北京大學國際漢學家研修基地

國際漢學研究通訊

Newsletter for International China Studies

第十一期
2015. 6

圖書在版編目(CIP)數據

國際漢學研究通訊. 第11期/北京大學國際漢學家研修基地編. —北京：北京大學出版社，2016.1

ISBN 978-7-301-26723-3

Ⅰ. ①國… Ⅱ. ①北… Ⅲ. ①漢學—研究—世界—文集 Ⅳ. ①K207.8-53

中國版本圖書館CIP數據核字(2016)第000337號

書　　名	國際漢學研究通訊(第十一期) GUOJI HANXUE YANJIU TONGXUN
著作責任者	北京大學國際漢學家研修基地　編
責任編輯	武　芳　翁雯婧
標準書號	ISBN 978-7-301-26723-3
出版發行	北京大學出版社
地　　址	北京市海淀區成府路205號　100871
網　　址	http://www.pup.cn　新浪官方微博:@北京大學出版社
電子信箱	zpup@pup.cn
電　　話	郵購部 62752015　發行部 62750672　編輯部 62756694
印 刷 者	北京大學印刷廠
經 銷 者	新華書店
	720毫米×1020毫米　16開本　30.75印張　488千字 2016年1月第1版　2016年1月第1次印刷
定　　價	78.00元

《國際漢學研究通訊》
Newsletter for International China Studies

目録

漢學論壇

論中國文學和文化的翻譯與傳播

張隆溪

自19世紀末以來,海外漢學和中國研究對於在歐美傳播中國文學和文化,做出了許多貢獻,取得了很大成就。然而漢學和中國研究畢竟不是歐美學術的主流,自身也難免有局限。近三十餘年來,隨着改革開放的深入,中國在經濟上取得極大發展,在國際政治方面也逐漸發揮更大作用,使得歐美和全世界越來越注重中國,也相應注重中國的文化傳統,於是漢學也越來越顯出其重要性。一方面,國外有越來越多的人願意瞭解中國,包括中國的歷史、哲學、文學和文化,學習中文的人也越來越多;另一方面,中國國內也希望把自己的文學和文化介紹給外部世界,把中國文學作品和人文研究的成果翻譯成外文,傳播到海外。可以説現在是漢學發展以及中國文化向外翻譯和傳播十分有利的時候。然而由於中外文化、社會和思想意識各方面都存在差異,要成功翻譯和傳播中國文化,並非易事,而會面臨不少問題和挑戰。如何瞭解和應對這些問題和挑戰,也就值得我們認真思考。

一、漢學與中國學術

讓我們首先考察漢學和海外的中國研究。所謂漢學,是從外部尤其是西方的立場和觀點看中國,而不是從中國内部看中國。這很容易使人想起"旁

作者單位:香港城市大學

觀者清、當局者迷”這句老話，或蘇東坡《題西林寺壁》所謂“不識廬山真面目，只緣身在此山中”那有名的詩句。這樣看來，作爲國際學術一部分的漢學，因爲是“旁觀”，有一定客觀和批評的距離，往往可以對中國和中國傳統有獨到的看法，對中國人理解自己和自己的思想和文化傳統，都有啓發借鑒的意義。西方漢學是西方學術的一部分，不僅其立場觀點不同於中國學者，而且其研究方法也不同。漢學家往往有西方社會科學方法的訓練，把這類社會科學方法應用於中國研究，也許有中國人自己不可能有的視角，也就可能看到中國人由於自己處於盲點而視而不見的東西。這當然就形成漢學研究不同於中國本土學術的特點，有時候也可能是其獨到的優點，但與此同時，也可能産生西方社會科學理論和方法的“傲慢與偏見”，即以爲中國和中國人只能提供研究的材料和對象，只有西方才可能提供研究的理論和方法。這其實是一種西方理論的傲慢，也因此往往産生其狹隘和局限。

所謂“旁觀者清、當局者迷”，只説到問題的一面，還有另一面，也可以用兩句老話來概括，那就是“隔霧看花”“隔靴搔癢”。從外部觀察固然可以有客觀距離，但也未嘗不會失之粗疏，没有直接經驗的真切體會。其實廬山之大，無論站在山中還是走到山外，都不可能窮盡其貌，這正是東坡要告訴我們的。東坡這首詩結尾兩句很有名，但也許恰恰因爲太有名，影響了後人對整首詩含義的理解。其實這首詩開頭兩句相當準確地説出了觀察理解任何問題的複雜多變，因爲東坡説得很明白：“横看成嶺側成峰，遠近高低各不同。”廬山面目是變化的，依據我們所處位置不同而呈現不同的面貌。把這個道理用來理解中國，我們就可以明白，要真正瞭解中國，就必須從不同角度看，把看到的不同面貌綜合起來，才可能接近於真情實貌。换言之，漢學和中國本土的學術應該互爲補充。上個世紀從50年代到70年代，漢學家們大多忽略中國本土的學術著作，認爲那基本上是受官方意識形態控制的學術，甚至是毫無學術價值的宣傳。但自“文化大革命”結束以後，中國學術已經發生了很大變化，中國學者們已經有獨立研究的意識，中國的學術著作也出現了許多優秀作品，其水準和以前不能同日而語。所以，今日海外的漢學家們已經不能繼續忽略中國學者的研究成果，中國學者也不能不瞭解漢學家的著述。中國和海外學術的互動交往已經越來越頻繁，所以我曾撰文説：“我們早已經該打破‘内’與‘外’的隔閡，拋棄‘社會科學模式’自以爲是的優越感，也拋棄西方‘理論複雜性’的自傲，融合中西學術最優秀的成果。只有這樣，我們才可能奠定理解中國及中國文化堅實可靠的基礎，在獲得真確的認識方面，更接近‘廬山真面目’。”①

西方漢學本身發展的歷史，對我們瞭解漢學與中國學術的關係不無幫助。最早從歐洲到中國、並在文化的互動交流方面產生影響的是基督教傳教士，尤其是耶穌會教士。利瑪竇（Matteo Ricci, 1552—1610）從意大利來到明朝末年的中國，在北京建立起基督教教會，與中國的士大夫階層多有交往，用中文撰寫了《天主實義》等多部著作，其影響不可低估。利瑪竇和其他傳教士又把中國的古代典籍用西文介紹到歐洲，以書信和著述陳述中國歷史文化和社會狀況，在啓蒙時代的歐洲產生了相當大的影響，使歐洲一些重要思想家對中國形成良好的印象。對於像伏爾泰那樣歐洲啓蒙時代的思想家們來説，中國没有教會，卻社會秩序井然，有自己的文化傳統，似乎正是他們所追求的以理性而非以宗教爲基礎的世俗社會。但利瑪竇死後，歐洲基督教教會内部産生了所謂“中國禮儀之争”，中國和西方的文化差異成爲争論焦點，不斷得到强調，而争論的結果，利瑪竇和耶穌會的“適應”策略被梵蒂岡教廷否定。在中國方面，康熙皇帝則禁止了西方傳教士在中國的活動。鴉片戰争後，中西之間的關係發生了很大變化，更多新教的傳教士來到中國，對中國的看法也與耶穌會教士很不一樣。無論如何，在文化方面最先瞭解中國的西方人是傳教士，西方大學裏最早聘請的漢學教授，也大多是傳教士。傳教士在中國社會，尤其在醫療和教育方面做出的貢獻，近年來已引起很多學者的研究興趣，得到肯定和嘉許。但“禮儀之争”中堅持原教旨和教義純正的一派，批評利瑪竇和耶穌會“適應”策略對異教的中國文化讓步過多，於是轉而强調中西語言和文化的差異，在中西之間設立起互不相通的對立。這些純粹派的思想觀念對中西跨文化理解和溝通的努力，到現在仍然還有一定影響 。

中西對立的觀念從“禮儀之争”時已初見端倪，只是在近代更具學術形

① 張隆溪，《廬山面目：論研究視野和模式的重要性》，《一轂集》，上海：復旦大學出版社，2011，228頁。此文原來用英文發表，見 Zhang Longxi, “The True Face of Mount Lu: On the Significance of Perspectives and Paradigms,” *History and Theory,* vol. 49, no. 1 (February 2010): 58-70.

式，往往上升到思想和哲理的層面。法國社會學家和人類學家列維－布魯爾（Lucien Lévy-Bruhl, 1857—1939）區分原始部落人與歐洲人的思維方式，認爲原始思維是審美和形象而非理性和邏輯的，只有歐洲人具有邏輯思維能力；與他交往甚密的漢學家葛蘭言（Marcel Granet, 1894—1940）就曾以此模式，著有《中國人之思維》一書。稍後法國漢學家謝和耐（Jacques Gernet, 1921—　）討論基督教傳教在中國之所以不成功，也歸結爲中國人思維方式與歐洲人根本不同，並斷定中國語言缺乏明確的語法，中國人的思維没有抽象能力，中國人缺乏對超越、精神和抽象觀念的理解能力。這類看法一直到今天仍然在西方學界有一定影響，如法國學者于連（François Jullien）就連篇累牘地著書立説，把古代中國與古代希臘描述爲絶然不同的兩種思維和兩種文化形態，中國乃是西方絶對的“他者”。這類看法在法國固然有一個連續幾代學人的傳統，但又不僅止於法國學者。美國學者尼斯貝特（Richard Nisbett）2003年出版了一部闡述文化差異的書，標題就把大意説得很清楚：《思想的地理學：亞洲人和歐洲人思維如何不同及其原因》（*The Geography of Thought: How Asians and Westerners think differently ... and why*）。曾在加州大學柏克萊分校任教多年、專門研究古代中國的漢學家吉德煒（David Keightley），按理説對中國古代典籍應該不會太生疏，但他在一篇文章裏爲了强調中國與希臘絶然不同，就以古希臘神話中詭計多端而巧言善詐的奥德修斯爲例，説古代希臘人認爲表面現象都是虚假不可信的，有所謂“認識論悲觀主義”（epistemological pessimism）；而與此相反，古代中國人則大多相信表面現象是可靠的，抱着所謂“認識論樂觀主義”（epistemological optimism）的態度[①]。可是《老子》七十章不是就抱怨説“吾言甚易知，甚易行。天下莫能知，莫能行”麽？《孟子·離婁上》不也説“道在邇而求諸遠，事在易而求諸難”麽？無論道家或儒家，老子和孟子都在抱怨説，人們往往連淺顯明白的道理都不懂，理解和行動都有問題。這哪裏有什麽中國人“認識論樂觀主義”的影子呢？中西思想、文化、傳統當然有這樣那樣的差異，但把兩者説得絶然不同，互相對立，就根本否定了文化交

① David N. Keightley, “Epistemology in Cultural Context: Disguise and Deception in Early China and Early Greece,” in Steven Shankman and Stephen W. Durrant (eds.), *Early China / Ancient Greece: Thinking through Comparisons* (Albany: State University of New York Press, 2002), p. 127.

往的可能。其實把文化差異絶對化的人都陷入一個邏輯上自相矛盾的困境，因爲無論東方還是西方,凡主張東西方互相之間不可能瞭解的人,都只可能或者是西方人,或者是東方人,按照他們自己的理論,就只可能或者瞭解西方,或者瞭解東方,而不可能跨越自己語言文化和思維方式的差異,瞭解或比較東西方之間的異同。可是他們卻恰恰宣稱自己似乎知道東西兩方面的情形,知道東西方人有兩種思維方式,截然不同,不可能相互理解。而按照他們自己的理論,卻恰恰是不可能有任何人能夠知道兩方面的情形,也就不可能知道東西方人的思維方式有什麽差異。過度强調文化之間差異的人,都的確不能不陷入這一個邏輯上的困境。我們要跨越文化差異,溝通中西思想和傳統,就必須批判這種絶對式的文化相對主義,真正站在平等的立場來理解對方,在東西方文化傳統的比較之中既見出中西文化各自的特點,又避免把差異簡單化、絶對化,形成文化之間的隔閡。

二、翻譯與文化理解

造成隔閡的原因很多,首先有一個語言差異的問題,但複雜文本的翻譯又絶不是一個簡單的語言問題。中文的確是很複雜的語言,有時候即使著名的漢學家和翻譯家,也難免出錯。亞瑟・韋利(Arthur Waley)大概是20世紀上半葉中國和日本文學最重要也最著名的翻譯家,他譯的《詩經》和《源氏物語》堪稱經典,他節譯《西遊記》中描寫孫悟空的精彩章節,題爲《猴王》,也十分成功。但就是這樣一位著名的翻譯家,也居然把“赤腳大仙”誤解爲“紅腳大仙”(red-foot master)。誤解很難完全避免,但翻譯往往不是一個簡單的語言問題,牽涉到文化傳統背景的理解,而涉及文化背景的錯誤就更值得我們注意。著名翻譯家華兹生(Burton Watson)曾翻譯《莊子》《韓非子》《史記》等中國重要的經典著作,很有成就。然而仔細察看之下,偶爾的錯誤也在所難免。《莊子・外物》結尾有這樣一段有名的話:“荃者所以在魚,得魚而忘荃;蹄者所以在兔,得兔而忘蹄;言者所以在意,得意而忘言。吾安得夫忘言之人而與之言哉?”華兹生的譯文讀來很流暢:

The fish trap exists because of the fish; once you’ve gotten the fish, you can forget the trap. The rabbit snare exists because of the rabbit, once you've gotten the rabbit, you can forget the snare. Words exist because of meaning; once you've gotten the meaning, you can forget the words. Where can I find a man who has forgotten words so I can have a word with him?[①]

這段文字看起來没有任何問題,但問題不在譯文的文字本身,而在原文意義的理解和把握。莊子原文前面幾個排比句,重點都在説明語言只是一種不得已而用之的工具,就像捕魚的竹籠、捉兔的羅網,一旦抓獲了魚、兔,就可以丟棄在一旁。接下來莊子發問道:我怎麼才找得到一位忘言之人,可以和他説話呢? 言下之意,世人大多只記得言,而不能得所言之意,所以莊子有此一歎。華兹生的譯文"a man who has forgotten words"固然是"忘言之人",可是英語裏的完成時態"has forgotten"表示此人已經忘了言,然而莊子尚未"與之言",那麼他忘記的就並非莊子之言,也就不是莊子要找的人。莊子希望找到的是能夠忘記他的言之人,因爲這個人才會得到他所言之意。這聽起來有點奇怪,但莊子的話往往都出人意料,超乎常理,要表達莊子這句話的原意就需把譯文稍做改動,把完成時態改爲將來時態:"Where can I find the man who will forget words so that I can have a word with him?"[②] 這看起來只是動詞時態的一個小小區分,但在意義上卻有根本的差别,關乎莊子哲理的準確表達。這裏舉出兩位翻譯名家的白璧微瑕,絶無意貶低他們的成就和貢獻,而只是想要説明翻譯,尤其文學、哲學之類複雜文本的翻譯,絶不是懂兩種語言就可以勝任,卻要求對原文的思想文化背景有相當深入的瞭解。許許多多翻譯成中文的作品,當中誤解和誤譯的地方也很不少,道理也一樣。

翻譯不是簡單的語言表達問題,因爲在一般語言表達能力之上,還要有文體、風格的意識,要熟悉並且把握學術語言規範等其他許多方面。學習外語達到能夠基本表達意思,並不是很難,但能夠做到像使用自己母語那樣使用外語,尤其寫作能達到正式發表出版的水準,不是不可能,卻也不是容易做

---

① *The Complete Works of Chuang Tzu*, trans. Burton Watson (New York: Columbia University Press, 1968), p. 302.

② 拙著《道與邏各斯》引用此段,就用將來時態來翻譯莊子這句話。见 Zhang Longxi, *The Tao and the Logos: Literary Hermeneutics, East and West* (Durham: Duke University Press, 1992), p. 20.

到的事。漢學家很少人直接用中文著書立説，中國人自己翻譯中文成外文，往往吃力不討好，都可以説明這一點。近年來我因爲編輯一套叢書，專門把中文學術著作譯爲英文出版，常常審讀譯稿，在這方面頗有些感受。翻譯不是自己寫作，而是"爲他人作嫁衣裳"，在大學環境裏往往不算自己的研究成果，於是教授學者們很少人願意做這樣的工作。能夠用英文寫作的中國人很少做翻譯，而且做也不見得就做得好，於是翻譯中文著作成英文，大多還是依靠母語是英語的譯者。我負責爲歐洲一家出版社編輯一套把中文學術著作譯成英文的叢書，到目前爲止已經出版了好幾部譯著，而除了一部是中國人翻譯之外，其他都是母語爲英語的譯者。英國漢學家葛瑞漢(A. C. Graham)有一本評價頗高的《晚唐詩選》，他在序言一開頭就説"翻譯中國詩的藝術是意象派運動的一個副産物"，又説"中國詩最好的譯者多半是詩人或愛好寫詩的人，依靠修改別人譯的初稿來翻譯"[①]。這當然是説，中國詩最好的譯者是以英語爲母語而又會寫詩的人，哪怕他們的中文水準不見得很高，但他們對詩的韻律有特別的敏感，依靠修改別人在傳達意義上也許更準確的初稿，就可以把中國詩翻譯成英美讀者更能接受，也更能領會其妙處的英語詩來。其實當年林紓翻譯西方小説，就正是這樣的情形。林紓自己不懂外文，但因爲他講究古文修養，文筆流暢而生動，就把懂外文的合作者口述的初稿下筆寫成優雅的文字，成功地把一百多部西方小説介紹到中國來。林譯小説儘管有不少誤解誤譯之處，但流暢可讀，尤其在清末民初那個時代，對於中國人開始瞭解西方，做出了很大貢獻。讓我們再回到葛瑞漢的那句話來。由於意象派以來英美現代詩首先注重意象(image)，如何傳達意象就成爲翻譯中國詩最重要的任務，而爲了意象，往往不得不捨棄原詩的格律形式和音韻。葛瑞漢把現代詩的感覺與西方讀者的期待講得很清楚，我們就可以明白爲什麽他認爲，翻譯中國詩最好由英美人來做。這話雖不能説百分之百正確，但也的確有一定道理。前不久北京大學許淵沖教授獲得國際翻譯家聯盟頒發"北極光"(Aurora Borealis)文學翻譯獎，這當然是中國翻譯家們值得引以爲榮的事，但許教授的翻譯在國内就曾引起不少争議，在國外本來幾乎没有人注意，獲

① A. C. Graham, *Poems of the Late T' ang* (Harmondsworth: Penguin, 1965), p.13.

獎之後反而引起一些從事文學翻譯的人激烈批評。當然,批評並不是壞事,但如果翻譯成外文的作品外國讀者大多不接受、不欣賞,翻譯就没有達到本來的目的,也就失去了意義。

三、出版信譽與合作的問題

翻譯作品在海外的接受,取決於好幾個因素,譯文品質當然重要,但也不是唯一決定性的因素。應該承認,由於政治和制度的原因,英美和歐洲許多讀者對中國出版的書往往還抱着一種懷疑態度,總是不大信任。這是中國文學和文化在海外翻譯與傳播面臨的問題,也是中國自己出版外文書刊必須面對的挑戰。在目前看來,最好的辦法是與國外在學界和讀者中享有良好聲譽的出版社合作,共同出版中國文學和文化的翻譯作品。我爲歐洲布里爾出版社(Brill)主編兩套叢書,在此就想以個人經驗爲例,談談相關的問題。布里爾在歐洲是一個有三百多年歷史的老牌出版社,1683年在荷蘭萊頓創立,出版各種學術著作、期刊、書目和參考書,内容廣泛,包括科學、法學和人文學科各個領域。自19世紀末的1890年以來,布里爾就一直出版一份專門研究傳統中國的學術刊物《通報》(*T'oung Pao*),在國際漢學界享有很高聲譽。布里爾近年來又出版一套專門把中文學術著作翻譯成英文的叢書,題爲"布里爾中國人文學術叢書"(Brill's Humanities in China Library),由我和德國哥廷根大學施耐德教授(Axel Schneider)共同主編。2007年以來,我們已經出版了七部中文學術著作的英譯本,包括洪子誠、陳平原、陳來、駱玉明、榮新江、葛兆光和羅志田等人很有影響、能夠代表中國當代學術研究成果的著作。不久還會有朱維錚、何懷宏、蔣寅等學者的著作陸續翻譯出版。我爲這套叢書寫了一篇總序,印在每部書前面,其中我提到中國在"文化大革命"後經過改革開放,經濟飛速發展,整個社會發生巨大變化,在國際上也越來越引人注目。隨着整個中國的改變,中國的學術研究也相應有很大變化,所以我説,在過去很長一段時間裏,西方學界幾乎完全忽略中國本土的學術,現在已經没有理由再繼續這樣無視中國學術,"現在時機已到,西方學者和其他感興趣的讀者應當接觸來自中國的觀點看法,而把重要的中文學術著作翻譯成英文出版,就是在這樣接觸中走出的重要一步"。由於布里爾是歐洲有名望的老牌出版社,出版的書基本上歐美大學圖書館都會收藏，所以我們這套中國人文研究叢書使中國學者優秀的著作進入了歐美大學的圖書館，也就進入了西方學術研究參考的範圍。這套叢書印製精美，在歐美學界已經産生一定影響，只要保持品質，長期堅持，對於在海外傳播中國文化和學術成果，就必定能做出不小的貢獻。保持品質一在選擇好書，再就是選擇好的譯者，這兩件事都不易，而尋找合格合適的譯者，尤其困難。在這方面，由於布里爾有自己人脈聯繫的網絡，可以找到願意從事中譯英工作的譯者，所以也相對而言比較順暢。

由於人文研究叢書的成功，布里爾出版社又在去年設立"東亞比較文學與文化研究叢書"（East Asian Comparative Literature and Culture），由我和波士頓大學的魏樸和教授（Wiebke Denecke）共同主編。雖然這套書出版直接用英文撰寫的書稿，但叢書性質是文學和文化的比較研究，而在這方面，我認爲錢鍾書先生的《七綴集》最具代表性，可以作爲研究典範，所以決定把《七綴集》的英譯本作爲這套叢書開頭的第一部出版。恰好一位新西蘭漢學家鄧肯（Duncan Campbell）已把此書譯完，他的譯稿經過進一步仔細修改，由我寫了一篇序，又得到楊絳先生同意，2014年初由布里爾出版了《七綴集》英譯本，並由此推出布里爾東亞比較文學與文化研究叢書。編輯這兩套叢書，尤其是把中文學術著作譯成英文出版的第一套叢書，使我體會到翻譯之重要，也認識到翻譯之困難。自古以來，翻譯就説明人們跨越語言和文化的隔閡和障礙，使不同民族可以互相瞭解、和睦共處。隨着瞭解的深入，對翻譯的需要、要求及標準也隨之提高。在我們這個時代，翻譯變得更爲重要。提高外語教學品質，培養更多更好的翻譯人才，加强國際合作交流，這就是促進中西方相互瞭解，推展文化傳播最根本的辦法。

中國古典文獻與宗教經典的會遇

羅明嘉

中國跨宗教對話的必要性

隨着中國成爲世界上越來越有影響力的經濟、文化、政治力量，它也正經歷着巨大的意識形態的挑戰和改變。宗教信仰體系，特别是儒教、佛教、基督教，迅速地赢得了許多新的歸信者，並在一定程度上填補了在價值、倫理和信仰方面的真空。

儘管中國有着豐富的宗教傳統，且這些宗教傳統正產生着越來越大的影響，但是，到目前爲止，仍然没有多邊的跨宗教對話，僅僅是一些近幾年才產生的零星的雙向對話，如學術研討會。而且，這類會議的大多數參與者都是世俗的、學術的宗教研究者，很少有不同宗教的代表。目前，有些國家已經爲多元宗教間對話做了初步嘗試，但在這方面中國仍然是一個處女地。

由於跨宗教間會遇的缺乏，關於中國跨宗教間對話的研究仍然非常少。在過去的二十年中，宗教研究在中國領先的大學裏和學術界已經成爲一種普遍的現象並且備受親睞，但遺憾的是，這種研究更多地關注個别的宗教，而不是多種宗教間的會遇。

作者單位：芬蘭赫爾辛基大學

跨宗教間對話的一個全球性的方法論

從全球性的角度來看,宗教對話的研究仍然是一個相對年輕的分支,僅僅只有半個世紀的歷史。目前這個領域迫切需要形成一種批判性的新理論和更好的實踐方式。如今,宗教間對話已經成爲西方自由主義的基督教學者和啓蒙思想主導下的純理論的、精英們的冒險。非西方人常常把它當作一種學術思潮,這種思潮推動了以家長式、個人主義、理性主義爲特徵的西方人文主義。宗教間對話的提案很大程度上是以宗教神學的各種基督教概念爲基礎的,而這些概念大多是從西方自由神學傳統借鑒而來(如:Swidler 1987, Hick 1995, Dupuis 2002, Knitter 2004)。

世界各地的宗教對話發展相對緩慢,其原因之一就是在其對話理論形成階段,常常受到西方的主導。西方自由主義和人文主義霸權對宗教對話具有潛在的危險,即宗教和意識形態並不能按照它們自己的標準來發展。因爲它們常常被某些公正和更加抽象的價值所利用,而這些抽象價值有悖於宗教的自我理解和認同。

非西方的宗教傳統信仰者對一個統一的宗教範式極爲不滿。因爲這種範式在重新闡釋這些宗教的真理時,卻讓個人信仰的真理信條做出了讓步。那些伊斯蘭教和東南亞宗教的信徒,不能接受這種方法。對於這些宗教,宗教對話的一個必要條件就是對宗教權威的認同。尤其是關係到他們各自的宗教信仰,這些不同的宗教之間仍然是相互對抗的,因此不同的宗教信仰不能夠被調和(Panikkar 1999, Heim 2001, D'Costa 2009)。

如今,西方啓蒙運動思潮面臨着後殖民主義、後世俗主義和後自由主義方法的挑戰。查理斯·泰勒警告道,占主導地位的北大西洋文明建立了"據説是無視差異的中立原則體系",但事實上,這些原則卻是"霸權文化的一種反映";因此被壓制的文化被迫採取其他的形式(Taylor 1994)。西方社會强調理性個體之間的跨宗教對話應建立在知識的基礎之上,這造成了他們對亞洲宗教傳統的歪曲理解(Welch 2011, 359)。因此,現在迫切需要建立一種跨信仰的方法論,這種方法將所有的宗教放在一個平等的地位上,因此能夠被所有的宗教都接受。在這種方法的指導下,西方傳統的首要性受到質疑,本國資源在與西方資源平等的基礎上被使用和解讀(Kwok 2005, 68)。

我們現在面臨的挑戰是要通過審視自己的價值評估體系,重新評價跨宗教對話的基本前提和範式。我們不能再盲目地認爲西方的認識論是優越的,因爲我們没有理由宣稱西方形而上學的敘述、學術化的語言以及認識論可以比我們自己所接受的宗教世界觀更加精准地描繪世界。文化需要通過與人們認爲是天經地義的或習以爲常的原則區分,來重新修訂自己的參照、規範和價值體系(Homi Bhabha, 轉引自 Welch 2011, 362)。

現在我們需要一種元對話:以最基本的對話概念爲基礎的對話。在跨宗教對話的背景下,理論研究自身需要更加處境化並且要更加關注非西方的認知和語言體系。它强調在宗教對話中"參與"的必要性,即强調從内部人士、某一宗教傳統和宗教生活體驗的參與者的角度作價值分析。這種參與的範式强調後殖民時期對認識論的重新評估,以具體化和内在性爲中心的後現代主義和女性主義思想和對靈性生活的研究以及宗教多元主義的不可化約性(Ferrer & Sherman 2008, 7-9)。

經文辨讀——跨宗教間對話的最具前景的新方法

經文辨讀是目前跨宗教對話文獻中人們最爲深入討論、且最具前景的方法。在過去的二十年裏,經文辨讀在猶太教徒、基督徒和穆斯林中贏得了廣泛的認同,並且獲得了世界範圍内學術界和宗教界越來越多人的關注(Ford & Pecknold 2006, Moyaert 2013, Higton & Muers 2013)。經文辨讀作爲一種亞伯拉罕宗教間的跨信仰對話,於20世紀90年代在美國和英國産生,由佛吉尼亞大學和劍橋大學進行了深入的發展,並在美國宗教學院的年度會議上被檢測。

在過去的幾年裏,中國的宗教學者已經展示了他們對經文辨讀的興趣:中央民族大學於2012年在北京建立了"比較經學與跨宗教間對話"創新引智基地;且這個基地已經在中國進行了經文辨讀的嘗試。此外,中國人民大學《基督教文化研究》也於2011年出版了關於經文辨讀的特刊。

經文辨讀是跨宗教間經典交叉閱讀與反思的一種實用的方法,目前僅在小群體中加以實踐,它是多邊的跨宗教對話的實踐,它使得公衆對宗教有一個更深刻的理性的認識。通常,經文辨讀的參與者,既是宗教信仰者,同時又是學術研究者和經文辨讀組織成員，當然，他們都是社會成員。經文辨讀也可在没有學界人員參加的情況下，由各種不同宗教的普通信徒來組織。

經文辨讀加深了參與者自己的宗教認同，並且使他們理解和尊重其他宗教經典中的信念。在參與者信仰相互交流的過程中，對他人信仰的理解可以加深參與者對自身宗教信仰的理解。經文辨讀能幫助參與者識别他自己或是對話對方的宗教身份。在認識自己和認識他人間，存在着一種相互性，認識自己就關係到認識他人。宗教認同並不僅僅意味着寬容，它還要求我們充分認識對方，承認他者的差異性(Hénaff 2010)。

與西方在抽象的啓蒙理性宗教信念方面達成一個最低的共識不同，經文辨讀在進行着一場深刻的、長期的、看似不能解決各宗教世界觀的差異的運動。經文辨讀被稱爲一種以智慧爲基礎的，一種後現代主義、後殖民主義、後自由主義、後保守主義的跨宗教間對話的方法。由此，經文辨讀就成爲反現代宗教的基要主義和現代自由主義之間的第三個區域了。它的目標不是達成某種共識，而是在一種更加清晰的層次上認識到差異，這將會提高論争的品質。可以説，經文辨讀提高了差異的地位，爲"獨特性"提供了空間(Sacks 2002, 49, 52, 61)。

經文辨讀遵循着開放性的原則，這並不是一項關於公共話語的優越理論的表面工作，"而是參與者享受他們一起互讀經典所帶來的成果"。這是實踐者的對話，他們不需要對爲什麼這樣做、在做什麼及經典的詮釋做出一致性結論。他們的對話不受任何理論的約束，並爲他們各自宗教傳統間的相遇形成了一個中立的領域(Higton & Muers 2012, 113-114)。經文辨讀意味着以一種求知的方式去做。作爲一個開放式的過程，它以"謙卑"(認識自身的不足)和"殷勤"(歡迎他者的出現，並將此作爲善和真理的潛在來源)爲基礎。經文辨讀創造了一種新的認知和行爲方式(Cornille 2008, 4-6)，它包含了對話的知識、情感、精神和社會維度，其所建立的宗教關係保護了差異性，並且加强了宗教之間的相互友愛和尊重。經文辨讀能夠在不同的宗教信徒間建立長久的友愛，因此它是在多元文化社會中實現跨文化交流最有效的方式。

中國背景下的經文辨讀

我們現在面臨的最大挑戰是要爲中國幾大主要宗教間的對話建立和發展一套最適合的方法論。我堅信經文辨讀會成爲中國幾大宗教間對話的最有效的方法。

目前,經文辨讀作爲亞伯拉罕三大傳統跨宗教的經典文本的一種跨宗教間的交叉閱讀方法,僅在小群體中被實踐。但就我自身的理解,它也可以推廣到中國幾大主要宗教的經典閱讀中。我認爲,儒家、道家和佛教的經典文本以及作爲經文辨讀一部分的亞伯拉罕的三大信仰傳統的文本可以有意義地結合起來,因此經文辨讀會成爲中國跨宗教間對話的一種卓有成效的方法。衆所周知,儒學是不是宗教一直存在争議。在Anna Sun最近的一本著作《作爲一種世界宗教的儒學》中,她有力地説明了儘管儒家帶有强烈的倫理取向,我們仍然有理由將儒學看作是一種宗教世界觀,因爲它填補了當代中國信仰和道德的真空(Sun 2013)。

分析跨宗教的概念和經文辨讀的實踐如何在中國得到具體的實現是非常有必要的。中國經文辨讀的參與者應當是中國宗教的真正代表:普通信徒、宗教專業人士、研究信仰傳統的學者。因此,經文辨讀的優勢之一是它的非精英主義和民主性:普通信徒,包括婦女和青年人以及宗教專業人士和學者都有平等的參與權。經文辨讀是一種現實的對話方法,因爲它的目標是提升差異的品質而不是相互排斥的抽象的世界觀之間一種人工的和諧。

中國經文辨讀的材料應當選擇六大宗教的經典文本:儒家經典“四書”“五經”,道家經典《道德經》《南華真經》《黄帝陰符經》,漢傳佛教《心經》《金剛經》《壇經》《阿彌陀佛經》《法華經》,伊斯蘭教的《古蘭經》和基督教的《聖經》(天主教和新教用不同的中國譯本)。每一期的議題由小組討論決定,關於議題的文本則由每一位元成員從自身的經典中選出。關於佛教,爲了簡單起見,經文辨讀首先以漢傳佛教爲主線。由於民間宗教缺乏合適的經典,因此很難將它納入到經文辨讀的活動中。

研究使得經文辨讀更有意義

在中國的社會文化背景下,經文辨讀這一實踐過程必須要通過對每一種宗教的自我理解以及與其他宗教的關係和認同的分析來得到深入發展。這一工作可以使我們獲得關於跨宗教間對話現實可能性的知識,並且能夠幫助經文辨讀在中國背景下更有效地實施。對中國跨宗教間對話的前提、條件和可能性進行分析使我們對經文辨讀中現實的、創造性的跨信仰對話有一個深刻的認識。這將加深我們對多邊宗教對話本質的理解。

在中國社會文化背景下,分析每一種宗教身份的自我理解最好應從以下三方面來入手:

1. 要分析中國每一種宗教的真理主張、信仰和教條,關注其信仰的歷史發展和哲學體系。

2. 要分析每一種宗教的美學的、視覺的、身體的表達,比如宗教藝術、儀式以及宗教經驗和靈修。

3. 要關注每一種宗教的倫理教條,特別是從跨文化的和全球的觀點來看道德合法性和研究每一種宗教是如何處理現代中國社會中迫切的政治、司法、經濟問題。

在對宗教進行這三方面的研究時,對中國的每種宗教的研究都應當從這樣一種視角出發,即在宗教身份的自我理解的基礎上,一種宗教如何理解它與其他宗教信仰的關係,它是如何認識中國的其他宗教的。認識自己和認識他人應當聯繫到一起。對一種宗教身份自我理解的分析要求採用一種參與式的方法:研究者不能將一種陌生的認識論範疇强加到信仰傳統上,相反要將傳統當作與活生生的宗教經驗相融合的參與者(Ferrer & Sherman 2008)。

宗教信仰

我們在研究中國宗教信仰時,需要研究中國背景下經文辨讀所涉及的宗教和哲學概念。我們還需要分析:在中國的歷史、文化、社會、政治背景下,中國的每種宗教是如何將自己認定爲一種宗教或信仰;是什麽使得一種宗教信仰區别於倫理的或是政治的世界觀;中國的宗教如何定義自身和其他宗教的關係以及如何作爲宗教世界大家庭的一部分;使得歸信者認定自己屬於某一特殊宗教的獨特信條是什麽；在哪種意義上，一個人的宗教身份是以信仰爲基礎的，在哪種意義上是以其他因素爲基礎，如文化、種族或生活方式？

"宗教"概念本身就是一個西方的發明，這也許説明非西方的概念可以將自身與西方標準描述或概念化區别開來（Welch 2011, 358-359）。"宗教是什麽"這個問題已經成爲中國知識分子討論的熱點話題。最新的所謂的"儒學是否是宗教的争論"就是産生於2000年到2004年間。

中國文化中，没有一種宗教已經取得了主導性地位。幾個世紀以來，各種宗教之間相互影響，使中國成爲了混合宗教身份的典型。我們需要關注中國人宗教身份的不確定性和含糊性；還需要研究宗教如何受着混合主義、實用主義、愛國主義、帝國中心主義以及民間信仰的影響；還需要關注種族身份和文化的多樣性，因爲，中國社會的多元文化主義與宗教傳統的多樣性有着緊密聯繫。

以美學—精神的方式進行對話

從本質上來講，中國文化是視覺文化。比如説，中國語言以形象思維爲基礎，這在實踐中深刻影響了各種宗教的自我表達。我們需要從宗教審美的維度來關注宗教的具體化，而且宗教正是透過宗教藝術、書法、建築、儀式、音樂、姿勢、運動和服飾等具體形式表達出來。對中國宗教中美學元素的分析和比較爲我們更深刻全面地理解跨宗教間對話提供了一種方法，這種方法將哲學思考與經驗、情感、精神和美學表達結合起來。事實上，美學—精神的跨宗教間對話方式可以稱作視覺的經文辨讀。

跨宗教間對話的研究主要和知識方面密切相關，它是通過比較多種信仰來尋找共同的理論認識。然而，如今由於這種方法與純知識比起來，没能考慮到人類宗教性的其他方面，而受到抨擊。越來越多的研究從更多的角度出發探討問題，如美學的、精神的或是神秘主義的角度（Cornille 2008, Cheetham 2010, Illman 2012）。當代全球的困境，如氣候變化和國際恐怖主義，似乎需要一種共存性的範式，這種範式不僅僅是建立在抽象的思維體系上，還是對個别宗教和它們的宗教生活方式的一個綜合瞭解。由於政治和權力結構的影響越來越突出，對話這一概念中積極的、建構性的本質日漸淡化（Amir-Moazami 2011）。

爲促使對話能夠超越知識與情感的分化，個人與政治的分化，精神與實踐的分化，我們需要從美學—精神的層面來分析對話。美學的對話作爲理性討論和論爭的補充越來越受到學術界的認同（Illman & Smith 2013）。對話的藝術在於轉變我們的思維方式，產生新的可能性；使我們停下來去反思、改變和解放我們自身（Cheetham 2010）。因此，想象力、具體的背景，如宗教視覺的、音樂的、神秘的表達都會爲我們理解、自我評價和共存打開新的渠道。想象力爲跨宗教間對話這一轉換性實踐和理解差異性提供了一個非常有意義的出發點。

跨文化的倫理和宗教慈善組織

隨着中國成爲世界上越來越具影響力的國家，在多元文化的背景下，對其倫理傳統轉變的探索是至關重要的。這類初步研究將會通過分析中國倫理傳統的基礎來完成。特別是如今具有影響力的儒家在與亞伯拉罕宗教和佛教的對比中，被重新評價。隨着西方國家的經濟化及其越來越依賴與中國的貿易往來，我們不禁要問：中國將會給世界帶來何種價值觀？我們有必要分析中國世界觀中價值與道德統一的元素以及中國社會文化中的宗教生活。同時，我們也要分析在中國背景下的倫理觀和世界上其他社會文化背景下倫理觀的相似性和差異性。換句話説，我們可以問：在根本不同的形而上學模式中產生共同的宗教責任的元素是什麽？我們還需要探索新的方法來面對以傳統倫理爲基礎的現代挑戰，如後殖民主義的身份、法律規則、公正、平等、社會福利、市民社會、民主、環境責任等。

在對中國跨文化倫理前景的分析中，20世紀90年代初期由孔漢思（Hans Kung）發起的"全球倫理項目"可以作爲一個出發點。孔漢思的方法獲得了中國學者的廣泛認同和讚賞。北京大學建立的世界宗教和全球倫理研究中心，在很大程度上就是受到孔漢思的推動。但我們仍然需要批判地分析和評價在改變中國形勢的過程中孔漢思方案的適切性。他方案的問題和局限仍然需要被指出並得到建構性的發展。

"走向全球倫理的宣言"（1993），被許多人看作是一份最基本的西方文件，它在多元立場上表達了理性的普遍主義和現代自由主義觀點。此宣言因其宣傳普世價值，而被認爲是西方道德哲學的嘗試。在孔漢思的"全球倫理方案”中：真正的人性是真正宗教的先決條件。但有些批評者認爲孔漢思的方案偷用了西方人文主義的概念來對抗宗教儀式的自我認同（Hedges 2010）。我們仍然需要特别關注後殖民主義範式，這種範式聲稱任何試圖建立普世價值的努力都注定是文化帝國主義。而真正的對話倫理應當源於環境的、文化的、宗教的理解。

另外，我們還需要直接關注實踐層面，特别是關注中國宗教神學和宗教慈善的實踐。如哪種神學主張（儒學、佛教、道教、伊斯蘭、基督教）堅持辦慈善機構，他們如何理解自身與相對應的其他慈善機構的區别？由於政府不能爲公民提供一個滿意的福利體系，從2008年以來，宗教慈善團體在中國得到合法的認可，他們在促進建立一個公正的社會中發揮着越來越重要的作用。目前，中國的非政府組織正在迅速增長，而且這些組織促進了中國社會福利的提升和公民社會的建立（Tao & Liu 2012）。我們有必要研究這類組織的宗教動機、倫理基礎和實踐模式；還需要特别關注宗教慈善組織間相互合作及這些組織與其他非政府慈善組織合作的可能性。

在當今全球化的中國，我們還需要建立一種新的倫理範式，這種範式應源於許多古老的傳統，並關注他們的多維度對話。我們不可避免地會接受形而上的宗教世界觀的排他性，但這並不能阻礙有責任心的社會合作，相反爲分擔社會責任釋放了能量。

經文辨讀對中國的意義

我堅信，在中國背景下，長期的經文辨讀，包括中國幾大主要宗教和每種宗教間的學術研究，會對中國跨宗教間對話産生巨大的影響。世界上主要的宗教，除印度教外，都在中國存在；中國跨宗教對話的實現將會給世界上其他地區的宗教對話提供一個有效的範式並産生積極的影響。在中國正在進行中的多邊宗教對話這一實踐將會成爲一個提升世界各地的對話品質的典範。

這種將一個富有成效的經文辨讀的實踐與意義深遠的學術研究的成果結合起來的方法，一定程度上可以加强中國宗教的自信心並培養中國社會的責任心和信仰自由。此外，成功的宗教對話還將會釋放道德能量，這對建立一個多元的、跨文化的、寬容的、和諧的社會是至關重要的，並且對公民社會、和平、正義、平等、福利和環境責任都具有重要意義。

參考文獻：

Amir-Moazami, Schirin: "Pitfalls of consensus-orientated dialogue: The German Islam Conference," *Approaching Religion* 1/1, 2011, 2-15.

Blue Book of Religions: Annual Report on China's Religions (2008-12) (in Chinese), Beijing: Social Sciences Academic Press, 2008-12.

Cheetham, David: "Exploring the Aesthetic 'Space' for Inter-religious Dialogue," *Exchange* 39, 2010, 71-86.

Cornille, Catherine: *The Im-Possibility of Interreligious Dialogue*, New York: The Crossroad Publishing Company, 2008.

Cornille, Catherine (ed.): *Criteria of Discernment in Interreligious Dialogue,* Eugene: Cascade Books, 2009.

Cornille, Catherine (ed.): *The Wiley-Blackwell Companion to Inter-Religious Dialogue*, Malden & Oxford: Wiley-Blackwell, 2013.

Cornille, Catherine & Conway, Christopher (eds.): *Interreligious Hermeneutics*, Eugene: Cascade Books, 2010.

Cornille, Catherine & Corigliano, Stephanie (eds.): *Interreligious Dialogue and Cultural Change*, Eugene: Cascade Books, 2012.

D'Costa, Gavin: *Christianity and World Religions: Disputed Questions in the Theology of Religions*, Malden & Oxford: Wiley-Blackwell, 2009.

Dupuis, Jacques: *Toward a Christian Theology of Religious Pluralism*, Maryknoll: Orbis Books, 2001.

Ferrer, Jorge N. & Sherman, Jacob H.: *The Participatory Turn: Spirituality, Mysticism, Religious Studies*, Albany: State University of New York Press, 2008.

Ford, David & Pecknold, C.C. (eds.): *The Promise of Scriptural Reasoning*, Malden & Oxford & Calton: Blackwell, 2006.

Goossaert, Vincent & Palmer, David A.: *The Religious Question in Modern China*, Chicago & London: The University of Chicago Press, 2011.

Hedges, Paul: *Controversies in Interreligious Dialogue and the Theology of Religions*, London: SCM Press, 2010.

Heim, S. Mark: *The Depth of the Riches: A Trinitarian Theology of Religious Ends*, Grand

Rapids & Cambridge: Eerdmans, 2001.

Hénaff, Marcel: *The Price of Truth: Gift, Money, and Philosophy*, Stanford: Stanford University Press, 2010.

Hick, John: *A Christian Theology of Religions: The Rainbow of Faiths*, Louisville: Westminster John Knox Press, 1995.

Hirvonen, Heidi: *Christian-Muslim Dialogue: Perspectives of Four Lebanese Thinkers*, Leiden & Boston: Brill, 2013.

Honneth, Axel: *The Critique of Power: Reflective Stages in a Critical Theory*, Cambridge (Mass.): The MIT Press, 1991.

Huang, Paulos: *Confronting Confucian Understandings of the Christian Doctrine of Salvation: A Systematic Theological Analysis of the Basic Problems in the Confucian-Christian Dialogue*, Leiden & Boston: Brill, 2009.

Illman, Ruth: *Art and Belief: Artists Engaged in Interreligious Dialogue*, London: Equinox, 2012.

Illman Ruth & W. Alan Smith: *Theology and the Arts: Engaging Faith*, New York: Routledge, 2013.

Knitter, Paul F.: *Introducing Theologies of Religions*, Maryknoll: Orbis Books, 2004.

Komulainen, Jyri: *An Emerging Cosmotheandric Religion? Raimon Panikkar' s Pluralistic Theology of Religion*, Leiden & Boston: Brill, 2005.

Küng, Hans: *Projekt Weltethos*, München: Piper, 1992.

Kuokkanen, Aleksi: *Constructing Ethical Patterns in Times of Globalization: Hans Küng's Global Ethic Project and Beyond* , Leiden & Boston: Brill, 2012.

Lin, Manhong: *Ethical Reorientation for Christianity in China: Individual, Community and Society*, Hong Kong: Chinese University of Hong Kong Press, 2010.

Kwok, Pui-lan: *Postcolonial Imagination and Feminist Theology*, Louisville: Westminster John Knox, 2005.

Panikkar, Raimundo: *The Intrareligious Dialogue*, New York: Paulist Press, 1999.

Ruokanen, Miikka: *The Catholic Doctrine of Non-Christian Religions, According to the Second Vatican Council*, Leiden & New York & Cologne Brill, 1992.

Ruokanen, Miikka & Huang, Paulos (eds.): *Christianity and Chinese Culture*, Grand Rapids and Cambridge: Eerdmans, 2010. (Also published in Chinese by China Social Sciences Press, Beijing.)

Sacks, Jonathan: *The Dignity of Difference: How to Avoid the Clash of Civilizations*, London & New York: Continuum, 2002.

Scriptural Reasoning: Journal for the Study of Christian Culture (in Chinese), Vol. 25, Beijing: Renmin University of China, 2011.

Sun, Anna: *Confucianism as a World Religion: Contested Histories and Contemporary Realities*, Princeton & Oxford: Princeton University Press, 2013.

Swidler, Leonard (ed.): *Toward a Universal Theology of Religion*, Maryknoll: Orbis Books, 1987.

Tao, Feiya & Liu, Yi (eds.): *Religious Charities and Social Justice in China* (in Chinese), Shanghai: Shanghai University Press, 2012.

Taylor, Charles *(et al.)*: *Multiculturalism: Examining the Politics of Recognition*, Princeton: Princeton University Press, 1994.

Timmerman, Christiane & Segaert, Barbara (eds.): *How to Conquer the Barriers to Intercultural Dialogue: Christianity, Islam, and Judaism*, Brussels: Peter Lang, 2006.

Wang, Yujie: *Religions and Society in Contemporary China*, Beijing: Renmin University Press, 2006.

Welch, Sharon D.: "Beyond Theology of Religions: The Epistemological and Ethical Challenges of Inter-Religious Engagement," in *The Oxford Handbook of Feminist Theology*, ed. by Mary McClintock Fulkerson & Sheila Briggs, Oxford *et al.*: Oxford University Press, 2011, 353-370.

Yang, Fenggang: *Religion in China: Survival and Revival under Communist Rule*, Oxford *et al.*: Oxford University Press, 2011.

Zhang, Zigang: *A Study of Contemporary Religious Conflicts and Interreligious Dialogue*, Beijing: Economic Science Press, 2009.

Zhao, Fasheng: *Humanism of Original Confucianism*, Beijing: China Social Sciences Press, 2012.

Zhao, Jianmin: *The Encounter between Christian Faith and Modern Chinese Culture*, Beijing: Religious Cultural Press, 2010.

Zheng, Xiaoyun: *Theravada Buddhism in China*, Beijing: China Social Sciences Press, 2012.

Zhuo, Xinping: *Theorien über Religion im heutigen China und ihre Bezugnahme zu Religionstheorien des Westens*, Frankfurt am Main *et al.*: Peter Lang, 1988.

The Chinese Classics Encounter the Canonical Texts of Religions

Abstract: As China becomes an increasingly influential economic, cultural, and political global power, it is also undergoing ideological changes. Religious belief systems, especially Confucianism, Buddhism, and Christianity, are winning new followers and, to some extent, filling the vacuum of values and faith. Yet in spite of the growing influence of religious traditions, no multilateral interreligious dialogue exists, and there is not yet much research on interreligious engagement in China.

Scriptural Reasoning (SR) could become a useful method of interreligious engagement among the followers of the major religions of China. SR is a new method of interreligious dialogue which is rapidly gaining attention in academic and religious bodies around the world. It was developed at the Universities of Cambridge and Virginia in the early 1990's. SR, so far practiced in small groups as an interfaith cross-reading of the canonical texts of the three Abrahamic traditions of Judaism, Christianity, and Islam, can be extended by including the representatives and sacred texts of Confucianism, Taoism, and Buddhism.

The practice of SR should be deepened by a historical-philosophical analysis of the identity of each Chinese religion. We need to ask: How, in the cultural and social context of China, does a religion understand its own identity in relation to the other faith systems of China, and how, on the basis of this, does it recognize the other religions of China? When analyzing this, we need to focus on the three dimensions of the Chinese religions: 1. truth claims or beliefs, 2. aesthetic-spiritual dimensions, and 3. ethical teachings, especially from the point of view of intercultural ethics.

This kind of approach, which combines a fruitful practice of SR with profound results of research, will open up new possibilities for dialogue in a pluralistic multicultural context, with implications for civil society, peace, justice, equality, welfare, and environmental responsibility.

英譯《文選》的疑難與困惑

康達維(David R. Knechtges)

《文選》英譯本第一册成書之後,我在序言中是這麽寫的:翻譯《文選》是“一項大膽,或許是不識實務的工作”①。自從第一册、第二册、第三册英譯《文選》先後出版之後,二十個寒暑轉眼之間就匆匆地過去了,今天重拾《文選》未完成的譯注工作,面對這項如此艱巨的工程,我似乎感到更加膽怯甚至氣餒了,如果現在重寫這篇序言,我會毫不猶豫地除去“或許”二字。我會這麽寫:“譯注《文選》是一項大膽,甚至不識實務的工作。”英文譯注的《文選》目前已經由普林斯頓大學出版社出版了三册,包括所有賦篇的翻譯與注解。注譯工作的進展是如此地緩慢,如果世界上有最慢翻譯家的稱呼,我想我應該可以當之無愧。

譯注《文選》的進展如此緩慢,主要有兩點原因:其一,《文選》本身所選的内容和題材;其二,歸咎於我的治學方法。《文選》組織龐大、内容豐富、文體繁多、作者複雜、包涵的題材廣闊,任何試圖翻譯整套《文選》的學者或專家,都將面臨極大的困難和挑戰。我想在這篇文章裏談談翻譯《文選》所遭遇的一些困難以及相應的解決之道。

《文選》共有六十卷,七百六十一篇作品,三十九種文學體裁,其中包括賦、詩、騷、表、檄、詔、頌、贊、論、史論、誄、碑、哀、策問、彈文等等。篇幅長短

作者單位:美國華盛頓大學

① David R. Knechtges, trans., *Wen xuan, or Selections of Refined Literature* (Princeton: Princeton University Press, 1981), p.xi.

不一，有的短詩僅有十行，有的賦篇或其他作品則長逾千行，當初編纂《文選》的主要目的之一，就是要彙編一套能夠代表當時所有重要文學作品的選集，因此翻譯《文選》就必先熟知各種不同的文體，充分了解每種文體形成的過程和發展的背景，並熟悉每種文體使用語言的特色。

翻譯《文選》主要面對的挑戰之一就是言語的艱澀與難解。《文選》是辭賦、駢體文的重要選集，賦篇佔了三分之一，辭賦、駢體文辭藻富麗、言語艱澀，僅是讀通、讀懂，就得大費周章，遑論注釋了。另外，《文選》也收集了大量的詩篇，奇文怪字，典故連篇，就以陸機、潘岳、謝靈運、謝朓等人的詩文爲例，即使有大量的注解和評注，還是難以真正讀懂，而在最需要註解的時候，注釋學家卻一筆帶過，略而不談，我想其中最主要的原因之一，就是歷來對最難解的辭句在注釋學家作注的時候，仍是不知所云，所以也就只好含糊其辭，一筆帶過了。

在21世紀的今天，賞讀《文選》還是如此地困難，原因之一就是語言環境和文化背景這兩道鴻溝，將我們和古代、中古世紀的文學隔開了。即使現代研究《文選》的中國學者跟研究《文選》的外籍學者相比，在詮釋《文選》內容方面，他們所遭遇的困難可能也不相上下。陸宗達教授曾經提到將《文選》翻譯成白話文"注不容易，譯恐怕更難，千餘年的文學作者的思緒，細微之處不易捕捉，獨特之處尤難表述"[①]。有些外國翻譯家在翻譯與他們自己類似的語言的時候，也有譯錯的時候。舉例來說，法國詩人夏爾·博德萊爾（Charles Baudelaire）將愛德加·愛倫·坡（Edgar Allen Poe）的小説《金甲蟲》（*Gold Bug*）翻譯成法文的時候，他就誤解了非裔美國人方言中"gose"的意思。在"as white as a gose"（像幽靈一般的慘白）短語中，"gose"是非裔美國人英語中的"ghost"（鬼或是幽靈），而他卻誤認爲是"goose"（鵝），把原文譯作"像白鵝一樣的潔白"[②]。如果愛德加·愛倫·坡的英文小說，爲法國譯者誤譯，那麼翻譯中國古代作品的學者不禁要問，中外古今年代相離更遠、文化差異更

① 陳宏天、趙福海、陳復興編輯，《昭明文選譯注》第一册，長春：吉林文藝出版社，1987，4頁。

② 見 Stephen Peitham, ed., *The Annotated Tales of Edgar Allan Poe* (New York: Avenel Books, 1981), p.267. 博德萊爾法文譯文讀"pâle comme une oie." 見 Charles Baudelaire, *Histoires extraordinaires par Edgar Poe* (Paris: Louis Conard, 1932), p.22。

大，使用的語言更是完全不同，那麽時、空、語言的隔閡和差異而造成的誤譯、誤解，將更難以計算。

在翻譯的過程中，閲讀古文最大的困難就是瞭解每字、每句、每行和整篇的内容含義，除了瞭解内容含義外，還必須瞭解當時所襯托的文化背景和社會環境。舉個例子，《文選》有許多描寫京都、城市的作品。爲了能夠準確翻譯這些作品的内容和含義，我曾花了大量時間閲讀關於長安、洛陽、建康、成都等古都的歷史和建築等方面的資料，審讀歷代的地理著作和新發掘的考古報告。在翻譯描寫皇宫和殿堂建築的句子時，我也儘可能多採用中國古代建築的辭彙和術語。除了建築術語之外，這類歷代都城的作品也充滿了對當時朝廷繁文縟禮的描繪。要把這些詞語都正確地翻譯出來，又必須全面閲讀、了解有關當時禮儀的著作和典籍，尤其是涉及朝廷繁瑣的典禮儀式和官服朝冠的著作。可説上至天文、下至地理，舉凡植物、動物、鳥獸、魚禽、礦石和星座的名稱，都必須有充分的認知才能着手翻譯這類作品。翻譯《文選》的賦篇，涉及的範圍可説包羅萬象，包括了中國的制度典章學、天文學、堪輿學、植物學、動物學、地質學、建築學、園藝學、城市規劃學等等知識領域。

我對翻譯的信念是：執着作品的原文和原意。我十分贊同俄裔美國作家和翻譯家納博科夫（Vladmir Nabokov）的金科玉言“最艱澀的逐字翻譯要比最流利的意譯更具有千倍的用途”（the clumsiest literal translation is a thousand times more useful than the prettiest paraphrase）[①]。譯文的流暢和可讀性是每位翻譯家所追求的目標，但是翻譯古代或是中古時代的中國古典文學作品，翻譯家就必須有勇氣表現出美國著名中、日語言學專家羅伊·安德魯·米勒（Roy Andrew Miller）所提出的箴言：翻譯必須具有“字字斟酌、探討語言和文字本義的勇氣”（lexical and linguistic courage）[②]。我個人認爲，在英譯的過程中，必須儘可能地、正確地傳達中文文本的原意，並且儘可能地保留或許會令讀者驚訝的比喻説法，甚至一些非比尋常的措辭用語。例如，西晉木華（生卒年不詳，約公元290年在世）的《海賦》，他用“天綱渤潏”[③]形容堯舜時代的洪荒大

① “Problems of Translation: ‘Onegin in English’,” *Partisan Review* 22 (1955), p.496.

② *Nihongo: In defence of Japanese* (London: The Athlone Press, 1986), p.219.

③《文選》，上海：上海古籍出版社，1986，卷第十二，543頁。

水。華滋生(Burton Watson)翻譯這句話的意思是"The Heaven-appointed waterways swelled and overflowed"(天定的洪水,波濤洶湧氾濫)[①]。這樣的譯文,儘管讀起來十分順暢,但是他錯譯了這句話的原義。"天綱"指的是"天之綱維"(mainstays of heaven),也就是維繫天體的繩網。這句話的意思當作:洪水氾濫高漲,連"天之綱維"都發泡起沫了——"the mainstays of heaven frothed and foamed"(渤潏 *bójué* 指洪水波濤的泡沫)。如果作直接的翻譯,這樣描述洪水的汎濫,或許令讀者有些驚訝,因爲這樣描繪洪水並不合乎邏輯,而且和我們今天對天文的認知也有所差異。但是就我個人而言,我認爲保留木華原來形容洪水氾濫的文字更爲重要,若以"意譯"來翻譯中國古典文學,那麽就很可能會失去原有文字滔天的效應和磅礴的氣勢。

儘管譯者努力追求譯文的準確性,但是有些時候,還是有許多詞語無法在翻譯的過程中充分地表達出來。因此附加詳細的注釋是必要的。我贊成納博科夫翻譯的方式,就是在翻譯中加入大量的注釋。納博科夫表述這一觀點,説得十分精彩:"I want translations with copious footnotes, footnotes reaching up like skyscrapers to the top of this or that page so as to leave only the gleam of one textual line between commentary and eternity."(我要在譯文中加入大量的注解,注解就像摩天大樓一樣向上攀升,達到這頁書或是那頁書的頂端,佔據了整頁篇幅的大部分,而在注釋與永恒之間,只留下少許的譯文)[②]。翻譯學術性的作品,更是應該負起這樣的責任,也就是在譯文中提供充分的注釋。這類注釋,事實上,就是一種評注,注明文中相關的詞句和語法、特殊辭彙、同字異音、特殊讀音、典故出處、字義辨明,並且討論罕見辭句的用法。

英譯《文選》我採用的方式如下:翻譯的英文正文在右頁,注解和説明列在左頁,注解部分可能遠超過翻譯的正文,以《西都賦》爲例,左面全頁的注解遠遠超過右上半頁翻譯的十行正文:

① *Chinese Rhyme-Prose: Poems in the Fu Form from the Han and Six Dynasties Periods* (New York: Columbia University Press, 1971), p.72.

② Nabokov, "Problems of Translation," p.512.

Zhang Heng in his "Western Metropolis Rhapsody" (*Wen xuan* 2.16b). I suspect that this pool, which the *Xinshi San Qin ji* (cited by Li Shan, 1.8b) says flowed into the White Deer Plateau (Bailu yuan 白鹿原) in the Lantian area, was simply part of the Kunming Pond (actually a lake), and that Ban Gu used the terms "sacred ponds and divine pools" as synecdoche for the entire Kunming Pond complex.

L. 132: Jiuzhen 九眞 was a Han commandery located in the area of modern Than Hoa, Vietnam. In 61 B.C. Jiuzhen presented Emperor Yuan a "strange animal" variously referred to as "a white elephant," "a colt with unicorn's color, and ox horns," or "a unicorn." See *Han shu* 8.259, *HFHD*, 2:240. Schafer considers it "a different species of rhinoceros" ("Hunting Parks in China," p. 330).

L. 133: Dayuan 大宛 is the name of the Central Asian kingdom usually identified as Farghana. E. G. Pulleyblank, however, has argued that Dayuan refers to the Tochari people of Sogdiana; see "Chinese and Indo Europeans," *JRAS*, 1–4 (1966), 22–26. In 101 B.C. the general Li Guangli 李廣利 defeated Dayuan after a four-year campaign. Upon his return he presented Emperor Wu with the "blood-sweating horses of Dayuan." See *Han shu* 6.202, *HFHD*, 2:102; *Shi ji* 123.3160, *Records*, 2.266; *Han shu* 96A.3894, *HFHD*, 2:132–5; Arthur Waley, "The Heavenly Horses of Ferghana," *History Today* 5 (1955):95–103; A.F.P. Hulsewé, *China in Central Asia* (Leiden: E. J. Brill, 1979), pp. 132–34, n. 332.

L. 134: The identification of Huangzhi 黃支 is tentative, but most scholars believe it is probably Kanchi (modern Conjeveram) in India. See Gabriel Ferrand, "Le K'ouen-louen et les anciens navigations interocéaniques dans les mers du Sud," *JA* 13 (1919):452–56; Fujita Toyohachi 藤田豐八, *Tōzai kōshō-shi no kenkyū, Nankai hen* 東西交涉史の研究：南海篇 (1930; rev. Tokyo: Ogihara seibun kan, 1943), pp. 124–30. For a more recent review see Su Jiqing 蘇繼廎, "Huangzhi guo zai Nanhai hechu" 黃支國在南海何處, *Nanyang xuebao* 7 (December 1951):1–5. The rhinoceros from Huangzhi was presented to Emperor Ping in A.D. 2; see *Han shu* 12.352, *HFHD*, 3:71.

L. 135: Various identifications for Tiaozhi 條支 have been proposed. Frederic Hirth, *China and the Roman Orient* (Shanghai: Kelly and Walsh, 1885), pp. 144–52 identified it as Chaldea. Shiratori Kurakichi 白鳥庫吉, *Seiiki-shi kenkyū* 西域史研究, in *Shiratori Kurakichi zenshu* 全書 (Tokyo: Iwanami, 1969–), 7:205–36 and "The Geography of the Western Regions Studied on the Basis of Ta-ch'in Accounts," *MTB* 15 (1956):146–60, considered it to be Mesena-Kharacene in the lower Euphrates valley. Fujita Toyohachi argues that Tiaozhi is actually Fars (Persis) in southern Iran; see *Tōzai kōshōshi no kenkyū, Seiiki hen* 西域篇 (1930, rev. Tokyo: Ogihara seitun kan, 1943), pp. 211–52. Miyazaki Ichisada, *Ajiashi kenkyū* アジア史研究. 5 vols. (Kyoto; Tōyōshi kenkyū kai, 1957), 1:151–84 claims that it is a transliteration of Seleucia in Syria. Tiaozhi was famous for its ostriches; see *Shi ji* 123.3163, *Records*, 2:268; *Han shu* 96A.3888; Hulsewé, *China in Central Asia*, p. 113.

L. 136: The Kunlun 崑崙 Mountains, which in Han times had almost mythical significance, stretched from the Qinling range across what is now northern Tibet to the Pamirs.

L. 140: This and the following lines reflect the cosmological aspect of Chinese architecture. The city and the buildings in it ideally were constructed as microcosms of the universe. Ban Gu's description of the palaces as imitations of Heaven and Earth probably refers to the cosmic house known as the Luminous Hall (Mingtang 明堂). According to the *Record of Rites of the Elder Dai* (*Da Dai li ji* 大戴禮記), it was "round on top and square on the bottom" (*Han Wei congshu* 8.19b), meaning that the roof of the central hall (Taishi 太室) was conical to conform with Heaven (which was considered round), and the base was square to accord with Earth (which was conceived of as square). See *i.a.* Wang Guowei 王國維 (1877–1927), "Mingtang miaoqin tongkao" 明堂廟寢通考, in *Guantang jilin* 觀堂集林, *Wang Guantang xiansheng quanji* 王觀堂先生全集 (Taibei: Wenhua chuban gongsi, 1968), 3.10–26; trans. by J. Hefter, "Ming-t'ang-miao-ch'in-t'ung-k'ao: Ausschluss über die Halle der lichten Kraft,

WESTERN CAPITAL RHAPSODY

Are located here and there.
Within the park:
There are unicorns from Jiuzhen,
Horses from Dayuan,
Rhinoceroses from Huangzhi,
135 Ostriches from Tiaozhi.
Traversing the Kunlun,
Crossing the great seas,
Unusual species of strange lands,
Arrived from thirty thousand *li*.

IV

The palaces and halls:
140 Their forms were patterned after Heaven and Earth;

114

除了上述《西都賦》,《文選》内其他的選文還包含了許多罕見的奇字或術語,要做通盤的研究才能翻譯成適當的英文。例如西晉潘岳(247—300)的《射雉賦》,潘岳用了幾個跟弓弩、雉雞有關的專用詞語,"捧黃間以密彀,屬剛罫以潛擬","黃間"和"剛罫"爲弓名和箭名,我把這句翻譯成:I raise the yellow crossbow and quietly bend it, /Fit the steel barb and stealthily take aim。

“摛朱冠之赩赫，敷藻翰之陪鰓。首藥綠素，身拖黼繪。青鞦莎靡，丹臆蘭綷”，則是形容雉雞身上如彩繪一般的羽毛，我翻譯爲：It displays the scarlet splendor of its vermilion comb, /Spreads the ruffled quills of its ornate plums. /Its head is enveloped in green and white, /From its body trails an embroidered design。這些難解的詞語，幸有劉宋時代徐爰（394—475）詳細的注解，徐爰與潘岳年代相去不遠，徐爰的注解解開了潘岳所用謎語似的辭句。例如：“爾乃摮場拄翳”（And then, I open up a clearing and erect a blind）一句，徐爰的注解作：“摮者，開除之名也。今傖人通有此語。射者聞有雉聲，便除地爲場，拄翳於草。”[①]由於徐爰的注解，今人才得知“翳”是在樹叢中搭起的“隱身之物”。

最近正在準備翻譯任昉的《奏彈劉整》——任昉彈劾劉整侵凌寡嫂，苛待孤侄之行。結語有這麽一段話：“令史潘僧尚議論，劉整應輒收付近獄測治。”“測”字在任昉時代具有特別的意思，作“刑訊”解，也就是“命囚犯站立稱測，拷打逼供爲罰……被囚禁的人，先斷其飲食三天，然後才容許家人送粥二升，如果是婦女、老人和小孩，可在一天半以後送粥，測罰滿十天停止”[②]。因此“測”字我譯作“subject to interrogation”。“測”字在這段行文中的用法與一般的用法不同，如果不仔細研究，很容易譯錯。

《文選》所包含的各種文體，賦體可説最具有挑戰性，司馬相如的賦以鋪張著稱，就以他的《子虛賦》爲例，他對丹青雌黄、赤玉昆吾、江蘺東蘠、龜甲玳瑁、楠木桂椒等都有着極其奢張的鋪陳，現在引用部分的原文作例：

In their soil:	其土
Cinnabar, azurite, ocher, white clay,	則丹青赭堊
Orpiment, milky quartz,	雌黄白坿
Tin, prase, gold, and silver,	錫碧金銀
In manifold hues glisten and glitter,	衆色炫耀
Shining and sparkling like dragon scales.	照爛龍鱗
Of stones there are:	其石

① 见 *Wen xuan* 9，p.417。

② 見姜小川，《中國古代刑訊制度及其評析》，《證據科學》2009年第17卷，522頁。

Red jade, rose stone, 則赤玉玫瑰
Orbed jades, vulcan stone, 琳瑉昆吾
Aculith, dark polishing stone, 瑊力玄厲
Quartz, and the warrior rock. 碝石碔砆
To the east there is Basil Garden, 其東則有蕙圃
With wild ginger, thoroughwort, angelica, pollia, 衡蘭芷若射幹
Hemlock parsley, sweet flag, 芎藭菖蒲
Lovage, selinum, 江蘺蘪蕪
Sugar cane, and mioga ginger. 諸柘巴且
To the south there are: 其南則有
Level plains and broad marshes, 平原廣澤
Rising and falling, splaying and spreading, 登降陁靡
Steadily stretching, distantly extended. 案衍壇曼
They are hemmed by the Great River, 緣以大江
Bordered by Shaman Mount. 限以巫山
The high dry lands grow: 其高燥則生
Wood sorrel, oats, twining snout, iris, 巴苞荔
Cadweed, nutgrass, and green sedge. 薛莎青薠
The low wet lands grow: 其卑濕則生
Fountain grass, marshgrass, 藏莨蒹葭
Smartweed, water bamboo, 東蘠雕胡
Lotus, water oats, reeds, 蓮藕觚盧
Cottage thatch, and stink grass. 菴閭軒芋
So many things live here, 衆物居之
They cannot be counted. 不可勝圖
To the west there are: 其西則有
Bubbling springs and clear ponds, 涌泉清池
Where surging waters ebb and flow. 激水推移
On their surface bloom lotus and caltrop flowers; 外發芙蓉菱華

Their depths conceal huge boulders and white sand.	内隱钜石白沙
Within them there are:	其中則有
The divine tortoise, crocodile, alligator,	神龜蛟鼉
Hawksbill, soft-shell, and trionyx.	玳瑁鱉黿
To the north there is a shady grove:	其北則有陰林巨樹
Its trees are elm, *nanmu*, camphor,	楩楠豫章
Cinnamon, pepper, magnolia,	桂椒木蘭
Cork, wild pear, vermilion willow,	檗離朱楊
Hawthorn, pear, date plum, chestnut,	樝梨梬栗
Tangerine and pomelo sweet and fragrant.	橘柚芬芳
In the treetops there are:	其上則有
The phoenix, peacock, simurgh,	鵷鶵孔鸞
Leaping gibbon, and tree-jackal.	騰遠射幹
Beneath them there are:	其下則有
The white tiger, black panther,	白虎玄豹
The *Manyan* and leopard cat.	蟃蜒貙犴[①]

這段短文,充滿了險僻生冷的詞語,辨認十分困難,感謝伯納德·里德(Bernard Read)[②]和近代學者[③] 英譯的《本草綱目》,使翻譯的工作容易多了。

---

①《文選》,第七卷,350—351 頁。

② *Chinese Materia Medica: Animal Drugs* (1931; rpt. Taipei: Southern Materials Center, 1977); *Chinese Materia Medica: Avian Drugs* (1932; rpt. Taipei: Southern Materials Center, 1977); *Chinese Materia Medica: Dragon and Snake Drugs* (1934; rpt. Taipei: Southern Materials Center, 1977); *Chinese Medicinal Plants from the Pen Ts' ao Kang Mu A.D. 1596* (1936; rpt. Taipei: Southern Materials Center, 1977); *Chinese Materia Medica: Fish Drugs* (1939; rpt. Taipei: Southern Materials Center, 1977); *Chinese Materia Medica: Insect Drugs* (1941; rpt. Taipei: Southern Materials Center, 1977); with C. Pak, *Chinese Materia Medica: A Compendium of Minerals and Stones* (1928; rpt. Taipei: Southern Materials Center, 1977).

③ Luo Xiwen 羅希文, trans. and annot., *Compendium of Materia Medica (Bencao Gangmu)*, 6 vols. (Beijing: Foreign Languages Press, 2003).

另外清代選學專家如張雲璈、胡紹瑛（1791—1860）、朱珔（1759—1850）[1]等的著作，以及中、西植物辭典等[2]，都對我的《文選》譯注工作有極大的幫助。在翻譯的過程中，若是有些詞語沒有現成的英文詞，我就按照中文的原意，尋找英文對等的詞語，自創新詞。例如上文提到的"昆吾"，原指火山頂上發現的一種富於銅和金的礦石，因此英文譯作"vulcan stone"（火山石）。"碔砆"一詞也很難確定是哪一種礦石，因此按照字義譯作"warrior stone"（武夫石）。有關植物的名稱，一般譯成英文俗名（通用名稱），附上拉丁學名，或是沒有合適的詞語，我就直譯拉丁文學名的含意，例如"苞"字拉丁學名作 *Rhynchosia volubilis*，我翻譯的英文"twining snout"就是從拉丁學名直接翻譯過來的。如果專家學者對辨認古植物名稱意見不一，不能統一，我就自創譯名，比方，"軒于"又稱"蕕草"，氣味奇特難聞，這種植物早在《左傳》時代就已經有了：僖公四年："一薰一蕕，十年尚有臭。"（這句話我認爲可以翻譯成：Now there is a fragrant plant, then a foul-smelling one; /In ten years there still will be a stench。）因此我將"軒于"譯成"stink grass"（臭草），并加注：*Youcao* 蕕草 is variously identified: *Digitaria sanguinalis (crabgrass), Caryopteris divaricata/nepetaefolia* (spreading bluebeard), or possibly a *Potamogeton*. See *Read, Chinese Medicinal Plants, 36, no. 143; Fèvre and Métailié, Dictionnaire Ricci*, 555。

中國神話中的奇人或是怪物也是翻譯家經常面臨的難題。有些常用詞語的英文翻譯已經廣爲接受，如鸞、鳳，前者作 simurgh，後者作 phoenix。但是也有許多難以翻譯的怪異動物名稱，我處理這類名稱的方式有兩種，或以拼音表示，或另組英文新詞，如梟羊作"roving simian"，蜚遽作"flying chimera"，翳鳥作"canopy bird"，而焦明作"blazing firebird"[3]。上文引用的

---

① 見《文選膠言》（1822年；台北：廣文書局，1966）；《文選箋證》（序 1858年；台北：廣文書局，1966；合肥：黃山書社，2007）；《文選集釋》（1836年；台北：廣文書局，1966）。

② 見 Hu Shiu-ying, *An Enumeration of Chinese Materia Medica* (Hong Kong: Chinese University Press, 1980); Hu Shiu-ying, *Food Plants of China* (Hong Kong: The Chinese University Press, 2005); Francine Fèvre, and Georges Métailié, *Dictionnaire Ricci des plantes de Chine* (Paris: Association Ricci—Les Éditions du Cerf, Paris, 2005); 高明乾主編，《植物古漢名圖考》（鄭州：大象出版社）。

③ 見 Knechtges, *Wen xuan*, Volume Two, 102, L. 329n, L. 330n, L. 334n, and L. 344n。

最後一行提到“蝘蜒”,這個詞語相當難解,郭璞認爲這是一種“大獸,似狸,長百尋”[①]。但是有些評注家則認爲這種大獸“長百尋”是作者誇大其辭,事實上這種動物只有“八尺長”[②]。蝘蜒是一種想象中的動物,究竟有多長,這不重要,或許郭璞是受了張衡《西京賦》的影響,張衡認爲蝘蜒爲“巨獸百尋”[③]。蝘蜒到底是一種甚麽樣的動物,漢學家(翻譯家)的看法不一,華滋生只譯作“leopard”(豹子),而没有作任何的解釋[④],法國漢學家吴德明(Yves Hervouet)認爲是“百尺之狼”(loup long de cent mètres—a hundred meter long wolf),並作了詳細的説明。他引用《説文解字》的説法、“獌”爲“狼屬”[⑤]。但是“獌”指的是“貙獌”或是“貙豻”,是一種狸貓,或是山貓一類的動物[⑥]。蝘蜒既然是一種想象中的動物,各家説法又不能一致,我的譯文就用“音譯”,另外作注來説明[⑦]。蝘蜒從名稱探究,很明顯的,是形容這種動物很長,而從字義上分析,當作蔓延(這種動物也寫作曼延),我曾經想譯作“behemoth”比蒙巨兽,但是後來放棄了這樣的想法,因爲西方的比蒙巨兽指的是“河馬”。既然欠缺合適的英文詞語,我只好採用音譯的方式來表達這種動物的名稱。

除了上面提到難以辨認的罕字奇文之外,最難翻譯的辭彙當屬描寫性的複音詞,我曾經寫了一篇文章專門討論這一類的翻譯[⑧],這篇文章也已經翻譯成中文了[⑨]。凡是具有兩個相同聲母或韻母的描寫性詞語,現代漢語通常把

---

① 見《文選》,卷七,351頁。

② 見高步瀛著,曹道衡、沈玉成點校,《文選李注義疏》,北京:中華書局,1985,卷七,1660—1661頁。

③《文選》,卷二,76頁。

④ 見 Watson, *Chinese Rhyme-Prose*, p.33。

⑤ *Le Chaptire 117 du Che-ki (Biographie de Sseu-ma Siang-jou)*, (Paris: Presses Universitaires de France, 1972), p.30.

⑥ 見 郝懿行,《爾雅義疏》,《四部備要》,卷六,6頁上。

⑦ 見 Knechtges, *Wen xuan*, Volume Two, p.63。

⑧ David R. Knechtges, “Problems of Translating Descriptive Binomes in the *Fu*,” *Tamkang Review* 淡江評論 15 (Autumn 1984—Summer 1985), pp.329-347; rpt. in David R. Knechtges, *Court Culture and Literature in Early China* (Aldershot: Ashgate Publishing Limited, 2002).

⑨《賦中描寫性複音詞的翻譯問題》,刊於俞紹初、許逸民編,《中外學者文選學論集》,北京:中華書局,1998,1131—1150頁;另見蘇瑞隆譯,《康達維自選集:漢代宫廷文學與文化之探微》,上海:上海譯文出版社,2013,135—156頁。

這些詞稱爲聯綿詞或是疊韻詞。聯綿詞或是疊韻詞這類的複音詞在早期的詩歌，尤其是《詩經》和《楚辭》中就很常見，而對後來的辭賦家來説，或許爲了展示他們的才華，他們特别喜愛採用這類詞語表達，而且越是冷僻就越好。

要了解這些詞語，自然免不了要參考許多注釋。然而，注釋家所作的解釋對於現代的讀者來説，總是覺得不夠詳細，不夠精確。例如對某一個聯綿詞的注解或作"高貌"，或作"亂貌"。如果精確一點的，也不外乎作"流水聲貌"這一類的解説。并没有提供精確的含義，只是簡單説明這個詞語在特定語境中所暗示的意義。例如，郭璞的《江賦》有兩句包含了四個聯綿字："潏湟淴泱。瀖潤瀾淪。"李善的解釋作"皆水流漂疾之貌"[①]。李善的注解雖然有助於了解這兩句話的詞義，但是他没有提供任何有關這八個字的個别注解。清代學者曾經作過這方面的研究，提供和該字有關的或體字，比方，胡紹瑛認爲"潏湟"或作"聿皇"，也就是揚雄"羽獵賦"中所稱的"水疾"[②]。諸如此類，不勝枚舉。

我翻譯這類聯綿詞的時候，儘量採用雙聲或是對等的疊韻詞來表達原文悦耳的諧音效果[③]。郭璞《江賦》的這兩句話。我的翻譯是:

> Dashing and darting, scurrying and scudding,
> Swiftly streaking, rapidly rushing.

這一類的翻譯，我會儘量在注解中解釋每個字翻譯的來源。舉例來説，雙聲字"瀖潤"一詞, 或是"倏"(sudden) 和"閃"(flashing) 的同義字。

近年來，我翻譯《文選》遭遇另外一個令人困惑的問題, 使得翻譯的進展更爲緩慢。這個問題就是《文選》"異字"的問題, 我當初開始着手翻譯《文選》的時候，以爲《文選》已是定稿，只要根據可靠的胡克家的版本就行了。事實不然，不久我就發現，《文選》有許多不同的版本，在準備翻譯之前，必須事先比較各種不同的版本，跟胡克家的版本相較，有些差異較小，有的差異就很大。如何處理這個問題呢?

---

① 《文選》，卷十二，560頁。

② 胡紹瑛，《文選箋證》卷十四，367頁。

③ Knechtges, *Wen xuan*, Volume Two, p.327.

限於篇幅,現在只舉謝靈運(385—433)的《述祖德詩》作爲一個例子説明。這首詩的敦煌版本現藏在俄羅斯聖彼得堡亞洲研究中心[①]。很明顯,這是一本唐代的手抄本,但是研究這首詩的學者對這首詩抄定的年代有不同的意見,我認爲北京大學傅剛教授的説法最可信,他將時代定在唐太宗年代[②]。《述祖德詩》是謝靈運爲贊頌他叔曾祖謝安和他祖父謝玄而作。這首詩的第一句話作:"達人貴自我。"達人指知命通達、豁達開放的人,指謝安和謝玄。李善引用《吕氏春秋》作注"陽 [=楊] 生貴己"[③]。陽生指楊朱,也就是拔一毛而利天下不爲的楊朱(《孟子》7A/26)。以自我爲中心的貴己/貴自我的觀念似乎與無私、豁通的達人正好相悖。俄羅斯聖彼得堡亞洲研究中心收藏的版本作"達人遺自我"。此處"貴"作"遺"字。不具名的評注家作注"墨翟貴己,不肯流意天下,故貴自我。作貴勝。遺,棄"。羅國威則認爲"墨翟……所著《墨子》一書,主張'兼愛'和'節用'……墨翟非'貴己不肯流意天下'之人,注非"[④]。敦煌本的評注家很可能見過原文作"達人貴自我"而非"達人遺自我",但是又如何解釋他曲解墨翟的中心思想來説明謝靈運的"達人貴自我"這句話呢? 以"貴己"或是"貴自我"來解釋"達人"似乎不合邏輯。但是如果從另一個角度來解釋"達人"爲"遺己"(舍己、忘己)或是"遺自我"的人,那麽聖彼得堡版本作"達人遺自我"就合乎邏輯了。類似這樣不同版本造成的困擾,就必須在注解中充分説明。

這篇短文,只能略談翻譯《文選》所遭遇的一些困難。希望所舉的幾個例子能夠明確説明,翻譯中國古典文學之前,研究中國的考據、考古、訓詁、聲韻等學科是同等的重要,如果不能正確地了解原文原意,何談正確地翻譯呢?

---

① 孟列夫 (L. N. Menshikov)、錢伯城主編,《俄藏敦煌文獻》,上海:上海古籍出版社,1992—2000,第四册,338—358頁。

② 見傅剛,《文選版本研究》(修訂增改版),北京:中華書局, 2014,285—295頁。

③ 見許維遹校注,《吕氏春秋集釋》(1935年),北京:中國書店,1985,卷十七,30頁下。

④ 羅國威,《敦煌本〈昭明文選〉研究》,哈爾濱:黑龍江教育出版社,1999,133頁。

From Primer to Second-level Reader: David Hawkes and Du Fu, An Appreciation

William H. Nienhauser, Jr.

This paper could be titled "Translating Du Fu is impossible." By that I mean that for me stand-alone translations of Du Fu rarely work. In class almost fifty years ago I recall that I only understood the rendition of "Chun wang" 春望 after my teacher, Liu Wuji 柳無忌 (1907—2002), had explained the historical background of the poem. This is not to say that stand-alone translations of Tang poetry cannot be handled by pens more skillful than mine, but just that for Du Fu, the "poetic historian" 詩史, it is much more difficult. Thus the manner in which William Hung 洪業 (1893—1980) and Chen Yixin 陳貽焮 (1925—2000) present Du Fu's poems within biographical frameworks seems to work best.[①] This is one of the reasons I was drawn to David Hawkes' masterful *A Little Primer of Tu Fu* (Oxford, 1967). Hawkes chose the thirty-five poems collected in the *Tang shi sanbai shou* 唐詩三百首 as his texts. Although he did not arrange them in chronological order, he did provide biographical contexts as part of his five-part analysis of each poem. These five parts were (1) the original text with a

---

Author Affiliation: University of Wisconsin, U.S.A.

① Stephen Owen, *An Anthology of Chinese of Chinese Literature: Beginnings to 1911* (New York: W. W. Norton, 1996), David Hinton, *The Selected Poems of Tu Fu* (London: Anvil Press Poetry Ltd., 1990), Erwin von Zach, *Tu Fu's Gedichte*, 2 v. (Cambridge, MA: Harvard University Press, 1952), and others follow a historical sequence in their translations of Du Fu.

Pinyin Romanization thereof, (2) the historical background in what he called "Title and Subject," (3) a section on "Form," (4) an "Exegesis" that includes a word-for-word translation, and finally (5) a prose rendition of the poem. What seems to be missing is the literary context for these poems, something one encounters immediately in Chinese commentaries both traditional and modern. What I plan to do in this paper is to engage with these Chinese commentaries on or of the poems from Hawkes' corpus as a means to what I hope will be a deeper understanding of each. I should emphasize that I have the utmost respect for David Hawkes' and his work,① and in my readings, to borrow Han Yu's 韓愈 (768—824) assessment of early critics of Du Fu, I am merely "an ant shaking a great tree" *pi²fu² han⁴ da⁴ shu⁴* 蚍蜉撼大樹.②

Let's begin with the fourth poem that Hawkes presented, "Yue ye" 月夜 "Moonlit Night."③ It is one of Du Fu's most famous poems.④ In "Title and Subject" Hawkes provides the circumstances in which the poem was written during autumn 756. That summer Du Fu had moved his family further away from the rebel-occupied capital of Chang'an to Fuzhou 鄜州, about 120 miles to the northeast. Du Fu then set out to join the new court of Emperor Suzong at Lingwu

---

① What has always impressed me with regard to Hawkes' scholarship was his courage. From what I know of his life, it seems he also lived courageously. While many Westerners avoid the major works of Chinese literature, Hawkes plunged into the classics with his dissertation and subsequent book on the *Ch'u Tz'u, The Songs of the South* (1959 and revised in 1985), followed that up with his *Little Primer of Tu Fu* in 1967 (revised and republished in 1987), turned next to *The Story of the Stone* (published from 1973-1986), and finally offered us *Liu Yi and the Dragon Princess: A Thirteenth Century Zaju Play* in 2003. All are works of superb scholarship and literary sensitivity. See *Little Primer of Tu Fu* (Oxford: Clarendon Press, 1967), *The Story of the Stone: A Chinese Novel by Cao Xueqin in Five Volumes* (Harmondsworth: Penguin, 1937-1986), *Liu Yi and the Dragon Princess: A Thirteenth Century Zaju Play* (Hong Kong: Chinese University Press, 2003).

② Han Yu, "Tiao Zhang Ji" 調張籍. Qian Zhonglian 錢仲聯, *Han Changli shi xinian jishi* 韓昌黎詩繫年集釋 (Shanghai: Gudian wenxue chubanshe, 1957), 9.435.

③ Hawkes, *Little Primer*, pp. 28-32; Qiu Zhaoao 仇兆鰲 (1638—1717), *Du shi xiangzhu* 杜詩詳注 (Beijing: Zhonghua, 1979), 4.309-10.

④ This poem has also been titled "Dui yue" 對月 (Facing the Moon; see Bian Lianbao 邊連寶 [1700—1773], Han Chengwu 韓成武 *et al.*, eds., *Du lü qimeng* 杜律啓蒙 [A Beginner's Guide to Du Fu's Regulated Verse], [Jinan: Qi Lu Shushe, 2005], p. 33).

靈武 another 180 miles to the northwest. Along the way he encountered a band of rebels who forced him to return with them to the capital, possibly as a porter. He must have been released when they got to Chang'an, but was unable to return to his family for some months. This poem was probably written, Hawkes notes, at or around the Zhongqiujie 中秋節 or Mid-Autumn Festival in 756.

Hawkes then briefly explains that the prosody of regulated verse like this *wu lü* 五律 pentasyllabic regulated poem) is too complex to be reviewed for his purposes (a textbook), but he does speak to the rhymes, noting that *kàn* and *hán* in lines 2 and 6 were not level tones. The first couplet is then presented in what his "Exegesis" as follows:

To-night Fu-chou moon
My-wife can-only alone watch
今夜鄜州月，閨中只獨看。

There is an explanation of the metonymical *guizhong* 閨中 as "wife." And at the end of the exegesis we have Hawkes' prose translation: "Tonight in Fu-chou my wife will be watching this moon alone."

Here I would like to suggest the first emendation to Hawkes' presentation: the importance of the rhyme which is here *han* 寒 "cold,"① a *ping-sheng* rhyme, thus 看 should not be read as *kàn*, but as *kān*.② Du Fu uses *kān* in a similar fashion in his "Shihao li" 石壕吏 (The Official at Shihao) which begins:

暮投石壕村，有吏夜捉人。Mu tou Shihao cun, you li ye zhuo[1] ren
老翁逾牆走，老婦出門看。Laoweng yue qiang zou, lao fu chumen kan

① Since Tang poets were accustomed to composing poems to set rhymes (it was one of the requirements of the *jinshi* 進士 examination), the idea of coldness (*han*) may color the entire poem.

② Qiu Zhaoao (4.309) points out in a note that 看 is here level tone 平聲. Satō Kōichi 佐藤浩一 notes that Qiu added 11,035 phonetic glosses in his annotation intending to guide the reader to the proper understanding of Du Fu's verse (Satō Kōichi, "Kyū Chōgō *To shi shōchū* ni tsuite" 仇兆鰲杜詩詳註の音注について, *Nippon Chūgoku gakkai hō* 日本中國學報, 58 [2006]: 171-87).

吏呼一何怒，婦啼一何苦。Li hu yihe nu, fu ti yihe ku.

At dusk I put up at Shih-hao Village,

An officer came in the night to seize people.

The old man fled over the wall,

The old woman came out to answer the door.

The officer's shouting at once how angered,

The cries of the woman how embittered.①

Kān in "Yue ye," again used as the final word of a pentasyllabic line, may connote the idea of "to take care of" or "look after"② as in Wang Jian's 王建 "Hanshi xing" 寒食行 (Lines on the Cold Food Festival Day), which begins:

寒食家家出古城，老人看屋少年行。③

Hanshi jiajia chu gucheng,

Laoren kanwu shaonian xing.

For the Cold Food Festival every family comes out of the old city,

The elders take care of the houses, the young gallants are on the move.

Although I don't think the *kān* in "Moonlit Night" means "to take care of," it suggests to me something more responsible in Du Fu's wife than merely looking at the moon. He watches the moon in Chang'an knowing that she *can only be watching* with the concern that *kān* may suggest.

The second couplet in "Moonlit Night," followed by Hawkes' word-for-word rendering, reads:

---

① Qiu Zhaoao, 7.528-29. Revised from Eva Shan Chou's translation *Reconsidering Tu Fu: Literary Greatness and Cultural Context* (New York: Cambridge University Press, 1995), p. 91.

② As Erwin von Zach, *Tu Fu's Gedichte*, V.8 (p. 162) translates: "Mein alter Wirt sprang über de Hofmauer und flüchtete; die alte Wirtin öffnete die Pforte, um nachzusehen, was es gäbe." William Hung paraphrases the line: "In the evening I found a lodging place at Shih-hao Village. / A recruiting officer came to take men at night. / My old host scaled the wall and fled; / His old wife went to answer the gate." William Hung, *Tu Fu: China's Greatest Poet* (Cambridge, MA: Harvard University Press, 1952), p. 141.

③ *Quan Tang shi* 全唐詩, (Taibei: Minglun Chubanshe, 1971), v. 5, 298.3374.

遥憐小兒女，未解憶長安。 Yao lian xiao er nv, wei jie yi Chang'an

In the word-for-word version:

Distant sorry-for little sons-daughters

Not-yet understand remember Ch'ang-an

Although Hawkes translates *yi* 憶 in line 4 literally as "remember," it disappears in his final prose translation which goes:

"I think with tenderness of my far-away little ones,

too young to understand about their father in Ch'ang-an."

In Du Fu's verse *yi* often means "to remember," but it can also mean "to miss" or "think of" as, for example, in the titles to three poems Du Fu wrote at about this time: "Yi youzi" 憶幼子 (Missing My Youngest Son; Qiu Zhaoao, 4.323) and "Yi di, er shou" 憶弟二首 (Two Poems on Missing My Younger Brothers, Qiu Zhaoao, 6.508-10).① A number of modern translators have, like Hawkes, understood the subject of *yi* to be Du Fu's children.② But traditional commentators are less sure. Wang Sishi 王嗣奭 (1566—1648), who Qiu Zhaoao called the best commentator on Du Fu, comments in his *Du yi* 杜臆: "the meaning is based on longing for his family, but by secondarily imagining his family members missing him, he has taken the poem to another level, until he seems to think of his sons and daughters not being able to miss him, and the poem reaches yet another level" 意本思家，而偏想家人思我，已進一層，至念及兒女之不能思，又進一層.③ Following a similar line of interpretation Huang

---

① See also Wang Li 王力 (1900—1986), *Wang Li gu Han yu zi dian* 王力古漢語字典 (Beijing: Zhonghua Shuju, 2000), p. 334, gloss 1.

② Stephen Owen, *The Great Age of Chinese Poetry: The High T'ang* (New Haven, CT: Yale University Press, 1980), "Tu Fu," p. 200 ("While faraway I think lovingly on daughters and sons, / Who do not yet know how to remember Ch'ang-an"); von Zach, III.9, p. 86: "In der Ferne sehne ich mich nach dem kleinen Sohn und dem Töchterchen, die es noch nicht verstehen, an ihren in Ch'angan zurückgebliebenen Vater zu denken."

③ Qiu Zhaoao, 4.309.

Sheng 黄生 (b. 1622) says: "Although the wife has her sons and daughters accompanying her, they do not understand why seeing the moon would cause one to miss Chang'an" 閨中雖有兒女相伴,然兒女不解見月則憶長安。我知閨中遠憶長安,對月獨垂清淚.[1] Bian Lianbao 邊連寶 (1700—1773) cites Yuan Shouding 袁守定 (1705—1782) who argued that [the lines] "I pity my little sons and daughters from afar, not understanding to miss Chang'an" mean that they "do not understand why their mother misses Chang'an" "遥憐小兒女,未解憶長安者," 未解其母之何以憶長安也.[2] This reading, although it strains traditional syntactical rules, is in concert with the image of Du Fu imagining his wife watching the moon. Yet a third possibility, and certainly not a primary meaning of the line, might be that Du Fu himself, returning to a capital city that had been sacked by the rebels, misses the former "Chang'an" as he knew it, for now that capital has indeed been destroyed: *guo po shanhe zai* 國破山河在.[3] Indeed, the poems of late 756 and early 757 all seem to focus either on Du Fu's family members or on current politics. For example, "Qian xing" 遣興, speaking of Du Fu's newly born youngest son,[4] and "Ai wangsun" 哀王孫 (Bemoaning a Royal Son; Qiu Zhaoao, 4.310—14), a poem which contains a summary of the military and political situation at the end of 756, follow "Yue ye" in many editions. Chen Yixin in his epic biography of Du Fu labels his chapter on these months *Riye geng wang guanjun zhi* 日夜更望官軍至 ("Day and night watching all the more for the arrival of the imperial troops"), a line from the Du Fu's "Bei Chentao" 悲

① *Du shi shuo* 杜詩説 (Hefei: Huangshan Shushe, 1994), p. 116.

② *Du lü qimeng* 杜律啓蒙, p. 33.

③ The well-known opening lines of Du Fu's famous "Chun wang" 春望 (Spring Prospects), thought to have been written a few months after "Yue ye" in early 757. Qiu Zhaoao, 4.320.

④ Du Zongwen 杜宗文 (Style name Jizi 驥子, Pony Boy); "Qian xing" 遣興: 驥子好男兒,前年學語時。問知人客姓,誦得老夫詩。世亂憐渠小,家貧仰母慈。鹿門攜不遂,雁足系難期。天地軍麾滿,山河戰角悲。儻歸免相失,見日敢辭遲。Qiu Zhaoao, 4.326.

陳陶 (Grieving over the Battle at Chentao).[1] Perhaps the straightforward reading of the line, with Du Fu's children too young to miss their father in Chang'an, is still the best solution, since it would imply that his wife has no one with whom to share her feelings for her husband.

The syntax of the third couplet of "Yue ye," *xiang wu yunhuan shi, qing hui yubi han* 香霧雲鬟濕，清輝玉臂寒, has also drawn attention from traditional commentators. Hawkes, literal translation is "Fragrant mist cloud-hair wet, / Clear light jade-arms cold." The final prose translation that Hawkes offers is "My wife's soft hair must be wet from the scented night-mist, / and her white arms chilled by the cold moonlight." Both lines suggest that considerable time has passed while Du Fu's wife has been gazing at the moon. Both the mist and the moon would have taken time to moisten (*shi* 濕) her hair and chill (*han* 寒) her bare arms. Thus the lines might better be understood as "In the scented mist her piled-high hair grows wet, / In the clear moonlight her bare arms turn cold." The term *xiang wu* 香霧 seems to have first been used by Du Fu. In later Tang poems such as Li He's 李賀 (791—817) "Qin Gong shi" 秦宫詩 (Poem on Qin Gong; J.D. Frodsham, *The Poems of Li He [791—817]* [Oxford: Clarendon Press, 1970], p. 163) it refers to the emanations from an incense burner. Qiu Zhaoao, however, argues that "the fragrance was emitted from the oil that Du Fu's wife used on her hair (p. 309). Although Du Fu's family was in straightened circumstances at this time, from the poem "Bei zheng" 北征 (Northern Journey) we know that she still carried powder and rouge: "In a minute powder and rouge cover their cheeks, / And the eyebrows are painted askew and too broad" 移時施朱鉛，狼藉畫眉闊 (Qiu Zhaoao, 5.400, William Hung, *Tu Fu*, p. 117). However we understand this couplet, it moves the poem from the metaphysical speculations of the first couplets to a physicality that calls to mind Li Bo's 李白

① Chen Yixin, *Du Fu pingzhuan* 杜甫評傳, 3 v. (Shanghai: Shanghai Guji Chubanshe, 2003), p. 283. Qiu Zhaoao, 4.314.

"Yujie yuan" 玉階怨 (Lament on the Jade Steps):

> 玉階生白露,夜久侵羅襪。卻下水晶簾,玲瓏望秋月。
>
> Yujie sheng bailu, ye jiu qin luowa,
>
> Que xia shuijing lian, linglong wang qiu yue.
>
> On the jade steps, white dew forms,
>
> As the night lengthens, soaking her silk stockings.
>
> She turns within, lowers the crystal curtains and
>
> Gleaming continues to gaze at the autumn moon.①

A major difference between the two poems are the objects of the poet's attention: Li Bo's is a more impersonal depiction, possibly an allusion to a woman abandoned in the Han dynasty, or a veiled self-reference, whereas Du Fu is speaking of his wife.

Nevertheless, there is a level of sensuality in the fifth and sixth lines that Hawkes' rendition, "my wife's soft hair must be wet from the scented night-mist, / and her white arms chilled by the cold moonlight," suggests. This suggestiveness continues in Du Fu's view of his wife's bare, jade-like arms, once warm, now chilled by her exposure to watch "their moon" from the window of her bed chamber.

The final couplet heshi yi xu huang, shuangzhao lei hen gan 何時倚虛幌,雙照淚痕乾 in Hawkes' word-for-word version becomes: "What-time lean empty curtain / Double-shine tear-marks dry." His polished prose translation reads "When shall we lean on the open casement together and gaze at the moon until the tears on our cheeks are dry?" The expression *shuang zhao* 雙照 seems lost here. Although both Du Fu and his wife are watching the moon alone (*du kan* 獨看), Du Fu hopes that the moon will link them by shining on them both. Some

① Li Bo, *Li Taibo ji* 李太白集 (Taibei: heluo Tushu Chubanshe, 1975), 5. 144-5. Perhaps Du Fu even echoes, intentionally or inadvertently, Xie Tiao's 謝朓 (464-499) "Yujie yuan" 玉階怨, a poem that portrays the separation of lovers from the lonely female's point of view.

scholars would also suggest, however, that it is either the reunited couple' s warmth or the moonlight itself that dries their tears,[①] and that the *xu huang* 虚幌 indicates a window covered in light silk or a sheer curtain. Beyond exploring the language of the poem itself, resonances with earlier well known poems such as "Yue chu" 月出 (The Moon Appears), Mao #143 in the *Shijing* are cited by some commentators as part of the context needed to fully understand "Yue ye." "The Moon Appears" is one of the earliest poems that metaphorically compares a lover to the moonlight and in the following translation by Martin Kern brings out the full erotic connotations that may also be operable in "Yue ye":

月出晈兮,佼人僚兮。舒窈糾兮,劳心悄兮。
月出皓兮,佼人懰兮。舒懮受兮,劳心慅兮。
月出照兮,佼人燎兮。舒夭紹兮,劳心慘兮。

The moon comes forth, how bright,
The beautiful girl, how adorable!
At leisure she is in her sensual allure—
My toiled heart, how anxious.

The moon comes forth, how brilliant,
The beautiful girl, how lovely!
At leisure she is in her beguiling charm—
My toiled heart, how troubled.

The moon comes forth, how radiant,
The beautiful girl, how vibrant!
At leisure she is in her enchanting appeal—

① Thus Stephen Owen ("Tu Fu," p. 200) rendered the last couplet: "When shall we lean in the empty window, / Moonlit together, its light drying traces of tears."

My toiled heart, how haunted.①

Du Fu's "Yue ye" has also reminded commentators of the nineteenth of the "Gushi shijiushou" 古詩十九首 (Nineteen Ancient Poems):

明月何皎皎,照我羅床緯。憂愁不能寐,攬衣起徘徊。
客行雖云樂,不如早旋歸。出户獨彷徨,愁思當告誰!
引領還入房,淚下沾裳衣。

The bright moon how pure and white,
Shines on my gauze bed curtains.
Worry and sadness won't let me sleep,
Pulling on a coat I get up to pace about.
Although traveling away from home is said to have its joys,
Far better early on to turn towards home.
I go out the door and wander about alone,
To whom can I tell these sad thoughts?
Looking out into the distance as I return to my room,
Tears fall soaking my gown.②

Although Du Fu's "Yue ye" is a well-known poem even in modern China, it offers a text laden with contexts that make translation difficult. Perhaps the poem is best understood by considering several of the available commentaries and translations. For me the driving emotion of "Yue ye" is a longing, both physical and emotional, replete with traditional associations. If I were to venture a new translation, it would run as follows:

---

① Martin Kern, "Lost in Tradition, the Classic of Poetry We Did Not Know," *Hsiang Lectures on Chinese Poetry*, Grace Fong, ed., volume 5 (McGill University, 2010), p. 43. Kao Yu-kung and and Mei Tsu-lin also point to the tactile nature of Du Fu's reference to his wife in the third couplet of this poem ("Syntax, Diction, and Imagery in T'ang Poetry," *HJAS* 31 [1971], p. 74).

② See also the translations by Stephen Owen, *Anthology*, p. 256, and Charles Hartman in Liu Wu-chi and Irving Yucheng Lo, *Sunflower Splendor: Three Thousand Years of Chinese Poetry* (Bloomington: Indiana University Press, 1990), p. 33.

今夜鄜州月，閨中只獨看。遥憐小兒女，未解憶長安。
香霧雲鬟濕，清輝玉臂寒。何時倚虛幌，雙照淚痕乾。

For the Fuzhou moon tonight,
My wife can only keep watch alone.
I pity my far away little ones;
They can't understand why Chang'an should be held dear;
A mist made fragrant moistens her piled high hair,
The clear moonlight chills her bare arms.
When will be able to lean at an open window
And let the moon shine on us both, drying the tracks of our tears.

The suggestions I've made in this essay are speculative. Of course, it is possible to read Du Fu without delving into the poems and the commentarial tradition. But this would be to read the poems in lower-case type, to skim meanings off the surface of deep pools. My reading of resonances is intended to supplement Hawkes' own exegesis and notes. They are intended to demonstrate both the complexity of Du Fu's verse and the meticulous accuracy and literary sensitivity of David Hawkes' comments on that verse.

盛唐山水詩文用語考證*

柯慕白(Paul W. Kroll) 著
楊杜菲 譯 童嶺 校

山川文化的漢學研究發端於約90年前愛德華·沙畹(Édouard Chavannes)關於泰山的具有里程碑意義的專著[①]。此後多年,許多重要的學術性論著相繼問世[②],與此相關且範圍更廣的傳統中國的"天人合一"理論,也從文化、政治、宗教等角度催生出大量文章與論著[③]。爲多方面研究該話題,幾部學術專

作者單位: 美國科羅拉多大學波爾得分校
譯者學習單位: 南京大學文學院

* 本文原載《通報》84.1-2(1998),後收入柯慕白《中古中國的文學與文化》(*Essays in Medieval Chinese Literature and Cultural History*, Ashgate Variorum, 2009)一書中,經作者授權翻譯爲中文。

① *Le T' ai chan : essai de monographie d' un culte chinois.* (Paris: Ernest Leroux, 1910).

② 比如 Michel Soymié, "Le Lo-feou shan: étude de geographie religieuse" , *Bulletin de l'École française d'Extrême-Orient* 54(1956), pp.1-132. 以及 Edward H. Schafer, *Mao shan in T' ang Times*, rev. 2nd edn. (Boulder: Society for the Study of Chinese Religions, 1989).

③ 主要的里程碑包括青木正兒(Aoki Masaru)1934年的文章《支那人の自然觀》("Shinajin no shizenkan"),收入《青木正兒全集》(*Aoki Masaru zenshū*)6卷,(Tokyo: Shunjūsha 1969—1975),卷2, pp.552-591; Richard B. Mather, "The Landscape Buddhism of the Fifth-Century Poet Hsieh Ling-yün", *Journal of Asian Studies* 18.1,(1958), pp.67-79; J.D. Frodsham, "The Origins of Chinese Nature Poetry", *Asia Major* 2nd ser., 8.1(1960), pp.68-104; Obi Kōichi, *Chūgoku bungaku ni arawareta shizen to shizenkan, chūsei bungaku o chūshin to shiru*, (Tokyo: Iwanami,1963)(譯者注:中譯本見《中國文學中所表現的自然與自然觀——以魏晉南北朝文學爲中心》,小尾郊一著,邵毅平譯,上海:上海古籍出版社,1989); Paul Demiéville, "La montagne dans l'art littéraire chinois", *France-Asie* 183(1965), pp.7-32, rpt. in Demiéville, *Choix d' études sinologiques* (1921-1970) (Leiden: Brill 1973), pp.364-389; 王國瓔,《中國山水詩研究》,台北:聯經出版社,1986; Tokura Hidemi, *Shijintachi no jikū: Kanpu kara Tōshi e*,(Tokyo: Heibonsha,1988)(譯者注:中譯本見户倉英美,《詩人們的時空——從漢賦到唐詩》);(轉下頁)

題論文集已經成稿[①],從而也擴充了該領域的參考文獻。《遠東亞洲叢刊》(*Cahiers d' Extrême-Asie*)爲"神聖地理學"的研究作出了不少貢獻[②]。如今人們很容易從中國期刊研究中找到有關中國山嶽的文章,頻繁的研究開始挑戰既有的題材,如王維的"山水詩歌"或者柳宗元的"田園散文"。我在山水上也花了很多心思,但是不知道研究的方向又該如何發展。

當然,我們應該對每一個幫助我們更仔細瞭解中國傳統文化的新事實與新數據充滿感激。當你已經看到了一座中國的山時,你並没有把所有的看完,但每座山的輪廓都是模糊地掩映在另一座山之後,它們的輪廓總是顯得太過相似。或許是時候重新審視我們關於中國山水的文學價值的主觀臆斷了。現在,一個研究唐朝的學者需要擦亮眼睛,才能辨清方向,因爲這是一個被深入研究且充滿吸引力的時代。因而,我們首先必須盡力回避那些影響我們判斷的被簡化的種類與陳詞濫調,以此來查明我們的結論,佐證我們的觀點。

因此我們不應該認爲:"自然詩"這個術語,只是一個輕率的没有批判價值的標籤。我們要承認中國詩歌——就像世界各地的傳統詩歌一樣——從一開始就喜用自然世界作爲意象的源泉;但這其實是一個顯而易見的事實。再進一步思考就會發現,我們對它的内涵其實一無所知。我們也不該在任何(甚至是最庸俗的)層面上確保自己找到了"宗教"與"文學"的文本之間毫無争議的差别,即便是研究唐代的學術精英,也自認爲他們的作品完全遵照了大學課程所規定的界限。相反,我們從最開始就接受了所被給予的,即文字的力量與神聖的存在之間不斷的融合與二者相互的吸引力。在這方面或許

(接上頁)最近的有 Donald Holzman, *Landscape Appreciation in Ancient and Early Medieval China: The Birth of Landscape Poetry* (台北:清華大學人文社會學院, 1996)。Jean-Pierre Diény 關於各種自然現象與想象之象徵的研究也不容忽視,比如"Pour un lexique de l'imagination littéraire en Chine: le symbolisme du soleil", *Bulletin de l' École francaise d' Extrême-Orient* 69 (1981), pp.119-152.在衆多研究動植物的著作中,最近很有趣的一篇是 Martin Kern, *Zum Topos "Zimtbaum" in der chinesischen Literatur:Rhetorische Funktion und poetische Eigenwert des Naturbildes kuei* (Stuttgart: Franz Steiner, 1994)。

① 包括本文1993年1月在加利福尼亞聖巴巴拉大學的早期譯本。

② 其中一個很好的例子是, James Robson 很有見地的文章"The Polymorphous Space of the Southern Marchmount (Nanyue): An Introduction to Nanyue' s Religious History and Preliminary Notes on Buddhist-Daoist Interaction ", *Cahiers d' Extrême-Asie* 8 (1995), pp.221-264。

有更多值得商榷的問題,動摇最根深蒂固的理念並非易事。

但是在談到本文的中心之前,我想强調兩方面的問題。我們可以從李白(701—762?)一首家喻户曉的絶句《山中問答》開始:

問余何意棲碧山,笑而不答心非閑。
桃花流水窅然去,别有天地非人間。①

通常認爲這首小詩表露了李白對世俗的輕蔑,反映了他對自然的熱愛,揭露了李白古怪性情中顯而易見的超然態度與玩世不恭,這種解讀儘管正確,也不過是老生常談罷了。敏鋭的讀者會注意到頸聯是對陶潛(365—427)《桃花源記》這一典故的引用,尾聯則是對道教洞天的暗示。也有賞析文章發現了詩歌標題中隱含的"對白"(山中問答,後文將提及)。其實根本没有真正的對話,因爲詩中人物没有用語言來回答②。然而,要是我們留意早期詩歌的歷史回音,像李白那樣,我們就會抓住一個更爲重要的暗示,這就是陶弘景(456—536)和齊高帝(479—483年在位)的相互問答。當齊高帝下詔讓這個有影響力的道教高人兼學者從静修處茅山進宫時,據説他當時頒了一道詔書,傲慢地質問陶弘景"山中何所有",不打算離開静修處的陶弘景,用以下一首絶句作了回答:

山中何所有,嶺上多白雲。
只可自怡悦,不堪持贈君。③

---

① 瞿蜕園、朱金城注,《李白集校注》,上海:上海古籍出版社,1980,卷19,1095頁;《全唐詩》(北京:中華書局,1960)卷178,1813頁。

② Stephen Owen, *The Great Age of Chinese Poetry:The High T'ang* (New Haven:Yale Univ. Press, 1981), p.136.(譯者注:中譯本見宇文所安著,賈晉華譯,《盛唐詩》,北京:生活·讀書·新知三聯書店,2014。)這篇出人意料地直截了當的文章忽略了詩歌語言的表達——即詩歌本身。當然,"問答"在任何實質意義上,幾乎都算不上對話,毋寧説文字。更糟糕的是,宇文所安把他對詩歌的理解很大程度上建立在這個晚出的、不定的題目上,儘管事實是所有李白最早的詩歌,包括8世紀中期的首部詩歌選集《河嶽英靈集》的修訂本,都把題目記載爲"答俗人問",明顯地表示這裏存在一個真正的回答。

③ 逯欽立編,《先秦漢魏晉南北朝詩》,北京:中華書局,1983,1814頁;《茅山志》(*HY* 304,譯者注:*HY* 爲哈佛燕京圖書館 Harvard-Yenching Library 縮寫)卷28,第2b頁;P. W. Kroll, *Meng Hao-jan* (Boston: G.K. Hall, 1981), p.101;參見《太平廣記》,台北:古新書局,1976,卷202,414a頁,其中奇聞異事引用了早期7世紀楊松玠的《談藪》。

李白詩的重點不在於人物没有回答，實際上他自己無聲地進行了答復，場景替他作了回答。所有可以用文字表述的事情，已經在兩個半世紀以前由陶弘景回答完畢[①]，很顯然詩人李白與這個笑而不語的人物截然不同，他字斟句酌，精心挑選意象，從而描繪出這幅場景。這裏提出的話不是從文字的角度看山水，而是一組文學意象，文字（而非眼睛）的第一性，以及文字隨着時間不斷積累的迴響，都是重新審視唐朝詩歌中的山水時需首要注意的。關於這點，我們將在下文進行深入的研究。

第二個方面是唐代人認爲理所當然的現實世界中的宗教力量，除非我們特别在文本或儀式中尋找，否則我們都易於做出這種錯誤的假設：山水對於他們（正如對於我們而言），不過是城市主導的生活中風景秀麗的背景而已。但是，對於所有不可否認的唐朝文明，在日常生活中，人們與自然環境有着我們難以效仿的親密關係。這需要想象與理解，才能使我們瞭解並融入到這樣一個世界，同樣我們必須有意識地强迫自己認識到印刷文化（或逐漸發展的後印刷文化）和與之相對的手抄文化之間的巨大差别。二者的差别如此之大以至於無論怎樣努力，我們一刻也不能掌控變換的意象，事實是我們對於這種持久的融合過於陌生了。舉個簡單的例子，對生活在8世紀長安的貴族或者平民而言，即使他們生活在城市中，都能在每一個晴朗的夜晚看到滿天繁星，而我們只有從充滿光污染和霧霾的城市搬離時才能看到這些景象。月亮每晚的光暈變化並非偶然，而是不可避免的、影響巨大的自然現象——這個事實並非毫無意義，這從根本上決定了一個人與周遭環境的關係。

正如本文一開始所説，近幾年許多學術研究都專注中國的"神聖地理學"，以至於太多的人説"中國山水最重要的特徵不僅是自然的物質，也是神

① 另一個來自過去的可能回應，被最早的李白評論家識别了，楊齊賢（1190年）："人問諸葛亮之志，亮抱膝，笑而不答。"原始材料見《三國志》注（北京：中華書局，1973）卷35，911頁，引自《魏略》。諸葛亮集詩人、傳説中的隱士及軍事家等身份於一體。見《分類補注李太白詩》（《四庫全書》）卷19，3a頁。楊齊賢的評論通常很到位，但遺憾的是（正如這個例子），並没有收進李白作品衆多的現在版本中，即使這些都聲稱是綜合的全面的研究。

聖的客體”[1]。因而很好地提醒我們,不管在政治領域還是宗教領域(這同本文所論述的一樣),這種見解是多麼地根深蒂固。例如,李隆基(玄宗,712—756年在位)執政期間,他對認可並授予其疆域内高山河流以力量産生了極端且持久的興趣,包括給五嶽(泰山、衡山、華山、恒山、嵩山)、四瀆(河、江、淮、濟)、四鎮(會稽、沂、醫無閭、霍)、四海授予新的封號,或給其他山川以神學意義上的組合稱號[2]。帝國對這些自然靈魂的有序供奉和安撫儀式由皇親國戚來執行,通常是正規血統的皇子[3]。唐玄宗於公元725年12月在泰山頂順利進行的封禪儀式是家喻户曉的,不過最有意思的是,當從這個角度看中國的疆域時,這片土地並非由地理力量固定並灌溉,而是或多或少地由當地有權勢的居民統治着。

或許你會認爲假如不是時代錯誤的話,這種觀點在唐朝就已經過時了。例如上面提及的這個莊嚴的儀式,已經很大程度上變成了一種正式的典禮。不過事實可能正好相反,這遠比僅僅祈願封號或是宣告模糊的(從現代的眼光看,令人欣慰的)理性主義傾向更加複雜。樂維(Jean Levi)的作品專門弄清楚了中古朝廷官員最主要的角色是國家信仰的代表,儒家“道”以天子爲中心,而非政治上的官吏禄蠹[4]。我們習慣於將唐朝的政府官員稱爲官僚主義者(對我們而言,難免有“權利之廊”、“統治之梯”的釋義),以至於把之前典型的活動都認爲是宗教使者所爲。就像道士那樣,他們試圖控制各種地方神,

---

① P. W. Kroll, “Verses From on High: The Ascent of T'ai Shan”, *T'oung Pao* 69(1983), p.223;修訂版收録於 *The Vitality of the Lyric Voice:Shih Poetry from the Late Han to the T'ang*, ed. Lin and Owen, (Princeton: Princeton Univ. Press, 1986), p.167。參見Gaston Bachelard(他最先把與天相對的地看作一個浩渺的使人敬畏的象徵),“La moindre colline, pour qui prend ses rêves dans la nature, est inspirée.”見其*La terre et les rêveries de la volonté* (Paris: Librairie José Corti, 1948), pp.378-402,尤其是pp.379-380和p.384的引用。

② 參見《册府元龜》(北京:中華書局,1980)卷33,7a-23b頁,一些與玄宗在此地活動相關的詔令;或《唐會要》(台北:世界書局,1960)卷47,834頁,以及《唐大詔令集》(台北:鼎文書局,1978)卷74,418、419頁。Kroll, “Verses From on High”, pp.236-237 n53 (pp.184-185 n53)對一些信息的概述。

③ 參考如《册府元龜》卷33,13a頁(日期爲開元二十五年十月初八,即737年11月4日的詔令);卷33,15b-16a頁(日期爲天寶三年四月初三,即744年5月19日的詔令);卷33,21b-22a頁(日期爲天寶十年三月十六,即751年4月16日的詔令);《令嗣鄭王希言分祭五嶽敕》,《唐大詔令集》卷74,418頁。

④ 特别見其“Les fonctionnaires et le divin:luttes de pouvoir entre divinités et administrateurs dans les contes des Six Dynasties et des Tang”, *Cahiers d' Extrême-Asie* 2 (1986), pp.81-106, 以及 *Les fonctionnaires divins* (Paris: Seuil, 1989).

以及其他地區通過非法崇拜(淫祠)的異教神。那些被稱讚的教化,不亞於通過修建學校和廟宇將狹隘的信仰替换成與儒家教義相關的教育,的確在中世紀,經過授權的儒家官方和道家規定在有影響的活動上有許多共同點,"或許在中國,朝廷官員和道家法師僅是神職的兩方面"[①]。

就這一點而言,我們可以通過同時代道家"十大洞天、三十六小洞天、七十二福地"的體系,來更好地欣賞以崇高的山河覆蓋的唐代神聖地理之圖,這些與佛教無關的記載在唐玄宗時期都由真人司馬承禎編成了法典[②]。我們甚至可以在這一時期看到皇室與道教的融合,當被司馬承禎説服相信五嶽真正的統治者是上清衆神(與獨有的培養皇帝的"山林之神"所不同)中的真人後,唐玄宗把他們從五嶽的洞府中請下來,卸除其在凡世的責任,還根據司馬承禎的意願,詔令在五嶽山頂各建真君祠一所,以表示對諸神的敬意[③]。

但山水並不是簡單的陸生地理,這是另一個經常被我們忽略的細節。所以它應該被提起,比如,"分野"或者"被分配的地域"這個傳統的概念在唐玄宗時期仍然流行,即因各種地理、政治原因,早期中國根據天上星次的位置來劃分地面上的位置,使二者相對應。讓我們回顧一下,道家的觀點認爲,在某些方面,可以把塵世的山水看做是天空現象的表現或者容納體,就像人體本身可以被看做縮影,是山谷、河流、太陽、月亮和星星等客體的具體化與超自然的身份。對於佛教徒而言,一行禪師這個佛教密宗的重要領袖,在原有的

---

① Levi,"Les fonctionnaires et le divin", p.106.尤其參見 K. M. Schipper, "Taoist Ritual and Local Cults of the T'ang Dynasty", in *Tantric and Taoist Studies, in Honor of R. A. Stein,* vol.3 (*Mélanges chinois et bouddhiques* 22; Brussels: Institut Belge des Hautes Études Chinoises,1985), pp.812-834,比如當地宗教信仰與道教"純"神學之間的衝突。關於中國文化的道教中心最好的論述是 Schipper, *Le corps taoïste: corps physique-corps social*, (Paris: Fayard,1982).更短篇幅的還有 Anna Seidel, *Taoismus, die inoffizielle Hochreligion Chinas* (Tokyo: Deutsche Gesellschaft für Natur-und Völkerkunde Ostasiens,1990).

② 其完整的佈局,可參見保存在11世紀類書《雲笈七籤》(*HY* 1026)第27章的《天地宫府圖》,這爲今天較爲熟知的杜光庭(850—933年)《洞天福地嶽瀆名山記》(*HY* 599)中的格局奠定了基礎。在其前言中,司馬承禎對這一神秘的衆仙所居之地進行了精彩的描述。有很多唐前的關於道教神聖地理的構想,比如6世紀的《無上秘要》(*HY* 1130)卷4,5b-10b頁,就對俗世地理做了恰當的概括。同時請注意早期關於二十八山的舉例,即葛洪(283—344)稱爲"可以精思合作仙藥"之地,參見王明,《抱朴子内篇校釋》,北京:中華書局,1988,增訂本,卷4,85頁。

③ 參考《雲笈七籤》卷5,15b頁中李渤於公元805年所撰《真系》的序言。《真系》中關於司馬承禎的記載是《舊唐書·司馬承禎傳》(北京:中華書局,1975,卷192,5127—5129頁)的藍本,後者記録了此事。

“分野”概念中添加了早期從印度天文學中傳入的更爲相關的二十八星宿,使已經成熟的體系更加完善。但我僅僅提及這些在研究中的問題,忽略了很多其他方面(比如對私人園林日益增長的興趣、歷史版圖的發展、佛教曼荼羅的傳播、以及持續性的對盆栽藝術的狂熱),作爲廣泛流傳的真理,儘管對於8世紀的作家而言很普遍,但我們仍需盡力將它們保持在我們眼前,或者至少不要拋得太遠,以免我們無意間對唐朝山水忽略而進行主觀揣測。

所有這些導致的是:儘管有記載完善的受南北朝時期影響的使山水文學柔和(的方式),儘管有受江南地區持續文化同化而帶來的對更加良好的自然環境的優雅鑒賞,我們仍然涉及了一種叫“自然的超自然主義”[①]的山水文學描寫,即使更深層次的庇護不再被當做“神聖的恐怖地帶”[②]而完全令人畏懼。

這些言論可能不够充足,但也是必要的準備工作,幾乎唐朝生活的每一個層面在一定程度上都與這個話題有關,而且很容易尋找研究方向。但是現在是時候思考一些文本了。

目前,我們暫且保持世俗的態度,讓李白來作試金石。首先看看兩首關於廬山瀑布的詩,其中以下這首七言絶句更爲知名:

日照香爐生紫煙,遥看瀑布掛前川。
飛流直下三千尺,疑是銀河落九天。[③]

尾聯享有許多盛譽,其把瀑布比喻成從天空傾瀉而下的銀河這個出人意料的構想,受到了蘇軾(1036—1101)特別的讚賞[④]。從詩的開始很難注意到

---

① 語出 M. H. Abrams 的名著 *Natural Supernaturalism: Tradition and Revolution in Romantic Literature* (New York: Norton, 1971),一本爲當前學科提供了很多有趣的參照點的書。

② Demiéville,“La montagne dans l’art littéraire chinois”15(372),其中描述了早期中世紀山的“古典觀”。參見 Marjorie Hope Nicholson, *Mountain Gloom and Mountain Glory: The Development of the Aesthetics of the Infinite* (1959; rpt. New York: Norton,1963),關於17世紀末到18世紀歐洲山水文學的相似變換的故事。

③《望廬山瀑布二首》第二首,《李白集校注》卷21,1241頁;《全唐詩》卷180,1837頁,10世紀的《文苑英華》(台北:新文豐出版公司,1979)卷164,4a頁首聯印刷失誤,現在所看到的首聯被放在第二句,前面用了一句“廬山上與星斗連”,這種開頭技巧在我看來,顯得太過單調而平實。

④ 王松齡點校,《東坡志林》,北京:中華書局,1981,卷1,4頁。

詩人是如何巧妙地醞釀這種構思的，我們首先看到的不是瀑布，而是廬山西北邊被太陽照射的峭壁——香爐峰[①]。和往常一樣，李白使用了一個地名的語義技巧，將香爐峰比喻成它的名字所表達的那樣，詩人向我們展現的第一幕是一只被迷霧籠罩的博山爐，蔓延的白雲是山中霧氣的痕跡。但是接下來，霧靄的顏色把可見的線條意象融入想象的整體，紫色的天際渾然一體，這裏建立了對詩歌結論中體現的超凡脱俗的最初偏見[②]。在最初支配性意象的促使下，頷聯、頸聯中體現的瀑布的飛流直下便顯得非同尋常，從而也引出尾聯來。

我們在這裏談論的例子，是一個運用想象和語言來刻畫的可能被認爲單純描寫的場景，並没有過多冗長的强調（我希望如此）。在繼續往下説之前，我們應該提出這首詩用五律寫的更長的姊妹篇[③]：

西登香爐峰，南見瀑布水。
掛流三百丈，噴壑數十里。4
欻如飛電來，隱若白虹起。
初驚河漢落，半灑雲天裏。8
仰觀勢轉雄，壯哉造化功！
海風吹不斷，江月照還空。12
空中亂潨射，左右洗青壁。
飛珠散輕霞，流沫沸穹石。16

---

① 廬山上有四個不同位置的香爐峰，所有唐以前的文獻中提到的都在西北邊，與佛教和尚慧遠有關。清代著名學者、李白詩注者王琦，在其論著中幾乎解決了所有關於李白詩中"香爐峰"之義的争論。然而，在最近出版的用紀傳體寫的關於李白詩作的書中，安旗（此前也有許多清代學者）認爲，大家所談的頂峰是更南的"香爐峰"（雖然詩歌最早的解釋是"西邊"）。可參見《李白全集編年注釋》，成都：巴蜀書社，1990，49頁。儘管安旗在地理學上的論據很有説服力，但這仍然只停留在李白時代就已出版的傳統文學表層。在6世紀《高僧傳》的慧遠傳記中，也指出在"香爐峰"西北附近也有一道漂亮的瀑布存在，參見《高僧傳》卷6，《大正藏》第2059部，第50册，358b頁。

② 關於紫色的象徵意義早已被Manfred Porkert總結爲："kosmische Ganzheit und Fülle, ungeschmälerte Macht, deshalb auch die wiedergewonne Einheit, die Rückkehr zum kosmischen Tao"，參見其"Untersuchungen einiger philosophisch-wissenschaftlicher Grundbegriffe und Beziehungen im Chinesischen"，*Zeitschrift der deutschen morgenländischen Gesellschaft* 110.2(1961)，pp.439-440.關於大氣的意象反映，參見Kroll, "Li Po's Purple Haze"，*Taoist Resources* 7.2 (1997)。

③《望廬山瀑布》第一首，《李白集校注》卷21，1238-1239頁。《全唐詩》卷180，1837頁。

而我樂名山,對之心益閑。

無論漱瓊液,且得洗塵顔。20

且諧宿所好,永願辭人間。22

儘管不如前一首出名,但這首詩也是一件值得研究的絶作。我們首先可以注意到,這首詩的前八句包含了在前一首絶句中出現的多數意象,只是在形式上有所擴展。在這首詩裏,“落下的銀河”這一意象把對瀑布的描繪刻畫引向了一個更深遠更完整的層面(第七、八句)。接下來韻腳一轉又迅速收回,並且用了一個祈使句和感歎句來讚美“造化功”,如此得精挑細琢且巧奪天工[①](此處應該注意,詩人極其貼切並巧妙地運用“轉”這個形容詞來描寫瀑布的浩大湧動之勢)。接着李白特别選取兩個細節來引人注意,每一個見過瀑布的人都應該知道第十一句所指爲何,在瀑布附近總有超乎想象的狂風在呼嘯。第十二句中的月亮,在水中若隱若現(瀑布在這裏被看作一條垂直的江河,相當於前面絶句中懸掛着的“川”),彷彿散射出一束光,而後又彌散在空中,這個美麗的景象與第五、六句[②]的光交相輝映。接下來變化的韻腳通過“頂針”這一修辭手法串起來,與第十二句月亮消散的“空”相連,第十三句一開始也是“空”的,就好像亂潨在空中四濺。在接下來的詩節中,水流直下沖洗兩邊青碧,這些濺起的小水珠(即第十五句中的“飛珠”)消失在周圍的彩霞中(宛若此前倒映的月光反照回空),而水流激起泡沫在山石中沸騰。

接着又换韻腳,表面的描寫部分告一段落,所以讓我們暫停下來,考慮一下李白是如何展開繁複的描述的。前面三節中,處處可見方位意象的上下交替使用,如垂直、水平,或者一些精心插入的水平蔓延(1上,2—4下,5—6上,7—8下,11水平,12上/下,13下,14水平,15—16向外/上/下),生動地模仿了瀑布“轉”的力量。人的眼睛總是被不同的風景吸引而上下移動,就好像這流動的瀑布自己以各種方式循環一樣。細酌每一處李白所强調的聽覺、觸覺和視覺,都會嚴重阻礙我們的討論進程,但是有兩處地方不得不提——第九、十

① 見 Edward H. Schafer, “The Idea of Greated Nature in T’ang Literature”, *Philosophy East and West*.15.2(1965), pp.153-160,關於此概念有一個粗略的看法。Gerard Manley Hopkins 關於“精粹”(instress)和“獨特”(inscape)的觀點具有值得深究的相似性。

② 這兩行受到學者高度讚美,包括胡震亨(1590)、王琦(相比前面的絶句,更喜歡這首)等評論家。

句——兩句插入式的聲韻一致的感歎語，這是在本詩中唯一没有换韻律的對句，也是比較常見的，這種有節奏的重複增强了詩的氣勢[①]。同時我們還可以注意到，第十三句中不尋常的半諧音（“空中亂潨射”），有意地把亂濺的聲音與水滴本身融合在一起。

在最後一節中詩人表達了他對這一場景的看法，表明了這一美景對他及其後來行爲的影響。廬山與佛教的聯繫，在慧遠（334—416）組織著名的蓮社時就開始了，這比其與道教聯繫更有歷史性的意義[②]。在這篇詩文中，李白撇開道教的“長生不老”教義，而選擇佛教的洗盡塵顔：瀑布把他世俗的污垢沖洗走了，使其得到净化與解脱（注意文學意象的恰當表述），他發誓永久地遠離人間世俗[③]。

最後，讓我們來看看更顯此詩技巧嫺熟的一點，那就是格律。在第一節的八句押仄聲韻：xAxAxAxA。第二節的四句押平聲韻：BBxB。第三節的四句押仄聲韻：CCxC。最後一節的六句（或者可以更恰當地，把它看作一個加了對句作爲結尾的絶句，最後補充的一組對句很明顯由第一個字“且”表明，與第二十句的第一個字“且”相呼應）押平聲韻，DDxDxD。這種不規則的押韻變化，可能是爲了表明詩中流動意象的“轉”的動作[④]。

現在我們必須抵制誘惑，避免對這兩首詩做簡單的推斷，很顯然可以看

---

① 這種現象不止在换韻的時候，而且在强調韻腳模式變换時也會用到，比如後面將談到的幾首AABA格式的絶句中開放式的押韻。當然，嚴格韻律模式所主張的“新體詩”不是這裏討論的話題。唐代詩歌中自覺的韻律模式對“舊體詩”産生影響的趨勢已經形成，尤其是李白，比如可以參見Elling O. Eide, “On Li Po” , in *Perspectives on the T' ang*, ed. Wright and Twitchett, (New Haven: Yale Univ. Press, 1973)，尤其是pp.377-383. 或者Kroll, “Li Po' s Transcendent Diction”, *Journal of the American Oriental Society*. 106.1, (1986)，特别是pp.115-117。

② 關於慧遠和廬山，可重點參看Erik Zürcher, *The Buddhist Conquest of China: The Spread and Adaptation of Buddhism in Early Medieval China*, 2vols, rev. edn. (Leiden: Brill, 1972), 1:208-11,241.（譯者注：中譯本見許理和著，李四龍、裴勇等譯《佛教征服中國——佛教在中國中古早期的傳播與適應》，南京：江蘇人民出版社，2005）。

③ 最後的不久將被遺棄的“人間”意象，可能會使我們想起“碧山”絶句的結尾處，“人間”這個詞對李白而言通常帶有輕蔑義。

④ 最後還有一個我認爲並非偶然的特别之處：在每一節中，不押韻的詩句中文字的聲調，通常與押韻的詩句相對，唯一例外的是第七句的“落”，此音節增强了“落”這個字的感歎語氣，從河漢意象的關鍵字“驚”也可看出來。我認爲李白或者其他所有的詩人，都没有繁瑣地把這些隱形的技巧碎片化，一個字獨有的發音，或者一個詞的韻律和振型（這是聽覺模式的本能意識），通常决定了一個作者在詞彙和語法概念上的選擇。參考本文附録裏用中古漢語的聲調類型對這首詩做的完整的音節重組。

出，較長那首具有更加形象而具體的描寫，事實也的確如此。在六百年後對此做出的評論中，韋居安（1368）除了一直稱讚其爲無與倫比的作品之外，還説道："除非一個人真正仰望過此景象，否則他將無法充分欣賞這首詩的完美。"[①]如果我們把這評論看成不只是個别讀者維護特權的洋洋自得的嘗試，那麽我們必須接受"要完全地理解這首詩，就需要在瀑布下真正的體驗"這種觀點。組成語言之外的東西更加複雜，我們無需找博洛尼亞或者巴黎的聖人來驗證或者混淆這一看法：關於語言和現實關係的深思，在傳統中國的孔子、莊子、老子和荀子等聖人那裏已經表達得很透徹了。衆所周知，這首詩對其他領域都有影響，特别是對山水畫理論。但還有一個關於"逼真"的老問題，作爲一種情景再現，"逼真"是如何進行動態描寫的？簡單來講，這個問題是毫無意義的。然而，當我們試圖使這個問題複雜化時，請牢記這一點。

李白當然不是第一個寫廬山詩的人。但在他725年或726年[②]作這兩首詩之前，廬山瀑布只是偶爾地在詩歌史上被提及。當然如果我們仔細尋找，就會發現早期詩歌中的措辭和意象預示了某些李白的用詞，即便是一些我們曾認爲是經典和獨特的詩句。與慧遠結交的中亞和尚支曇諦（卒於411年）的《廬山賦》，遺留下來的三十句詩保存在七世紀的《藝文類聚》中，李白肯定也曾見過，其最後一組對句寫道：

香爐吐雲以像煙，甘泉湧霤而先潤。[③]

有人對李白廬山絶句中最開始的意象産生過疑問嗎？這個從香爐峰上升起的紫煙有它的根本原型，這在於香爐峰與瀑布的相應關係而非其顏色。此處的重點，至少在通常意義下不是對典故的引用，而是用了一個詞法循環，正如支曇諦《廬山賦》中的詩句在一定程度上影響了李白如何看待並再現這

---

①《梅澗詩話》，（*TSCC*）卷1，3頁。

② 安旗在《李白全集編年注釋》第49頁認爲是725年；但是詹瑛（另一個研究李白的當代權威學者）則認爲是726年，參考其著作《李白詩文体繫年》（北京：作家出版社，1958），5頁；或者《李白集校注》卷21，1240、1241頁。然而黄錫珪認爲這些詩寫於李白隱居廬山期間的756年，參考《李太白年譜附李太白編年詩目録》（1906年首印；北京：作家出版社1958年重印），75頁。

③《藝文類聚》（台北：文光出版社，1974）卷7，134頁；《全晉文》卷165，16b頁，收入《全上古三代秦漢三國六朝文》，嚴可均編（1893年初刊）。

樣的場景。

我們也可在《藝文類聚》裏的其他兩首關於廬山的詩中，找到相同的與李白詞彙共鳴的根源。首先是鮑照(414?—466)的一首詩：

千岩盛阻積，萬壑勢回縈。[①]

這與李白絶句的第三句(飛流直下三千尺)產生了隱約的共鳴，同樣，長詩的第八句(半灑雲天裏)也如此。第二首江淹的詩更有趣：

絳氣下縈薄，白雲上杳冥。
中坐瞰蜿虹，俛伏視流星。[②]

這裏的"絳氣"就是李白詩中的"紫煙"嗎？這裏的"蜿虹"是"白虹"(五律的第六句)的原型嗎？最能體現技巧性的地方在於，江淹從山上俯視流星(奔跑或湧動的星星)這個意象，可能有助於觸發李白催生出"銀河落九天"這個著名的關於瀑布的想象。

或許這些理解有助於我們一窺詩人的創作手法。雖然重構詩人閱讀所積累的文本，我們也未能解釋詩的神秘性或者抓住詩歌的魔力，但是我認爲，通過關注其思想庫，我們已經深刻領悟了李白及其他唐代詩人。即使我們不能像洛斯(J. L. Lowes)探究柯勒律治(Coleridge)那樣徹底地去挖掘[③](在其他唐以前關於廬山的作品中，應特別注意鮑照著名的登大雷岸寫給其妹的書

---

①《藝文類聚》卷7，134頁，《登廬山詩二首》第二首；《先秦漢魏晉南北朝詩》，1282頁。

②《藝文類聚》卷7，134頁，《從冠軍建平王登香爐峰》；《先秦漢魏晉南北朝詩》，1557頁；《文選》(台北：文津出版社，1987)卷22，1058頁(這是一本李白——像所有唐代詩人一樣——熟知的詩文選集)。

③ 參考洛斯的名著 *The Road to Xanadu: A Study in the Ways of the Imagination* (1927; rpt., Boston: Houghton Mifflin, 1964)。或許從這個方面研究李白最好的是阮廷瑜《李白詩論》的"詩語淵源"章(台北：編譯館，1986)，183—225頁，但其中忽略了佛教和道教的影響。葛景春《李白思想藝術探驪》對這個課題做了許多深刻而中肯的評論(鄭州：中州古籍出版社，1991)，這是近十年關於李白詩歌研究最卓越的著作。

信，這奠定了後來很多詩歌的基礎[①]）。

從本質上講，儘管像"原始資料研究"一樣令人滿意，我們也不能受制於此，這説明過多假設山水描寫的現實主義是很危險的。至少在詩歌中，只有極少數觸及詳細場景的描寫。比如王維（701—761）的山水田園詩，多數都是相對受限的基礎詞彙組成的普通意象，很少有可見的特性[②]。與王維相比，其他的唐朝詩人描述更加確切。對於他們來講，詩歌創作規律暗示其對選擇出來的意象進行重組，而非對場景一絲不苟的描寫，我喜歡肯尼斯·伯克（Kenneth Burke）的説法，這也適用於中國詩歌：

> 一個人不可能一直敘述意象而不涉及其象徵意義。詩人的意象從另一個具有象徵性的關聯物而來，一旦考慮到這點，我們就可把一個物體的意象轉爲其象徵意義，不僅物體本身，而且對關係的結構也起作用。[③]

這些把詩歌組成一個整體（如果可行的話）的"象徵關係"，賦予文字説話的權力，而不只是字面意義。作爲讀者，我們一直對構成"關係結構"所用的技巧進行評判，這種結構也暗示了傳統的偏見和個人的呻吟（我稱之爲過去的低吟和現在的怒號）。這裏我們可以引用劉勰的觀點，他對每一個文學話題都有精闢的論述，以下是其《物色》中的相關章節：

> 然物有恒姿，而思無定檢，或率爾造極，或精思愈疎。且詩騷所標，並據要害，故後進鋭筆，怯於爭鋒。莫不因方以借巧，即勢以會奇，善於適要，則雖舊彌新矣。是以四序紛迴，而入興貴閑；物色雖繁，而析辭尚簡；使味飄飄而輕舉，情曄曄而更新。古來辭人，異代接武，莫不參伍以

---

① 錢仲聯，《鮑參軍集注》，上海：上海古籍出版社，1980，卷2，84、85頁。Kang-i Sun Chang, *Six Dynasties Poetry* (Princeton: Princeton Univ. Press, 1986, p.89)（譯者注：中譯本見孫康宜著、鍾振振譯，《抒情與描寫：六朝詩歌概論》，上海：三聯書店，2006），零星地以詩歌的形式爲這篇文章作了選譯，其中有太多不足，因此不能作爲一個譯本推薦。文中她對鮑照所見日落景象的準確描述的評論，忽視了鮑照不可能從他所站的位置看到其描述的場景這個事實，這種描述更像是富於抒情性和想象力的絶技。同時我認爲，這一章節中孫對鮑照細微、逼真意象的其他評論也過激了。

② 詳見Kroll, *Meng Hao-jan*, pp.99-100。

③ *Attitudes toward History* (1937; 2nd edn., Los Altos, Calif: Hermes, 1959), pp.281-282.

相變，因革以爲功，物色盡而情有餘者，曉會通也。若乃山林皋壤，實文思之奥府，略語則闕，詳説則繁。然屈平所以能洞監風騷之情者，抑亦江山之助乎！[①]

這與我們話題的主旨相關，包含了對文學的秩序、前人的經驗以及山水的靈感來源三個方面的關注。

要想徹底領悟李白在廬山瀑布中的觀點，我們只需看看其曾被艾龍(Elling Eide)[②]深入探究過的《廬山謡寄盧侍御虚舟》中的一組對句：

銀河倒掛三石梁，香爐瀑布遥相望。[③]

這裏又出現了銀河，而且還是和香爐峰一起，這兩個意象在《留别金陵諸公》的一組對句中又同時出現，李白在詩中把廬山説成是他此行的目的地：

香爐紫煙滅，瀑布落太清。[④]

天空中的紫氣再次出現了，隨着從天際飛流而下的瀑布而消失不見[⑤]。很明顯，這些意象一旦被詩人偶遇，就成爲一種象徵，比如廬山就不可避免地預示着某種象徵意義。如果我們認同安旗和詹瑛的時間推測，就會發現二十多年來這些意象對於李白來説是合理的，實際情景也符合此觀點。然而黄錫珪推斷的時間，把所有詩歌的創作時間歸結到相同的一年裏，我們由此可以推斷詩人是如何不斷改變初衷的。

---

①《文心雕龍斠詮》，李曰剛輯(台北：編譯館，1982)，卷43，1903頁。《物色》篇的標準次序在第46，但是范文瀾(《文心雕龍注》，台北：開明書店，1958年重印，卷10，2a頁)認爲其原本位置並不在此，李曰剛受到啓發，改至50篇中的第43篇，從而充實了其著作。李這個詳盡的版本，與大陸學者周振甫的《文心雕龍注釋》(台北：里仁書局，1984年重印，引文參看卷46，846頁)一様，也成爲目前供學者選擇的堪與范注相媲美的版本。

② “On Li Po”, pp.379-387.

③《廬山謡寄盧侍御虚舟》，《李白集校注》卷14，863頁；《全唐詩》卷173，1773頁。此詩創作時間，安旗(1580頁)和詹瑛(143頁)認爲是760年，黄錫珪(77頁)認爲是756年。

④《留别金陵諸公》，《李白集校注》卷15，926頁；《全唐詩》卷174，1784頁。此詩創作時間，安旗(914頁)和詹瑛(76頁)認爲是750年，黄錫珪(75頁)認爲是756年。

⑤“太清”是天的領域，在地球以上40里，根據葛洪《抱朴子内篇校釋》卷15，275頁。在中古時期發達的道教宇宙觀裏面，其上依次是太極、上清、玉清。當然，“太清”在中古詩歌中也通常被看成宇宙的總稱。

但是至此爲止,我們僅僅討論了李白的詩,如果仔細品味他的其他作品,可以發現另外一篇非常有意思的可與前面相提並論的文章。這是一篇寫於8世紀中葉某個時間[①]的題爲《秋於敬亭送從姪專遊廬山序》的古文,像李白多數古文一樣鮮爲人知。文中的感情非常吸引人並且值得拜讀,所以我全文引用,我們大致看看,但要特别注意所有關於廬山的句子,讀起來會非常熟悉:

余小時,大人令誦《子虛賦》,私心慕之。及長,南遊雲夢,覽七澤之壯觀[②]。酒隱安陸,蹉跎十年[③]。初,嘉興季父謫長沙西還,時余拜見,預飲林下。專乃稚子,嬉遊在傍。今來有成,鬱負秀氣。吾衰久矣!見爾慰心,申悲導舊,破涕爲笑。

方告我遠涉,西登香爐。長山橫蹙,九江卻轉[④]。瀑布天落,半與銀河争流;騰虹奔電,潨射萬壑,此宇宙之奇詭也。其上有方湖石井,不可得而窺焉[⑤]。

羡君此行,撫鶴長嘯。恨丹液未就,白龍來遲[⑥],使秦人着鞭,先往桃

---

① 安旗認爲是757年(1968頁),詹瑛(94頁)是753年,黄錫珪(93頁)是754年。

② 參考《子虛賦》,《文選》卷7,349頁;David R. Knechtges, *Wen xuan, or Selections of Refined Literature*, vol.2 (Princeton: Princeton Univ. Press, 1987), p.55。

③ 約727—735年的安陸,在湖北省北部,是李白與第一任妻子許氏的居住地,在這裏不爲人知地隱居了多年,尤見前野直彬(Maeno Naoaki), "Anriku no Ri Haku"(《安陸の李白》), *Chūgoku koten kenkyū*,(《中國古典研究》)16(1969),pp. 9-22。

④ 九江,在唐代也稱"潯陽",廬山北面,傳説是長江的九條分支。

⑤ 這個位置及有赤鱗嬉戲的湖是廬山當地傳説的一部分,根據慧遠《遊廬山記》記載,當地農夫看到香爐峰上湖中的景象,都驚歎不已。亦可參見爲李白所熟知的劉峻(462—521)《世説新語》第十注釋的部分引用,徐震堮,《世説新語校箋》,北京:中華書局,1984,卷2,314頁;Richard B. Mather英譯本:*Shih-shuo Hsin-yü: A New Account of Tales of the World* (Minneapolis: Univ. of Minnesota Press, 1976), p.288. 亦可見嚴可均《全秦文》卷162,7b頁。

⑥ 李白把自己描述成一個隱藏的能手:他撫鶴卻不能乘鶴(享受崇高榮譽的坐騎)離開,只得像孫登那樣長嘯來排遣苦悶,他等待着煉丹爐裏長生不老的丹藥早日煉成,期盼着白龍來帶他去超然的境地,過陵陽山上棄官的竇子明那樣的隱居生活。

花之水[1]。孤負夙願，慚歸名山[2]，終期後來，攜手五嶽。情以送遠，詩寧闕乎？[3]

顯然，這裏的廬山是李白的一個記憶，腦海中曾經描繪讚美廬山的詞語再次逐漸被喚回，用在這篇散文裏，重提這些相同的意象也證明了其象徵性的地位。如果混亂難免要與時俱進，那麼回憶就會越來越亂，或許我們可以把象徵物的積累（不管是有意或無意的）看成定量。其他苛刻的闡述可能會批評這種意象的重提是偷工減料，除非這種"陳腔濫調"是詩人自己特有的藝術手法。因而對這個問題最糟糕的反擊，可能就是言語上的自我陶醉，對每一位詩人而言，這顯然是反對他們的可控訴的罪行。儘管對新鮮事物充滿現代派熱情，我們仍然欣賞李白用其符號鍵在不同環境和場景中切换自如的能力，這些意象都在他的掌控之中。簡單地說，李白第一次表述廬山的語言判斷，就將他對廬山的感受和領悟永記於心了。

是不是一來到廬山，這些李白銘記於心的關於廬山和瀑布的想象，就從他同時代的其他人那裏以不同的順序呈現呢？對此我們没有時間來詳盡瞭解。所以我們需要另外兩個重要作家的例子來滿足我們的求知欲與好奇心。第一個是張九齡（678—740），他是最具影響力的詩人之一，尤其還是開元年間（713—742）最具權力的政治家之一。他深受粵北人民的敬仰，是8世紀南方"克里奥爾人"的典型代表，在漢族的中心地帶逐漸變得重要並且被接受。很遺憾的是他的詩很少被研究，因爲缺乏一些美麗的景色和有趣的特點。張九齡對其南方出身很在意（這是不常見的）卻又引以爲傲，他許多出色的作品都是讚美此前未被北方人欣賞的本土地域的光環之處，並且試圖重新

---

① 如果朝廷中的政治困難加大，（懷着對廬山的敬畏或回憶）他將不告而别，做一個如秦朝混亂時期桃花源裏那樣的居民。"秦人着鞭"，賈誼（前201—169）《過秦論》對秦始皇的描述，用長鞭將桃花源的人召回。嚴可均《全漢文》卷16，5b頁。

② 他仍然希望在像廬山那樣的名山上，漫遊在自由世界裏，但現在又感到不值得這樣做。不同的記載在《文苑英華》卷721，5b頁（"慚未歸於名山"），這裏通過意思完全相同卻出人意料的運用，提供了一個很好的實際上讀起來相互矛盾的例子，同時也說明，《文苑英華》的改變是編者想弄清僅僅被他誤解的一個點的嘗試。

③《秋於敬亭送從侄耑遊廬山序》，《李白集校注》卷27，1566頁；《全唐文》（台北：大通書局，1979）卷349，6b—7a頁。"敬亭"指敬亭山，在安徽東南的宣城北郊。

定位傳統的區域偏見,比如著名的《荔枝賦》[①]。在其他以南方爲背景的創作中,他有時顛倒外界對非漢地區事物既定的一致看法。因此在一些詩中我們發現,平常長臂猿悲切的哭號出人意料地趕走了南方詩人的陰鬱感:對他而言反而是熟悉同伴的歡迎聲[②]。換句話講,當他認爲有必要的時候,張九齡並不怯於在詩中跟着自己的感覺走(這方面他和李白很像,是另一個領域的局外人)。

在8世紀中期,廬山並不屬於典型的"南方",但是因爲張九齡對南方山水之美極其敏感的偏好,他過多地把瀑布邊蒼翠繁茂地帶看作值得沉思的地方,也就不足爲奇了。有兩首關於此場景的詩,第一首叫做《湖口望廬山瀑布水》(這湖當然指的是彭蠡湖,即現在的鄱陽湖)[③]:

萬丈紅泉落,迢迢半紫氛。
奔流下雜樹,灑落出重雲。4
日照虹霓似,天清風雨聞,
靈山多秀色,空水共氤氲。8

這首緊湊的律詩成功地把我們引入了這個瀑布的場景——即使(請記住題目)詩人是從很遠的距離遠望的。如果逐字逐句分析這些意象,我們可以發現這與李白關於廬山瀑布的律詩的前十六句中一系列意象具有相似之處。然而,張九齡選擇用更受限制的形式來完成這首詩[④],這是必然受到讚美和欽佩的。例如,第三、四句中的排比句充滿了活力,還有第五、六句中建立的有效張力。即使張九齡對每一個意象的描述風格與李白不同,我們也不得

---

①《荔枝賦》,《曲江張先生文集》卷1,11a—12b頁;《全唐文》卷283,1b—3b頁。

② 一些材料見筆者關於張九齡記載的前部分,*The Indiana Companion to Traditional Chinese Literature*, ed. Nienhauser (Bloomington: Indiana Univ. Press, 1986), pp.207-209。

③《湖口望廬山瀑布水》,《曲江張先生文集》卷4,8a頁;《全唐詩》卷48,590頁;《文苑英華》卷264,3b頁。第二和第三句我根據的是《曲江張先生文集》和《文苑英華》,第四句則是《文苑英華》。(我認爲張九齡不可能在一首律詩中的四句中用兩個"落"字。)

④ 現在一些漢學家繼E. H. Schafer之後,也把所有的律詩看成"雙絶句"("double quatrains"),不得不承認,這種用詞是不恰當的。"雙絶句"應該更傾向於指:(1)由不同韻律的兩組絶句構成的兩個詩節的詩歌,這並不是律詩的例子;或者可能是(2)特殊的律詩,其中前四句的聲調類型與相應的後四句完全匹配,這種精確的重合與兩首絶句聲調序列的鏡像標準相反。眼前這首張九齡寫的詩其實就屬於第二類。

不注意那曾在李白詩中出現過的紫煙、霓虹、清風。如果談到影響力的話，因爲我們首先讀了李白的詩，而且他是二者中更有名的，所以可能更傾向於承認他的優先地位。但可以肯定的是張九齡這首詩（及下一首詩）寫於727年的春天，當他南下洪州（彭湖西南的延伸地帶）[1]任職經過此地時，如果安旗和詹瑛關於李白作詩時間在725年或726年的推斷準確，則張九齡是在李白之後創作的。然而，李白當時還未出名，他的詩還没流傳開來，無法推測張九齡是否看過這個年輕人關於瀑布的詩作。當然，如果黄錫珪關於李白756年的作詩時間推測是無誤的，理解就完全不同了，因爲到那個時候，李白肯定被認爲讀過張九齡的詩。但是我認爲可能最保險的是，不要在作品中摻入任何方面的直接影響。兩位詩人都是獨立創作出這些意象的，他們的想象，無疑都從收入標準總集如《文選》或《藝文類聚》中的4、5世紀（之前提到的）先哲的作品而來[2]。這種技巧重振了歷史遺承，這些個人的文章對整體都有備受矚目的貢獻："傳統與個人天賦"可以被看成孔子格言"温故而知新"的文學變體，或者與受約束的一流爵士樂的産生、印度拉格的演奏者類似。

現在再來看張九齡的第二首長度是第一首兩倍的詩，題目爲《入廬山仰望瀑布水》[3]：

絶頂有懸泉，喧喧出煙杪。
不知幾時歲，但見無昏曉。 4
閃閃青崖落，鮮鮮白日皎。
灑流濕行雲，濺沫驚飛鳥。 8
雷吼何噴薄，箭馳入窈窕。
昔聞山下蒙，今乃林巒表。12
物情有詭激，坤元曷紛矯。
默然置此去，變化誰能了。16

這是從另一個角度所見到的瀑布，前面十句只是描寫自然現象，這次並

① 楊承祖，《唐張子壽先生九齡年譜》，台北：台灣商務印書館，1980，52頁。
② 參見孟浩然《彭蠡湖中望廬山》的對句："香爐初上日，瀑水噴成虹。"Kroll, *Meng Hao-jan,* p.73.
③《入廬山仰望瀑布水》，《曲江張先生文集》卷4，8b頁；《全唐詩》卷47，573、574頁。

没有表達任何有關遠離世俗的情感。但是儘管如此，山水非凡的特徵比之前想象的更加顯眼地出現在詩人面前。比如"詭激"（注意此處對"激"原意的適當改動：水受阻遏）這個例子，偶然從大地多產本能（坤元[①]）所形成的世間種種自然現象中而來。當然詩人也懷疑這些事物的形成都是有源可溯的，但是創造和改變它們的强大力量的意志是不言而喻、並且很難被我們看透的。

這是一個關於自然山水令人驚訝並喚起記憶的看法，而不是山水的特别之處。這是一種當看到令他驚奇的自然景色時，張九齡慣用的表達方式。比如這首關於湞陽峽（位於廣東中部的北江邊，大約在張九齡的故鄉曲江和廣州之間）的詩[②]：

行舟傍越岑，窈窕越溪深。
水闇先秋冷，山晴當晝陰。4
重林間五色，對壁聳千尋。
惜此生遐遠，誰知造化心。8

當談到最後一組對句時，研究唐代文學的人都會不由自主地聯想到柳宗元（773—819）《小石城山記》[③]結尾處的哲學思考。被貶謫到南邊，抑鬱的柳宗元困惑造物者（即張詩中的"造化者"）爲何把無與倫比的美景安放在南方這没有文明人欣賞的蠻荒之地。如我們所見，柳宗元這個著名的沉思顯然受惠於二十多年前張九齡的言辭（毫無疑問柳宗元是熟悉張九齡作品的）。但我們偏題了，我們談論的是張九齡的作品，在上面這首詩中，像之前的那首一樣，張九齡很巧妙地把外部景色描寫轉爲内心情感的抒發（情景交融）。可以

---

①"坤"或者這裏所説的"坤元"，具有潛藏、承受的本能，象徵着大地，是萬物生長的根源。"乾"，或者"乾元"，象徵着天。這種古老對應背後的物力論，可能很好地使我們想起希臘神話，坤對應蓋亞，乾則對應烏拉諾斯。

②《湞陽峽》，《曲江張先生文集》卷4，11a頁；《全唐詩》卷48，590頁。關於湞陽峽的傳説，參見《水經注疏》（南京：江蘇古籍出版社，1981）卷38，3186頁。這部彙集晚清學者楊守敬和熊會貞珍貴評論的精印本，是真正的集大成之作。

③《柳河東集》（台北：世界書局，1963）卷30，第317、318頁；《全唐文》卷581，16a/b頁。這篇文章多次被翻譯，但是E. H. Schafer在其 *The Vermilion Bird: T' ang Images of the South* (Berkeley: Univ. of California Press, 1967), p.117的翻譯是無人能及的（譯者注：中譯本見薛愛華著，程章燦、葉蕾蕾譯，《朱雀：唐代南方的意象》，北京：生活·讀書·新知三聯書店，2014）。

斷言，至少在兩首詩中，這裏肯定不僅僅有對材料的組成和異常事物的擺放那種感覺上的目的，即使這個目的已超過我們的理解範圍：

我感到有物令我驚起，它帶來了崇高思想的歡樂，一種超脱之感，像是有高度融合的東西。（譯者注：華兹華斯《丁登寺旁》）

但是帶有“落日的餘暉，浩瀚的海洋和清新的空氣，蔚藍色的天空”的寓所並沒有對“人的心靈”敞開。這與李白關於廬山瀑布的詩形成鮮明對比。當然，即使文學傳統在措辭和意象上常常有一定程度的一致性，也没有必要期待每個人都是同樣的觀點。

最後再看看我們現在已經很熟悉的廬山瀑布，李華（735年進士，逝於769年？）的《望瀑泉賦》，創作於他在安史之亂後自願歸隱期間[①]，所以是在李白、張九齡還有許多其他盛唐詩人拜訪廬山、並作詩於此之後所寫的。我們可以認爲，這篇賦比我們之前所遇到的體裁更廣泛地處理描繪了所見景物，作者在黄昏時分從彭蠡湖的船上看到了這一場景[②]：

曙無雲兮川無波，泛余舟於空碧。
彼廬山浮重湖之上兮，巖極天之峻壁。4
凝黛色之深明，噫林嶺之岑寂；
何神造之杳冥，躍騰泉於山脊。8
孤流皎皎於蒼梁，翠淙千仞兮懸帛；
玉繩縋於寥天，銀河垂於廣澤。12
春風雷兮簁霜雪，穿重雲而下射；
白龍倒飲於平湖，若天地之初闢……[③]16

這應該足夠表明這一文段所傳達的感情了，在措辭上比此前的詩體更加自由和深遠，李華《賦》的這一選節具有豐富的意象和象徵。詩人儘量選擇富

① 756年，李華被迫與叛軍政府勾結去奪取長安，這成爲直到臨死前還一直縈繞他的恥辱。

② 改變了首行縮進，改變了原文中詩的長度，引文没有韻律變化，因此也没有分節。有節奏的助詞“兮”代表破折號。引入的兩三個字爲一行的，不計入行數。

③《全唐文》卷314，10a頁。

於暗示的方式來描寫瀑布，而且他的一些措辭的確很引人注目，尤其容易被這懸掛在千英尋高的亮藍色絲綢（“瀑布”一詞使人想起紡織物這一意象）驚住。從天空懸吊下來的玉繩、被篩選的霜雪、特别是罕見的白龍倒飲，這些與李華文中落下的銀河、厲風和與之相隨的雲朵交融在一起。當天地初闢時，這些生動的事物使人有了創作的靈感。在這裏，瀑布成爲語言技巧的動力，獲得了處理詩意現實的全新視角。

通常，我們必須首先處理兩個基本因素：世界和文本，世界和詞彙。在最字面的（或許也是最抽象的）層面上，我們應該像處理文字一樣處理世界。宇宙的或者地球的“山水”這一概念，作爲一個解讀的重要文本，在伏羲通過觀察天地間的象、法以及鳥獸紋路來發明《易經》中八卦的神話中，擬古地被表現出來。如果不與“口述第一性”相比較，我們只需要看看中古道教展示的關於永恒聖典的思想，以及在物質世界創造之前，人們最初在熾熱的光下的口頭語[①]。從宇宙起源論以及認知科學來説：文字創造了世界。

如果往下至人類精神、歷史時間、語言（不只是語言，而且是巧妙組織的語言），成爲使精神時刻在未來永存或者永恒的方式，希望在這點上，我已經闡述清晰了，因爲這是另一個可能巧妙卻致命的使我們從唐朝人的論點上轉移的方式。我們生活在一個文化全球化的時代，從字面上説，這是非常矛盾的。然而我們可能希望點别的，這個世界充滿了傳播媒體和技術的混亂，不需要詩歌或者藝術作爲必需品。在這個文化中，不是所有人都支持和欣賞、或者陶醉、深入瞭解藝術，但即使是這些人，他們大多數都不能存活於一個没有藝術裝飾的世界。六十多年前，保爾·瓦雷里（Paul Valéry）把這類人定義爲真正的資産階級，對藝術毫無感覺：“他們中並非所有人都像我們所説的對藝術充耳不聞，他們只是不被‘只存在於遺忘的存在’的理念折磨，也不被‘豐富

① 關於這點，有很多需要説的，但這離我們討論的焦點“唐”有點遠了。感興趣的讀者可以閱讀 Isabelle Robinet, *Méditation taoïste* (Paris: Dervy-livres, 1979) , pp.29-44. 這本書的英譯本 *Taoist Meditation: The Mao-shan Tradition of Great Purity*, tr. Pas and Girardot (Albany: State Univ. of New York Press,1993)，準確度稍欠佳，應該謹慎閱讀。

的精神世界是生活的必需品'這樣一種强烈的欲望所困擾。"[①]但是對於中國中古受過教育的精英來講,他們是一千多年以文本爲中心的文化的繼承者,根據傳統,的確知道如何操縱詞彙、語法和韻律的語言藝術的命令和持續運用,扮演了必要的生活核心。對於創作我們研究作品的精通文字的少數派而言,其他任何東西都是次要的。這就是説,基本的觀點是通過文字看世界,或者世界被看成文字(毫無疑問,被流放遠離皇權的落後地區,希望渺茫,頂多有一些相同境遇的同伴,遭受着可怕的刑罰:孤立之感必將成爲不可思議的摧毁力量)。

分析唐代山水的意象和場景時,我們要説什麽呢?我們通過同情的(希望是正確的)理解、對唐代作家的幻想充滿想象力的(希望是真誠的)重建來展開研究。關鍵是"幻想",因爲這些作品是他們自己想象的重構,並不是一絲不苟的描寫。這就是傳統的重要性,詩人們在創作時,潛移默化地受到了前人表述的制約和鼓舞。我們對待山上的煙霧、瀑布中的薄霧,這些風景激起了我們或感人的、或個人的、或心靈的、或無形的視覺感受,但是我們精心考慮和部署的詞藻,依然是前人在面對其他風景時使用過的。我們不僅看眼睛看到的事物,也看心靈的眼睛强加給我們的東西,由此引申,則是"五蘊"的消散,不然就是文本的解構。

當回頭看那些獨特的山水詩篇時,我們可以肯定的是,大多數的這些考慮都會隨着我們的閱讀而慢慢消失,就像瑪格麗特鷄尾酒杯口抹的一圈鹽一樣。我們很願意相信文章中所描述的情景,就是作者眼前看到的風景,多數情況也許會是這樣。讓我們來聽聽華萊士·史蒂文斯(Wallace Stevens)的意見吧,這是一個着重關注實體和現象轉换的詩人:

> 詩是生命的想象,一首詩是生命的一個細節,被思考良久以至人的思想已經成爲它不可分割的一部分,或者是一個被强烈感覺到根深蒂固的生命之細節。因此,當我們説世界是真實事物的濃縮,與想象的非真實事物毫無區别以至難以分辨時,譬如説,當我們提及藍天時,我們可以

① "The Necessity of Poetry", in Paul Valéry, *The Art of Poetry*, tr. Folliot (rpt., Princeton: Princeton Univ. Press, 1985) , p.219.

> 肯定被提及的總是某個這樣的事物，不論通過思想還是感覺，即便我們毫無察覺，這已經成爲我們生命體驗的重要部分。很容易想到，很少有人意識到那一時刻，我們所有人都會碰到，當我們第一次看到藍天時，就意味着，我們不僅僅看到了它，也望着並體驗着它，同時第一次感覺到，我們就生活在物質詩歌的中心，一種要不是存在於其中的非地理學的話將不可容忍的地理學中心。没有多少人會認識到，他們所看到的是其自身思想的世界和自身感覺的世界。①

劉勰曾云："物色盡而情有餘者，曉會通也。"因此，在文學的世界裏，"愉悦山"像廬山一様存在，繁星點點的"上清"之路與美國大峽谷的裂隙一様真實。

一篇奇妙的文章需要用心去理解。比如我們的文章主題與美國文化有關時，我們可以選擇讀赫爾曼·梅爾維爾(Herman Melville)的心理小説 *Pierre: or, The Ambiguities*，書中用了三大段華麗且優美的文字來描述位於伯克郡的格雷洛克山，這是我見過的唯一一部對山有貢獻的書。無獨有偶，盛唐時代也有一篇山被人格化的文章，這篇文章確實存在，也許有人已經猜到了，它就是李白寫的《代壽山答孟少府移文書》。壽山是安陸一帶的一座小山，文中李白代壽山説話，停留安陸期間(約727—735)，李白隱居在此。這篇"書信"是李白寫給一位當地少府(少府是替代尉的官職，在縣一級擔當調解者)的回文。在之前的移文中，少府用責備和詼諧的口吻來貶低壽山，説它不夠資格庇護那些本應該投身政治的隱士。回文的前三分之二是壽山的自我辯解，這也是文章妙趣横生的地方。文章剩下的三分之一，介紹了與它一様的隱居者李白，尤其寫了對李白無拘無束美德的讚美，正如梅維恒(Victor Mair)解釋的那様，李白在這裏成爲了焦點②。在這篇半詼諧半嚴肅的文章的前三分之二部分，壽山(代指李白)試圖澄清山與國家的恰當關係。作爲一名有文學修養或精通古典文學的作家，李白寫作對仗工整、詼諧幽默，又靈活多

① 選自"The Figure of the Youth as a Virile Poet", in *The Necessary Angel: Essays on Reality and the Imagination* (New York: Vintage, 1951), pp.65-66。

② 見其"Li Po's Letters of Political Patronage", *Harvard Journal of Asiatic Studies* 44 (1984), pp.141-142。

變,能與之媲美的恐怕只有孔稚珪(447—501)的《北山移文》[①]了。原文寫道:

淮南小壽山謹使東峰金衣雙鶴,銜飛雲錦書於維揚孟公[②]足下曰:

僕包大塊之氣[③],生洪荒之間[④]。連翼軫之分野[⑤],控荆衡之遠勢[⑥]。盤薄萬古,邈然星河。憑天霓以結峰,倚斗極而横嶂。頗能攢吸霞雨,隱居靈仙。産隋侯之明珠[⑦],蓄卞氏之光寶[⑧],罄宇宙之美,殫造化之奇。方與崑崙抗行,閬風接境[⑨]。何人間巫、廬、台、霍之足陳耶?[⑩]

昨於山人李白處見吾子移文,責僕以多奇,叱僕以特秀,而盛談三山五嶽之美[⑪],謂僕小山無名無德而稱焉。觀乎斯言,何太謬之甚也!吾子豈不聞乎:無名爲天地之始,有名爲萬物之母[⑫]。

假令登封禋祀,曷足以大道譏耶?然皆損人費物,庖殺致祭,暴殄草木[⑬],鐫刻金石,使載圖典,亦未足爲貴乎?且達人莊生,常有餘論,以爲斥鷃不羨於鵬鳥,秋毫可並於太山[⑭]。由斯而談,何小大之殊也?

---

① 參見 James Robert Hightower 的譯本, “Some Characteristics of Parallel Prose”, in *Studia Serica Bernhard Karlgren Dedicata* (Copenhagen: Ejnar Munksgaard, 1959), pp.70-74。

② 維揚是揚州的舊稱,可能是孟公的故鄉。

③ “大塊”即大地,象徵着大自然,參看《莊子集釋》,郭慶藩撰(北京:中華書局,1961)卷2,22頁;卷6,110頁;卷6,119頁。

④ 洪荒,統一的混沌時代,地球還是一個龐大而未分化的荒野。

⑤ “翼”和“軫”,兩種星宿名,分别指巨爵座和烏鴉座裏的星,共同控制古楚地(即湖南及以南)的天數,參見 E. H. Schafer, *Pacing the Void: T' ang Approaches to the Stars* (Berkely: Univ. of California Press, 1977) , pp.76-77。

⑥ 即荆州和衡州,或者按照另一種説法,指荆山與衡山。

⑦ 隋侯,曾經救了一條受傷的蛇,後來蛇給他一顆産自西亞的明珠作爲回報。參看《淮南鴻烈集解》,劉文典撰(北京:中華書局,1989)卷6,198頁,高誘212年注。

⑧ 卞和獻玉,在其原型中,兩任君主都没有認出玉的價值,相繼以欺君之罪砍去他的雙腳作爲懲罰,第三任君主使璞玉顯世發光。與此前所述的隋侯珠齊名。區别認識三位君主的各種來源,可參看高誘的注;《韓非子》(《百子全書》)十三,卷4,4a頁;以及《後漢書》(北京:中華書局,1965),卷80B,2633頁,李賢(651—684)注。

⑨ 宇宙論中西方的崑崙山,閬風(岩崖上的風)是崑侖三巔之一。

⑩ 中國四大神山,與壽山一樣,不能與崑崙相提並論。注意李白再次對人間的輕蔑。

⑪ “三山”指東海中的蓬萊、方丈和瀛洲三座仙島。

⑫ 引自《老子》第一章。

⑬ 即爲祭壇準備場地。

⑭《莊子集釋》卷1,14頁;卷2,39頁。

又怪於諸山藏國寶、隱國賢，使吾君榜道燒山[①]，披訪不獲，非通談也。夫皇王登極，瑞物昭至，蒲萄翡翠以納貢[②]，河圖洛書以應符[③]。設天網而掩賢[④]，窮月窟以率職[⑤]。天不祕寶，地不藏珍，風威百蠻，春養萬物。王道無外，何英賢珍玉而能伏匿於巖穴耶？所謂榜道燒山，此則王者之德未廣矣。

昔太公大賢，傅説明德，棲渭川之水，藏虞虢之岩，卒能形諸兆朕，感乎夢想[⑥]。此則天道闇合[⑦]，豈勞乎搜訪哉？果投竿詣麾，舍築作相，佐周文，贊武丁。總而論之，山亦何罪？乃知巖穴爲養賢之域，林泉非祕寶之區。則僕之諸山，亦何負於國家矣？[⑧]

因此即使這些山是政治的體現，但更大程度上代表的是它們自己，只有李白借山之口爲他自己説話。其他時候，他和與其字相同的太白山對話，二者可看成崇高的二重體[⑨]。"山水文化"是多樣的，山水並不全是唐朝文化的一個話題，史蒂文斯認爲，這是存在於自然地理詩中的"非地理性"，正如他在另一處所説："一個詩人的詞語屬於没有它們就不存在的事物。"[⑩]

但是最後的一詞對於唐代的作家來説應該被保留，讓我們從吴筠(卒於

---

① 孫惠，在3世紀早期，爲躲避朝廷紛争而隱居山林，東海王通過放榜尋找，才使其出世，《晉書》(北京：中華書局，1974)卷71，1883頁。此前一百年，阮瑀用同樣的方法躲避曹操的召喚，但是不久，曹操爲逼出阮瑀，就放火燒了其隱藏的山林，《三國志》卷21，600頁，裴松之(372—451)注引3世紀的《文士傳》。放火燒山以求隱士入朝的原型是公元前7世紀，晉文公燒綿山來找介子推，這段歷史載入《新序》(《漢魏叢書》)卷7，14b-15a頁，劉向(前77-6)在結尾記載，介子推拒絶從大火中出來，最後被活活燒死。這個不幸的結局並非《左傳》和《史記》關於介子推事跡的一部分。

② 即來自中亞和遥遠的南邊的外來品。

③《河圖》和《洛書》，象徵着聖人和賢人的到來。

④"天網"，天上用星星編成的網，因此天子就是地上的權威。

⑤ 月窟，都是寒冷的月亮所在地，在地球上即遠西的土地。

⑥ 太公即吕尚(最初的姜子牙)，被找到時正在渭河北岸釣魚，文王在出獵前占卜曰："所獲非龍非螭，非虎非羆；所獲霸王之輔"《史記》(北京：中華書局，1972)卷32，1477、1478頁，吕尚以太公身份成爲文王的主要謀臣。傅説，是商朝武丁時期的丞相，在虞、虢之間的傅巖修路的一堆囚徒中被發現，他被囚禁於此。在找到他之前，武丁做了一個夢，夢見自己受一位名爲"説"的聖人相助，《史記》卷3，102頁。

⑦ 闇合，即命中註定的巧合。

⑧《代壽山答孟少府移文書》，《李白集校注》卷26，1521—1524頁；《全唐文》卷348，16b-18a頁。安旗(1851頁)和詹瑛(9頁)認爲的創作時間是727年，黄錫珪(92頁)認爲是731年。

⑨ 參見Kroll, "Li Po's Transcendent Diction", pp.113-117。

⑩ "The Noble Rider and the Sound of Words", in *The Necessary Angel*, p.32.

778年)的賦來進一步看看,這是一位與李白同時代的詩人、道教名人。吴筠近年從文學歷史的邊緣浮出,全得利於薛愛華(E. H. Schafer)的關注,他寫了兩篇主要關於吴筠道教主題詩作的長文[①]。然而,吴筠的賦仍未被翻譯,我想用其中的一首《巖棲賦》來結尾,這會給我們另一種劉勰所謂"江山之助"的盛景。文章的題目來自於謝靈運(385—433)著名的《山居賦》序的第一句:"古巢居穴處曰巖棲,棟宇居山曰山居。"[②]這是對山水價值可變性的充分運用。吴筠在自然界創造了一種超凡的内在性,因此他没有打破世俗的自由進入一個更高的層面(這樣他完全能夠寫出其他的文章),而是欣喜地使自己沉浸於山水景物之中。或者反過來,像他做的那樣,最後發現整個自我(變得沉着),内心和諧平衡、平静於這純净的非地理的非物質性,在這裏他可以隱藏他的光芒(隱藏自我本質),淹没在黑潭死水或者遁世的漩渦中。

巖棲賦[③]

感玄聖之垂訓[④],悟已親而名疎[⑤]。
言可放而從默,身應卷而勿舒[⑥]。4
爰鷦鷯之巢林,在一枝而有餘[⑦]。
性所悦而難違,託兹山以結廬[⑧]。8

---

① "Wu Yün's 'Cantos on Pacing the Void'", *Harvard Journal of Asiatic Studies* 41 (1981), pp.377-415. 以及"Wu Yün's Stanzas on 'Saunters to Sylphdom'", *Monumenta Serica* 35 (1981-1983), pp.309-345. 薛愛華的第一篇文章中有關於吴筠的生平梗概,但有幾處觀點欠妥,按照傳統的説法,吴筠與李白742年被召回朝廷脱不開干係,但是參看郁賢皓1981年的文章《吴筠薦李白説辨疑》,載《李白叢考》,西安:陝西人民出版社,1983,65—78頁,文中謹慎考察了與此相關的時間,以及吴筠人生中的其他事件。

②《宋書》(北京:中華書局,1974)卷67,1754頁。嚴可均《全宋文》卷31,1a頁。

③《巖棲賦》,《宗玄先生文集》(*HY* 1045)卷1,1a-2a頁;《全唐文》卷925,3b-4a頁;《文苑英華》卷98,2a-3a頁。

④"玄聖"指老子。

⑤ 隨着對現實的"悟",詩人在世間的名也漸漸消散。

⑥ 默默地"卷"而不是在他人面前展示或者"舒"。

⑦ 參考《莊子集釋》卷1,13頁,"鷦鷯巢於深林不過一枝",關於此話題的全面解釋,可參見張華(232—300)《鷦鷯賦》,嚴可均《全秦文》卷58,3a/b頁,序言:"鷦鷯,小鳥也,生於蒿萊之間,長於藩籬之下,翔集尋常之内,而生生之理(《莊子集釋》卷6,115頁)足矣。"

⑧ 此句强烈暗示了陶潛詞的用語。

果棲遲而我愜[1],即逍遥之靈墟[2]。

觀其繚崇巒,横峻谷。

激泌泉,羅森木。12

後巍峩以縈紆,前參差而聳伏。

追陰壑之夏涼,偃陽崖之冬燠[3]。16

美勁節於松筠,翫幽芳於蘭菊。18

虚籟清耳[4],閑雲瑩目。20

因海鶴以警夜[5],任鵾雞以知旭[6]。

慮静於無擾,神恬於寡欲[7]。24

於是歌考槃於詩人[8],諷嘉遯於太易[9]。

遠浮俗之難險,消毁譽之損益。28

蹈方外之坦途[10],信可免於兢惕。

既即陰以息影,由不行而滅跡。32

雖區中之末計,實世表之長策[11]。

人所棄而己收,故處約而恒適。36

---

① 參考《詩經》第138首(譯者注:《詩經·陳風·衡門》),第1、2行:"衡門之下,可以棲遲。"

② "逍遥"來自《莊子》第一章。

③ 這兩句提供了一個很好的陰陽互補的例證。

④ "虚籟"近似於天籟之音(普通耳朵是聽不見的),《莊子集釋》卷2,22—24頁。

⑤ 參見鮑照《秋夜詩》兩首第二首的第9、10行:"霽旦見雲峯,風夜聞海鶴。"《先秦漢魏晉南北朝詩》,1308頁。

⑥ 參見"大招":"鵾鴻群晨,雜鶖鶬只",《楚辭補注》(北京:中華書局,1983)卷10,224頁。

⑦ 參見"見素抱樸,少私寡欲。"《老子》,第十九章。

⑧ "考槃",《詩經》第56首(譯者注:《詩經·衛風·考槃》)的前兩字,也是題名,這是一首讚美隱士之樂的頌歌。孔穎達(574—648)所作的唐代權威注本《毛詩正義》把"考槃"解釋爲"成樂",也正是此處吴筠之意。《毛詩正義》卷3,53c頁,載《十三經注疏附校勘記》,北京:中華書局,1979。

⑨ 參考《周易正義》(《十三經注疏附校勘記》)卷4,36b頁,韓康伯(卒於約385年)注以及孔穎達正義,關於詞語"嘉遯"在第33卦的第五句(遯),象徵一種值得稱讚的自我意願的退隱。

⑩ "方外",參見《莊子集釋》卷6,121頁,孔子對於遊於世俗之外的人的解釋。

⑪ 參考第79、80句,陸機(261—303),《歎逝賦》,《文選》卷16,727頁:"精浮神淪,忽在世表。"但是,與陸機相反,吴筠意在遠離俗世,《文苑英華》:"……世途之長策。"

覽無見以收視,聽無聲以黜聽[①]。

和匪專於旨酒,樂奚必於絲桐。40

焚清香以練氣[②],啓玉檢而擊蒙[③]。

期遣滯於昭曠[④],庶近真於感通[⑤]。44

筌太虚之有象,覆妙用之非空[⑥]。

朝天甚簡,採藥多暇。48

形猶資於吐納[⑦],意已迸於將迓[⑧]。

知道無廢興,而物有存謝。52

故挹生本而常生,體化宗而不化[⑨]。

蕭蕭絶塵,誰與爲鄰。56

跡遠而朋遊益廣,機忘而鳥獸可馴。

---

①《文苑英華》和《全唐文》作"黜聽",以便與第40句的"桐"押韻,*HY*作"逃默",其中"默"與第36句的"適"押韻。參考第175、176句"遠遊"("遠遊",《楚辭補注》卷5,174頁):"視儵忽而無見兮,聽惝恍而無聞。"參見Kroll, " On 'Far Roaming' ", *Journal of the American Oriental Society* 116.4(1996), p.663, p.669;《莊子集釋》卷11,173頁亦有引用:"無視無聽,抱神以靜。"或參考陸機《文賦》(《文選》卷17,763頁)的第15句:"其始也,皆收視反聽。"

② 參見"怨清香之難留",謝靈運《山居賦》,《全宋文》卷31,4b頁。詩人所謂"練氣",是希望净化肉體以達到更高的境界。

③ "玉檢"是詩人在顯示其隱退的文本的漩渦中"真"(第44句)的聖典。"蒙",詩人戰鬥因而使人想起《易經》的"蒙卦",即坎(水)下艮(山)上,傳統認爲(象)這是"山下有險,險而止"的表現,《周易正義》(《十三經注疏附校勘記》)卷1,8b頁。故此未嘗不是一個富於文化聯想的典故。

④ 參考《莊子集釋》卷12,198頁,描述"神人":"上神乘光,與形滅亡,此謂昭曠。致命盡情,天地樂而萬事銷亡,萬物復情,此之謂混冥。"這又與謝靈運的文章相關,在《富春渚》的詩句中用了"昭曠"這個詞(《文選》卷26,1240頁),他爲做了一個辭官隱世的決定而感到高興:"宿心漸申寫,萬事俱零落。懷抱既昭曠,外物徒龍蠖。"

⑤ "真"是居住在最高處的物體。"draw near"這個詞同時滿足了《文選》的"延"(taken causatively)以及《文苑英華》和《全唐文》的"近"(whether causative or active)。"感通",相互在精神和態度方面的共鳴。唐代宗教和世俗的文章,都顯示建立在兩個世界之間通過精神的探訪回應人類虔誠行爲的聯繫。

⑥ *HY*用"筌"和"覈"分别作爲第45、46句的第一個字;《文苑英華》和《全唐文》則是"鑒"和"覆"。從同義詞角度考慮,正確的配對應該是"鑒—覈"或者"筌—覆",我用第二組做了校改。兩句話中起物體作用的五個字(太虚之有象/妙用之非空)都是巧妙地在"有—無"與"無—有"觀念上構成對立。

⑦ "吐納"指由呼吸控制的引導一個人"氣"的練習,在這過程中,排出舊的氣息,再吸入新的不可或缺的氣息。此外,參考《莊子集釋》卷15,137頁,對"道引之士"、"養形之人"的描述。

⑧ 第三字《文苑英華》和《全唐文》作"迸",*HY*爲"屏"。"將迓",指穩定地接受來來往往的世間變换,正如隨後詩句詳盡闡述的那樣。參考《莊子集釋》卷6,115頁:"殺生者不死,生生者不生,其爲物無不將也,無不迎也,無不毁也,無不成也。"

⑨ 詩人已經達到與"旋轉世界的静止點"同等的境界。

韻靡叶於當時,心常依於古人。60

仰巢由[①]浩浩之逸軌,詠羲農默默之化淳[②]。

師黄老之玄奥[③],友松喬之道真[④]。64

慙無功之逮物,良獨善於吾身[⑤]。

只所幸其自得,敢韜精於隱淪。

附　録

李白《望廬山瀑布二首》其一,中古漢語對此詩聲調(O代表平聲韻,X代表仄聲韻)與韻律模式的重組。

	西登香爐峰,	OOOOO	
	南見瀑布水。	OXXXX	rA
	掛流三百丈,	XOOXO	
4	噴壑數十里。	OXXXX	rA
	欻如飛電來,	XOOXO	
	隱若白虹起。	XXXOX	rA
	初驚河漢落,	OOOXX	
8	半灑雲天裏。	XXOOX	rA
	仰觀勢轉雄,	XOXXO	rB
	壯哉造化功。	XOXXO	rB
	海風吹不斷,	XOOXX	
12	江月照還空。	OXXOO	rB

① 指古代名人巢父和許由,他們都選擇隱逸,不與世人往來。

② 指文明始祖伏羲和神農,在史前的迷霧中,伏羲首先教會人們漁獵的方法,神農則教人們農事和藥理。《文苑英華》和《全唐文》第61、62這兩句作"仰由皓之逸軌(即商山四隱士,見《漢書》,北京:中華書局,1975,卷72,3056頁)詠羲農之化淳"。

③ 黄帝和老子。

④ 指赤松子和王子喬,二人作爲仙人合稱至少可以追溯到《遠遊》(譯者注:《楚辭》卷5),第23、24、54、61-74行。見Kroll, "On 'Far Roaming'"。

⑤ 參見《孟子》卷7A,第9章(譯者注:《盡心》上):"窮則獨善其身,達則兼濟天下。"

	空中亂潨射，	OOXOX	rC
	左右洗青壁。	XXXOX	rC
	飛珠散輕霞，	OOXOO	
16	流沫沸穹石。	OXXOX	rC
	而我樂名山，	OXXOO	rD
	對之心益閑。	XOOXO	rD
	無論漱瓊液，	OXXOX	
20	且得洗塵顏。	XXXOO	rD
	且諧宿所好，	XOXXX	
22	永願辭人間。	XXOOO	rD

元稹悼亡詩《夢井》新釋*

——以中國古代井觀爲視點

山崎藍

前言

當代因自來水的普及,井失去了用武之地,但在日本奈良縣等地,至今仍在舉行祭祀廢井的活動,活動中,必須在井口留下縫隙,或設置通到地面的管道,以使井神呼吸①。另外,從《繼子和井》的情節、講述小野篁經井通往冥府的《小野篁赴冥土之井》的故事中也可看到,井被視爲通往異界的通道②。

中國文言小説中,不僅出現了井神及棲息在井中的龍(井龍)③,還有不少

作者單位:日本明星大學人文學部

* 本文根據日文論文《元稹悼亡詩《夢井》新釋——中國古代における井户觀の一側面》,《東方學》第116輯,74—91頁翻譯並增訂。本項研究接受了JSPS科研費26770130的贊助。

① 參看日色四郎,《日本上代井的研究》,橿原:日色四郎先生遺稿出版會,1967。大島建彦,《水的民俗和信仰》,《民俗信仰的衆神》,東京:三彌井書店,2003;山本博,《井户的研究》,京都:綜藝舍,1970等亦有提及。

②《繼子和井户》的故事有兩種版本,一是淘井日繼子掉入井底最終遇救,另一是繼子爲了尋找掉入井底的碟子而進入井底,給住在裏面的老婆婆揉肩搓腰,獲得禮品而返回。請參照《日本昔話事典》,東京:弘文堂,1977,871頁。關於小野篁的故事,請參看田中嗣人,《小野篁傳説考》,《華頂博物館學研究》第10號,2003,16—31頁。

③《太平廣記》卷二三一《陳仲躬》、卷四二〇《井龍》等中有井龍的描寫,《太平御覽》卷一八九引《白澤圖》、《廣博物志》卷七引《緝柳編》及《子不語》卷一七《井泉童子》等中有井神的描寫。關於井龍、井神,吴裕成,《中國的井文化》,天津:天津人民出版社,2002;堀誠,《八角井異聞——井中之怪》,《早稻田大學教育學部學術研究——國語·國文學編》第41號,1993,105—116頁;《井中奇聞——死生之命和生殖》,《中國文學研究》第19期,1993,66—81頁;《井中餘聞——鏡、夢和神靈》,《中國文學研究》第22期,1996,77—91頁中亦有提及。

作品是作爲連接人世和異界之“境界”來描述井的[①]。但是文言小説中那種井的意象的代表物——“境界之井”,卻未在詩歌中得到描繪。在詩歌與小説中,對井的關注之處、着重之點互不相同。

本文關注因領域不同而形成的各異的井像,討論涉及井的資料。特别指出中唐詩人元稹創作的悼亡詩《夢井》是一篇有意識地利用文言小説中描寫的“境界之井”的構思而創作出的重要作品,希望由此揭示中國古代“井觀”的一個側面。

首先,筆者舉幾個具有代表性的例子,來説明人們對“井”這一空間持有何種認識。

一、清潔感——修德養民之井

《釋名》卷五《釋宫室》中稱:“井,清也。泉之清潔者也。”井是湧出清潔之水的場所。《井》還是《易》的一卦,《易·井》“巽下坎上”辭如下:

> 井,改邑不改井。無喪無得。往來井井。汔至亦未繘井。羸其瓶,凶。

對於“井”,孔穎達有如下表述:

> 井者,物象之名也。古者穿地取水,以瓶引汲,謂之爲井。此卦明君子修德養民,有常不變,終始無改,養物不窮,莫過乎井,故以修德之卦取譬名之井焉。

由這些表述可見,井是始終不變的,它被視爲君主修德養民的象徵。此外,《初學記》卷七“井第六”引《風俗通》稱:

> 《風俗通》云:井者,法也。節也。言法制居人,令節其飲食,無窮

---

① 關於文言小説中的井,除了上述列舉的各論文之外,還有大室幹雄,《圍棋之民俗話》,東京:岩波書店,1995;中野美代子,《天井和地井——境界之幻想空間》、《有關世外桃源的風水——竪坑、井和“天井”》,收《奇景之圖像學》,東京:角川春樹事務所,1996,194—223頁等。這些研究以文言小説爲主要材料,以具有“境界”之作用的井爲中心進行了分析。在日文中,“境界”一詞有連通兩個以上不同的空間、領域之“領域”的涵義。下文中使用的“境界”一詞均指上述意義。

竭也。

這可以説與稱“井,法也”的《經典釋文》卷一“井”的記述同出一轍。由此可見,存在着一種將井與清廉、德、節制相聯,認爲它給人們帶來恩惠的“井觀”。

二、閉塞感——藏污納垢的井

另一方面,井亦被作爲狹小封閉的場所來描寫。《九歎》是追思屈原忠信之節氣,讚頌其品德的作品。其中有如下一節:

> 菀蘼蕪與菌若兮,漸藁本於洿瀆。淹芳芷於腐井兮,棄雞駭於筐簏[①]。(劉向《九歎·怨思》)

對此文,王逸注中有如下表述:

> 言積漬衆芳於污泥臭井之中,棄文犀之角,置於筐簏而不帶佩,蔽其美質,失其性也。以言棄賢智之士於山林之中,亦失其志也。

按王逸注,第四句是感歎自己(即衆芳)懷才不遇,被棄置於“腐井”中的狀況。如同《莊子·秋水篇》裏的“井鼃”的故事,這裏描述的井是窄小而令人窒息的空間[②]。作者爲了表達不爲君主所用的鬱悶感,把自己比喻爲被棄置井底的香草。下述詩也是相同的例證:

> 不見山巔樹,摧扤下爲薪。豈甘井中泥,上出作埃塵。(作者不詳或劉孝威《箜篌謡》一部)[③]

《井中泥》或許出自《易·井》卦初六“井泥不食。舊井無禽”這一典故。王

---

①《詩經》《楚辭》所收的漢代以前的作品中没有涉及井的作品。

②《莊子》相關部分的原文如下:“井鼃不可以語於海者,拘於虚也。”

③《文苑英華》卷二一〇“上出作埃塵”作“時至出作塵”。逯欽立《先秦漢魏晉南北朝詩》(北京:中華書局,1983,1871頁)在《梁詩·箜篌謡》中,有“本集。《文苑英華》二百十失名,次劉孝威後;《樂府詩集》八十七作無名氏,詩紀八十八”一文。又云:“逯案:《御覽》引此篇起首四句,題作古“歌辭”。細玩其辭,亦不類六朝人作。今分别列入漢詩及此集。”

弼注曰：

> 最在井底，上又無應，沈滯滓穢，故曰井泥不食也。井泥而不可食，則是久井不見渫治者也。久井不見渫治，禽所不向，而況人乎。一時所共棄捨也。

由此可以推定《箜篌謡》中的"豈甘井中泥"一句表達的應是作者自己不甘爲井底之污泥，被世人遺棄的思想。提供清澈之水的井，若不適時清掃，井底立即就會積澱淤泥，變爲污穢空間。

三、作爲通往異界的通道之井

正如本文開頭所述，六朝志怪以來的文言小説中常描述"作爲境界之井"。如《太平廣記》卷一九七《張華》（出 自《小説》）中有如下記述[①]：

> 又嵩高山北有大穴空，莫測其深。百姓歲時每遊其上。晉初，嘗有一人悮墜穴中。同輩冀其儻不死，試投食於穴，墜者得之爲糧，乃緣穴而行，可十許日，忽曠然見明。又有草屋一區，中有二人，對坐圍碁。局下有一杯白飲。墜者告以飢渴，碁者曰，可飲此。墜者飲之，氣力十倍。碁者曰，汝欲停此不。墜者曰，不願停。碁者曰，汝從西行數十步，有一井。其中多怪異，慎勿畏。但投身入中，當得出。若飢，即可取井中物食之。墜者如其言。井多蛟龍，然其墜者，輒避其路。墜者緣井而行，井中有物若青泥，墜者食之，了不復飢。可半年許，乃出蜀中。因歸洛下，問張華。華曰，此仙館。所飲者玉漿，所食者龍穴石髓也。

正如"井多蛟龍"所述，此井中棲息有蛟龍。此作品中可見如下故事，井是從異界返回人間的通道，有人耳聞井中傳來的"雞犬鳥雀聲"，躍入井中，便

---

① 前述大室幹雄、中野美代子認爲典故出自《搜神後記》，但汪紹楹在《搜神後記》(北京：中華書局，1981，2頁）指出"本條未見各書（指唐宋的類書——筆者補注）引作《繼搜神記》（指《搜神後記》——筆者補注）"。李劍國也在《新輯搜神後記》(北京：中華書局，2007)中，因上述原因而未輯録該條。《北堂書鈔》卷一六〇、《初學記》卷五中指出典故出自《世説》，《太平廣記》卷一九七中指出典故出自（梁）殷芸《小説》。本文採用了《太平廣記》所收的原文。

到了宛如世外桃源之地[①]。

而在《太平廣記》卷三九九《王迪》(出《祥異集驗》)中有如下記述:

唐貞元十四年春三月,壽州隨軍王迪家井,忽然沸溢,十日又竭。見井底,有聲,如嬰兒之聲。至四月,兄弟二人盲。又一人死。家事狼狽之應驗。

不僅有"雞犬鳥雀聲",從井中還會傳來異界之音,嬰兒之聲則是死與不祥的徵兆[②]。(梁)宗懔《荆楚歲時記》中寫道:"正月未日夜,蘆苣火照井廁中,則百鬼走。"這一想法,梁朝以後也被繼承下來,據説現在的荆州依然流傳着"井廁好藏百鬼"這一諺語[③]。"井廁"這一詞反映了地坑結構的廁與井被視爲同類空間。從人們點火來驅逐"井廁"之幽靈(鬼)的行爲中可看出,存在着像井或者廁等"陰暗場所"中有幽靈(鬼)的認識。《太平廣記》有不少故事描寫了井的異常現象,以及由此造成的死亡、投井自殺、殺人藏屍於井中的情節。這表明通過井抵達的異界不一定都是世外桃源。井亦是與死亡和怪異密切相關的空間。

四、故鄉的景象和井桐——時光流逝與憂愁

自漢代以來詩歌中出現的井的形象中,以下兩種具有代表性,一是"留在故宅的井",一是"思念異地的戀人而汲水的井"。

首先,舉故宅詩中的井爲例。

兔從狗竇入,雉從樑上飛。中庭生旅谷,井上生旅葵。(作者不詳《古詩三首·其二》一部)

---

①《太平廣記》卷二〇《陰隱客》(出《博物志》)。前述中野美代子引用了《陰隱客》及《張華》等,指出:"井也和通往桃花源的洞穴一樣,其功能完全相同。"中野《有關世外桃源的風水——豎坑、井和"天井"》,208頁。

② 有學者從日本民俗學的觀點出發認爲,嬰兒的哭聲中包含着穿越人世到達異界的信息。請參照森下みさ(米沙)子,《居於境界的孩子和老人》,赤坂憲雄編《作爲方法的境界》,東京:新曜社,1991。

③ 王毓榮,《荆楚歲時記校注》,台北:文津出版社,1992,89-90頁。

這是十五歲從軍，到了八十歲才得以重返故鄉的老兵之歌。在日夜想念的故宅中已不見親人踪影。棲居着的是雞、兔，以及無人培育卻枝葉繁茂的穀物和旅葵。

草深斜徑没，水盡曲池空。林中滿明月，是處來春風。唯餘一廢井，尚夾兩株桐。（元行恭《過故宅詩》一部）

這是元行恭暫返家鄉時吟詠的作品。生長在井端的桐樹——"井桐"，因其落葉飄零的景象，常被作爲令人聯想到凋落的植物，出現在自六朝至唐代的詩中[①]。對於《過故宅詩》中的"廢井""兩株桐"的意象，原田直枝氏指出，這是"借用了井和桐樹等景物來象徵故鄉、故宅的表現方式"，由此體現了造成這一系列荒蕪的時間之綿延，吟詠者離開故鄉期間逝去的光陰之漫長[②]。

留在故宅的井之所以被描繪爲植被叢生的樣子，是由於井是不可或缺的生活器具，只要有人過日子，井上就不會長滿野草。正如第一章引用的《易》中稱"井，改邑不改井"，井是一直留在原地的。詩歌將作爲汲水場所的角色已經終結的、無法遷移而逐漸腐朽的井，和"旅葵"及"桐"等至今仍枝葉繁茂的植物，作爲一組對比的事物而加以描寫，使故鄉、故宅的荒蕪情景和時光流逝更爲鮮明[③]。

其次舉描寫思念身在異地的戀人，在井邊汲水的人物之作品爲例。

已泣機中婦，复悲堂上君。羅襦曉長襞，翠被夜徒薰。空汲銀牀井，誰縫金縷裙。所思竟不至，空持清夜分。（庾丹《秋閨有望詩》一部）[④]

歸飛夢所憶，共子汲寒漿。銅瓶素絲綆，綺井白銀牀。雀出豐茸樹，蟲飛玳瑁梁。離人不相見，争忍對春光。（庾丹《夜夢還家詩》）[⑤]

---

① 關於中國文學中的桐樹的形象，有俞香順，《中國文學中的梧桐意象》，《南京師範大學文學院學報》2005年第4期等的研究。俞氏指出"井桐"的落葉表示秋天的到來，落葉是引起悲愁的觸媒。

② 原田直枝，《"江南爲瘴癘之地"以及故鄉》，《中國文學報》第71册，2006，19-20頁。

③ 故宅井詩，另外還有何遜《行經范僕射故宅詩》、江總《南還尋草市宅詩》等。

④《先秦漢魏晉南北朝詩》中，"銀牀"作"銀床"、"空持清夜分"作"持酒清夜分"，依據《玉台新詠箋注》卷五進行了更改。

⑤《玉台新詠箋注》卷五中，"争忍對春光"作"難忍對春光"。

"銀牀",除上述兩部作品之外,也常出現在描寫井的詩中[1]。如《秋閨有望詩》那樣的描寫女子佇立井邊汲水之身姿的詩,從六朝至唐代多有出現。井被作爲描寫與心上人相隔千里的女子之離愁的最佳舞台,常被詩人所用[2]。《夜夢還家詩》則是歌詠一位男子借夢與遠隔千里的妻子相會,夫妻用銅瓶從井中汲水之情景的罕見作品。男主人公在夢中與愛妻一同汲水,夢中醒來,妻子的身影悄然不見。將男子帶回現實世界中的窗外春光,更添悲愁之情。

詩歌中的井呈現出了既不同於"作爲境界之井",也不同於使人聯想起清廉與品德的井的景象。日常生活中不可或缺的設施之井,被描述爲使人真切感知日常生活之崩潰、應有的幸福之不在的場景[3]。

五、元稹《夢井》的定位

(一)《夢井》原文及先行研究的整理

元稹爲了追思元和四年(809)死去的妻子韋叢,創作了三十三首悼亡詩,《夢井》被認爲是韋叢死後翌年的元和五年(810)創作的[4]。首先引用其全文。

夢上高高原,原上有深井。登高意枯渴,願見深泉冷。徘徊遶井顧,自照泉中影。沉浮落井瓶,井上無懸綆。念此瓶欲沈,荒忙爲求請。遍入原上村,村空犬仍猛。還來遶井哭,哭聲通復哽。哽噎夢忽驚,覺來房舍静。燈焰碧朧朧,淚光凝冏冏。鐘聲夜方半,坐卧心難整。忽憶咸陽

---

① 所引用的作品之外,還有庾肩吾《九日侍宴樂遊苑應令詩》、(梁)簡文帝蕭綱《雙桐生空井》等中有"銀牀"。"銀牀"出《玉台新詠箋注》卷七吴兆宜注,而(南宋)吴曾《能改齋漫録》中引用了作者不詳的《淮南王》的一部分等,吴曾以《山海經》"海内崑崙墟,在西北,帝之下都。高萬仞,面有九井,以玉爲檻。郭璞注曰,檻,欄也"的記載爲依據,認爲:"蓋銀牀者,以銀作欄。猶《山海經》所謂以玉爲欄耳。"

② 例如有(梁)簡文帝蕭綱《雙桐生空井》、張籍《楚妃怨》、陸龜蒙《井上桐》等。女子思念戀人時提及井的詩更多。

③ 李白《長干行二首·其一》第四句"遶牀弄青梅"的"牀"一説是床,一説是井,若是後者,可解釋爲以井爲背景的少男少女的初戀詠唱。但筆者認爲,六朝爲止的詩歌中尚未出現類似的作品。

④ 對於《夢井》的編年,參考了冀勤點校,《元稹集》,北京:中華書局,1982;楊軍箋注,《元稹集編年箋注》,西安:三秦出版社,2002;花房英樹,《元稹研究》,京都:匯文堂書店,1977。楊軍、花房英樹均認爲《夢井》是元和五年創作的。

原，荒田萬餘頃。土厚壙亦深，埋魂在深埂。埂深安可越，魂通有時逞。今宵泉下人，化作瓶相警。感此涕汍瀾，汍瀾涕沾領。所傷覺夢閒，便覺死生境。豈無同穴期，生期諒綿永。又恐前後魂，安能兩知省。尋環意無極，坐見天將昞。吟此夢井詩，春朝好光景[①]。

關於元稹的悼亡詩有很多先行研究[②]，其中山本和義舉出《夢井》原文進行了分析[③]。另外，高橋美千子注意到元稹創作了許多寫夢的詩，從夢詩作品之一的角度探討了《夢井》[④]。

山本氏指出："對男女離別，以落瓶來歌詠的自古就有"，將（齊）釋寶月《估客樂》及李白《寄遠》與《夢井》進行了比較（後文詳述）。高橋氏針對《夢井》指出："明確區分了夢境和夢醒時分，凸顯了夢境部分。並在夢醒後，按自己（指元稹——筆者補注）的意願解釋了夢，並同時梳理了自己的情感。在此所描寫的夢，是採用了古有的比喻的幻想情景。在該詩（指《夢井》——筆者補注）中，描述了夢中靈魂自由漫遊的情景。夢境中雖不見妻子，但元稹解釋爲妻子的靈魂化身爲'瓶'出現。"她還舉出山本氏注中只提及詩題的白居易的《井底引銀瓶》，考察了其與《夢井》的不同之處。

山本氏所舉的釋寶月《估客樂》原文如下：

有信數寄書，無信心相憶。莫作瓶落井，一去無消息。（釋寶月《估客樂·其二》）[⑤]

① 以《元氏長慶集》(用弘治元年楊循吉景宋傳鈔本影印，北京：文學古籍刊行社，1956，158-159頁)爲底本。底本中，"相警"作"相誓"，依據《新刊元微之文集》(北京圖書館藏宋蜀刻本影印，上海：上海古籍出版社，1994，245頁)進行了更改。前述《元稹集》和《元稹集編年箋注》也依據《新刊元微之文集》作"相誓"。對於"便覺"，《新刊元微之文集》和《元稹集》作"便隔"。

② 本稿未涉及的有關悼亡詩的先行論文有：入谷仙介，《關於悼亡詩——從潘岳到元稹》，《入矢教授小川教授退休紀念中國文學語學論集》，京都：筑摩書房，1974；陳寅恪，《元白詩箋證稿》，上海：上海古籍出版社，1978；張樾暉，《元稹〈夢井〉詩賞析》，《古典文學知識》1998年第5期；陳翀，《代摯友亡妻賦詩的白居易——元稹妻韋叢之死及悼亡唱和詩》，《日本中國學會報》第59集，2007年等。

③ 山本和義，《關於元稹的艷詩及其悼亡詩》，《中國文學報》第9册，1958，81-83頁。

④ 高橋美千子，《對元稹之夢的探討》，《中國文學報》第32册，1980，57-59頁。

⑤《玉台新詠箋注》卷十中無作者名，而作《近代西曲歌五首·估客樂》。另外將"有信數寄書"作"有客數寄書"。

李白《寄遠》也和《估客樂》相同，也用落瓶比喻戀人的杳無音訊[①]。下面舉出高橋氏引用的白居易《井底引銀瓶》的開頭六句：

> 井底引銀瓶，銀瓶欲上絲繩絶。石上磨玉簪，玉簪欲成中央折。瓶沈簪折知奈何，似妾今朝與君别。（白居易《井底引銀瓶》一部）[②]

《井底引銀瓶》是新樂府三十首之一，開頭六句以沉瓶折簪來比喻女子清晨與情人分别後的境遇。文章中描述了委身於所愛之人，但因未明媒正娶而被男方雙親趕出家門，無處可歸而不知所措的女子的情景，以此勸誡婦女應行事慎重。

山本氏認爲這兩首詩（指釋寶月《估客樂》和李白《寄遠》——筆者補注）中歌詠的落瓶較單純。而元稹詩（指《夢井》——筆者補注）中卻是動態的，而且具有結構性意義。詩中生動地表現了對命運撕心裂肺卻徒勞無功的反抗心情，“元稹在自古使用的比喻中註入了全新的生命力。這種新的生命力正來自於建立在親身體驗上的强烈情感，以及表現該情感的小説式敘述之巧妙”。另外，據高橋氏分析，比起白居易《井底引銀瓶》，《夢井》中覺醒時比夢中更爲傷悲，這表明“明確夢之因果關係，使人更深刻地直面現實，思緒涉及生死問題。瓶爲妻子的化身，彷彿在勸誡自己（指元稹——筆者補注），並通過夢使自己對人生不安的意識呈現出來”[③]。

山本、高橋兩氏所比較、探討的是釋寶月《估客樂》、李白《寄遠》、白居易《井底引銀瓶》三首詩，而中國民俗學家吴裕成在收集關於井的各種資料時，引用了《易》之《井》卦的一部分和《墨子·非儒下》的一部分，以及《夢井》的一部分，指出《夢井》中“瓶”爲元稹亡妻，以及黄泉和井有關，瓶可能具有象徵

---

①《寄遠十二首·其八》（據詹鍈主編，《李白全集校注匯釋集評》，天津：百花文藝出版社，1996）頭四句如下：“憶昨東園桃李紅碧枝，與君此時初别離。金瓶落井無消息，令人行嘆復坐思。”

② 據謝思煒撰，《白居易詩集校注》，北京：中華書局，2006，419頁。《全唐詩》中，詩題作《井底引銀缾 止淫奔也》。“缾”爲“瓶”的異體字。以下略同。

③ 對於《井底引銀瓶》的頭六句，高橋氏指出：“這個夢是悲慘的……到達悲哀的頂點後從夢中醒來。”認爲這是對夢的描寫。可是這實際上應該是將與男子離别的女子的境遇比喻爲提取時繩斷而沉入井底的瓶，以及因爲被磨細而從中折斷的玉簪。高橋氏認爲是夢中發生的事情，這種説法有欠妥當，筆者不能贊同。

性。就筆者所見,該文是利用詩歌以外的資料分析《夢井》的唯一論著[1]。

吴氏引的《墨子·非儒下》如下:

其親死,列屍弗斂,登屋窺井,挑鼠穴,探滌器,而求其人焉……

據此記載,當時作爲一項招魂儀式,有窺井行爲。但有人認爲"窺井"以下文字不見於《禮記》、《儀禮》等有關喪葬禮儀的記述中,屬於謾語[2],但考慮到在日本各地至今仍保存着這一風俗:爲唤醒失去意識的人,邊窺井邊喊其人姓名,其人就能生還。與此相參看,《墨子·非儒下》的記載稱得上是頗有意義的資料[3]。

吴氏除了《墨子·非儒下》記載之外,作爲賦予瓶吉凶徵兆的例證,還引用了《易》的《井》卦一節:"井汔至,亦未繘井,羸其瓶,凶。"但是正如本文第一章所述,該文字應爲"汔至亦未繘井,羸其瓶,凶"。在吴氏論著中,時而可見在引用和使用資料方面令人質疑之處。另外,雖然廣泛徵引了有關井的衆多資料,但遺憾的是,未對元稹《夢井》詩整體進行分析,與吴氏論著之前的山本氏和高橋氏的研究成果相比較爲簡單。

本文將參考這些研究成果,並利用其他資料,從不同的角度進一步分析《夢井》[4]。

(二) 瓶的意義——黄泉和瓶

正如先行研究中已經指出,《夢井》的特點之一是,繼第七句至十句"沈浮

---

① 參照前述吴裕成,《中國的井文化》,307-308頁。

② 孫詒讓,《墨子閒詁》,北京:中華書局,2001,288頁。吴氏未提及此附註。

③ 參照中山太郎,《井神考》,《日本民俗学(神事篇)》,東京:大和書房,1976,121-138頁;以及前述大島建彦,《水的民俗和信仰》。

④ 劉航、李貴兩氏在《白居易〈井底引銀瓶〉的民俗學問題》中,引用揚雄《酒箴》(論考中爲《酒賦》但此爲錯誤)、王昌齡《行路難》、元稹《夢井》,指出本文中所引用的《井底引銀瓶》的開頭四句表現了中國古代的民間禁忌、水崇拜、井崇拜、瓶崇拜及簪的象徵自身之意(《文史知識》2001年1期,62-66頁)。筆者執筆拙稿《元稹悼亡詩〈夢井〉新釋——以中國古代井觀爲視點》(《東方學》第116輯,74-91頁)時該論著雖已發表但因筆者當時不知未能引用。但該考論與前述吴氏同樣未對拙稿《元稹悼亡詩〈夢井〉新釋——以中國古代井觀爲視點》的主要題材元稹《夢井》進行詳細的分析。劉航、李貴兩氏還引用了張籍《楚妃怨》等,作爲表現"瓶"本身被作爲崇拜對象,以"瓶"是否落入井中占卜凶吉的作品。但筆者對"瓶"本身爲崇拜對象的觀點不敢苟同,對劉航、李貴兩氏的很多觀點都難以贊同。

落井瓶，井上無懸綆。念此瓶欲沈，荒忙爲求請”的描述之後，第二七句至二八句明確提到了“今宵泉下人，化作瓶相警”，也就是妻子化作瓶在夢中出現這一點。作爲提及瓶的文獻可以舉出吴氏也曾指出的《易》之《井》卦“汔至亦未繘井。羸其瓶，凶”（第一章既述）。可見對瓶破抱有不祥之感。關於瓶，《太平御覽》卷七五八《缾》寫道：“雜五行書曰，懸缾井中，除邪鬼。”瓶被認爲具有辟邪功能。

此外，東漢揚雄《酒箴》中描寫的井中之瓶也頗有意思。《酒箴》記載在《漢書》卷九二《遊俠傳》之《陳遵》中。因才能受王莽器重的陳遵，嗜酒如命，他常對“時時好事者從之質疑問事，論道經書”（《漢書》卷九二）的友人張竦説起《酒箴》。《酒箴》是揚雄當“黄門郎”時爲“諷諫成帝”所作，“其文爲酒客難法度士”，揚雄爲了譏諷法度之士，運用了下述比喻：

> 子猶瓶矣。觀瓶之居，居井之眉，處高臨深，動常近危。酒醪不入口，臧水滿懷。不得左右，牽於纆徽。一旦叀礙，爲瓽所轠。身提黄泉，骨肉爲泥。自用如此，不如鴟夷。鴟夷滑稽，腹如大壺，盡日盛酒，人復借酤。常爲國器，託於屬車，出入兩宫，經營公家。由是言之，酒何過乎。①

陳遵將飽讀經書的清廉之士張竦喻作爲繩所繫，不自由且與危險相伴的瓶，將性格豪放飲酒爲樂的自己喻作鴟夷。及王莽敗，張竦爲賊兵所殺。井底與黄泉相通，張竦，即“瓶”被投入黄泉，喪失性命。井既如第一章所述是像徵清廉、法度，又如第三章所述也是與死亡、怪異相伴的“境界”。

小南一郎氏的研究也頗有助於理解《酒箴》。小南氏提出瓶或壺可作爲靈魂的附身。根據小南氏的研究，三國（東吴）至西晉時期，長江下游流域墓葬中經常供有叫做“神亭壺”的壺，死者之魂通過該神亭壺到達祖先靈魂所在的世界，反之，爲死者招魂之際，該壺就成爲靈魂的附身，靈魂憑藉它返回人世，據説浙江省一帶至今仍將這種壺叫做“魂瓶”。另外，從東漢楊氏墓中出

① 顔師古注有如下記載：“眉，井邊地。若人目上之有眉。”“纆徽，井索也。叀，縣也。瓽，井以甎爲甃者也。轠，擊也。言瓶忽縣礙不得下，而爲井瓽所擊，則破碎也。”“提，擲也。擲入黄泉之中也。”“鴟夷，韋囊以盛酒，即今鴟夷（勝）［幐］也。”

土的壺上的朱書中記載有靈魂靠“瓶”赴來世，而且還有稱相同的壺爲“神瓶”及“解注瓶”等例子[①]。另外，據鍾方正樹氏的研究，在新石器時代至六朝時期的中國湖北省和河南省、奈良時代以後的日本各地，多見從井底出土瓶等大概爲汲水器皿的完整的土器。與之相關，锺方氏還舉出事例，試圖説明由於水之神性，井和壺密切相關，他舉出了源賴義將在八幡神的佑護下得到的水裝入壺中帶回，並將它埋在新挖的井的底下的所謂“壺井水傳説”。另外，還介紹了日本水占卜之一的“依瓶之水”，指出有可能是原爲裝神水的容器之瓶，其後轉變爲神靈之附身[②]。

如此在日中兩國，包括瓶在内的壺狀容器被作爲一種道具，它具有與靈魂相隨的功能。如前所述的楊氏墓那樣，在後漢墓中隨葬着題寫朱書或墨書的壺，這些遺物在陝西省至河南省西部一帶常可見到。在揚雄出生的蜀川至今没有發現同類遺物，但是至少在長安，揚雄已經認識到瓶作爲靈魂附身的功能，可能正是這種認識構成了《酒箴》構思的背景。高高懸掛在通向黄泉的井上、反复升降、不知何時會破碎的瓶，如將它視爲盛放靈魂的容器的話，即使是當代的讀者，也會因此而感到不安吧。

把瓶作爲靈魂的附身來看的話，還存在一些更有趣的作品。如王昌齡《行路難》，其文如下：

> 雙絲作綆系銀瓶，百尺寒泉轆轤上。懸絲一絶不可望，似妾傾心在君掌。人生意氣好遷捐，只重狂花不重賢。宴罷調箏奏離鶴，回嬌轉盼泣君前。君不見眼前事，豈保須臾心勿異。西山日下雨足稀，側有浮云無所寄。但願莫忘前者言，剉骨黄塵亦無愧。行路難，勸君酒，莫辭煩。美酒千鍾猶可盡，心中片愧何可論。一聞漢主思故劍，使妾長嗟萬古魂。[③]

---

① 小南一郎，《壺型之宇宙》，《東方學報》第61册，1989，165-221頁。朱書原文如下(依照《靈寶張灣漢墓》(河南省靈寶縣出土，《文物》1975年11月，80頁)：“天帝使者，謹爲楊氏之家，鎮安塚墓，謹以鉛人金玉，爲死者解適，生者解罪過，瓶到之後，令母人爲安宗，君自食地下租歲二千萬，令後世子孫，士宦位至公侯富貴，將相不絶，移丘丞墓伯，下當用者，如律令。”

② 參看鍾方正樹，《井的考古學》，東京：同成社，2003。

③ 對於“勸君酒，莫辭煩”，《全唐詩》卷一四一注有“一作勸酒莫辭煩”。另外“心中片愧何可論”的“愧”上有“一作恨”的注釋。

第一句至第四句説的是用兩根絹絲做井繩，垂掛銀瓶，百尺之下的清涼之水就可用轆轤打上來。一旦割斷繩子，瓶就不復存在。任轆轤與繩擺佈的瓶就如同我的心掌握在你的手掌心。

因爲夫君變心而不得不離開婆家的女人，將自己的心情與不知何時會掉入無底深淵的井中之瓶重疊起來。揚雄作爲表達"法度之士"的命運而描繪的井中之瓶的危險處境，在此則用來表述一名女子的命運。同樣的構思在顧況《悲歌六首·其三》中也能見到，他把女子的心聯繫到繩上，描寫了隨着轆轤的轉動而"惆悵"的情景[①]。前面例舉的白居易《井底引銀瓶》，也認爲是以相同的構思爲背景創作的[②]。據上述小南一郎氏的考證，從唐代至元明，在廣東、湖南、江南各地的墓葬中都有形狀不一的壺被發掘。由此可見壺型容器被視爲靈魂之附身的觀念，在中國古代廣爲流傳，並繼承至今。

最後一句"使妾長嗟萬古魂"雖覺唐突，但筆者也希望根據井中之瓶上附有女性之靈魂這一想法來解釋。如第一章引用的《易》之《井》卦中"井，改邑不改井"，井是不能遷移而一直存在着的，而瓶則要聽憑繩和轆轤的擺佈。所有的井中都可能長眠着無數無可奈何落入井底深處的瓶。主人公正是聯想到附在這些瓶上的每一個靈魂，悲歎自身也是其中之一，才寫出"使妾長嗟萬古魂"之句的吧[③]。

（三）《夢井》中的井——井的用法

基於以上研究，再回到元稹的《夢井》。至於在該詩中元稹的個人體驗有

---

① 原文如下："新系青絲百尺繩，心在君家轆轤上。我心皎潔君不知，轆轤一轉一惆悵。"據王啓興、張虹注，《顧況詩注》，上海：上海古籍出版社，1994，103頁。《全唐詩》卷一九中的詩題作《短歌行》。

② 白居易《井底引銀瓶》中，第一句、二句爲"井底引銀瓶，銀瓶欲上絲繩絶。"與此相對，第三句、四句爲"石上磨玉簪，玉簪欲成中央折。"對於"簪"，《太平御覽》卷六八八中有"夢書曰，簪者爲身，簪者己之尊也。夢著好簪身歡喜"。和魂的附身之瓶欲上升，繩子卻斷了相同，玉簪也是從中折斷，這種描寫似乎也是基於玉簪所含有的民俗意義的。參看前述劉航、李貴，《白居易〈井底引銀瓶〉的民俗學問題》。

③ 筆者所見，王昌齡的校注本中，提及"使妾長嗟萬古魂"的只有李國勝校注，《王昌齡詩校注》，台北：文史哲出版社，1973，249頁。對於"一聞漢主思故劍，使妾長嗟萬古魂"這兩句，李氏引用了《漢書》外戚傳，説明文中"故劍"指舊妻，並指出："按此言天意回复，己或可冀重蒙恩眷，然猶嘆萬古忠義逐臣，去而不返。"可是該詩是棄婦詩之一，而這兩句應該是悲嘆被棄婦人遭遇的。對於李氏的説法筆者不能贊同。

多重要，元稹是否做了這夢等問題，我們已無法考證，但我們可以從詩歌是如何創作出來的這一角度來對其進行分析。筆者分析如下：

第一句“夢上高高原”至第一六句“覺來房舍静”。在夢中，登上高原，喜逢一深井。登上高處感覺到口渴，希望深深井底有清澈的冷泉。環繞井端，俯看井底，只見自己的身影和沈浮不定的瓶，不知何故不見吊瓶的繩子。男子以爲此瓶會沉入井底深處，突然感到非常焦急。走遍平原上的村莊，欲尋求幫助，到處不見人煙，只聽見犬吠聲。正如陶淵明《桃花源記並詩》中說的“荒路曖交通，雞犬互鳴吠”，犬聲是只能在悠閒的田園生活中才聽得到的，而在此詩中卻回盪在渺無人煙的村莊中。男子返回井邊，抽泣起來，哭聲彷彿是唤回死者的“招魂”。這聲音傳到井底，從抽泣至慟哭，不甚傷悲，幾近窒息而從夢中醒來。

第一七句“燈焰碧朧朧”至第三〇句“丸瀾涕沾領”。青白的燈光下，清晰地映現出淚痕。鐘聲告知已是夜半，但坐卧不寧難以平静。忽然想到咸陽的平原。萬餘頃荒田展現在眼前，土厚墓亦深。（亡妻的）靈魂埋在深穴底處，怎麽能穿越如此深厚的土層。但靈魂卻時而能如願返回人世。今宵在黄泉下的已亡之人化作瓶來告知自己。爲此深受感動的主人公淚流難禁，沾濕了衣襟。

第三章例舉的《太平廣記·張華》中，經過井回到蜀地的一男子，向張華講述起自己在井中所見，張華爲他“解謎”，講清了男子的經歷的意義所在。《夢井》的第一七句至第三〇句就相當於該“解謎”部分。彷彿仍在夢境的男子分析了夢中所發生的事情及心境。雖然妻子的肉體深埋在地下，但其靈魂仍能夠自由來往於陰陽界。妻子將自己的靈魂附在靈魂附身的工具——“瓶”中，來跟自己傾心交談。就元稹的夢詩而言，僅以高橋氏所舉的詩例來看，在元稹的作品中，如此詳細地“解夢”的，僅有《夢井》。《夢井》中“解夢”部分具有重要的意義。經長時間詳細“解夢”之後，夢境中的男子漸漸醒來，更深的悲情湧上心頭。

第三一句“所傷覺夢間”至最後一句“春朝好光景”。從夢境覺醒後最爲傷感，因爲由此意識到了生死“境界”。夫婦同入墓穴的時節終究會來，但是，留在人間的我獨自生活的時間太漫長。更擔心的是，在死後的他界即使兩人相遇，是否還能相認。左思右想也理不清思緒，起身時天空已呈魚肚白。就讓我在春朝的好光景中，吟此夢井詩吧。

如第四章所述，以往詩中提到井時，所描寫的一般都是令人想起時光流逝的故宅之井及井桐，或是思念遠在異地的戀人而汲井水的人。如果沿用這種形式，以詠井來追思妻子的話，就該是妻子曾經汲水的井在妻子逝去的如今已變成一片荒蕪，或者是，當年與你一起在那井邊汲水，如今安在等形式。但在元稹的《夢井》中，前半是夢，中部是“解謎”，其結尾部分描寫了長時間“解夢”之後引來的更深的悲歎。構思之複雜不見先例。

另外，夢醒時分“鐘聲夜方半”一句，在詩中流露出現實的時間，最後以即將破曉的天空來收尾。通過明確指出時間，體現了對夢反省所費的時間之長和悲傷之深切，這種手法可以説是極近於小説的描寫方式。

並且，我們還應注意到，基於將瓶作爲靈魂之附身這一認識，使用文言小説中描寫的“作爲境界之井”這一構思，來創作追悼亡妻詩的創作手法。揚雄《酒箴》和王昌齡《行路難》描寫了瓶不知何時掉落井底的危險境況。釋寶月《估客樂》、白居易《井底引銀瓶》中，沉入井底的瓶意味着決定性的斷絶。看到將要沉入井底的瓶而焦慮，爲了求援而奔走的《夢井》中的主人公的姿態，正是在這些作品的基礎上又增添了小説式的描寫方式創作出來的。

那麼妻子的靈魂是從何處來到這個高原之井的呢？雖有第二六句“魂通有時逞”，但靈魂是用何種方法化爲瓶的呢？我認爲由於當時在墓葬中將壺作爲隨葬品相當流行，井通往黄泉的觀念也很普遍，因此在元稹的構思中，靈魂化爲瓶應該不會是向上飛往天空，而是向下通過井來實現的。文言小説中的主人公們通過井到達了理應到不了的地方。我想《夢井》正是利用這個構思，藉助成爲靈魂之附身的瓶，讓亡妻短暫地回到人世的作品。從咸陽的平原穿越厚厚土層，來到不知爲何處的高原之井的妻子之靈魂，正是利用“井”才使這一旅程變爲可能的，這就是《夢井》中最重要的創造性之處。

川合康三氏指出：“至當時爲止共有的世界觀在中唐解體，這同時也意味着從舊習束縛中擺脱出來。爲文學奠定基礎的一些觀念發生了質變，中唐文人以各自的方式認識世界，構築了具有獨特風格的文學。”[1]元稹《夢井》與以往的詠井詩明顯屬於不同類型，也可以說是攝取小說中各種形式的井而創作出的悼亡詩。《夢井》或許就是在以中唐爲文學史轉折點的文學史中誕生出的一種新型創作手法的作品之一吧。

六、結語

以上筆者利用中國各種文獻中關於井的描述同時分析了《夢井》。闡述了小說和詩中對井的描寫有明顯區別，而中唐以後的詩的新趨向深受小說的影響。另外，如果將揚雄《酒箴》、王昌齡《行路難》以及白居易《井底引銀瓶》一起來考慮的話，可見元稹《夢井》是以小說中描寫的井爲舞台，並基於作爲靈魂的附身之瓶的意義而創作的。同時指出，雖然本稿中提及的只是瓶，但通過考慮道具所持有的民俗意義，可能更深入地理解作品[2]。

本文對詩歌中“轆轤”、“桔槔”、“壺”等瓶之外與井有關的道具的描寫方式，以及“井龍”、“井神”記載中所見的涉及井的信仰等尚未進行研討[3]。另外關於井，還有一些以《井賦》爲題，以賦的形式吟誦的，其中郭璞的《井賦》是可窺見其獨特世界觀的作品。《井賦》是否描寫出了與文言小說、詩歌不同的世界呢？這一研究課題有待筆者日後探討。

A New Interpretation of Yuan Zhen's Poem of Mourning "Meng jing": One Aspect of the View of Wells in Ancient China

As well as featuring gods of wells and dragons living in wells, there are quite a number of Chinese works of literary fiction in which wells are described as marking the boundary between this world and the spirit world. But the well

---

① 川合康三，《終南山的變容——盛唐至中唐》，《中國文學報》第50册，1995，67頁，後收入劉維治、張劍、蔣寅譯，《終南山的變容》，上海:上海古籍出版社，2007。

② 前述“簪”也可以說是其例之一。

③ 關於詩歌中“轆轤”，請參照拙文《李賀〈後園鑿井〉——六朝、唐代中有關井的描寫》，《六朝學術學會報》第11輯，2010，69-92頁。

that thus serves as a boundary is not depicted in poems, and there arose a situation that could be described as a form of segregation regarding the treatment of wells in literature. In this article, I draw attention to differences in the image of wells due to differences in genre and examine material relating to wells. In particular, I suggest that "Meng jing 夢井"(trangslated by Arthur Waley as "The Pitcher"), a poem of mourning composed by the mid Tang 唐 poet Yuan Zhen 元稹, is a work that makes use of the idea of the well as a boundary, and I attempt to clarify one aspect of the view of wells in ancient China.

In addition to being associated with integrity, virtue, and law, wells were also depicted as spaces that caused feelings of entrapment or confinement. In literary fiction, wells also marked boundaries adjoining death and supernatural phenomena. In the case of wells appearing in poems from the Han period onwards, one finds depictions of wells attached to old houses and paulownias standing beside wells, which evoke the passage of time, and of people drawing water from a well while thinking of a loved one in a far-off land.

Meanwhile. Yuan Zhen's "Meng jing" has a complex structure,describing a dream in the first half, providing an interpretation of the dream in the middle section, and ending with a description of the poet's grief, which has increased as a result of his interpretation of the dream. In addition. Yuan Zhen's use of the technique of time spent in reflecting on the dream and the depth of his grief is very similar to the techniques employed in fiction. Furthermore, against the background of the fact that the pitcher could serve as a receptacle for the soul, he has used the idea of the well as a boundary, found in literary fiction, to compose a poem of mourning. When one considers this poem together with Yang Xiong's 揚雄"Jiu zhen"酒箴,Shi baoyue's 釋寶月 "Gu ke le" 估客樂 , and Bai Juyi's 白居易 "Jing di yin yin ping" 井底引銀瓶 , it could be said that the figure of the protagonist of "Meng jing", who frets as he watches the descending pitcher and rushes about seeking help, was created by applying a fictional mode of depiction on top of this accumulation of earlier poems. But the notion of the soul of the deceased wife coming from the plains of Xianyang 咸陽 to a well in some highland or other is original to "Meng jing." Characters in literary fiction used wells as passages for going to places that, properly speaking, they ought not to have been able to visit. The poem "Meng jing" uses this idea and has the deceased wife make a fleeting visit to the world of the living by means of a pitcher, which served as a receptacle for the soul. It is this that constitutes the greatest achievement of "Meng jing."

經學史論述的日語翻譯與有關問題
——以清代《尚書》今古文言説爲中心

橋本昭典

一、前言

皮錫瑞(1850—1908)的《經學歷史》是記録從儒學誕生到清朝的兩千多年的解釋經書的歷史大作。筆者和茨城大學井澤耕一教授一起,花了二十多年的時間將這部著作翻譯成了日文。現在已經翻譯完成。像這樣的經學著作,只是翻譯是不能完全理解的。特別是爲了幫助現代的日本讀者理解,必須加上注釋。《經學歷史》已有周予同(1898—1981)的很詳細的注釋。現在我們如果没有他的注釋,無法讀懂《經學歷史》。但是這個注釋至今已數十年,其間世界的經學研究也在發展。其中一項重要的發現是近年來出土的戰國時期的竹簡資料。這樣的出土資料展現了古代經書的原始面貌,是研究中不可缺少的參考資料。現在的經學史著作的翻譯注釋必須參考這些出土資料。另一方面,就是隨着時代的不斷變化也有千古不變的問題,例如,訓詁、佚文、佚書、典故、今文古文、解釋的問題等。這些問題,在研究經學的兩千多年歷史中,很多學者探討過。但是,之中很多問題是還没解決的。從而不可避免地會帶給我們翻譯者讓人煩惱的問題。筆者在此舉個例子進行探討。

皮錫瑞的《經學歷史》現在有三種翻譯版本。日本宫本勝的日文翻譯(到

作者單位:日本奈良教育大學

第七章),美國華盛頓大學Stuart. V. Aque的英文翻譯,韓國李鴻鎮的韓文翻譯[①]。這三種翻譯基本上是按照周予同的注釋翻譯過來的。我們的翻譯工作是以要超過前人爲目標的。

二、判斷佚文引用的出處

《經學歷史》第二章"經學流傳時代"第三節有如下一文:

> 劉向稱荀卿善爲《易》,其義略見非相、大略二篇。是荀子能傳《易》、《詩》、《禮》、《樂》、《春秋》,漢初傳其學者極盛。

周予同把注釋附在"劉向稱荀卿善爲《易》,其義略見非相、大略二篇"這個句子之後説;

> 見劉向《校荀卿敘録》。

劉向之語在《别録·校荀卿敘録》中可見。《别録》是早就散佚的書。雖然遺失,這個部分因附在宋本《荀子》中,所以流傳至今。周予同的注釋附在"非相、大略二篇"這個句子之後,讀者認爲劉向之語是至"非相、大略二篇"。

在此,要把前人兩種翻譯參看。

宫本日譯:

劉向は、荀子が善く易を修得していたと述べ、その内容の概略は非相· 大略の二篇に見ることができると称している(『荀子』序録)。[②]

Aque英譯:

Liu Xiang stated that Xun Qing was good at *the Changes*, and that this can been in the Feixiang and Dalue chapters(自注). [③]

---

① 宫本勝,《皮錫瑞〈經學歷史〉譯注(1)~(7)》,《北海道教育大學紀要》1991—2002。Stuart V. Aque, *Pi Xirui and Jingxuelishi,* University of Washington digital library (2004). 李鴻鎮,《中國經學史》,首爾:同和出版社,1984。我們翻譯草稿見井澤耕一、橋本昭典,《皮錫瑞〈經學歷史〉譯注(周予同注 皮錫瑞〈經學歷史〉全譯)(0)~(23)》,《千里山文學論集》48~76,1992—2006(現在準備出版)。

② 宫本勝,《皮錫瑞〈經學歷史〉譯注(二)》,《北海道教育大學紀要》人文科學編,42(2),1992,43頁。括號裏的是宫本自注。

③ Stuart V.Aque, *Pi Xirui and Jingxuelishi,* University of Washington digital library (2004), p.489.

從這兩段譯文能看出兩個翻譯都把劉向之語看作到"非相、大略二篇"。把這個句子如此解釋,按照周予同的注釋,是當然的結果。

那《别録》是怎麽記載呢?《别録》的佚文裹找不到同文[①],只有"孫卿善爲《詩》《禮》《易》《春秋》"之句。根據此文,皮錫瑞可能會表現爲"劉向稱荀卿善爲《易》"。這又是不是爲"劉向云",而是爲"劉向稱"的理由。然則"其義略見非相、大略二篇"這個句子,我們要認爲不是劉向之語,而是皮錫瑞之語。上面兩種譯文都給讀者帶來此文整體都是劉向之語的誤解。所以嚴格地來説,應該把此文解釋爲:

> 劉向稱"荀卿善爲《易》"(注:見劉向《校荀卿敘録》)。其義(注:荀卿易學)略見非相、大略二篇。

但是,解釋此文還有問題。其實此處不是皮錫瑞之語,而是汪中(1744—1794)的《述學·荀卿子通論》的引用。從而"其義略見非相、大略二篇"是汪中之語。梁啓超(1873—1929)的《清代學術概論》稱:"汪中之荀卿子通論,我輩今日讀之,誠覺甚平易,然在當日,固發人所未發,且言人所不敢言也。"荀卿易學略見《荀子》非相、大略二篇,這個見解是汪中提出的。在此皮錫瑞没寫汪中的名字。

周予同有一篇文章,《從孔子到孟荀——戰國時的儒家派别和儒經傳授》[②],之中有"劉向又稱荀卿善爲《易》(注),其義亦見非相、大略二篇"一文。這篇文章裹周予同把注釋附在"善爲易"這個句子之後。這樣注釋可以避免誤解劉向之語到何處。但是,引號到最後,並且又没寫汪中之功。清末民初,汪中的荀卿學之功可能屬常識,描寫出荀卿傳經之功,不用把他的名字寫出來。不過,現代我們翻譯這個句子,在此一定有必要加注釋。

三、關于翻譯今文説、古文説的問題——以《尚書》爲例

《經學歷史》第三章"經學昌明時代"第三節有如下一文:

---

① 姚振宗輯録,鄧駿捷校補,《七略别録佚文 七略佚文》,上海:上海古籍出版社,2008。
② 朱維錚編校,《周予同經學史論》,上海:上海人民出版社,2010,570頁。

> 試舉《書》之二事證之。伏生《大傳》以大麓爲大麓之野，明是山麓。《史記》以爲山林，用歐陽説，《漢書·于定國傳》以爲大録，用大夏侯説，是大夏侯背師説矣。伏生《大傳》以孟侯，爲迎侯《白虎通》朝聘篇用之，而《漢書·地理志》，周公封弟康叔，號曰孟侯，用小夏侯説，是小夏侯背師説矣。

在此舉出關于傳授經説，弟子違背師説的兩個例子。一是《尚書·舜典》："納于大麓。烈風雷雨弗迷"的解釋；一是《尚書·康誥》："王若曰，孟侯，朕弟，小子封"的解釋。首先探討前者。

1. 關于"大麓"的解釋

"伏生《大傳》以大麓爲大麓之野，明是山麓，《史記》以爲山林，用歐陽説，《漢書·于定國傳》以爲大録，用大夏侯説"。那麽此文怎樣翻譯成日語？直譯的話，讀者肯定不能理解其意義，因而有必要加上注釋。怎樣解釋注釋纔好呢？此短文其實有非常複雜的背景。

首先，《舜典》是僞古文的篇名，在今文《尚書》中這個文章是堯典篇之文。《孔傳》認爲："麓録也，納舜使大録萬機之政，陰陽和，風雨時，各以其節不有迷錯愆伏。明舜之德合於天。"要理解"大麓"之語的議論，首先要參照《十三經注疏》。《孔傳》中爲："麓録也。"既然《孔傳》有"麓録也"，讀者就會認爲是古文説。但是皮錫瑞云是今文的"大夏侯説"。

其次，要知道兩個解釋的關係，就要調查孫星衍的《尚書今古文疏證》。此書列舉了《史記·五帝本紀》等"山麓之説"之例，稱："此俱孔氏安國古文説。"

皮錫瑞把伏生《尚書大傳》和《史記》的"大麓"作爲"山麓"的學説稱爲歐陽説；把"大麓"作爲"大録"的學説稱爲大夏侯説。這都是今文説。但是，孫星衍《尚書今古文疏證》把"山麓之説"當作孔安國的古文説，"大録之説"是"夏侯、歐陽等今文説也"。要理解這解釋的背景，調查有關書籍，會越加混亂。

要解決這個問題需要一點知識。關于孫星衍(1753—1818)的《尚書今古文疏證》，吕思勉(1884—1975)《經子解題》認爲："其時今古文之派别尚未大明，誤以司馬遷爲古文，實爲巨謬。"皮錫瑞《經學通論》還評價："孫星衍《尚書今古文注疏》,於今古説搜羅略備,分析亦明。但誤執《史記》皆古文。致今古文家法大亂。”據此,孫星衍把“山麓之説”作爲古文説是錯誤的。

《尚書大傳·虞夏傳》:“故堯推尊舜而尚之,屬諸侯焉。致天下于大麓之野。”皮錫瑞説:“伏生《大傳》以大麓爲大麓之野,明是山麓,《史記》以爲山林。”《尚書大傳》和《史記》都是“山麓之説”。《尚書大傳》的鄭玄注:“山足曰麓,麓者,録也。古者天子命大事,命諸侯,則爲壇國之外。堯聚諸侯,命舜陟位居攝,只天下之事,使大録之。”皮錫瑞《尚書大傳疏證》:“鄭,謂麓取録之義,本之《漢書·于定國傳》萬方之事大録君。”《尚書今古文疏證》:“鄭注大傳用博士付會之説,殊非伏生之旨。”鄭玄把歐陽説和大夏侯説並在一起。因此,王先謙(1842—1917)的《尚書孔傳參正》認爲:“立説未免兩岐”,“疑,麓者,録也以下數語,爲後人羼入”。還有,《經典釋文》引用鄭玄《尚書注》:“麓,山足也。”從這裏來看,鄭玄用的是歐陽的今文説。

另一方面,有些學者認爲伏生《尚書大傳》從古文説。段玉裁(1735—1815)的《古文尚書撰異》云:“馬、鄭注《尚書》皆云麓山足也。雖缺佚不完,而《釋文》以别于王,云麓録也,則知馬、鄭注古文不爲大録之解。”根據《經典釋文》引鄭注“麓,山足也”的記載,段玉裁認爲鄭玄從“山麓之説”,而且認爲是古文説。《史記》也是“山麓之説”,段玉裁斷定還是古文説。即是,他把“大録之説”作爲今文説,把“山麓之説”作爲古文説。其根據是《漢書·儒林傳》:“司馬遷亦從安國問故。遷書載堯典、禹貢、洪範、微子、金縢諸篇,多古文説”的記載。

段玉裁《古文尚書撰異》又云:“鄭注《大傳》云,致天下之事,使大録之。”他從鄭玄《尚書大傳注》把“麓”解釋爲“録”而推論到《尚書大傳》是“大録之説”,因爲《尚書大傳》是今文,所以他提出“大録之説”是今文説,“山麓之説”是古文説的結論。但是,這樣理解有“可知注古文與《大傳》注迥殊”的矛盾。在此段玉裁的主張是,“鄭意必謂處艱難而裕如也”,即是鄭玄的意圖不過是想説不管舜處于何種困難的處境都表現出悠然的態度。

近人程元敏《尚書學史》司馬遷的尚書學引段玉裁《古文尚書撰異》云:“訓大麓爲大山足,古文尚書説也,與今文《尚書大傳》異,《大傳·虞夏傳》,堯推尊舜,屬諸侯,致於大録之野。鄭玄注山足曰麓,麓者録也……鄭釋字本義爲山麓，在國郊，又借爲録，申義爲大録萬機，即總攝天下政事。爲壇祭告上天，是堯薦舜爲天子於天也。段《撰異》，堯本紀……此條説大麓蓋安國説也。”[①]

那麽，《尚書大傳》到底是“山麓之説”還是“大録之説”？關于此問題，皮錫瑞《今文尚書考證》所説的“若以麓爲録，何必加之野二字耶”，是接近正確的。陳喬樅（1809—1869）的《今文尚書經説考》也講“鄭注《大傳》，二義而綜釋之，此合古今文説而一之也。段氏《古文尚書撰異》不達此義”，可參看。

《尚書大傳》和鄭玄注、鄭玄《尚書注》都是佚文。這個問題上也可以懷疑曾經發生過文獻上的混亂。但是，現在暫且如段玉裁所云，能夠相信《經典釋文》“王云録也。馬、鄭云山足也”的記載和寫法，即使如此，段玉裁把“山麓之説”作爲古文説是否正確呢？皮錫瑞《今文尚書考證》指出過：“段玉裁以山麓之説爲古文，大録之説爲今文。蓋徒見今文説之誤者解爲大録，不知今文説之不誤者正解爲山麓。”皮錫瑞堅持認爲今文家大夏侯違背了今文家歐陽的師説。

2. 怎樣判斷師説的授受

接着，探討關于“山麓之説”是爲歐陽説，“大録之説”是爲大夏侯説的根據。

皮錫瑞《經學通論》：“漢武帝立博士，《尚書》惟有歐陽。太史公《尚書》學，不言受自何人。考其年代，未能親受伏生，當是歐陽生所傳者。”[②]

陳喬樅《今文尚書經説考》有更加詳細的論述：“《論衡·吉驗篇》之言與《史記》合，此蓋亦歐陽《尚書》説也。《論衡·書解篇》，于書家獨有歐陽公孫，是歐陽《尚書》之明證。太史公時，《尚書》獨有歐陽博士，知《史記》所據《尚書》是歐陽氏學。”

司馬遷《史記》中所見把《尚書》的解釋作爲古文説的根據論述如下：

> “司馬遷亦從安國問故。遷書載堯典、禹貢、洪範、微子、金縢諸篇，多古文説。”（《漢書·儒林傳》）

① 程元敏，《尚書學史》，上海：華東師範大學出版社，2013，682頁。

② “論伏傳之後，以《史記》爲最早，《史記》引書多同今文，不當據爲古文”。

如上所述，經學的師説及其授受是從傳世文獻上片斷的記述推論而得出的。因而把《史記》所見的《尚書》解釋看作是今文説也可以，同樣，把《史記》所見的《尚書》解釋看作是古文説也可以。

馬融《書序》云“逸十六篇，絶無師説”。根據這個記述，能判斷“山麓之説”、“大録之説”都是今文説的可能性比較大。

還有，皮錫瑞《經學通論》説“論治《尚書》，當先看孫星衍《尚書今古文注疏》，陳喬樅《今文尚書經説考》”，但是，又説孫書“誤執《史記》皆古文，致今古文家法大亂”，陳書“誤執古説爲今文，以致反疑伏生，違初祖”。此記述表現了學習《尚書》的困難之處。

《經學歷史》之中“伏生大傳以大麓爲大麓之野，明是山麓，《史記》以爲山林，用歐陽説，《漢書·于定國傳》以爲大録，用大夏侯説”這一句裏存在着這麽複雜的背景。只有理解這樣的背景，纔能正確地翻譯這一句。注釋也是爲了幫讀者理解此文的，所以盡量簡潔地寫出較好。但是有限的文字裏面解釋這個背景還是不容易的。

3. 關于“孟侯”的解釋

第二個記述：“伏生《大傳》以孟侯，爲迎侯，《白虎通》朝聘篇用之，而《漢書·地理志》，周公封弟康叔，號曰孟侯，用小夏侯説，是小夏侯背師説矣。”

這是關于《尚書·康誥》“王若曰，孟侯，朕弟，小子封”的解釋。

《尚書大傳·略説》云：“天子太子年十八曰孟侯。”由此可知，“孟侯”是十八歲的太子之意，是指十八歲的成王。因此，《康誥篇》是寫周公告之十八歲的武王之子成王和自己的弟弟康叔之語。另外，《漢書·地理志》“周公誅之，盡以其地封弟康叔，號曰孟侯，以夾輔周室”，顔師古注云：“康叔亦武王弟也。孟，長也。言爲諸侯之長。”據此，孟侯是“諸侯之長”之意，是指諸侯之長的康叔。由此，《康誥篇》是記述周公代替年少的成王告之成爲諸侯之長的弟弟康叔之語。像這樣，這一句裏存在着完全不同的解釋。關于此解釋，與“大麓”之語同樣存在師説、今文古文説及其授受的問題。在此不提這個問題，只論述關于翻譯的難點問題。

“王若曰，孟侯，朕弟，小子封”一文，翻譯爲“王這樣説，孟侯，我的弟弟，小伙子封”。這樣翻譯的話，“王”是誰?“孟侯”是什麽意思？讀者不大明白。

于是要嚴格地翻譯,今文説古文説的問題是不可避免的。對于王是誰,有兩個解釋。"孟侯"也一樣。根據伏生説(今文説)嚴格翻譯的話,是"周公這樣説,十八歲的太子成王和我弟弟小夥子康叔";根據小夏侯説(或古文説)嚴格翻譯的話,是"成王這樣説,又是諸侯之長,又是我弟弟的小伙子康叔"。這兩種翻譯都不像《尚書》的文章,缺少又簡潔又有深度的《尚書》文章的魅力。還是對于原文忠實地翻譯而加上不同解釋的注釋比較好。

四、翻譯經學史的意義

以上以今古文説爲例,論述了關于《尚書》的解釋中翻譯的困難。如此,閲讀經典及經學書籍需要相當長的時間和勞力。

《經學歷史》有以下記述:

> 分文析字,煩言碎辭,學者罷老且不能究其一藝。(《漢書·劉歆傳》)
>
> 古之學者耕且養,三年而通一藝,存其大體,玩經文而已,是故用日少而畜德多,三十而五經立也。後世經傳既已乖離,博學者又不思多聞闕疑之義而務碎義逃難,便辭巧説,破壞形體説五字之文,至於二三萬言後進彌以馳逐,故幼童而守一藝,白首而後能言;安其所習,毁所不見,終以自蔽。此學者之大患也。(《漢書·藝文志》)
>
> 桓譚《新論》云,秦近君能説堯典篇目兩字之誼至十餘萬言。但説曰若稽古三萬言。

以上的記述都是漢代書籍的引用。如"學者罷老且不能究其一藝","白首而後能言"所説,鑽研經學需要長期的時間。如"曰若稽古三萬言"所説,鑽研經學需要大量的勞力。這是兩千年前就已經指出的。

現在學術界浸透着成果主義,學者的業績在數量上競争。專攻經學,進行經典及經學關係的翻譯,需要勞力較大,相對成果較少,並且較難得到專攻經學以外人士的理解。在這樣的情況下,進行經學的翻譯到底有什麽意義呢?最後想從《經學歷史》的翻譯問題上探討其意義。

《經學歷史》經學昌明時代,第二節引用《史記·儒林傳》如下云:

但云“武安侯田蚡爲丞相，絀黄、老刑名百家之言，延文學儒者數百人，而公孫弘以春秋白衣爲天子三公，封以平津侯，天下之學士靡然鄉風矣。公孫弘爲學官，悼道之鬱滯，迺請爲博士官置弟子五十人。郡國縣道邑有好文學、敬長上、肅政教、順鄉里者，詣太常，得受業如弟子。一歲皆輒試，能通一藝以上，補文學掌故缺。其高第可以爲郎中者，太常籍奏。即有秀纔異等，輒以名聞”。此漢世明經取士之盛典，亦後世明經取士之權輿。史稱之曰“自此以來、則公卿大夫士吏彬彬多文學之士矣”。方苞謂古未有以文學爲官者，誘以利禄，儒之途通而其道亡。案方氏持論雖高，而三代以下既不尊師，如漢武使束帛加璧安車駟馬迎申公，已屬曠世一見之事。欲興經學，非導以利禄不可。古今選舉人才之法，至此一變，亦勢之無可如何者也。

《史記·儒林傳》記載，公孫弘的上奏被武帝認可，之後起用經學者爲官吏，因此經學開始興隆。《史記》評價這樣的情況説：“自此以來，則公卿大夫士吏斌斌多文學之士矣。”關于此事實，方苞（1668—1749）持否定的評價：“弘之興儒術也，則誘以利禄……由是儒之道汙，禮義亡，儒之途通而其道亡。”對此皮錫瑞云：“方氏持論雖高……。”

其實這裏也有非常複雜的背景。《儒林傳》與《史記》其他部分的寫法不同，司馬遷在《史記·儒林傳》以“贊”開頭，即是以“太史公曰：餘讀功令，至於廣厲學官之路，未嘗不廢書而歎也”這一句開始的。司馬遷所説的“功令”，一般被當作公孫弘提出來的教育法規。他看到當時的教育法令而感歎了。但是，“歎”這個字有“贊歎”和“悲歎”相反的兩個意思，看取哪一個意思，會産生相反的意思。對“誘以利禄”，司馬遷是“贊歎”還是“悲歎”？從後面“公卿大夫士吏斌斌多文學之士矣”的記載來看，司馬遷對這種狀況是贊歎的，這是一般的説法。但是方苞没有這樣考慮。

方苞之論在《方望溪文集》中“書儒林傳後”及“又書儒林傳後”可見，如下所云：

蓋歎儒術自是而變也。古未有以文學爲學官，以德進，以事舉，以言楊，詩、書、六藝特用以通在物之理，而養其六德，成其六行焉耳……其以文學爲官，始于叔孫通弟子以定禮爲選首，成于公孫弘請試士太常，而儒術之汙隆，自是而中判矣。

古代學習經書，通理、養德、爲實踐而行。其結果是被推舉當官吏。但是，公孫弘鑽研經書走上官吏之路以後，雖然儒學興起，但是世人以獲取地位和金錢爲目的而學習經書。方苞爲此而進行批駁。

是書敘儒術至漢興……驟觀其辭，若近於贊美，故"廢書而歎"，皆以爲歎六藝之難興也。然其稱歎"興於學"也，承太常諸生之爲選首，稱"學士鄉風"，承公孫弘以白衣爲三公，稱"斌斌多文學之士"，承選擇備員，則遷之意居可知……而復正言以斷之曰"學官弟子行雖不備，而至於大夫、郎中，掌故以百數"。其刺譏痛惜之意，不亦深切著明矣乎。

方苞云："子長所讀功令，即弘奏請之辭。"他寫的這篇文章把《儒林傳》視爲司馬遷的譏刺。司馬遷對公孫弘開闢的"誘以利禄"之路堅持否定，《史記·儒林傳》是評價武帝的儒學獎勵政策的文章。但是，實際上司馬遷是在暗中批判。總之，可以説《史記·儒林傳》是一種"微言大義"。所以爲了理解皮錫瑞寫的"方氏持論雖高"一文，要考慮到以上这些内容。

在此又出現一個問題。皮錫瑞接着説："史稱之曰'自此以來，則公卿大夫士吏彬彬多文學之士矣'。"周予同注："語見《漢書》儒林傳。"此語確實可見《漢書·儒林傳》。但是，《史記·儒林傳》也可見："自此以來，則公卿大夫士吏斌斌多文學之士矣"，幾乎是相同的句子，不同之處是"斌斌"和"彬彬"，這兩個詞在意義上没有不同。如果考慮方苞如此批評的話，就是其文章裏有"斌斌"和"彬彬"的不同，應該認爲皮錫瑞引用了《史記·儒林傳》，"史稱之"的"史"應該指《史記》[1]。經典解釋史論述的文章里經常存在這樣的讓人難懂的背景。

回到正題，探討皮錫瑞如何考慮"誘以利禄"。方苞否定"誘以利禄"，皮錫瑞説"方氏持論雖高……"，此"高"是"清高"之義，皮錫瑞對于這個想法評

---

① Aque 譯《經學歷史》把注釋附在這一句説，"Zhou Yutong comments: For the original passage,see the Rulinzhuan chapter of the *Hanshu*"*Pi Xirui and Jingxuelishi,* p.544.

價爲“清高”。但是,考慮到現實狀況,而判斷爲“欲興經學,非導以利禄不可”,更視爲“勢之無可如何者也”。在此我們看得出来皮錫瑞的重視現實的態度。

一般視皮錫瑞堅持今文經學派的立場的看法比較多,實際上皮錫瑞對不同立場的想法給予一定的評價。這樣的記述在《經學歷史》裏也常見。皮錫瑞本來持有應該堅持古代以學問目的爲修養自己的“清高”之路的想法,但是由于時代變化的原因,而有把“導以利禄”視爲“勢之無可如何者也”的現實態度。

鑽研學問是否應該與“利禄”相結合?關于像這樣在現代也相通的問題,通過翻譯經典解釋的歷史,能夠追尋先賢者們宏大的考慮問題的軌跡,这也可以算是“治經致用”了。

地理環境決定論的輸入與近代的先秦學術及文明建構

潘静如

晚清之際,很多歷史人物都不約而同發出千古大變局的概歎。學術界一度把劇變主要歸因于西方的介入。近幾十年來,隨着研究的深入,人們越來越傾向于提出更具體的解釋。比如,何以往這個方向變而不是别的方向?何以某一西方學説在中國風行很久,而别一學説迅即歸于消歇?顯然,這背後有本土的自我選擇。這就是説,雖然西方的介入最初是通過武力達成的,但西方"文明"征服並根植于中國,是要經過互相選擇的過程的(當然二者並不處于平等的位置)。那麽,中國的傳統文化在這種選擇中扮演的角色就很值得重新考量。近代對先秦學術史和文明的建構就是一個極佳的例子,既與西方地理環境決定論的輸入相關,又與本土的歷史傳統有着密不可分的聯繫。本文以此爲例,展開論述。

一、引子:南學北學之辨

近代有關南學、北學的討論,發端于梁啓超。早在20世紀30年代初,就有學者作了總結:近代研究南北文化的發展和差異,最有功勞的,首推梁啓

作者學習單位:北京大學中文系

超[①]。梁啓超的關注點不限于區區南、北之辨，但要理解近代以地域論學風氣的興起，以此切入十分必要。需要説明兩點。第一，南學、北學兩個詞涵義很複雜甚至紊亂，約略來説，有時候是泛指，有時候是專稱；撇開《周禮》、《大戴禮記》等典籍中關于"五學"的記載，南學、北學作爲專門術語來使用，始于南北朝隋唐之際，分指南、北朝的學術（主要是經學）[②]。第二，中國文化由東西互動而變爲南北互動，是在東漢以後的事，重心由北而南轉移，則始于東晉六朝，一般認爲成于安史之亂後[③]。如果以南、北分説六朝以後的政治、軍事以逮學術，那倒算不上孤明先發，但是以南、北來辯章先秦的學術思想，梁啓超則有大輅椎輪之功。

1902年，梁啓超的《論中國學術思想變遷之大勢》開始在《新民叢報》第五號上連載。其中關于先秦（諸子）思想學術的誕生及特色，他有一段影響深遠的論述：

> 欲知先秦學術之真相，則南、北兩分潮，最當注意者也。凡人群第一期之進化，必依河流而起，此萬國之所同也。[④]

地理尤其是河流與特定文明的關係，在這裏被强調，可與他差不多同時撰述的《中國地理大勢論》所謂"文明之發生，莫要于河流……自周以前，以黄河流域爲全國之代表，自漢以後，以黄河揚子江兩流域爲全國之代表，近百年來以黄河、揚子江、西江三流域爲全國之代表"相參看[⑤]。因而，"北地苦寒磽瘠，謀生不易""南地則反是，其氣候和，其土地饒，其謀生易"也就成了他解釋和分析先秦學術思想的基礎，從"北派崇實際，南派崇虚想"以下共十一條。論及諸子時，又稱"北派之魁，厥惟孔子；南派之魁，厥爲老子"，"孔、老分雄於南北，而起于其間者有墨子焉。墨亦北派也，顧北而稍近于南。墨子生于宋，

---

① 陳序經，《梁啓超的南北文化觀》，《南北文化觀》，《嶺南學報》1934年第3卷第3期，26頁。

② 皮錫瑞，《經學歷史》，北京：中華書局，1959，170—192頁。

③ 劉師培，《中國地理教科書》，揚州：廣陵書社，2013，41頁。錢穆，《國史大綱》上册，北京：商務印書館，1996，457頁。

④ 梁啓超著，夏曉虹編，《論中國學術思想變遷之大勢》，上海：上海古籍出版社，2001，25—27頁。

⑤ 梁啓超，《中國地理大勢論》，《飲冰室合集·飲冰室文集之十》，78、79頁。原載《新民叢報》1902年第6號，23頁。

宋,南北要沖也,故其學于南、北各有所采,而自成一家言"[①]。這一思路,甚至在他20年後的著述中還能看到。

用南、北地域的差異來分析先秦學術思想,很快成爲士人效仿的典範。錢基博《中國輿地大勢論》、劉師培《南北學派不同論》,並受梁的影響,史學界早有定論,不待贅言。事實上,梁的影響遠不止于此,這裏舉兩個學術界不太注意的例子。

比如,1906年,王國維撰《屈子文學之精神》,開頭就説:"我國春秋以前,道德政治上之思想,可分爲二派:一帝王派,一非帝王派。"[②]表面看來,王進而把北方派和南方派定義爲帝王派和非帝王派,似與梁説無涉,但梁啓超早稱北學精神爲"則古昔,稱先王",南人"不崇先王"[③]。復次,王以老孔爲代表的南北二分法不出梁説的範疇,並且南北二派一連串的對比,大抵可以在梁氏分疏的十一條南北不同特色中找到對應點。他把屈原論説成南人而具北性者,遂而滋生出"歐穆亞(humour)之人生觀",進而發爲感情充沛的而又富于執着精神和想象力的騷賦,讓人有足夠的理由懷疑一方面受梁啓超對墨翟的分析的影響,一方面直接繼承梁啓超"屈原,文豪也,然其感情之淵微,設辭之瑰偉,亦我國思想界中一異彩也""屈原生于貴族,故其國家觀念之强盛,與立身行己之端嚴,頗近北派;至其學術思想,純乎南風"之説。

再如,江瑔《新體經學講義》,1917年刊于《新體師範講義》第一至四期,次年由上海商務印書館出單行本,後有再版[④]。書中説:"大抵北方之地,土厚水深,民生其間,多尚實際;南方之地,水勢汪洋,民生其間,多尚虛無。尚實際,則其學不外乎記事析理二端;尚虛無,則其學多言志抒情,往往寄想于冥漠,此南學、北學之所由分也。"這一論述,與梁啓超若合符契,連用語都頗多雷同。最妙的是,江瑔在分析完孔、老爲代表的北學南學之後,又説:"惟蘇張縱

---

① 實際上,早在1901年梁啓超撰《中國史敘論》就已强調"地理與歷史,最有緊切之關係","故按察中國地理,而觀其歷史上之變化,實最有興味之事也"。參見梁啓超,《中國史敘論》,《清議報》1901年第90期,3頁。

② 王國維,《屈子文學之精神》,《王國維遺書》第五册,北京:商務印書館,1940,31—32頁。

③ 按,梁啓超、王國維之説並非全是後天的建構,《荀子·非十二子》已經把當時學派分出六派,其中兩派是"不法先王"和"略法先王"。見王先謙,《荀子集解》,北京:中華書局,1988,93—95頁。

④ 張京華,《〈讀子卮言〉出版弁言》,江瑔《讀子卮言》,上海:華東師範大學出版社,2011,序言页。

橫,師事鬼谷,北學而近于南。而墨翟宋人,界南北之中,是以南北並學,既重力行,亦明天鬼。"[①]不消説,是來自梁啓超的啓發。

重新考察近代的學術史書寫,梁啓超的南學、北學之辯確實開啓了新的範式。這不僅應當歸功于梁啓超的健筆,還應看到時人對西學的普遍接受以及本土的固有傳統。

二、他山之石:地理環境決定論的輸入、傳播與接受

晚清之際,西方的地理環境決定論輸入中國,激起了巨大反響。所謂地理環境決定論,英文寫作 environmental determinism,有時也寫作 climatic determinism 或 geographical determinism,是一種物質決定論。它的含義在不同的學者那裏有廣狹之分,大體上它强調自然環境對人們的生理、生活習慣、文化特點乃至社會制度等有着決定性作用。遠在古希臘時代,比如希羅多德(Herodotus)、希波克拉底(Hippocrates)、柏拉圖、亞里士多德就有近似決定論的樸素表述。到了16世紀,法國的讓·波丹(Jean Bodin)在所撰《易于理解歷史的方法》這一史學理論著作中,不但流露出決定論傾向,並且實際上已經依據南方民族、中部民族、北方民族因環境的不同而造就的文明差異把人類歷史分爲三個時期(每個時期長達2000年)[②]。但作爲系統的理論,則始于孟德斯鳩的《論法的精神》,暢于黑格爾、巴克爾(H. T. Buckle),大成于拉采爾(Friedrich Ratzel)及其美國學生森普尔(E. C. Semple),成了19世紀末20世紀初最爲流行的學説之一[③]。20年代以後,此説逐漸衰微[④]。然而,只能説是衰

---

① 江瑔,《南學北學之別》,《新體經學講義》,北京:商務印書館,1918,56—57頁。

② 張廣智,《西方史學史》,上海:復旦大學出版社,2000,114頁。

③ 李旭旦,《近代人生地理學之發達及其在我國之展望》,管理中英庚款董事會十週年(1941年)紀念論文。宋正海,《地理環境決定論的發生發展及其在近代引起的誤解》,《自然辯證法研究》1991年9期。其實,把某某歸入環境決定論是危險的做法,例如拉采爾與之前的決定論者已略有不同,比較地看重自然與人文的互相聯係,儘管還没有完全脱去決定論色彩。再比如巴克爾,他也非一個純然的地理環境決定論者,有些表述恰恰走到了這一理論的反面。本文爲論説指稱方便,也只能從俗。

④ 例如法國白菱漢(Jean Brunhes,現在通譯白吕納)就很不贊同,他説:"人類地理學所倡之'必然論'(determinism),已爲今之學者所否認。人類地理學者以爲人類之發達,完全受土壤、氣候,與其他環境之影響,故其解釋,莫不系統秩然,因果歷歷不爽,而不知人類雖在宇宙覆載之中,盡有自由活動之余地。"雖然他所謂"已爲今之學者所否認"偏指黑格爾、巴克爾、李特爾(Karl Ritter)、(轉下頁)

微，而不能説是消歇，所以，争議並未停止，尤其是隨着新環境主義(neo-environmentism)的不斷修正，相關問題仍在討論之中[①]。明乎此，才能理解這個學説爲何在近代中國一度成爲幾近真理的存在，因爲它在學理上確實有一定的説服力。

儘管如此，學理上的説服力卻不是這一學説得以風行中國的主要原因。考察決定論在近代中國的風行，不能忽略它潛在的因素。它之所以在19世紀末風行于西方，是與當時興起的帝國主義、社會達爾文主義緊密相聯的。環境決定論是社會達爾文主義者的意識形態在地理學上的體現，它爲帝國主義的世界統治提供了一個自然主義的解釋[②]。反觀當時的中國，知識界剛好獲睹了嚴復譯出的《天演論》(1897年)，無條件地接受了進化論，逐漸把它擴展、運用到每一個領域中去。正是在這個意義上，"物競天擇適者生存""落後就要挨打"成爲了一個普遍真理。在不少知識分子那裹，西方文明優于中國文明彷彿是不證自明的，而且從某種意義上講，"武力/國力=文明"成爲了他們的思維基點之一。學習西方也就成了自救的唯一出路。因此，環境決定論背後隱藏的邏輯，中國知識界往往不會察覺或懷疑，或者即使察覺，也仍然奉爲真理。像海洋文明優于陸地文明、歐洲中心論、中國人種與文明西來説等，得到一些"先進知識分子"的贊同就是這種精神的體現[③]。

環境決定論的最初輸入當然與梁啓超密不可分。梁啓超1899年在《清議報》上連載《飲冰室自由書》，其中有《孟的思鳩之學説》一篇就涉及環境決定

(接上頁)科爾(J. G. Kohl)這些人，對拉采爾也能給予較高的贊美，但無疑，拉采爾是他的批評對象，他本人持的是地理或然論(probability或possibility)。見白菱漢，《人生地理學》，張其昀譯，商務印書館，1930，3—4頁。

① Gwilym Lucas Eades,"Determining Environmental Determinism", *Progress in Human Geography* 36.3 (Jun 2012), pp.423-427.

② Richard Peet, "The Social Origins of Environmental Determinism", *Annals of the Association of American Geographers*, (09/1985),Vol 75, No.3, pp.309-333.

③ 這種情況在民國期間似乎愈演愈烈。例如，傅斯年1918年撰寫《中國學術思想家之基本謬誤》，摘引了一篇幾年前英國雜志《十九世紀與其後》(*The Nineteenth Century and after*)上的一篇論東方民族性的文章，聲稱"東方學術，自其胎性上言之，不能充量發展""所謂近世文明者，永無望其出于亞細亞人之手；世間之上，更不能有優于希臘、超于羅馬之政化"。傅斯年雖然"憤其狂悖"，但不得不承認作者所論"昭信不誣"。見《中國古代思想與學術十論》，桂林：廣西師範大學出版社，2006，189頁。此文最初載1918年4月15日《新青年》第4卷第4號。

論思想;1902年他又在《新民叢報》上發表《法理學大家孟德斯鳩之學説》來介紹孟德斯鳩的政治思想。1900年,《論法的精神》節譯本《萬法精理》連載于江蘇留日學生創辦的《譯書彙編》,根據小序"今所譯者即何氏本也"[①],可知這個節譯本所據爲日本人何禮之的日譯本;同年秋,張相文也開始着手翻譯何禮之的日譯本,只是出版是好些年以後的事了。正如學者指出的那樣,這幾件事並没有促成環境決定論的盛行,其影響甚微;它的東漸有一個過程[②]。類似的,像對黑格爾、李特爾、拉采爾有關地理環境決定論的零星介紹,也應作如是觀。所以,本文不擬考察這些人的理論在何時何地始有何人介紹,而是側重于這一理論在1900年代的實際影響。可以順帶指出,當時對這一理論的發展脈絡和代表人作了集中介紹的有《漢聲》1903年6期佚名的《史學之根本條件》,據考證,此文也是舶來品,譯自日人坪井馬九三的《史學研究法》[③]。因此,可以説,"新史學"的面向很多,但强調地理是極其重要的一環。

真正讓這一理論風行于世的還得推梁啓超的實踐和運用,他先後發表的系列論文,很快掀起了運用這一理論研究問題的狂潮。雖然筆者懷疑這些文章與徐勤那篇早出的很少有人關注的《地球大勢公論(並序)》多少有點關係[④],但是,毋庸置疑,從日本接受的西方地理環境決定論才是這些文章最重要的理論基礎和靈感淵藪。

全面考察它在1900年代尤其是上半期的輸入、傳播與接受,似乎應當分爲這樣兩個層次:第一,西方人或日本人的著述,不論是否譯出,只要晚清士人確曾檢閲過並受其影響就行;二,中國人的譯介、接受和運用。現據學界的成果和筆者的考察,排比當日影響較大的幾部著述,列出一表(見下頁),同時附論相關著述,各以類聚。此表意不在全,只是提供一個"瞭望臺",闕陋甚

---

① 無名氏,《萬法精理》譯序,《譯書彙編》1900年1期,35頁。

② 郭雙林,《西潮激蕩下的晚清地理學》,北京:北京大學出版社,2000,42頁。唐文權,《二十世紀初孟德斯鳩地理決定論東漸述論》,《唐文權文集》,武漢:華中師範大學出版社,2013。

③ 俞旦初,《二十世紀初年中國的新史學思潮初考》,《史學史研究》1982年3期。另,关于这些人及相关理论的输入情况,參郭雙林《西潮激蕩下的晚清地理學》,内有很詳尽的考证。

④ 徐勤,《地球大勢公論》,海上釣鼇客(梁啓超)編《維新政治文編》(第四册)卷二十,上海中西譯書會,1901年。徐勤這篇文章開頭便論述了幾大文明與河流的關係。《文編》既是梁啓超編定,他看過此文無疑。此外,《文編》卷三中華强士《亞非兩洲將興説》、卷二十徐勤《中國盛衰關于地球全局》、卷二十一劉振麟《地運趨于亞東説》等篇,梁啓超當也有汲取。

多，但即使不闕陋，也不能准確反映當時的傳播與接受情況：比如前面提到的關于孟德斯鳩的若幹篇文章，其直接影響力遠不如後來梁啓超的一篇雄文；再比如巴克爾的《英國文明史》或浮田和民的《史學通論》雖然對環境決定論有詳細的敘説，但其在中國本土産生巨大影響很可能反而後于梁啓超。不過無論如何，這有助于我們更直觀的理解。

從列表中可以看出：第一，地理環境決定論輸入的渠道和滲透的領域是非常多的，比如地理學、史學、政治學、社會學、人種學等等；第二，日本著作在當中扮演了重要至極的角色；第三，不管是日本還是中國士人，甚至包括孟德斯鳩、黑格爾、巴卡爾、拉采爾等人（在當時中國士人的認識和理解層面上），在强調地理或自然環境的重要性上是一致的，但各家是否果真主張或贊同決定論，則很難作詳細的界定。

不論如何，晚清士人對自然環境的重視是明確無疑的，而且很大程度上是受了地理環境決定論的影響。但是，這並不是説，這是晚清士人論説的惟一基礎或來源。

1900年代地理環境決定論的傳播、接受與運用簡表

著述	版本、内容	接受、運用及其他
浮田和民《史學通論》	1903年和衆書局李浩生中譯本等6種譯本。第五章《歷史與地理》引及黑格爾、巴克爾的學説，重天然之力。	（1）未譯出前，梁啓超已閱讀過，其《地理與文明之關係》譯撮自此書。 （2）《新民叢報》刊《章太炎來簡》論此書“于修史有益”。 [附論] 梁啓超在《清議報》和《新民叢報》上的系列文章影響很大，論點大都帶有幾分決定論色彩。一時並起的同類文章很多： 1. 梁啓勳《論太平洋海權及中國前途》。 2. 錢基博《中國輿地大勢論》。 3. 蔣觀雲《中國興亡一問題論》。 4. 佚名《地理與國民性格之關係》。 5. 吴民《江蘇與漢族的關係》。 6. 佚名《史學之根本條件》。 7. 佚名《豫省近世學派考》。 8. 熊秉穗《中國種族考》。 9. 陳黻宸《地史原理》。決定論色彩極濃。

續表

著述	版本、内容	接受、運用及其他
志賀重昂《地理學講義》	1901年金粟齋薩瑞中譯本。 卷一《論地理學之必要》强調地理對商業、軍事、海航以及歷史研究的重要性。	(1)梁啓超《亞洲地理大勢論》《歐洲地理大勢論》譯撮自此書。 (2)無名氏彪蒙書室本《繪圖中國白話地理》鈔撮自此書甚多。 [附論] 晚清尤其是20世紀初譯行的日人地理學著述極多,這些著述往往都是新近之作,有的明確聲稱采用黑格爾、巴克爾等人的學説,有的則未聲稱,但往往都具一定程度的環境決定論色彩。比如: 1. 鳥居龍藏《人種志》。人種與地理關係。 2. 佚名《地理人文關係論》。人文與地理關係。 3. 世界語言文學研究會編譯《最新人生地理學》。一依黑格爾海洋文明、平原文明之説。 4. 牧口長三郎《人生地理學》。詳下欄。 5. 守屋荒美雄《國際地理學》。極其强調地理之重要性。但末章所附譯者識曰:"凡得地勢交通之便利者,可以爲地球獨一無二之雄國,(亦)可以使爲公共瓜分之孱國。"決定論色彩變淡了。
木[牧]口長三郎《人生地理學》	1906年蘇屬學務處江蘇師範生中譯本。 内容詳右欄。	(1)梁啓超《中國思想變遷之大勢》第八章《近世之學術》自注稱此書"舉日本全土風俗、政治種種發達之差異,而悉納之于地理,旁引泰西各國以爲證",所據當是日文本。 [附論] 晚清教科書編纂盛行,最初大都轉販自日本,不獨地理一科爲然。當時國人編纂的地理教科書至少有157種。現列幾本當時的帶有環境決定論色彩的地理教科書,以見傳播與接受之一斑: 1. 鄒代鈞《京師大學堂中國地理講義》。 2. 張相文《新撰地文學》。 3. 臧勵龢《新體中國地理》。 4. 謝洪賚《瀛寰全志》。 5. 屠寄《中國地理學教科書》。論亞洲各國"國體及政體"頗帶決定論色彩。

續表

著述	版本、内容	接受、運用及其他
那特硜《政治學》	1900年《清議報》66册刊玉瑟齋主(麥孟華?)部分譯文;1902年廣智書局馮自由譯本;1902年商務印書館胡翼翬、王慕陶合譯本。 第一章即論氣候、地勢對于民族、政治及國家之影響。	(1) 劉師培《南北學派不同論》明確稱山國之民崇尚實際、澤國之民崇尚虚無之説本自此書。 (2)《新民叢報》上刊有廣智書局《政治學》廣告。 (3) 那特硜(Karl Rathgen)在日本講授的《政治學》最精彩的地方在于對國家概念的界定,自1891至1897間有多種日人譯述本風行于世,影響奇大。除了中文譯者而外,與此書發生過極大關係的還有:梁啓超、章太炎、鄧實、陳曾矩、陳天華。 [附論] 拉采爾1897年著《政治地理學》,也極有影響。晚清立憲運動興起之際,關于政治學的書籍多譯自日人著作,强調地理重要性的有: 1. 劉鴻鈞等編譯日人《政治地理》。決定論色彩不濃,但時時論及地理與政治的關係。 2. 小野塚喜平次《政治學大綱》。梁啓超在《新民叢報》上發表的《國家原論》譯自此書第二編"國家原論"。又按,第三章第二節《國家之競争力》甚重氣候、土地(包括面積、地位、大勢、境界)及動植物材料。
巴克爾《英國文明史》	1903年南洋公學院譯本;1903年《藝政通報》第15號林廷玉譯本(轉譯自日文);王建祖譯本;1906年至1907年《學部官報》魏易譯本。 論氣候、土壤、食品、萬有現象對人類的影響,有決定論傾向。	(1) 梁啓超通過浮田和民《史學通論》征引、宣傳其論。 (2) 除譯本而外,介紹巴克爾或該文明史的刊物及著述有:《政藝通報》、《中外日報》、《新世界學報》、《美國留學報告》、《泰西政治學者列傳》等。 (3) 1900年代,除相關譯者而外,曾讀過或征引此書的有:楊蔭杭、湯壽潛、陳懷、陳黻宸、汪榮寶、嚴復等。 [附論]浮田和民《史學通論》而外,日人史學著作被譯爲中文的或中國人自撰的史學著作,用及其地理環境決定論的有: 1. 家永豐吉、元良勇次郎合著《萬國史綱》。 2. 汪榮寶《史學概論》(1902年)。且謂"利用實際之山河以解釋歷史之事實,英人帕克爾(按即巴克爾)論高山大河之勢力影響于人間之心理,此在今日已成老生長談"。按,此書參照了坪井馬九三、浮田和民等人著作。 3. 翼天氏《中國歷史》。第二章《地理略説》。 4. 吕瑞庭、趙徵璧《新體中國歷史》。敘論强調"歷史與地理,如精神與肌膚……所以然者,蓋因地理影響于歷史上物質之文明,而遂成爲歷史地理之公例也"。第四章《歷史與地理之關係》。 5. 桑原騭藏《東洋史要》。第一篇有地勢、人種兩章。 6. 白河次郎、國府種德《支那文明史》。漢族來自西亞説。 7. 陳黻宸《京師大學堂中國史講義》。强調史學與輿地、物理之關係,但"天然力""人力"並重。

三、傳統資源

《禹貢》爲中國的地理學之祖。儘管秦漢以後,不管史志敘述裏還是士人心目中的"中國",常常是以本部十八省(有時得加上四大邊疆)的形式而存在的,但《禹貢》及《夏官·職方氏》《吕氏春秋·有始覽》《爾雅·釋地》的"九州"記述[①],雖互有抵牾,依然有其長遠的影響。古史辨運動興起之際,顧頡剛等人一度把《禹貢》九州説的形成延遲至戰國時期[②]。但有意思的是,到了1980年代,邵望平先生的考古活動依然緊緊圍繞着"九州"來展開,把它當成天然的"區域類型"來使用,而且自認爲證明了"九州"區域在公元前第三千年間已經形成[③]。考古學當然可以而且應該通過考古活動來劃分(新石器時代的)區域類型,但是誰能説假如没有九州的記載,邵先生會得出這樣的結論?再譬如,一位西方學者指出,《管子·水地》"齊之水道躁而複,故其民貪粗而好勇;楚之水淖弱而清,故其民輕果而賊"的這段論述,正與希波克拉底的説法一樣,帶有地理環境决定論的色彩[④]。這兩個例子,前者表明雖然後人可以借助于全新的視角、理論或方法來考察古代社會甚至史前社會,但是傳統經典的引導——不管是陷阱還是津逮——始終是存在的;後者則表明"東海西海,心同理同",换言之,在中西的久遠傳統中,各自有類似的人類體驗、想象或認知,而這正是當某種認知成爲體系化的理論之後,可以風行世界的基礎——按照勞埃德的意見,不同年代和地域的人們對外部世界的探索呈現出的多樣性,只

---

① 本世紀初刊出的上博竹簡《容成氏》也有"九州"的記載。參馬承源編,《上海博物館藏戰國楚竹書(二)》,上海:上海古籍出版社,2002。蘇建洲,《上海博物館藏戰國楚竹書(二)校釋》,台北:花木蘭文化出版社,2006。

② 顧頡剛,《詢〈禹貢〉僞證書》,《古史辨》第一册,海口:海南出版社,2005,175頁。當時,王國維的《古史新證》援據經典舊説及彝器銘文,明確表示《禹貢》九州説絶不晚于東周,甚或在西周乃至西周以前。現在由于燹公盨的發現,王説得到了進一步的支持。參李學勤《論燹公盨及其重要意義》、裘錫圭《燹公盨銘文考釋》、朱鳳瀚《燹公盨銘文初釋》、李零《論燹公盨發現的意義》,均載《中國歷史文物》2002年第6期。

③ 邵望平,《〈禹貢〉九州的考古學研究——兼論中國文明起源的多元性》、《禹貢九州風土考古學叢考》,分别見《九州學刊》第2卷第1期(1987年)、《九州學刊》第2卷第2期。

④ W. Allyn Rickett, *Guanzi: Political, Economic, and Philosophical Essays from Early China: A Study and Translation. Volume II.* Princeton University Press, 1998, p.106. 實際上,這一點中國學者早有發現和論述,參宋正海《地理環境决定論的發生發展及其在近代引起的誤解》。

是與研究風格的不同相關,而不是本質上有不可通約性(incommensurability)[①]。

的確,晚清士人還有一個傳統資源。這裏應分兩個維度:一個是用新理論來發現或詮釋舊説,比如梁啓超對于劉獻廷"地文學"的揄揚[②],就屬此類,即梁啓超本人的中國學術、思想史建構並没有直接受其影響,只是在書寫時發現了它與新理論或新學科暗合;另一個是明確受到傳統論説的影響,以劉師培《南北學派不同論·總論》爲典型。不過,我們强調,兩個維度之間並没有百分百的界限,可以是並存的。

梁啓超在《論學術思想變遷之大勢》説:"古書中言南、北分潮之大勢者,亦有一二焉。《中庸》云:'容寬以教,不報無道,南方之强也';'衽金革,死而不厭,北方之强也'。《孟子》云:'陳良,楚産也,悦周公、仲尼之道,北學于中國。北方之學者,未能或之先也。'是言南、北之異點,彰明較著者也。"[③]點明了其論點的部分來源。在《中國地理大勢論》中,他又成段摘抄《貨殖列傳》來分疏南北風俗。劉師培《南北學派不同論》前的一段總論,征引了從《王制》"廣谷大川,民生其間者異俗"到《貨殖列傳》《地理志》等篇籍中的内容:它們既是闡釋對象,也是話語資源。

正如近代西方學者的地理環境決定論關涉種族、語言文字、思想風俗、制度等層面的内容一樣,傳統經典的書寫也藴涵着同樣的精神。除了上引《水地》《王制》而外,還有:

> 堅土之人肥,虚土之人大,沙土之人細,息土之人美,耗土之人醜。[④]
>
> 以土會之法,辨五地之物生:一曰山林,其動物宜毛物,其植物宜早物,其民毛而方。二曰川澤……其民黑而津。三曰丘陵……其民專而長。四曰墳衍……其民皙而瘠。五曰原隰……其民豐肉而庳。[⑤]
>
> 土地各以其類生[人],是故山氣多男,澤氣多女……暑氣多夭,寒氣

① 勞埃德,《古代世界的現代思考:透視希臘、中國的科學與文化》,鈕衛星譯,上海:上海科技教育出版社,2008。

② 梁啓超,《論中國學術思想變遷之大勢》,108、109頁。

③ 梁啓超,《論中國學術思想變遷之大勢》,27頁。按,《中庸》所云"南""北",未必是以長江爲界,梁啓超有所誤讀。

④ 孔廣森,《大戴禮記補注·易本命第八十一》,北京:中華書局,2013,251頁。

⑤ 孫詒讓,《周禮正義·地官·大司徒》,北京:中華書局,1987,699頁。

多壽……堅土人剛，弱土人肥，壚土人大，沙土人細，息土人美，秏土人醜……東方川谷之所注，日月之所出，其人兑形小頭……長大早知而不壽……南方……其人修形兑上……早壯而夭。西方……其人面末僂……勇敢不仁。北方幽晦不明，天之所閉也，寒水之所積也，蟄蟲之所伏也，其人翕形短頸……其人惷愚，禽獸而壽。[1]

匈奴……居于北蠻，隨畜牧而轉移……逐水草遷徙，毋城郭常處耕田之業，然亦各有分地。毋文書，以言語爲約束……其俗，寬則隨畜，因射獵禽舍爲生業，急則人習攻戰以侵伐，其天性也……苟利所在，不知禮義。[2]

它們彷彿只是客觀描述，至少並未使用直接表示因果關係的詞語，但叩詢紙背之義，決定論傾向不難發現。就像近代歐洲的決定論隱含着歐洲中心論思想，上述引文的後兩條同樣隱含着"華夏中心論"。但這不是本文要展開的問題。上舉的前三條頗强調地理對人的生理的影響。我們還可以看到傳統經典强調地理對政治、軍事以及歷史研究的影響，司馬遷"作事者必于東南，收功實者常于西北"[3]的那段分析就是典型的例子，儘管還摻雜了一些神秘色彩。

諸多帶有決定論色彩的表述，班固完成了總結："凡民函五常之性，而其剛柔緩急，音聲不同，系水土之風氣；好惡取舍，動静亡常，隨君上之情欲，故謂之俗。"[4]他一面强調天然環境的影響，一面也注重人文教化。更上溯之，漢初伏生作《虞夏傳》已説："律樂者，人性之所自有也，故聖王巡十有二州，觀其風俗，習其性情，因論十有二俗。"[5]這有春秋時代禮樂教化論的影子，但表述的扼要醒豁，是頗值得注意的。不過，筆者之所以强調《地理志》，卻並不因上引的那句總結，而是基于班固在下文對各地（國）自然環境以及相應的性情、風俗、文學乃至學術思想的描述。比《貨殖列傳》以敘述各地天然産物爲主、

① 劉文典，《淮南鴻烈集解·地形訓》，北京：中華書局，2013，168—175頁。

②《史記·匈奴列傳》，北京：中華書局，1982，2879頁。

③《史記·六國年表》，686頁。

④ 班固，《漢書》，北京：中華書局，1962，1640頁。

⑤ 皮錫瑞，《今文尚書考證》，北京：中華書局，1989，57、58頁。

間及風俗更進一步,《地理志》完成了一個極有譜系的建構。譬如,在論及巴蜀時,《地理志》説:"巴、蜀、廣漢本南夷,秦並以爲郡,土地肥美,有江水沃野,山林竹木蔬食果實之饒……民食稻魚,亡凶年憂,俗不愁苦,而輕易淫泆,柔弱褊陒。"[①]排除流露出的感情色彩,梁啓超等人論南學時的用語與之相當一致。在以後的典籍比如《隋書·地理志》《通典·州郡篇》中,這個譜系得到了繼承。

魏晉以後,因地理、政治的共同作用,南、北分裂加劇。隋唐以後,雖統一的時間遠大于分裂的時間,但南北風俗、文化的不同,士人多有論述。因此,以梁啓超、劉師培爲代表的晚清士人在接受地理環境決定論時受到傳統的牽引無疑。梁啓超更是申明道:"欲以觀各家所自起,及其精神之所存……請據群籍,審趨勢,自地理上、民族上放眼觀察,而證以學説之性質。"[②] 概言之,梁啓超是根據既有的典籍,從地理、民族的角度來闡述或者説建構先秦的學術思想和文明體系。

四、全新的先秦文明建構範式:以章太炎、蒙文通、傅斯年、王國維爲中心

梁啓超在論南學北學的時候,並没有忽略唐以後政治地理的影響,但天然地理似乎仍是決定性的。從哲學、經學、佛學、詞章到美術音樂,他畢竟是不分時代、熔于一爐的,大抵在他標舉的"北俊南孅,北肅南舒,北强南秀,北僿南華"範疇之中[③]。據筆者所知,第一個給予熱烈贊美的是陳黻宸。他説:"偉哉,人之論中國文明者,切切于南北派之辨,而曰北人崇實際,南人貴理想,北人尚氣節,南人重辭華。鎔孔老于一冶,彙百家而同歸,大端斯舉,條流畢貫,折衷至當,我何間然。"[④]這篇文章載于《新世界學報》1902年第5期,反應之快,令人咋舌。他彷彿發現了"新大陸",中國文明尤其是先秦文明原來可

① 班固,《漢書》,1640—1645頁。

② 梁啓超,《論中國學術思想變遷之大勢》,25頁。

③ 梁啓超,《中國地理大勢論》,《飲冰室合集·飲冰室文集之十》,84—87頁。

④ 陳黻宸,《地史原理(下)·文明原始表》,《陳黻宸集》,北京:中華書局,1995,600頁。當然,陳黻宸對梁啓超的説法並不是全盤接受。

以這樣詮釋。所以他想到的第一件事就是依據地理環境重新做一份"文明原始表"。

從地域的角度來詮釋中國文明，是一個新視角。影響之大，無遠弗屆。梁啓超和時人確乎開啓了新的範式。

1910年，章太炎刊行了他的《國故論衡》，下卷《諸子學·原學》説：

> 世之言學，有儀刑他國者，有因仍舊慣得之者。細征乎一人，其巨征乎邦域。荷蘭人善行水，日本人善候地震，因也。山東多平原大壇，故周魯善頌禮；關中四塞便騎射，故秦隴多兵家；海上蜃氣象城闕樓櫓，恍賁變眩，故九州五勝怪迂之變在齊稷下。因也，地齊使然也。[①]

所謂"因也，地齊使然也"，就是地理決定論的變相表達。考察章太炎最早有類似表述的是重訂本《訄書·原學第一》(1904年)："視天之郁蒼蒼，立學術者無所因。各因地齊、政俗、材性發舒，而名一家……故正名隆禮興于趙，並耕自楚，九州五勝怪迂之變在齊稷下。地齊然也。"[②]這是初版《訄書》(1900年)裏所没有的。"地齊使然也""地齊然也"，頗類梁啓超不斷强調"地勢然也"[③]。章太炎重訂《訄書》是從1902年開始的，至1904年完成並在東京出版。《訄書》出版不到兩年，就着手重訂，是必有因。他在《自訂年譜》中説："余始著《訄書》，意頗不稱。自日本歸，里居多暇，復爲删革傳于世。"[④]這一年(1902)，他剛好因張園國會案避居日本三個月，甫一回國，便删革新著，嫌其"不稱"；直到晚年，他還説"舊著《訄書》，多未盡理"[⑤]。這與章太炎眼界變得開闊有關。1902年二月初九，他剛到日本不久，致吴君遂書云："鄙人東行已二十日，初寓新民叢報社，後入東京，寓牛込區天神町六十五番支那學生寓中。(有湘人朱菱西爲東道，任公之弟子也)屏居多暇，仍爲廣智删潤譯稿，間作文字登《叢

---

① 章太炎，《國故論衡》，上海：上海古籍出版社，2011，101頁。

② 章太炎，《訄書》，上海：中西書局，2012，114頁。

③ 譬如，梁啓超稱："名家言者，其繁重雜博似北學，其推理詼詭似南學，其必起于中樞之地，而不起于齊魯秦晉荆楚者，地勢然也。其氣象頗小，無大主義可以真自立，其不起于大國而必起于小國者，亦地勢也。"《論中國學術思想變遷之大勢》，30、31頁。

④ 湯志鈞，《章太炎年譜長編》上册，北京：中華書局，1979，128頁。

⑤ 章太炎，《自述爲學次第》，《中國現代學術經典·章太炎卷》，石家莊：河北教育出版社，1996，655頁。

報》中，以供旅費而已。”[①]這裏有三個關鍵詞：新民叢報，任公，廣智。梁啓超這一年在《新民叢報》發表了好幾篇帶地理決定論色彩的文章，自不待言。“廣智”即廣智書局，梁啓超募集華僑資本所創設，所謂“譯稿”當即馮自由1901年開始翻譯的那特硜《政治學》（此書對地理的强調参見上表）；太炎旅居日本潤色此書一事，馮自由曾經提及[②]。同是在這一年，章太炎還爲廣智書局譯述了日人岸本能武太的《社會學》[③]。也是在這一年，章太炎與梁啓超、吴君遂探討學術甚歡，最着力的就是關于重修中國通史的問題。用不着多説，章太炎這次“意外”的避居日本，雖爲時極短，但對于他的學術思想的演變，大有裨益。雖然他並没有完全沿着這個思路來展開他的先秦學術史構建，而是意在自創一個全新的體系（《國故論衡·諸子學》便是）[④]，不過，無論如何，他强調地域對文明的發生有着重大影響是無可争辯的事實。

沿着地域這個思路來建構先秦學術體系的，不得不提蒙文通。在蒙文通之前，廖平、劉師培之間的今、古文之争，如治絲而棼，漸入死角，彼此不約而同采取淡化今、古之争，而歸于齊、魯之辯。在蒙文通看來，“兩師言齊、魯學雖不同，而舍今、古而進談齊魯又一也。廖師又言：‘今學統乎王，古學帥乎霸。’此皆足導余以先路而啓其造説之端”[⑤]。本此，他從所謂的今古文裏尋出魯學、齊學、晉學，來建構東周六國時的學術體系：齊學本是百家言，晉學本是古史家，魯學是孔子嫡傳即六藝之學；齊學晉學爲霸（伯）制，魯學爲王制[⑥]。另一方面，他把諸子學也統轄于齊、晉、魯三系[⑦]。爲了使眉目清楚，蒙文通又據地域考察，以道家和詞賦源于荆楚，是南方派；法家、古史盛于三晉，是北方派；墨翟及别墨的名家起于鄭宋，便是中央派；《詩》《書》《禮》《樂》傳習者並燕

---

① 湯志鈞，《章太炎年譜長編》，130頁。

② 馮自由，《吊章太炎先生》，《制言》1936年第25期。馮自由，《革命逸史》初集，北京：中華書局，1981，54、55頁。馮自由，《革命逸史》二集，33、34頁。

③ 湯志鈞，《章太炎年譜長編》，138頁。《社會學》有光緒二十八年（1902）上海廣智書局印行本。

④ 章太炎之所以選擇另建體系，除了他不喜隨俗、别有所見之外，還與他的認識有關：“地齊限于不通之世。”也就是説，當世界日漸發達，地域對于思想文化的影響力漸趨衰微。

⑤ 蒙文通，《經學抉原》序，《經學抉原》，上海：上海人民出版社，2006，54、55頁。

⑥ 蒙文通，《經學導言》，《經學抉原》，21—34頁。

⑦ 在蒙文通這裏，推尋晉學、魯學、齊學的本源分别是舊法世傳之史、詩書禮樂、百家之言。後兩者屬于新派，前者舊法世傳之史屬于舊派。而舊派有二，一是舊法世傳之史，一是皇帝道家。

齊方士，是東方派；西方無派[①]。這樣一來，不免雜糅而失之棻亂。有時，他又把孔墨作爲東方之學，而荆楚（南）三晉（北）的道、法二家統統作爲西方之學，申言“中國學術，固有南北之殊，尤見東西之别，一愛人、一貴己，一文一質，其判較然也”[②]。這裏還是有疑竇，比如詩書禮樂之學與燕齊迂怪之説何以並爲東方之學？他于是開始從時期上來分析，認爲燕齊陰陽家之流屬東方的前期文化，儒墨是東方的後期文化。换言之，東方文化中心，前期在齊（爲海洋性文化），後期在魯（爲陸地文化）[③]。很明顯，他脱去今、古之争，直接上溯六國，建立晉學、齊學、魯學三系，一面固是承廖平、劉師培的余緒，一面也是有了清楚的地域分布意識。他論説諸子學從地域上分東西南北中各派，則衍自梁啓超以來的新説。

蒙文通最負盛名的卻是主文化多元論的《古史甄微》。筆者以爲，蒙文通不是考古學出身，然而卻能建立這樣一個體系，不光是天才式的冥想使然，也是由近代學術思潮的內在邏輯（inner logic）引逗出來的；當然，這個邏輯並非先天的或自在的，很大程度上乃是經過了蒙文通的預設。本來，蒙文通已經梳理出六國學術體系，繼續探源，則齊學百家言、魯學六藝是新派，晉學古史是舊派。舊派又有古史與黄帝道家兩系的不同。他自然想弄清古史的脈絡，卻恰恰發現古史傳説不一，有明顯的地域分布差異[④]。他在各篇序言或雜著中都强調自己的《古史甄微》與《經學抉原》《天問本事》二書“循環相通”；由于各篇的初稿、改稿、發表時間有錯亂，所以在描述上不免紛紜，此不具引[⑤]。從思維邏輯上講，《經學導言》（《經學抉原》前身）先出，立齊學、魯學、晉學三目，

---

① 蒙文通，《經學導言》，《經學抉原》，37頁。蒙文通《經學抉原》，《經學抉原》，91頁。蒙文通説法前後或有遊移，比如他又把“儒墨東方之化”與“法家西方之治”並舉，原因不外是法家起于三晉、盛于秦，與吴楚相較則爲北，與齊魯相較則爲西，見蒙文通，《周秦民族與思想》，《經學抉原》，141頁。

② 蒙文通，《經學抉原》，《經學抉原》，87頁。

③ 蒙文通，《周秦民族與思想》，《經學抉原》，144頁。需要指出的是，此篇作于1938年以後，與《經學導言》《經學抉原》不完全是同一時間的同一系列了。

④ 除了《古史甄微》而外，他同時撰述的文章也交代了這一點。詳蒙文通，《論先秦傳述古史分三派不同》，《成大史學雜志》1929年1期，33—39頁。

⑤ 蒙文通，《古史甄微》，上海：商務印書館，1933，序2、18頁。蒙文通《經學抉原》序，55頁。蒙文通，《天問本事序》，《史學雜志》（南京）1929年1卷4期，20頁。按，《古史甄微》最初從1929年起連載于《史學雜志》（南京）。

是爲了解決今、古學的提法各有窒礙；次撰《古史甄微》，因古史傳説的不同，立太古民族地域分布爲三系，即海岱（齊魯/東方/泰族），河洛（三晉/北方/黄族），江漢（吴楚/南方/炎族）；于是復詳考江漢一系文明，以爲補充，故有《天問本事》之撰[①]。這樣，三書各有偏重，卻"循環相通"：《經學抉原》澄明齊魯六藝之學，《古史甄微》探原三晉古史之學，《天問本事》發明吴楚之學；從時期上講，《經學導言》建立了六國學術體系，而《古史甄微》《天問本事》則建立了太古以逮兩周的文化體系。蒙文通作《古史甄微》從想法到方法上所受的影響必非單一[②]，與20世紀初引入的史地理論是緊密相聯的。他自己説："寫《古史甄微》時，就靠讀書時學過些西洋史，知道點羅馬、希臘、印度的古代文明，知道他們在地理、民族、文化上都不相同。"[③]蒙文通在1906年隨伯父入四川高等學堂分設中學，那時正是編纂文明史和新式教科書的井噴期。淵源所自，可得而推。在細節方面，比如他在《古史甄微》中説："尚忠，北方之質也，此黄族之崇實用，好剛勁之習也。尚敬，南方之惑也，此炎族之好逸豫、信鬼神之習也。尚文，此東方人之智也，此泰族之重思考、貴理性之習也。"可説有梁啓超的直接影響，而又試圖進一步從種族角度加以完善。

蒙文通主華族"自東而西"説，分太古文明爲海岱、河洛、江漢三系，很容易讓人想到傅斯年的《夷夏東西説》。有學者認爲，《夷夏東西説》雖然出現在《古史甄微》之後，但二者並無沿襲的痕迹[④]。若是説觀點不盡相同，可以同意，若説後者没受到前者的影響，則還可以討論。先看一下傅斯年的《戰國子家敘論》，最初在1930年刊出，第六節標題作《論戰國諸子之地方性》，其中説道："近人有以南北混分諸子者，其説極不可通。蓋春秋時所謂'南'者，在文

---

① 王承軍《蒙文通先生年譜長編》："一九二七年……春，先生撰《古史甄微》，是爲先生成名之作。第二年，又成《經學抉原》、《天問本事》，並教授于成都大學、成都師範大學、成都國學院、成都敬業學院。"《經學抉原》前後修改多次，其主體實際上是曾經的《經學導言》，只是有所修正，因而本文在論定撰寫次序上仍以《經學導言》爲第一，《經學抉原》爲附。參王承軍，《蒙文通先生年譜》，北京：中華書局，2012，71頁。

② 筆者注意到蒙文通在同時撰的另一篇文章很可能受王國維《殷周制度論》啓發，見蒙文通，《周初統治之法先後異術遠近異制考》，《中央大學半月刊》1930年1卷9期。《古史甄微》的思想資源亦然。

③ 蒙文通，《治學雜語》，蒙默編《蒙文通學記》，北京：三聯書店，2006，2頁。

④ 王汎森，《從經學向史學的過渡——廖平與蒙文通的例子》，《近代中國的史學與史家》，上海：復旦大學出版社，2010，93頁。

化史的意義上與楚全不相同，而中原諸國與其以南北分，毋寧以東西分，雖不中，猶差近。”[①]這是批評梁啓超及其追隨者無疑[②]。前面説過，蒙文通已强調“中國學術，固有南北之殊，尤見東西之别”，部分地修正了梁説，傅説與之相合。傅又説：“在永嘉喪亂之前，中國固只有東西之争，無南北之争。所以現在論到諸子之地方性，但以國别爲限不以東西南北等泛詞爲别。”[③]梁啓超在《論中國學術思想變遷之大勢》提到過北派、北東派、北西派、北南派、南派等説法，傅斯年所謂“東西南北等泛詞”也許指的就是這個，如果不是，則很可能是針對蒙文通的，如前所述，蒙文通那裏有所謂東、西、南、北、中央各派的分法。無論如何，傅斯年注意戰國諸子的地方性，分爲齊（燕附）、魯、宋、三晉及周鄭、南國、秦國六個地域來論述，不能説不是在衡量梁啓超、蒙文通等人得失的基礎上而來的。

接下來考察他的《夷夏東西説》。這篇文章最初是《民族與中國古代史》中的三章，大致寫于1931年春天。1933年傅撮合三章爲《夷夏東西説》發表。他强調“歷史憑借地理而生”，“以考察古地理爲研究古史的一個道路”來證明“三代及近于三代之前期，大體上有東西兩個不同的系統”，“這兩個系統因對峙而生争鬥，因争鬥而其混合，因混合而文化進展；夷與商屬于東系，夏與周屬于西系”[④]。傅依此疏通傳統文獻及零星的出土甲骨，最後從地理上加以綜合分析：黄河下游及淮河流域一帶爲東平原區，太行山及豫西群山以西的地域爲西高地系。在他看來，這兩個區域有根本的地形差異。同在東區之中，商與夷不同，諸夷以淮濟間爲本土，商人自北（東北渤海一帶）而南；同在西區之中，諸夏與周又不同，夏以河東爲土，周以岐渭爲本。他强調這些的意義在于：“因地形的差别，形成不同的經濟生活，不同的政治組織，古代中國之有東

---

① 傅斯年，《戰國子家敘論》，《中國古代思想與學術十論》，103頁。

② 傅斯年的批評並不怎麽對題，他念念不忘秦漢以前只有東西之争，没有南北之争，此是就政治（當然確與文化相關）而言，忽略了梁啓超所謂南北，只是强調南北人因居處環境的不同而造成的心理、世界觀、人生觀的不同，遂影響于學術、思想而已。惟梁啓超引《中庸》“南人之强”“北人之强”云云，以爲就是跟他以江漢、河洛來區分南北是一回事，則傅斯年駁之甚是。

③ 傅斯年，《戰國子家敘論》，《中國古代思想與學術十論》，103頁。

④ 傅斯年，《夷夏東西説》，《中國古代思想與學術十論》，1、2頁。

西二元,是很自然的現象。”[①]他在其他文章中也認爲“殷商時代以前,黄河流域及其鄰近地帶中,不止一系之高級文化”[②]。他在國外的學習使他有了系統的思路和方法,但在整個近代史的脈絡上,仍可視爲與梁啓超等人倡導的地理環境決定論和“新史學”是一脈相承的。一個較有説服力的事實是,他頗受到巴克爾《英國文明史》中(其重要性参見上文表格)地理史觀的影響,一度想把它譯成中文[③]。至于種族本來就與地理密切相關,傅斯年一直有自己的思考,只須看他對王國維《鬼方玁狁考》的贊許[④],就可瞥見關懷和矚目所在。

王國維除了《鬼方昆夷玁狁考》而外,還撰有《西胡考(上、下)》、《西胡續考》、《胡服考》等。他在《鬼方昆夷玁狁考》中説:“我國古時,有一强梁之外族。其族自西汧隴,環中國而北,東及太行常山間。中間或分或合,時入侵暴中國,其俗尚武力,而文化之度,不及諸夏遠甚。又本無文字,或雖有而不與中國同,是以中國之稱也,隨世異名,因地殊號。”[⑤]此文詳考鬼方的流變,不僅從音韻上加以證明,更“自史事及地理觀之”。這裹不妨指出從1901年梁啓超發表《中國史敍論》開始,種族、地理及文明的關係已逐漸進入人們的視野。梁啓超《中國史敍論》特别講人種、地勢,《新史學》還專論“歷史與人種之關係”“地理與文明之關係”。當時,受此影響的論著就有熊秉穗《中國種族考》、吴民《江蘇與漢族的關係》等文章(參上表),可見影響之大。其時,一同輸入中國的還有日人和西人著的中國文明史或世界文明史。對于王國維而言,他的接受還要更早,早在1899年,他就爲樊炳清譯的桑原騭藏《東洋史要》作過序,稱其“簡而賅,博而要”[⑥];這本書對亞洲東方民族的論述,是盡人皆知的。

如前所述,王國維1906年撰寫《屈子文學之精神》就深受梁啓超啓發。正是這一年,他寫出了久負盛譽的《殷周制度論》。其中心論點即“中國政治與

① 傅斯年,《夷夏東西説》,《中國古代思想與學術十論》,49頁。

② 傅斯年,《諸子天人論導源》,《春秋策:先秦諸子與史記評述》,北京:中國華僑出版社,2013,6頁。

③ 原信存台灣“中研院”近代史研究所檔案館。參王汎森,《王國維與傅斯年——以〈殷周制度論〉與〈夷夏東西説〉爲主的討論》,《學術思想評論》第三輯,沈陽:遼寧大學出版社,1998。

④ 傅斯年,《〈新獲卜辭寫本後記〉跋》,《傅斯年全集》第三册,台北:聯經出版事業公司,1980,262頁。

⑤ 王國維,《鬼方昆夷玁狁考》,《觀堂集林》上册,北京:中華書局,1959,583頁。

⑥ 王國維,《東洋史要序》,桑原騭藏《東洋史要》,東文學社,1899。

文化之變革,莫劇于殷周之際”“自五帝以來,政治文物所自出之都邑,皆在東方,惟周獨崛起西土”[①]。三代制度尤其是殷商制度的不同,古人時有論及,大抵見于對經典的隨文疏證[②]。然而他論説殷、周制度的差異,與經典義疏不同,而是從地理上的東、西分布着眼。他還聲稱:“殷周間之大變革,自其表言之,不過一姓一家之興亡與都邑之轉移,自其裏言之,則舊制度廢而新制度興,舊文化廢而新文化興。”[③]何以有這麼大的變革?在他看來,是因商、周分處東、西,所以文化、制度不同。由于夏商周三代王權的直線型更迭體系是有史可稽的,因而他似乎有一個三代同族的預設,在論述殷周制度和文化的劇變時,始終没有朝種族方面設想。到了傅斯年那裏,他把這幾個問題結合起來考慮了。正如我們看到的,《夷夏東西説》把地理與種族、文化融爲一爐來論述。

通過梁啓超、章太炎、王國維、蒙文通乃至傅斯年的例子,我們可以清晰地看到晚清之際開始輸入的地理環境決定論是如何一步步滲透到一流學者的先秦學術、思想或文化史建構之中的。它對近代學術産生的影響之大,是毋庸置疑的。

餘論

用地理環境決定論來代表或指稱一個人的學術主張,是很危險的事。歐洲各國及美國的所謂決定論者,可以列出一長串的名單來[④],但不論是他們自己,還是旁人,並不一定贊同這種指派或標簽,因一般人在使用這個詞語的時候,容易以概念來替代一個人的具體論説,並且他們在“決定”的對象、程度以及原因上,有着不同的主張。不過,就一般意義來説,不管各家所治的是歷史學、政治學,還是文化學、人類學,格外强調地理的重要性是一致的。

---

① 王國維,《殷周制度論》,《觀堂集林》上册,451、452頁。

② 柳詒徵,《傳疑之制度》,《中國文化史》上册,上海:東方出版中心,1988,105—111頁。

③ 王國維,《殷周制度論》,《觀堂集林》上册,453頁。

④ 以美國爲例,這種情形最爲突出。參 Gerald Larson Hardin, *Environmental Determinism: Broken Paradigm or Viable Perspective?* East Tennessee State University, ProQuest, UMI Dissertations Publishing, 2009, pp.28-121。

1900年代興起的新史學頗關涉這一學説，在理解上，我們也應持一種開放態度：知道它只是一個概念，並且學人在接受它時深受傳統經典敘述的影響。地理學、考古學不用説，那時的歷史學、政治學、社會學無不强調地理的重要性，其所以强調地理，來源不必單一，即不必受所謂地理環境決定論的影響，但都匯聚爲一個時代的聲音，影響着學術的思路和方法。

由梁啓超開啓的先秦學術史或先秦文明史建構，其軌轍也不必單一，我們只注意其關注地理的一面。從梁啓超到王國維、蒙文通、傅斯年，決定論是被逐漸淡化了的，但地理分布的差異和地理環境的不同而導致的族群、政治、文化的不同，始終是學人的關注點，即他們並不完全强調文明或思想被決定，而是强調不同地域的不同制度、風俗、思想、文化特點或彼此之間的互動。這一點，其實從梁啓超《論中國學術思想變遷之大勢》分論南、北諸子學以及論"諸派之初起，皆各樹一幟，不相雜廁；及其末流，則互相辯論，互相熏染……其時（戰國時）學界大勢，有四現象：一曰内分，二曰外布，三曰出入，四曰旁羅"[①]就已經開始。雖然，嚴格來説，他們進行的是古史研究，並不完全等同于我們今天所謂的地緣文化學，但對現實的地域文化觀和地緣文化研究頗有影響。所以，柳詒徵要憤憤地説"有清之季，海内人物並無南北之分，自梁氏爲此説，而近年南北人乃互分畛域，至南北對峙，迄今而其禍未熄，未始非梁氏報紙論説之影響也"[②]，而陳序經在20世紀30年代撰《南北文化觀》要專辟一節《梁啓超的南北文化觀》，來作爲現代探討地域文化的第一人。

The Acceptance of Environmental Determinism and the Modern Construction of the Pre-Qin Academic and Civilization

Abstract: Liang Qichao was the first modern scholar who studied the Pre-Qin acdamic and clarified the differences between Northern and Southern ancient China with environmental determinism which was imported from the West and became rather popular in the fisrt decades of the 20th century. The

① 梁啓超，《論中國學術思想變遷之大勢》，33頁。

② 柳詒徵，《論近人講諸子之學者之失》，宋洪兵編《國學與近代諸子學的興起》，桂林：廣西師範大學出版社，2010，249頁。

Japanese books and periodicals were the most important carrier and sources for the new theory. Besides, some English and French book should be mentioned. As it corresponded to China's inherent idea, the impact of the theory was rather profound. Lots of scholars such as Zhang Taiyan, Meng Wentong, Wang Guowei and Fu Sinian started to reconstruct the Pre-Qin academic and civilization , which should be partly attributed to the new theory.

小議域外漢籍的"外"

黄雅詩

前言

隨着國際影響力的增長,中國對於海外的各方面研究也在逐步發展,有廣布各國的孔子學院,也有近年新興的研究領域如"域外漢籍"者。與"對外漢語"的"外"字相同,"域外漢籍"的"外",也經常被用作指稱相對於中國國內的其他地方。然而隨着研究範圍的日漸擴大,相關争議也屢見討論。稍加觀之,可知各方討論焦點其實不在明白指稱"漢字書籍"的"漢籍"一詞,而是"域外"二字,此間卻非簡單"疆域之外"可以説明。

從其英文翻譯觀之,此一研究關鍵詞的常見翻譯有以下幾種:overseas Chinese books①、the documents of Chinese literature in Asia②、Chinese classics③,也有 foreign books written by Chinese④。其研究指涉的地理範圍,從海外、他國或直言亞洲者,研究對象涵括各類古今漢籍文本。至今,"域外漢籍"一詞

作者學習單位: 北京大學中文系

① 鄒振環,《"華外漢籍"及其文獻系統芻議》,《復旦學報(社會科學版)》2012年第5期,114頁。

② 王曉平,《亞洲漢文學文獻整理和一體化研究——以韓國寫本〈兔公傳〉釋録爲中心》,《天津師範大學學報(社會科學版)》總第208期,2010年第1期,58頁。

③ 孫文,《重構東亞古典文明的理論嘗試——"漢籍比較文獻學"研究》,《河南大學學報(社會科學版)》第53卷第3期,2013年5月,135頁。

④ 朴貞淑,《關於中國"域外漢籍"定義之我見》,《長春大學學報》第18卷第4期,2008年7月,58頁。

仍未出現確定的英文翻譯,反映出此一名詞所包含的廣泛領域,及其研究內容的複雜性。上述四種翻譯,多以"國界之外"作爲定義,然而現今大部分研究所針對的地理範圍,卻並非如斯廣泛。這一現象,既是"域外漢籍"作爲新興學術名詞所需面對的問題,也是其亟須解決的內在矛盾。

作爲四大古文明中唯一尚且通行於世的語言,漢語的影響已非車載斗量所能計算。藉漢語爲載體的文化交流,延續數千年的朝貢體制,在東亞地區日積月累地形成一穩定而各有特色的文化圈,隨着歷史發展,這份深厚長遠的文化基礎成爲今日東亞國家的特色。然而,這般以文化爲主要聯繫的國際關係,在面對以民族國家爲單位的現代外交體系時,此中差異便自呈於世——從"域外漢籍"之名,及其諸多相關討論,便可探知一二。相關問題並非只是單純的學術考量,還包括各國本身自有的民族意識、國際定位,甚或對文化霸權的憂慮等。是以,"域外漢籍"這一廣泛使用的學術名詞,則僅囿於中國國內,東亞諸國則各有其稱。

一、"域外漢籍"的幾種定義

"域外漢籍"的"域外"二字,最爲直白的解釋即爲"疆域之外",但此一定義又因涉及夷夏問題,而顯得別有深意。"域"者,其邦、界、區等特定範圍的指稱,然"外"者,則是相對於"內"而言之。其用於直稱夷狄者,見於《公羊傳·成公十五年》:"春秋,內其國而外諸夏,內諸夏而外夷狄。王者欲一乎天下,曷爲以外內之辭言之?言自近者始也。"此處先將夷夏以內外別之,又從天下角度加以區分:相對於自視文明程度較高的中原地區,夷狄在政治歸屬和地理性質兩方面,都不屬於此一範圍,故以內外別之,而非上下、左右等其他方位綴詞。對此,《白虎通·禮樂》更爲具體地指出:"作之門外者何?夷在外,故就之也。夷狄無禮義,故不在內。《明堂記》曰:'九夷之國,在東門之外。'所以知不在門內也。"夷狄之所以不被納入中原,乃是因其缺乏禮義之道,故"九夷"是爲拒之門外、以待禮樂教化者。由此觀之,則"域外"一詞,隱含着尚未開化或社會文化水平低下的貶義,該處不僅不如中原,甚至需要中原施以援手。此般按照文化作爲夷夏內外區別的寓意,正是傳統中國朝貢制度的核心理論,然在當今講求平等的國際關係裏,採用此一名詞指稱他國,或有所不妥。

今以“域外”爲稱者,或爲直接借用魯迅《域外小説集》之義,然該書所收皆爲歐美之作,其各方面皆與“域外漢籍”所涉者相去千里。雖然,“域外漢籍”一詞,表面上即爲“中國以外的漢字古籍”,但如此直白簡單的定義,並無法滿足學術上深度探討的需要,而足以放諸四海皆準的精確定義,學界至今仍莫衷一是。或有以地區爲別者,如王瑞來云:“中國域外漢籍,其義有二:一是中國典籍在域外的翻刻;一是域外學者以漢字撰寫刊行的著作。”[①]此説雖簡潔扼要,然其對於“域外”仍無詳細定義。又如金程宇從收藏地加以界定,云:“主要指域外所藏中國古籍(包括域外刻本)以及域外人士撰寫的漢文典籍。”[②]此將世界各大圖書館、博物館等地所藏之各類漢籍並皆納入,恐失之廣泛。況且,此一情況即爲“國際漢籍”,何須别稱“域外”一詞,再加上,若直接按照收藏地點加以稱之,則可更爲明確。又或如鄒振環從刊刻地和撰著者國籍的角度出發,以“華外漢籍”爲主,延伸出“外刻外著漢籍”、“華刻外著漢籍”、“外刻華著漢籍”、“海外古籍佚書”和“漢外籍合璧本”等五類[③]。此一分類比起“漢語撰寫之典籍”更爲具體、完備,且有補於前二者之説,然對於内容和收藏地等,並無多加規限。此分類雖可稱周到,然其“華外”與“國際”實爲同義,且以“華—外”相對,恐更易引起東亞各國的文化敏感。

除地區之外,或有從“漢文化圈”出發者,如《域外漢籍珍本文庫》之定義,便以“與漢文化有關”者爲定義。又有王勇曾云:“若從‘漢籍乃中華文明結晶推演,‘域外漢籍’應定義爲凝聚域外人士心智的漢文書籍,是在中華文明浸潤下激發的文化創新,構成東亞‘和而不同’的獨特文明景觀。”[④]今暫不論“凝聚域外人士心智”之語,王論所指雖過於博泛,卻可視爲内地“域外漢籍”研究的内在核心。同樣從“中華文明”出發,張伯偉“站在中國人的立場上”指出:“‘域外漢籍’指的就是在中國之外的用漢字撰寫的各類典籍,其内容大多植根於中國的傳統學術。”其《域外漢籍研究入門》一書將“域外漢籍”分爲三類:

---

① 王瑞來,《緣爲書同文,異口論漢籍——東京第七届中國域外漢籍國際學術會議追記》,《中國典籍與文化》1993年2期,123頁。

② 金程宇,《近十年中國域外漢籍研究述評》,《南京大學學報(哲學·人文科學·社會科學)》2010年3期,111頁。

③ 鄒振環,《“華外漢籍”及其文獻系統芻義》,104—114頁。

④ 王勇主編,《東亞坐標中的書籍之路研究》,北京:中國書籍出版社,2013,133頁。

歷史上域外文人用漢字書寫的典籍、中國典籍的域外刊本或抄本,以及流失在域外的中國古籍(包括殘卷)。針對首類典籍的作者,張書明確指出其身份:"朝鮮半島、越南、日本、琉球、馬來半島等地,以及17世紀以來歐美的傳教士等人。"[①]觀其所述各類,即爲"外國人"或"國際人士"者。加之,該書於開篇處曾指出若以國别爲稱,將易疏漏歐美地區漢籍及傳教士漢文著述,故以"域外漢籍"一詞包納之。然若以"國際"一詞稱之,既可解決此疏漏,也可統納作者身份的區别。相對於研究内核的王論,張書以"漢文化圈"爲研究範圍,對朝鮮半島、日本、越南的諸般資料,進行廣博深入且系統性的梳理,實可作爲近年相關研究的全方位概述。

從上述幾説觀之,今日普遍使用的"域外漢籍"一詞,雖屢言"域外" 爲"中國之外",實際上卻是將"域外"此一範圍廣大的名詞,限定(或等同)爲東亞的特定國家。如此,既然其研究重心皆在"漢文化圈"或相近領域,則爲何不直言"東亞"或各國國名,而須另起别稱。或有言以此區别歐美所藏漢籍,然既以國界爲限,則又回到前述的"國際漢籍"問題。此間糾纏並非一、二家所獨有,乃爲"域外漢籍"此一複合詞的名實爭議所在,然現今許多研究成果,多於此習焉不察。

有鑒於此,或有試圖從其他角度切入者,如以最爲相關的文獻學出發,有陳正宏云:"古籍版本學上的域外漢籍,是指古代中國周邊受漢文化深刻影響的幾個國家以漢文(主要是漢語文言文)撰寫、刊行的書籍。"[②]此説雖言版本,其重點與範圍實與張書相同,皆將漢譯佛教、敦煌文獻等文獻排除。然,既已有特定的"幾個國家",則或直言國名可更利於分别。又或有劉玉才所言:"中國典籍流傳至周邊地區,除原本得以存藏以外,通常再以傳鈔、翻刻的形式擴大輻射面,延生出日本本、高麗本(朝鮮本)、安南本;其後又以注疏、諺解、選編、評點、翻案等形式融入各民族文化元素;然後接受中國文化薫陶的域外文人又仿作或創作出本土漢文著作。以上存藏或産生於中國之外的各類漢文

① 張伯偉,《域外漢籍研究入門》,上海:復旦大學出版社,2012,1頁。

② 陳正宏,《域外漢籍及其版本鑑定概説》,《中國典籍與文化》2005年1期,1頁。

典籍，我們統稱之爲域外漢籍。"[①]此説着重於成書形式，並直接明指"周邊地區"這一地理限制，且在最大程度上囊括各類文本内涵，故其既可與前述張書之論互爲補充，也可作爲此一研究在文獻學方向的總結。

"域外漢籍"一詞的議論並非只在國境之内，被列爲"域外"的諸國家，也對此頗有微言。首先，此一指稱以中國爲出發點，故對於其他國家而言，以本國人之研究著述卻必須以"外"之，或恐難以接受；其次，對於"域外"的定義懸而未決，或與"國際"同義，或爲"東亞"代稱，又或是特定國家的合稱，分歧尚大。"域外漢籍"一詞的研究重心既爲"漢籍"，若以撰寫語言爲其定義，則當如王勇所建議："'漢籍'爲不分時代、不别國籍、不拘種類、不囿内外之總稱，中國人原創稱'中國漢籍'，日本人原創曰'日本漢籍'，以此類推。"[②]如此稱呼，既回歸最基本的"漢籍"之義，又在客觀中立的情況下，指明各文獻的區别，或爲明簡之法。

二、"域外漢籍"的研究範圍和目的

今言及"域外漢籍"研究，皆以台灣聯合報文化基金會於1987年12月所召開之研討會爲肇端，而該會論文集當中所使用的"中國域外漢籍"一詞，則在省去前二字的情況下，沿用至今。此後，隨着各方研究成果紛見於世，這一研究領域也成爲備受矚目的新興學科，但是針對"域外漢籍"這個名詞，或有以爲譁衆取寵者，或有以爲無須詳細定義者。是以，今日大多數研究成果，多是在模糊未明的定義範圍下，進行書籍流傳的文獻考補。又，因着研究領域的擴大和深入，原本爲了方便而簡省的詞組，已然不足以學術研究之用，其準確含義有待進一步地詳細定位。若欲完整定義"域外漢籍"，從其研究範圍和目的起論，或可究其根本。

---

① 劉玉才，《東亞漢籍版本之考察——以〈寒山子〉詩集爲個案》，《北京論壇文明的和諧與共同繁榮——新格局·新挑戰·新思維·新機遇："文明的構建：語言的溝通與典籍的傳播"語言分論壇論文及摘要集》，2012年11月，197、198頁。

② 王勇，《從"漢籍"到"域外漢籍"》，《浙江大學學報(人文社會科學版)》第41卷第6期，2011年11月，10頁。

從研究範圍觀之,國際上與中國相關的研究領域有二:“漢學”和“中國學”。白永瑞曾對此二詞提出分别:“漢學主要是指對中國語言、文學、歷史、哲學等的研究,即以研究中國古典世界的人文學爲主,西方的 Sinology 翻譯的漢學就具有這種意義。與此不同,中國學,即用英文爲 China Studies 或者 Chinese Studies,主要是指中國的政治、經濟等社會科學方面的研究,即屬於所謂區域學領域的學問。”[①]由此言之,則漢籍與“漢學”之相關性,應當遠高於其與“中國學”,而從其集中於近代以前文史領域的相關學術成果觀之,實可爲證。又,若直接以漢籍的内容及本身性質觀之,則其與關注時事、偏重政經區域學的“中國學”,其間出入更是顯而易見。

而以“中國古典世界的人文學”作爲“漢學”定義,或嫌廣泛。嚴紹璗曾經這樣定義:“就其學術研究的客體對象而言,應是中國的人文學術,諸如文學、語言、歷史、哲學、藝術、法律、宗教、考古等等,實際上這一學術研究本身就是中國人文學科在域外的延伸。”[②]又或有閻純德之言:“漢學的歷史是中國文化與異質文化交流的歷史,是外國學者閱讀、理解、研究、詮釋中國文明的結晶。作爲外國人認識中國及其文化的橋樑,漢學是中國文化和外國文化撞擊後派生出來的學問,實際上也是中國文化另一種形態的自然延伸。”[③]從“中國人文學科的外在延伸”到“中國與外國文化的交流”,“漢學”一詞所蘊涵的意義落實到具體層面上,前者是“域外漢籍”的内容,後者是“域外漢籍”的用途。

閻純德形容漢學的學術意義“在於漢學家以自己的本土文化語境作爲觀察中國文化的基點。”[④]而張西平在談及漢學未來方向時,曾言:“探討中國文化在域外的傳播和接受只是它的一個維度,這只是漢學研究目的的一個方面。我覺得還有另一個維度尚未被關注,那就是中國近代學術的變遷。”[⑤]這兩段話和張伯偉、葛兆光等人所提出“異域之眼”,實爲異曲同工。且觀王勇

---

① 白永瑞,《對話互動之相映——全球本土視角下的台灣與韓國之漢學研究》,《漢學研究通訊》第31卷1期(總121期),2012年2月,23頁。另,相同觀點的進一步説明可參見錢婉約的《從漢學到中國學》(北京:中華書局,2007)。

② 嚴紹璗,《我對國際Sinology的理解和思考》,《國際漢學》2006年,7頁。

③ 閻純德,《異名共體之漢學與中國學》,《國際漢學》2012年2期,29頁。

④ 閻純德,《異名共體之漢學與中國學》,29頁。

⑤ 張西平,《三十年來的中國海外漢學研究略談》,《國際漢學》2009年2期,66頁。

對“域外漢籍”的定義:“‘域外漢籍’應定義爲凝聚域外人士心智的漢文書籍,是在中華文明浸潤下激發的文化創新。”[①]由此可見,在這一層面上,“漢學”與“域外漢籍”實互爲表裏。作爲文化的結晶,“域外漢籍”彷若中國人在異鄉婚嫁後所孕育的後代,其差别只在於父母和出生地而已。

從研究目的觀之,針對“域外漢籍”的相關研究多採用宏觀角度,力圖進行古今中外的各種比較、探尋,正如王小盾所言:“它(域外漢籍研究)的實質是在更大的視野中關注事物的内在關係,特别是關注事物的起源和來由。從這一角度看,研究‘域外’,可以説是深入研究中國事情的前提。”[②]除了往内追溯,也可在外旁觀,如張伯偉形容“域外漢籍”爲:“它是漢文化之林的獨特品種,是作爲中國文化對話者、比較者和批判者的‘異域之眼’。”[③]

正是出於此般文化比較的目的,因此,所謂“域外漢籍”的“域外”其實更多時候是某些特定國家的代稱。静永健曾指出:“‘域外漢籍研究’所要探尋的不只是‘文化背景之不同’的結論,它更希望進一步揭示出‘同一文化大背景之下爲何會産生這種不同的接受形態’。”[④]在“同一文化”的前提下,“域外漢籍”的範圍便圈定在使用漢字的地域,如王勇以“構成東亞‘和而不同’的獨特文明景觀”之語,論述域外漢籍[⑤]。又如黄俊傑在《儒學與東亞文明研究叢書》中,便有多處直接以朝鮮半島、日本、越南與中國進行比較,並以此作爲“東亞文明”的研究主體。然而,在已有特定研究對象的情况下,使用“域外”二字或過於廣泛,是以目前這一研究領域的稱呼尚莫衷一是。比起中國所採用的“域外漢籍”一詞,東亞諸國大多以“東亞漢文學”爲代稱,因“域外漢籍”研究“會直接涉及三方面的研究,一是漢字文學研究,二是東方文學研究,三是比較文學研究”[⑥]。此論明確顯示出相關研究的内在深意:既有“漢文之學”,又兼有“漢文文學”之意。然而,比起漢籍所包含的各種跨學科信息而

---

① 王勇,《東亞座標中的書籍之路研究》,北京:中國書籍出版社,2012,133頁。

② 王小盾,《探索中的域外漢籍研究》,《中國社會科學報》2010年4月27日,第7版。

③ 張伯偉,《〈域外漢籍研究叢書〉總序》,《中華讀書報》2007年7月11日,第6版。原文爲:“它(域外漢籍)是漢文化之林的獨特品種,是作爲中國文化對話者、比較者和批判者的‘異域之眼’。”

④ 静永健,《中國學研究之新方法:域外漢籍研究》,《中國社會科學報》2010年4月27日,第7版。

⑤ 王勇,《東亞座標中的書籍之路研究》,133頁。

⑥ 朴貞淑,《關於中國“域外漢籍”定義之我見》,《長春大學學報》第18卷第4期,2008年7月,1頁。

言，劃爲文學便已是一種局限。因此，與其以文學代稱，便有王曉平提出“亞洲漢文學”、陳慶浩提出“漢文化研究”等，較爲宏觀的角度。

三、“域外”實爲“東亞”

如前所述，現今“域外漢籍”研究的地理範圍，大多集中於朝鮮半島、日本、越南等地。造成此一現象的主因，正在於朝貢制度——這個塑造出東亞文化圈的制度，它同時也是導致“域外漢籍”無法明確定義的根源。

作爲傳統中國沿襲數千年的外交系統，朝貢系統具有典型的中國思想特色：“化”[①]。以中國爲核心，加以深厚文化的優越性，朝貢系統歷經漫長歲月，建立起以漢字爲主的文化圈。然而，此看似運作良好的體制，實際上卻是：對外能强、能專而不能弱，對内能化、能容而不能獨。面對外來的各方“他者”，上至荆楚吴越，下至遠渡而來的西方諸國等，中國必須在文化、武力和經濟等層面佔據主導地位，才能行使“萬國來朝”和“懷柔遠人”的大國風範，一旦顯出弱勢，則或如南宋爲陪臣，或如晚清爲魚肉。另一方面，由於地形、交通、經濟等因素，華夏自古便非單一民族的天下。況且，在與“他者”融合、交流的歷史進程中，不論思想文化、風俗信仰抑或民族血統，傳入者大多始於民間，融於主流，最後成爲中國文化的一部分。若論朝貢制度對文化傳播之力，則以單向輸出文化思想爲主，在政治上更是要求臣服朝拜。在此框架下，只有“我者－他者”的角度，和“内－外”的相對區别，而無法産生如“東亞”、“亞洲”等具有“亞洲意識”的詞句。

“東亞”這一名詞作爲地理指稱時，便已包含了相對於“西方”的比較意義，而近年“東亞共同體”的積極開展，更加深了東西相對的意涵。這個實際上應當包含東北亞和東南亞的廣泛地理名詞，卻經常被借用爲朝鮮半島、日本和越南的特定代稱。對此，或正如孫歌所言：“我們之所以對現在的東亞論述不太滿意，是因爲現行的思路把東亞的不同區域作爲對等的民族國家來考

---

① 趙汀陽，《天下體系：世界制度哲學導論》，南京：江蘇教育出版社，2005，9頁。“中國思想的基本能力不僅僅在於它能夠因時而‘變’，更在於它什麽都能夠‘化’之。”

慮，並且試圖實體性地整合出一個完整的東亞。”[①]所謂“對等的民族國家”，即是西方國際體系的基礎，然而“一個完整的東亞”，卻又只見於往昔朝貢制度的格局，力求將兩種南轅北轍的觀念合爲一體，正是“朝貢制度”與“東亞”並列時的隔閡。越是力求整合，便越容易陷入其間歷史糾纏與諸般紛擾，至而多生滯礙，疑問叢生。此一距離最爲明顯的例子，即是甲午戰争之後，中國仍將日本視爲藩屬的一份子，僅以其爲早一步學習西方者，卻未曾注意到日本已經“脱亞入歐”，擺脱了朝貢制度及其相關制約，轉而進入以民族國家爲主的西方外交體系[②]。正是因爲忽略國際關係的運轉已然採用西方體系的架構，反而習慣性地採用朝貢制度爲視角，故而造成了無法彌補的慘敗局面。

一直以來，朝貢體系宛若貼身軟甲般，層次分明且堅固柔韌地拱衛着華夏文化，在這安心舒適的環境裏，中國無需捍衛本已高高在上的宗主地位，只需每年加意恩賜遠人即可。但在西方主導的全球化時代，外交規則建立在民族國家的基礎上，自主獨立的國家在尋求定位的同時，以符合世界格局的宏觀視野，規劃和執行各類相關政策。是以，若是仍舊依循朝貢體系的思路，不僅僅會對鄰近國家在外交、政治或軍事的方面形成壓迫，更容易産生文化霸權。從歷史觀之，“東亞共榮圈”的形成與後果，足堪佳例。也因着這般令人難以忘懷的經驗，至今提起“東亞”時，諸國或是憂慮着霸權再臨，或是極力避免再度發生。

雖然東亞在“朝貢制度”的影響下，有儒家文化權作共通性基礎，但稍加研究便不難發現，此一共通性僅在表面，各國實皆有其特色。即使是以漢字作爲文化基礎，各國文獻的獨立性及衍生研究，也各有偏重。因此，與其陷入“朝貢制度”與“民族國家”的拉扯之中，或是陷入古今中西的衝擊對立之下，“域外漢籍”之名，或許使用地域作爲方向冠詞，更可突顯其學術層面的獨立與客觀。

① 白永瑞、陳光興、孫歌，《關於東亞論述的可能性》，《書城》2004年12期，38頁。

② 楊念群，《近代以來中日韓對“亞洲”想象的差異及其後果》，《清華大學學報（哲學社會科學版）》2012年第1期（第27卷），48、49頁。

小結

從研究成果、範圍和目的觀之,域外漢籍與漢學存在一定程度的重疊。此一現象導致域外漢籍研究的興起或有譁衆取寵之嫌,然而,若從學科内涵論之,這般情況也可視作新興學科開創期的局限性。漢學早自19世紀便已開展,作爲研究文本的"域外漢籍"卻近乎晚了一個世紀才被視作關鍵詞,此間差距有着政治因素和歷史背景的影響,另一方面也是時常被提起的"一時代有一時代之學術"之況。然而,因着"域外漢籍"涉及的不同層面,其學術性不可避免地沾染上政治和意識形態問題,並進而導致諸多争論。

在"域外漢籍"研究肇端之後三年,"東亞共同體"一詞被提上國際舞台,但遲至今日仍未見進展。即使是儒家薰陶最深的日、韓兩國,其合作進展一如構築東亞共同體的遲滯緩慢,遑論各國先後加入的東盟,其各自爲政、難求同一的紛亂,長期不得解決。然而,與窒礙難行的政治共識相比,"東亞文化圈"卻是廣受承認。並非因其無涉於利益,更主要是因爲有延續兩千餘年的朝貢制度。"如果説接受'朝貢'概念即是一種'文化認同',那麼朝貢關係凸現了東亞國際的共性。"[①]透過朝貢制度,東亞各國在漫長歲月裏,吸收中國的文化和思想,並進而發展出各自的特色,但當此一制度瓦解崩頹,共享的價值體系也隨之分崩離析。在東亞浸淫於儒家思想長達千餘年之後,滿清入主紫禁城一事卻在東亞文化圈(或言朝貢制度)的同心圓上,劃了難以彌補的一刀。遑論晚清受制於砲火條約的攻擊、割據,又有接連而來的兩次世界大戰、越戰和冷戰等國際戰争,更造成東亞至今仍遭受歷史傷痕的桎梏[②]。是以,東亞文化圈的成員們在軍事、外交方面固然多有齟齬,但仍可藉由相近的文化基礎,尋求政治、經濟方面的共識。此時,作爲朝貢制度及文化交流的有形遺産,域外漢籍便可作爲補充。

然而,漢字雖出於中國,卻非中國獨享,東亞多國皆有漢字書籍、文學作品,現以漢語爲官方語言者也非中國獨一家。則將漢籍别之以内、外的稱呼,其所隱含的意識形態,極易被視爲天朝上國優越感的痕跡。對此,東亞地區

① 周方銀、高程主編,《東亞秩序:理念、制度與戰略》,北京:社會科學文獻出版社,2012,124頁。
② 詳見沈丁立,《中美關係、中日關係以及東北亞國際關係》,《當代亞太》2009年2期,5—19頁。

在20世紀下半葉所興起的"去中國化"運動,便是對此類思維的反動和擺脱。另一方面,域外漢籍的關注方向多在"傳播"二字,且是由"内"而"外"的輸出方向。不論是嚴紹璗、陳慶浩、大庭修、黄仕忠等,或是王勇、張伯偉、夫馬進等,大部分討論追究的對話主題,多聚焦於東亞文化圈内中國典籍往外流傳的情況研究:或爲各國所存之版本、目録和收藏,或如小説、戲曲之文本内容改寫或沿襲,又或論中國思想、制度等往外輻射之影響。比之論中國輸出者,反論鄰國往中國輸入者寥寥,清代《論語》回傳中國實屬少見之例,近年所整理出版的東亞藏本也多被視爲補充。這些晚近踏入學術潮流的他山群玉,仍被固定在被動接受、影響的定位上,而非國與國之間的文化交流成果。這般或囿於朝貢框架的研究思維,看在關注相關學術動態的日、韓、越等國學者眼中,便不難理解他們並未採用"域外漢籍"一詞,而是提出"漢文學"、"東亞文化圈"、"漢字文化圈"或"儒家文化圈"等概念,力求擺脱文化陰影。或有言,根據不同研究主題而有不同代稱,然中國與鄰國之各有偏好,已足可爲證。

而這一切,或許可以從"域外漢籍"的明確定義開始。正如趙汀陽所言:"中國的思想假定的是,對於任何他者,都存在着某種方法能夠將它化爲和諧的存在。"[①]中國曾經透過思想傳播,凝聚出一個高度近似的文化圈,維持了東亞的和諧穩定,雖然朝貢制度今不復存,但思想仍在、書籍仍在。如今,透過研究這些往昔的文化遺産,進而從中尋求彼此的文化異同與交流足跡,正是東亞共同體諸國努力的方向,以及建立共識的開端。

---

① 趙汀陽,《天下體系:世界制度哲學導論》,15頁。

現代漢語歐化語法現象:20世紀初西方翻譯小説有關情況的考察*

——論“被”字句以及動態助詞“着”使用頻率的增加

朧　蘭(Lara Colangelo)

引言　本研究的領域、限制和結構

幾十年來,有關現代漢語誕生的研究已能夠證明,在漢語形成過程當中,西方語言通過數類外來詞的引入在詞彙方面起着重要作用。到目前爲止,已有数部介紹西方語言與漢語接觸在詞彙方面所産生的結果的著作問世。但該種接觸的作用有可能不僅涉及詞彙層面,也涉及語法層面。虽然有關西方語言對現代漢語語法所産生的影響的研究,在學術界无疑是具有相当大的意義和價值的,但是目前相關的研究却仍然寥寥無幾,遠遠不及詞彙領域的有

作者單位:意大利羅馬第二大學

* 本文起源於筆者的博士論文(參見:L. Colangelo, *L' influenza delle lingue europee sulla morfologia e la sintassi del cinese moderno: la 'grammatica europeizzata' (ouhua yufa* 歐化語法*) nelle traduzioni di narrativa occidentale all' inizio del XX secolo*, 2012)。論文涉及的有關歐化語法現象的範圍遠遠大於本文所論。論文中所論的歐化語法成分共有八個,即:位於名詞之後用來表示複數的後綴“們”使用頻率的增加、“一個”/“一種”名詞性標記使用頻率的增加、動態助詞使用範圍的擴大、“被”字句使用頻率的增加、結構助詞“de”在書面上的分化(的/地/得)以及用法上的變化、第三人稱代詞“ta”在書面上的分化(他/她/它)以及用法上的發展、介詞“當”使用頻率的增加以及連詞化和“從句後置”現象使用頻率的增加。本文聚焦於上述兩種現象(即“被”字句使用頻率的增加和動態助詞“着”使用範圍的擴大)作爲對各種歐化成分所進行的分析的例子。

關研究。鑒於此種情形，本文試圖部分地填補研究上的這種空白來進一步考察所謂“歐化語法現象”的性質和特點。此外，至今，漢語歐化語法現象所屬的語言接觸的種類本身很少有學者去研究。大多數有關語言接觸的研究所關注的爲其直接形式（即所謂“直接語言接觸”）並集中於雙重語言這一現象或者操不同語言的兩個社會團體的接觸等各種人與人之間的接觸。西方語言對漢語語法的影響則屬於間接語言接觸，因爲它是因以華語爲母語的人與西方語言文字之間的接觸而產生的。因此，本文同時也希望能爲間接語言接觸這一領域提供一些新的認識。

作爲本研究的前提，這裏也需要指出另一點，即漢語詞法和句法中的歐化成分没有詞彙中的顯著，外來詞可以很容易地被認定爲是由外語引進的成分，而歐化語法成分則恰恰相反，很難百分之百地確定爲“外部”因素。詞彙上的歐化指的是，漢語中原來没有，後來由於對西方語言的模仿而出現的新詞語，這些新詞語具有使人毫无疑问地將之認定爲外來詞的語音上或結構上的特點[①]。因此，學者們就可以通過對原始文獻（尤其是翻譯作品）的仔細分析去尋找某個外來詞在漢語中的第一次出現，從而將其比較肯定地歸納爲外來詞。而漢語語法上的歐化，大多數情況下指的是一些虚詞和句式，這些虚詞和句式在漢語中虽然原本存在，但較少見，後來作爲對西方語言中類似成分的模仿而開始普遍起來，並且有時也会出現用法上的變化。换言之，漢語中，語法上的歐化一般表現爲某種語法成分使用頻率的增加[②]。此種使用頻率的增加也包括該語法成分新用法的出現在内。因此，恐怕無法毫無疑義地證明此種語法成分的歐化性質。基於此種考慮，本研究選擇觀察某些語法成分在某段具體的時間内使用頻率和用法上的變化，而這些語法成分是我們認爲很有可能可以被理解爲歐化成分的。

就分析的資料而言，既然歐化成分被引入漢語的主要途徑爲西方翻譯小

---

① 比如“鐵路”和“磅”這兩個詞分别爲德語單詞 *eisenbahn* 的仿譯詞和英語單詞 *pound* 的音譯詞，不難將其認定爲漢語詞匯上歐化的典型例子。可見：F. Masini, “The Formation of Modern Chinese Lexicon and Its Evolution Toward a National Language: The Period from 1840 to 1898,” *in Journal of Chinese Linguistics*, 1993, pp. 158/200.

② 賀陽，《現代漢語歐化語法現象研究》，北京：商務印書館，2008，36頁。

説，那么本研究就選擇聚焦於這種資料。值得注意的是，雖然這種資料在歐化語法研究方面具有很大的價值，但是，到目前爲止，研究該方面的學者尚未比較系統地對其進行分析。20世紀初，文學期刊在中國紛紛創刊，翻譯小説的連載是這些期刊的一個主要内容，且這些連載往往會被出版成書，所以，就分析的時間而言，我們的研究將聚焦於20世紀前三十年。此外，除了大量的翻譯小説外，爲了能夠將所分析的翻譯小説中有關歐化語法的研究，與當時最有代表性的幾部中國原創小説的語言情況進行比較，本研究的資料也包括少量的原創小説。更具體地说，本研究所進行的分析工作主要是依靠觀察20世紀前三十年這兩種不同的資料中歐化語法成分的使用頻率和用法上的變化来完成的。

大多數有關歐化語法的已有研究都强調漢語中所謂歐化語法成分的出現或者普遍使用只有在"五四"運動之後才開始。衆所周知，在漢語的歷史演變中，1919年確實是格外重要的階段，因此諸多學者也認爲"五四"運動與現代漢語的誕生有着密切的關係。但實際上，早在20世紀初（"五四"之前）中國文壇中对西方文學的翻譯活動已經很活躍，由此推断，歐化語法現象也很有可能在那时就已經開始出現了。基於此，我們認爲，很多學者有可能过度地重視"五四"這一階段，而忽略了"20世紀前二十年"这一阶段對漢語歐化語法現象出現的重要性，因此本研究在觀察20世紀前三十年的同時，特别注意考察這段時間裏1919年前的階段。

本文共分爲三部分。在第一部分中我們將對20世紀初中國語言文學背景提供小綜述，但僅限於論述那些有可能促使了歐化語法現象出現的因素和動力；在第二部分中我們首先對所謂漢語歐化語法現象下定義，其次再介紹有關該現象的已有的主要研究成果；在第三部分中我們先提供有關"被"字句使用頻率的增加以及動態助詞"着"使用範圍的擴大這兩種歐化語法現象已有的研究成果的綜述，再通過列出摘自原始文獻的若干具體的例子，詳細闡述在本研究使用的原始文獻中所觀察到的有關這兩種歐化語法現象的情況。

一、20世紀初中國語言文學背景簡述

(一) 20世紀前三十年:中國語言及翻譯活動背景

20世紀前三十年對現代漢語的誕生是極其重要、充滿事件的階段。在本部分中,我們將會提及許多這一階段歷史上的大事件與大人物,尤其是與"五四"運動這一時期有關的部分。但我們不會对这些内容提供詳細的介紹,因爲這將超出本研究的範圍。我們只會重点讨论那些爲确定小説這一文學類别的價值以及西方文學翻譯的重要性做出过很大貢獻的某些人物、雜志等,因爲,如前所述,我們認爲漢語語法的歐化過程主要是起源於翻譯活動。

19世紀末清政府正處於内外交困的狀況之中,中國已陷入政治、經濟、社會以及文化等各方面的危機。在此期間,少數较爲開明的文人開始意識到,進行文化改良與救國、實現民族復興有着密切的關係,因此要達到救國與實現民族復興的目標不僅需要實現政治維新,也需要進行文化改良,而中國的這種文化更新不得不涉及語言的层面。基於此種思考,梁啓超創造了所謂"新文體"並强調小説对于改造社會能起到的作用及其價值,并通過發表相關文章影響到了多數當時較爲進步的知識分子。梁啓超等人之所以提倡"小説界革命",是因爲他們認爲,能夠担負起改革社會重擔的不是傳統小説,而是那些"新小説",即西方小説以及以西方小説爲榜樣的日本新小説,因爲中國傳統小説雖然影響力巨大,但帶來的影響卻主要是負面的,甚至是"吾中國群治腐敗之總根源"[①]。

與此同時,西方小説翻譯活動出現了前所未有的發展。多數文學期刊不僅出版中國作家的原創小説,而且也大量出版翻譯小説。當時最有代表性的從事出版文學著作(尤其是小説)的文學雜志主要有《新小説》《繡像小説》《新新小説》《小説世界》《月月小説》《小説林》《小説時報》《小説月報》等期刊。如同许多中國學者已经指出的那样,"這些都促使了翻譯文學興盛。文學翻譯成爲文壇時尚,清末民初的新小説家們都熱衷於進行翻譯活動,翻譯小説歷

① 王宏志,《重釋"信、達、雅"——20世紀中國翻譯研究》,北京:清華大學出版社,2007,162頁。

年增多,在1902年到1907年間,翻譯小説的數量甚至超過了創作小説"[①]。據統計,清末民初的翻譯小説多達4000餘種,譯介最多的作家爲柯南·道爾、哈葛德、大仲馬等人,但除這些著名作家之外,其他的西方作家也經常被譯介。而從國别上看,在翻譯小説中,被翻譯最多的是英國小説,其次是法國、日本、俄國等國的小説。在翻譯小説的類型上,主要是偵探小説、政治小説、言情小説和科學小説。由此可見,當時的文學翻譯産物具有相當大的多元性[②]。

清末民初的外國文學翻譯爲"五四"時期的翻譯活動提供了借鑒,爲迎接翻譯文學高潮的到來起到了很大的准备作用。到了"五四"時期,20世纪前二十年一直在進行的翻譯活動進一步發展,而且這種發展與这一时期新的文學社的創立及其文學期刊的創刊有着緊密的聯繫。《新青年》于1915年創刊,1915至1921年間該雜志不斷出版多數外國著作的中文譯本。1915年起以及20年代期間新創建並具有不同文學主張的文學社團及其機關刊物促使了新文學翻譯高潮的到來。這段時間的翻譯文學以內容的多元化以及類型的豐富性爲主要特點,而且這段時間的譯者與20世紀初基本上不會外語的最早的譯者如林紓、嚴復等人有所不同,因爲這一代的中國譯者往往對外語有一定的認識,或是曾經到外國留过學,或是與外國人有所接觸。這一階段出版文學著作的最有代表性的期刊主要爲《小説月報》《創造季刊》《創造周刊》等刊物。翻譯領域上起較大作用的是文學研究會的主要陣地、由茅盾主編的《小説月報》[③]。該期刊以將世界文學介紹給中國讀者爲主要目的之一。因此,1921至1924年間先後推出了多期外國文學專號,如俄羅斯文學專號、法國文學專號以及著名作家專輯,如陀思妥耶夫斯基、拜倫、莫泊桑、易卜生、安徒生、羅曼·羅蘭等作家專輯。此外,《小説月報》還主持出版了百餘種"文學研究會叢書",較系統地翻譯介紹了英國、法國、德國、西班牙、俄國、日本、匈牙利等國家以及北歐的文學作品及文學史,同時也致力於將全世界在不同時期

① 謝天振、查明建主編,《中國現代翻譯文學史(1898—1949)》,上海:上海外語出版社,2004,16頁。

② 謝天振、查明建主編,《中國現代翻譯文學史(1898—1949)》,18頁。

③ 文學研究會於1921年在北京正式成立,發起人爲鄭振鐸、沈雁冰、周作人等著名作家。其機關刊物爲十一年前(1910年)已創刊的《小説月報》。

出現的各種文學派别一同介紹給中國讀者。同樣在介紹外國文學和外國文學史上做出較大貢獻的另一文學社團爲1921年成立的創造社。1922至1929年間其刊物不斷介紹西方各種文學思潮,如浪漫主義、象征主義、未來主義等,翻譯得較多的主要爲浪漫主義文學。在其社團成員中,郭沫若于翻譯歌德、雪萊等浪漫主義作家的作品方面起到了較大作用①。在從事文學翻譯的其他社團當中,較有代表性的有:由魯迅組織發起、成立於1925年、在翻譯領域主要從事俄羅斯文學翻譯的未名社,主要翻譯外國民間文學和諷刺幽默文學的語絲社,主要致力於翻譯羅曼·羅蘭和尼采作品的沉鐘社以及主要出名于翻譯莎士比亞著作的新月社。當時的許多著名作家大多活躍于各種不同的社團中,比如茅盾、周作人和鄭振鐸活躍于文學研究會,郭沫若和郁達夫活躍于創造社,梁實秋和徐志摩爲新月社的成員,等等。所有這些文學家,除從事原創寫作(即自己編寫小説等作品)之外也從事外國文學作品的翻譯工作,並憑藉自己的聲譽使其受到中國讀者的歡迎。

從20年代中後期開始,尤其是在中國左翼作家聯盟成立(1935年)之後,由於政治上的原因,在翻譯領域里,對蘇聯文學作品關注得相對比較多,但同時對西方其他國家文學作品的翻譯工作也在繼續進行,尤其是二三十年代英美文學的翻譯也出現了日漸繁盛的趨勢,其中譯介最多的作家有夏洛蒂·勃朗特 、哈代、辛克萊等人。創刊于1934年的《譯文》則專門以介紹外國文學爲目的,先後翻譯了包括英、法、俄、德、意大利、日本、丹麥、匈牙利等國家在内的外國文學作品。

從1937年起,抗日戰爭期間的翻譯活動在很大程度上受到了政治格局的影響,在此期間翻譯家們多傾向於翻譯反映反法西斯戰爭題材的作品,尤其是蘇聯文學以及歐美現實主義文學。

如前面所提及的那樣,當時最有權威的作家的寫作工作與各種不同文學社團的活動及其出版刊物有着緊密的關係。值得注意的是,這些作家在從事原創寫作的同時也致力於外國著作的翻譯,因爲他們認爲西方文學的翻譯在很大程度上能促使中國文學與語言的更新。換言之,中國的大多數影響力較

① 謝天振、查明建主編,《中國現代翻譯文學史(1898—1949)》,19頁。

大的作家,不僅是著名作家,而且也是積極的翻譯家。不僅如此,在多數情況下,他們的寫作活動往往是從翻譯外國作品開始的,即他們一生當中是先從事翻譯工作,然後才開始從事原創寫作的。相比其他國家文學史,這是一個比較突出、中國獨有的情況。考慮到西方著作對中國作家及其文學修養的積累過程有可能產生的影響,這是值得注意的一種現象。聯繫此種背景,便不難理解爲何在當時多數作家的著作中歐化成分會比較多(無論是詞彙上的歐化還是語法上的歐化或者文體上的歐化)。郭沫若、郁達夫、魯迅、老舍、巴金只是文學作品顯示出較爲明顯的外國影響的衆多作家中的幾位[①]。

(二)主張漢語歐化的作家兼翻譯家

就翻譯著作的類型而言,19世紀末的翻譯活動關注的主要是科學技術類著作,因爲一般認爲西方國家的"優越性"(即其現代化)主要就在於其擁有科學和技術方面的廣泛知識。而到了20世紀初,如前面論及梁啓超時所提到的,翻譯家們開始慢慢地將自己的興趣轉向小説以及其他文學著作。

在此一新的時代背景之下,這一特點表現得最爲突出的作家之一就是魯迅。在開始原創寫作之前,魯迅與其弟周作人一直在進行翻譯工作。最初魯迅提倡并積極進行科學小説的翻譯工作[②],提出借翻譯科學小説以"改良思想,補助文明"的觀點[③]。當時,魯迅主張"寧順而不信"的翻譯理論,他這一時期的翻譯方法屬於"意譯",不大忠實於原文,按照當時的翻譯習慣對原文進行大刀闊斧的變動及改寫,使其符合當時讀者的口味。

魯迅與其弟周作人一同翻譯、並於1909年出版的《域外小説集》這一文集在魯迅的翻譯活動上可視爲一個轉折點。之所以這樣説,首先是因爲本集子

---

① 外語的影響在不同程度上表現在不同作家的著作當中,有的作家歐化風格較爲明顯,有的則不太明顯(比如,巴金的著作相對而言更爲歐化)。同樣,同一作者的不同著作歐化性質也會不一致,其文體有時更爲傳統,有時則更爲歐化。

② 魯迅最早的翻譯作品應該是1903年發表的《斯巴達之魂》。雖然魯迅多次自己稱該書爲譯作,但從未提到過其原著,而且目前爲止研究魯迅的學者也没能夠查出它的原著,因此,無法肯定該作品是翻譯還是創作。如果不將其看作譯作,魯迅最早的譯作爲1903年翻譯的《哀塵》(V. Hugo, *L' origine de Fantine*)、《月界旅行》(J. Verne, *De la terre a la lune*)、1906年翻譯的《地底旅行》(J. Verne, *Voyage au centre de la terre*)等作品。

③ 李寄,《魯迅傳統漢語翻譯文體論》,上海:上海譯文出版社,2008,75頁。

包含俄羅斯與北歐作家編寫的短篇小説(而魯迅在此之前翻譯的主要爲科學和政治著作),其次是因爲魯迅從此就開始使用“直譯”的翻譯方法,即争取“逐字譯”或者最起碼“逐句譯”。但是,《域外小説集》在當時没有受到讀者歡迎。該文集失敗的原因不僅是因爲當時的發行量非常小,而且也是因爲魯迅這個時期的“直譯方法”實際上仍采取的是在讀者看來可能過於難懂、不適合表達近現代人生活及思想感情的先秦語體。魯迅當時所使用的語言,在古代漢語的基礎上添加了許多外來詞及其他歐化成分(主要爲地名和西方標點符號),讀起來不順,不夠自然而且遠遠没有中國當時流行的林紓及屬於桐城派的其他文學家翻譯外國小説時使用的語言優雅[①]。

在《域外小説集》之後的其他翻譯作品中,魯迅選擇使用的則是大衆更爲易懂的白話文,但同時他仍然繼續主張“直譯”翻譯理論並追求對原文更高的忠實度。就這一時期的翻譯活動而言,魯迅除試圖輸入原文内容之外,也力圖輸入其形式,即儘可能模仿原文的句法與語氣。他對自己當時的翻譯方法給出了下面的描述:

> 文句仍然是直譯,和我歷來所取的方法一樣;也竭力想保存原書的口吻,大抵連語句的前後次序也不甚顛倒。[②]

因這種翻譯方法,魯迅受到多數文人的公開指責。當時文學界的許多人物稱魯迅翻譯風格生澀深奥,將它稱之爲“硬譯”。魯迅曾在兩篇文章中給這種語體選擇予以解釋。第一篇文章《“硬譯”與文學的階級性》(1930年),是用來反駁梁實秋在1929年發表於《新月月刊》的《論魯迅先生的“硬譯”》一文中的指責。在該文章中,梁稱魯迅的譯文難懂,“讀這樣的書,就如同看地圖一般,要伸著手指來尋找句法的線索位置”,甚至説魯迅“‘硬譯’基本上與‘死譯’無大區别”。第二篇文章《關於翻譯的通信》(1931年),則是寫給瞿秋白的。1931年瞿秋白以J. K.爲筆名,給魯迅寫了一封公開信,發表於《十字街頭》第1期。信中除贊揚魯迅對翻譯事業的貢獻之外,還列舉了其譯文的一些

① 李寄,《魯迅傳統漢語翻譯文體論》,133頁。

② 這是魯迅在《苦悶的象徵序》中所寫的(可見:王宏志,《重釋“信、達、雅”——20世紀中國翻譯研究》,242頁)。

缺點乃至錯誤。對此，魯迅在高興、感謝之餘，回信告訴對方，自己選擇使用"直譯"方法的原因在於當時漢語本身的缺陷：

中國的文或話，法子實在太不精密了，作文的秘訣，是在避去熟字，刪掉虛字，就是好文章，講話的時候，也時時要辭不達意，這就是話不夠用。[①]

因此，魯迅認爲通過"逐字譯"以及對原文的模仿可以改善漢語，增强其表達能力。他主張的不僅是詞彙上的模仿，而且也是語法上、句法上的模仿。其實"硬譯"一説，爲魯迅本身所創造而後來由其反對者所使用，因爲魯迅知道自己的譯作確實較爲難懂而不順[②]。但他"寧信而不順"，也就是説，他認爲在最初階段"不順" 、不易懂的直譯是可以接受的，以便豐富和完善漢語，使其變成更爲完整的國語。在他看來，最初讀起來很彆扭、甚至不像中文的那些詞語、句式等，會逐漸地被中國讀者"消化"並開始被理解爲漢語的組成部分：

一面儘量的輸入，一面儘量的消化、吸收，可用的傳下去了，渣滓就聽他剩落在過去裏。但這情形當然也不是永遠的，其中的一部分，將從"不順"而成爲"順"，有一部分則因爲到底"不順"而被淘汰，被踢開。[③]

此後幾年，魯迅多次强調漢語歐化的重要性，稱其有時甚至爲面對當時漢語精密性缺乏的唯一方法：

竭力將白話做得淺豁，使能懂的人增多，但精密的所謂"歐化"語文，仍應支持，因爲講話倘要精密，中國原有的語法是不夠的，而中國的大衆語文，也決不會永久含胡下去。譬如罷，反對歐化者所説的歐化，就不是中國固有字，有些新字眼，新語法，是會有非用不可的時候的。[④]

① 王宏志，《重釋"信、達、雅"——20世紀中國翻譯研究》，245頁。
② "硬譯"這種説法，是魯迅在《文學與批評》(1929年)後記中創造的。
③ 摘自《關於翻譯的通信》，王宏志，《重釋"信、達、雅"——20世紀中國翻譯研究》，246、247頁。
④ 摘自魯迅1934年的文章《答曹聚仁先生信》，可見：彭建華，《論魯迅翻譯策略》，2011。

在1934年的《康伯度先生答文公直》一文中，魯迅指出連反對歐化並批評魯迅歐化文字的文公直寫的信也含有一些歐化成分：

> 中國語法的歐化並不就是改學外國話，但這些粗淺的道理不想和先生多談了(……)不過還要説一回：我主張中國語法上有加些歐化的必要。這主張，是由事實而來的。中國人"話總是會説的"，一點不錯，但要前進，全照老樣卻不夠。眼前的例，就如先生這幾百個字的信裏面，就用了兩回"對於"，這和古文無關，是後來起於直譯的歐化語法，而且連"歐化"這兩個字也是歐化字；還用着一個"取消"，這是純粹日本詞；一個"瓦斯"，是德國字的原封不動的日本人的音譯。都用得很愜當，而且是"必要"的。①

魯迅並不是那個年代唯一一個對漢語表達能力缺陷表示不滿並指出迫切需要將其更新的知識分子。清末民初中國處於政治與文化不穩定的情形之中，許多文人已經表示出了此種擔心，早就意識到語言文學改良與救國有着緊密的關係。魯迅與瞿秋白的討論以及魯迅與梁實秋的筆戰都發生在"五四運動"之後，此時已正式"廢棄"文言文，但新生的白話文還需要經歷一個摸索成長的過程，需要明白如何改良白話文，如何使其能夠作爲與外語表達能力一樣强的規範化新國語。在這個過程中，很多權威人士强調漢語書面語表達不夠精密，主張向西方語言學習，在翻譯領域公開支持直譯。在這樣一種情形下，基於這種改良語言的迫切需要，瞿秋白等其他幾位知識分子甚至在考慮將漢字換成拉丁字母文字系統的可能性。瞿秋白對當時的漢語持有極爲消極的看法，在他看來，漢語中用於交流和信息傳播的語言也顯得非常有限和貧乏，很多情況下處在一種極不發達的"原始狀態"。以下片段摘自瞿秋白1931年致給魯迅的《論翻譯：給魯迅的一封信》②：

> 中國的語言(文字)是那麼窮乏，甚至於日常用品都是無名氏的。中

① 魯迅，《康伯度先生答文公直》，《申報·自由談》，1934年第8號。

② 瞿秋白給魯迅的這封信曾以《論翻譯》爲題發表於1931年12月11日、25日《十字街頭》第1、2期。

> 國的語言簡直没有完全脱離所謂"姿勢語"的程度——普通的日常談話幾乎還離不開"手勢戲"。自然,一切表現細膩的分别和複雜的關係的形容詞,動詞,前置詞,幾乎没有。①

就這一點,魯迅與瞿秋白的觀點相同。而且,兩位均認爲翻譯活動在漢語改良過程當中能起到極其重要的作用:

> 翻譯——除出能夠介紹原本的内容給中國讀者之外——還有一個很重要的作用:就是幫助我們創造出許多新的字眼,新的句法,豐富的字彙和細膩的精密的正確的表現。②

由此可見,瞿秋白認爲外國文學的翻譯不僅可以讓中國讀者了解到新的内容,而且也會有助於將新的詞彙成分以及使漢語更精密、更具有表現力的語法成分引進到漢語中。瞿秋白與魯迅觀點的區别在於有關"直譯"以及白話文性質本身的看法。瞿秋白雖然也以"白話"來指稱他所主張的翻譯語言,但卻不是"五四"意義上的白話。作爲致力於無産階級革命文學及其理論建設的倡導者,瞿曾多次强調漢語書面語應該是"口頭上能夠説得出來"的。"他所説的白話,更强調'聽得懂',他對於五四式白話的不滿主要就在於'讀出來是不能懂得的,非看着漢字不可'"。而且,"他所主張的'聽得懂'並且口頭上'能夠説得出來'的白話,不僅是他鼓吹普洛的革命文學的必然要求,也是爲他最終要實現漢字拼音化的設想做鋪墊的。正是出於這樣的革命與革命文學的目的,瞿秋白認爲魯迅翻譯的世界無産階級革命文學的名著《毁滅》做到了'正確',但還没有做到'絶對的白話',更不贊成魯迅極而言之的'我是至今主張寧信而不順'的觀點。他認爲,魯迅所説的'現在可以容忍多少的不順',是'没有着重的注意到絶對的白話本位的原則'。如果使'寧信而不順'成爲一種傾向就會造成很壞的結果:'一般青年的翻譯因此完全不顧群衆的需要。'由此可見,瞿秋白認爲,魯迅雖然翻譯了無産階級革命文學的作品,但其

① 王宏志,《重釋"信、達、雅"——20世紀中國翻譯研究》,245頁。
② 該片段也摘自瞿秋白《關於翻譯的通信》一文。

譯文還不是絶對的白話"[1]。但無論如何,如前面所提到的那樣,魯迅和瞿秋白都認爲翻譯會有助於改善漢語,並非常看重翻譯在輸入新的表現法方面所起的重要作用。而且,這是在當時許多先進的知識分子當中較爲普遍的看法。對於本研究而言,這是非常值得注意的情況,因爲這意味着,外語對漢語的影響不僅不是當時作家和翻譯家没有意識到的一種自然現象,而且甚至可以説是由他們專門地、有意識地推進的一種過程。换言之,當時許多權威人士都曾公開主張通過對外語語法成分的模仿與引進,來豐富和完善自己的語言。其實,嚴厲批評魯迅翻譯方法的梁實秋自己也非常重視翻譯,而他打擊魯迅的原因與其説純粹涉及文學方面,不如説很可能也涉及意識形態方面[2]。

積極提倡漢語中引進外語成分——包括語法成分——的另一位權威人士爲傅斯年。以下片段摘自傅斯年1919年編寫的《怎樣做白話文》,能夠明確説明他的這一觀點:

> 我們主張新文學,自然也得借經於西洋的新文學。劈頭便要創造,便不要榜樣,正合了古人説的"可憐無補費精神"。只可惜我們歷史上的白話產品,太少又太壞,不夠我們做白話文的憑借物。(……)這高等的憑借物是什麽,照我回答,就是直用西洋文的款式、文法、詞法、句法、章法、詞枝(……)一切修詞學上的方法造成一種超於現在的國語、歐化的國語,因而成就一種歐化國語的文學。[3]

由此可見,傅斯年公開支持漢語歐化,認爲創造新白話文必須在很大程度上模仿西方語言的各個方面,無論是詞彙方面還是詞法和句法方面或者修

---

① 李今,《翻譯的政治與翻譯的藝術——以瞿秋白和魯迅的翻譯觀爲考察對象》,《河北學刊》,2007年2號。

② 王宏志還指出:信仰了新人文主義的梁實秋,認定了文學活動是"有紀律的,有標准的,有節制的",而文學的内容更是基本的普遍的人性。這樣,它不可能不跟主張文學是具有階級性的左翼以至黨員作家發生沖突……梁實秋在《文學是有階級性的嗎》裏,完全否定了文學的階級性,他認爲,文學是最基本的人性的表現,在這基本的人性上,資本家和勞動者並没有什麽分别……"把文學的題材限於個别階級的生活現象的範圍之内,實在是把文學看得太膚淺,太狹隘了",而當時已受到馬列主義思想影響的魯迅當然不能接受這一觀點。他認爲,文學家雖自以爲超階級,但他們也無意識地受本階級的階級意識所支配。(王宏志,《重釋"信、達、雅"——20世紀中國翻譯研究》,267、268頁)

③ 趙家壁主編,《中國新文學大系(1917—1927)》,上海:良友圖書印刷公司,1935,217—227頁。

辭方面,因此他也主張最適合引進外語成分的"直譯"方法。

同樣,當時許多著名作家和傑出知識分子,如胡適、陳獨秀等人,雖然没有如此公開地主張漢語語法的歐化,但也强調模仿西方語言的好處。比如,胡適曾公開指責傳統白話過於簡單,譴責其缺點並多次强調在廢棄文言文的同時也需要翻譯外國現代文學傑作,因爲西方文學會爲中國新國語的創建樹立榜樣[①]。

魯迅、胡適、傅斯年只是强調外國文學翻譯對創造新國語能起到重要作用的中國文學家中的幾位。郭沫若、高植等許多其他著名作家也認爲西方文學的翻譯是完善和豐富漢語的主要途徑[②]。如前所述,在研究現代漢語歐化語法的過程當中,這是值得强調的重要因素,這些作家的此種觀點和理論,由於他們的名聲與權威而促使了他們實際的翻譯工作本來已自然引起的歐化語法現象的進一步發展,而這能夠進一步地證明所謂"漢語歐化語法成分"的歐化性質。

二、現代漢語歐化語法現象定義與已有的研究成果綜述

本文中所謂歐化語法是指現代漢語在西方語言尤其是英語的影響下産生或發展起來的語法現象。這種現象既涉及漢語的詞法,也涉及漢語的句法。如前所述,"五四"期間已有多位作家討論過漢語模仿和吸收西方語言成分這種問題,但專門從語言學的角度研究漢語歐化語法現象的第一位語言學家是王力。其《中國現代語法》專門包含"歐化的語法"一章[③],而且此後他在自己的其他著作中也曾多次論及這一主題[④]。在《中國現代語法》中王力列出他認爲是從"五四"運動開始普遍起來的一系列新的語言現象,比如:通過添加詞綴創造新複音詞的趨向上升、句子的延長、被動式使用頻率的增加、連結

① 胡適,《中國新文學大系·導言》,《中國新文學大系(1917—1927)》,第一卷,1—32頁。

② 謝耀基,《現代漢語歐化語法概論》,香港:光明圖書公司,1991,19—200頁。

③ 王力,《中國現代語法》,北京:商務印書館,1943(上册)/1944(下册),下册,334—374頁。

④ 比如:王力,《中國語法理論》,北京:商務印書館,1944(上册)/1945(下册)(本文參考的是王力,《中國語法理論》,《王力文集》,濟南:山東教育出版社,1984,第1卷,433—503頁,下同);王力,《漢語語法史》,北京:商務印書館,1989,327—339頁。

成分使用頻率的增加、模仿西方語言中不定冠詞的“一個”/“一種”名詞性標記用法的興起[①]、表示進行貌的動態助詞“着”以及添加在名詞之後用來表示複數的後綴“們”使用範圍的擴大[②]、人稱代詞“他”字面上的分化(分爲陰性、陽性和中性)[③]等現象。在1958年出版的專著中[④],王力對上述歐化現象提供了進一步的介紹並論及以前專著中未提及的歐化現象,如:新興的聯接法、新興的平行式等現象。王力在其最早論及歐化語法現象的《中國現代語法》中早就强調,所謂“西方語言的影響”主要是指英語的影響,因爲當時會英語的中國人在人數上遠遠多於會其他西方語言的,而在西方翻譯著作中,最多的也是翻譯自英文作品的譯著,其數量遠遠高於其他語言的著作,所以漢語與英文的間接語言接觸比起漢語與其他西方語言的要强烈。基於此,他認爲與其説漢語的“歐化”不如説漢語的“英化”:

> 所謂歐化,大致就是英化,因爲中國人懂英語的比懂法、德、意、西等語的人多得多。[⑤]

除王力之外,也有其他學者持同樣的觀點[⑥]。然而,我們認爲,“歐化語法”這一種説法雖然不夠精密並過於泛指,但同時也具有更爲靈活、更有概括性這樣一種好處。儘管在20世紀初的翻譯著作中英文譯著最多,但也有不少其他歐洲語言的作品被翻譯成中文,因此不能絶對地排除其他歐洲語言在某種程度上對漢語也有影響。

王力之後,有關漢語歐化語法的研究停滯不前,要等到20世紀80年代才有新的發展。唯一例外的是20世紀50年代北京師範學院中文系編撰的《五四以來漢語書面語言的變遷和發展》一書的出現[⑦]。該書是有關中國1919年

---

① 王力將這種現象講述在“新稱述法”一段落中。

② 這兩種現象講述在“記號的歐化”一段落中。

③ 這種現象講述在“新代替法”一段落中。

④王力,《漢語史稿》,北京:科學出版社,1958(本文參考的是《王力文集》,濟南:山東教育出版社,1988,第9卷,607—621頁)。

⑤ 王力,《中國現代語法》,下册,334頁。

⑥ 可見:謝耀基,《現代漢語歐化語法概論》,2頁。

⑦ 北京師範學院中文系漢語研究組編,《五四以來漢語書面語的變遷和發展》,北京:商務印書館,1959。

後幾十年間語言情況的重要參考，雖然不是有關歐化語法現象的專著，但該書對漢語在“五四”之後發生的變化進行了詳細的描述，並多次强調在這種過程中西方語言對漢語所起到的重要作用。

有關漢語歐化語法現象的第一部專著是由美國作者Cornelius Kubler編寫的[①]。該書來自Kubler的語言學碩士論文，其研究主要基於作者對巴金小説《家》1931年的原版和1957年的修訂版所進行的比較分析。巴金從少年時代起就愛讀原文外國小説，尤其是英國和法國小説，他于1927年赴法國留學，留學期間曾積極從事外國文學翻譯。巴金與西方文學的接觸以及其對外語較高的熟悉程度對其寫作風格造成了較大的影響。到了20世紀50年代，過於歐化的語體變得不符合當時的政治形勢，因此巴金因自己的寫作風格而受到批評，被迫作出自我批評並改寫自己的小説，在改寫過程中巴金對作品中的歐化成分進行了較多的删除和改寫[②]。通過對其兩種不同版本的小説進行對比分析，Kubler試圖考察像巴金這樣以中文爲母語並懂外語的中國作家兼翻譯家將何種語言成分歸納爲歐化成分。换言之，在Kubler看來，凡是修訂版中被修改或删除的成分可理解爲歐化的成分。基於此種分析Kubler發現的歐化現象主要爲：“們”使用頻率的增加、做狀語的結構助詞“的”/“地”使用頻率的增加、動態助詞“着”使用範圍的擴大、人稱代詞“他”在書面上的性分化、主語的增加、被動式的增加、副詞“在”的增加(表示進行時態)、由介詞“在”引出的介詞結構後置現象(位於謂語之後)、由介詞“當”引出的時間從句使用頻率的增加、定語的延長等現象。

在Kubler的著作出版幾年之後，香港學者謝耀基出版了有關歐化語法的專著《現代漢語歐化語法概論》[③]。該書先聚焦於詞彙，以介紹不同類型的外來詞爲主，然後致力於描寫一系列有可能可被認定爲歐化的語法成分和句式。這些成分中的大多數是王力和Kubler的書中已論及過的，但與之不同的是，謝耀基除列出所謂歐化成分之外還試圖將其分爲善性的歐化和惡性的歐

---

① C. Kubler, *A study of Europeanized Grammar in Modern Written Chinese*, Taipei: Student Book co. Ltd., 1985.

② C. Kubler, *A study of Europeanized Grammar in Modern Written Chinese*, pp. 36-43.

③ 謝耀基，《現代漢語歐化語法概論》，香港：光明圖書公司，1990。

化兩類。在謝耀基看來,該成分積極的或消極的性質取決於其使用的必要性程度,換言之,凡是純出於對西方語言的"機械性"模仿、在漢語句中不起實際作用的多餘成分可被理解爲惡性歐化,而凡是出於對西方語言的"選擇性"模仿並有助於使漢語句子更精密、更清楚的成分,則可被理解爲善性的歐化。

在谢耀基之後,對有關西方語言對漢語影響的研究做出較爲重要貢獻的是美國學者 Edward Gunn 的 *Rewriting Chinese: Style and Innovation in Twentieth-Century Chinese Prose*(1991)一書[①]。該書試圖考察從清末起一直到20世紀90年代(尤其是1918年至1986年這段時間)漢語書面語發生的主要變化。作者認爲在這些變化當中,有一部分與西方語言和日語對漢語的影響有關。Gunn所使用的資料主要是通過隨機抽樣挑選出的現當代文學作品的片段。該書關注的主要是20世紀在漢語小說和散文中所發生的修辭方面、文體方面的變化而不是語法方面的變化。雖然作者的研究聚焦於新興的修辭格(如:回指、語句間隔反覆、連詞疊用等等)及寫作風格上的變化,但還有一小部分是專門介紹語法上的一系列新現象的,如:被動式的增加、"們"使用頻率的增加、從句後置現象的出現、做狀語的結構助詞"的"/"地"使用頻率的增加等現象。

此後,幾乎有二十年左右的時間,有關漢語歐化語法的研究有所停止[②],直到2008年賀陽出版《現代漢語歐化語法研究》一書時才有新的發展[③]。現任中國人民大學文學院副院長的賀教授在這一專著中較爲系統、完整地分析了現代漢語中主要的歐化成分中相當大的一部分,介紹的歐化成分涉及所有詞類,也涉及語序,其數量遠遠大於以前其他學者所論及的(比如"N的V"結構、"NV"結構和"PP的V"結構爲以前很少有人論及的歐化成分[④])。其所使用的

---

① E. Gunn, *Rewriting Chinese:: Style and Innovation in Twentieth-Century Chinese Prose*, Stanford, California: Stanford University Press, 1991.

② 論及漢語歷史和演變並順便提到西方語言對漢語影響的著作當然也有,但都不是比較系統地研究漢語歐化語法現象的專著(比如:向熹,《簡明漢語史》,北京:高等教育出版社,1993;徐時儀,《漢語白話發展史》,北京:北京大學出版社,2007,等)。

③ 賀陽,《現代漢語歐化語法現象研究》,北京:商務印書館,2008。

④ "N的V"結構 、"NV"結構和"PP的V"結構都是以謂詞爲中心語的定中結構,實例分別爲:"價格的增加" 、"企業管理"和"對這些問題的考察"。這三種結構均爲舊白話中已有的,但在賀陽看來,"五四"以來使用頻率的增加及發展與外語影響有關。

資料可分爲兩種:一種是作爲未受西方語言影響而代表漢語語法傳統樣本的14到19世紀末的明清小説,另一種則是作爲受到西方語言影響的“五四”以來具有代表性的中文作品[①]。作者没有將使用的所有著作全部分析,而是通過計算機軟件進行隨機抽樣而摘出部分片段。通過對這兩種資料的對比分析,賀陽試圖確定哪些語法現象是“五四”之前已經存在的而哪些則是“五四”之後才出現或普遍起來的,並認爲“五四”以前的舊白話中没有、“五四”之後突然出現的語法成分即可被理解爲歐化語法成分。爲了解釋自己的這種觀點,賀陽强調,與語言的自然演變有關的語言變化(尤其是語法層次上的)一般需要某種漫長、逐漸的過程才能出現,而賀陽所觀察到的那些語法成分在漢語中的出現或普遍使用則發生得較爲突然、迅速,因此很有可能與外語的接觸及影響有關:

> 如果没有特殊的原因,語法的自然發展通常采取漸變的方式,如果一個語法演變在一個很短的時間内完成並通行開來,我們就有理由懷疑這一演變可能是一種接觸性演變。[②]

本研究同意賀陽的這種看法,也認爲某些語法成分的使用頻率的突然增加很有可能與外部原因有關(即與外語影響有關)。但與賀陽和大多數其他學者不同的是,本研究主張漢語中所謂歐化語法成分或至少其一部分不是“五四”以後才開始普遍起來而是“五四”之前(20世紀初)就已經開始普遍起來的[③]。换言之,許多學者認爲“1919年”在歐化的過程當中起了決定性的作用,將“五四運動”視爲漢語歷史上的某種轉折點,因此也將其理解爲漢語中是否有歐化語法成分的分水嶺,而本研究則力圖證明1919年之前某些歐化成分在西方翻譯小説中已經開始流行開來。

---

① 採用這兩種資料進行對歐化語法的大規模的分析,其實是Kubler在專著“今後的研究方向”一章早就指出的研究方向(C. Kubler, *A study of Europeanized Grammar in Modern Written Chinese*, p.146)。

② 賀陽,《現代漢語歐化語法現象研究》,36頁;此外可見:賀陽,《現代漢語歐化語法現象研究》,《世界漢語教學》2008年4期,16—32頁。

③ 賀陽在其專著中多次説明指出這一點(可見:賀陽,《現代漢語歐化語法現象研究》,1、2、36、285、293等頁)。

賀陽的專著在介紹一部分歐化語法成分時,也將其所使用的書面資料中有關歐化語法成分的數據與北京話口語語料庫中的有關數據進行了對比分析。他的這種將書面資料與口語資料進行比較的做法是前所未有、具有原創性貢獻的。通過此種對比分析,賀陽得出的結論是,歐化語法現象只涉及漢語書面語,不涉及漢語口語:

> 五四以來發生重大變化的是漢語的書面語而不是口語,漢語語法的歐化是間接語言接觸的結果,其産生有比較嚴格的語體限制,即它們一般都只限於書面語,而不會深入日常口語。所以在當今日常口語中不説而舊白話也未見的語法現象就有可能是歐化的結果。[①]

其實,王力在其1944年專著中已指出:

> 歐化的語法往往只能出現於文章裏,和口語没有關係。[②]

賀陽再次强調這一點,並通過分析口語資料試圖對它予以充分的證明。由於資料的豐富性(舊白話小説和"五四"之後的小説)以及研究方法的系統性,賀陽的著作對本研究是重要的參考,因此我們在分析各種所謂"歐化語法成分"時經常會以他的研究結果(尤其是他收集的有關舊白話的數據和使用的例句)爲主要參照系。

最近幾年除賀陽上述著作之外,基本上没有出版過其他有關漢語歐化語法現象的新專著,但有幾篇有關英語對香港漢語所引起的影響的文章也值得參考[③]。雖然香港漢語與大陸漢語(或者説普通話)在某些方面有所不同,但這些文章論述了有關英語對漢語引起的影響的幾種有意思的現象,比如香港

---

① 賀陽,《現代漢語歐化語法現象研究》,36頁。

② 王力,《中國語法理論》,465、466頁。

③ 其中較有代表性的文章爲:石定栩、朱志瑜,《英語對香港書面漢語句法的影響——語言接觸引起的語言變化》,《上海外國語大學學報》1999年4期,2—11頁;何自然、吴東英,《内地與香港的語言變異和發展》,《語言文字應用》1999年4期,82—87頁;朱志瑜、傅勇林,《英漢翻譯的影響與香港書面漢語的結構變異》,《外語與外語教學》2002年10期,55—60頁;石定栩、朱志瑜、王燦龍,《香港書面漢語中的英語句法遷移》,《外語教學與研究》2003年1期,4—13頁。

漢語在英語的影響下發生的詞彙轉類現象[①]。這種現象也許已經開始或將來會從香港漢語波及大陸漢語(由於英語對大陸普通話的影響或者由於香港漢語對普通話的影響),從而形成漢語語法歐化的一種新形式——即使是直接的。

三、20世紀初西方小説中文譯文中的兩種歐化語法成分

本部分我們將會介紹本研究所使用的原始文獻中兩種歐化語法現象的有關情況。我們將會對這兩種不同的資料進行對比分析,即將本文所分析的西方翻譯小説中有關歐化語法的研究結果,與同時期最有代表性的幾部中國原創小説的語言情況進行比較。此種對比分析的主要目標,是力圖進一步證明這兩種歐化語法現象的歐化性質,並考察兩種現象的出現是否早於1919年。爲此,我們將會試圖爲每一個語法成分回答以下五個問題:

1. 在選擇分析的時間段内(20世紀前三十年)所研究的被假定爲"歐化語法"的成分其使用頻率是否有增加?

2. 所研究的語法成分的普遍使用先出現在翻譯小説中還是先出現在原創小説中?

3. 所研究的語法成分在用法上是否産生了變化?

4. 所研究的語法成分在用法上的變化先出現在翻譯小説中還是先出現在原創小説中?

5. 所研究的語法成分使用頻率的增加和用法上的變化是在五四運動之前開始出現還是五四運動之後開始出現的?

(一)選擇原始文獻時所遵守的原則

在第一部分中我們已經指出以下幾點:1) 20世紀初中國對西方著作的翻譯和引進達到了前所未有的發展;2)所有這些翻譯著作也變成了將一系列

① 比如説,爲了模仿漢語中本没有相應形容詞的英語形容詞,直接用漢語中的名詞或動詞來當作形容詞使用。典型的例子有"很時尚"(等於 fashionable)、"非常吸引"(very attractive)等説法 。

歐化語法成分引入漢語的主要途徑;3)當時多數最有代表性的作家兼翻譯家在其翻譯(或編纂)的作品中曾有意識地、專門創造並普遍使用一些歐化成分和句式來彌補漢語表達能力的不足。由此可見,爲了研究漢語歐化語法現象,20世紀初的翻譯作品絶對是非常重要的資料。儘管如此,如引言中所述,以前有關歐化語法現象的研究從未聚焦於此種資料。部分地分析翻譯著作的唯一學者爲賀陽,但其使用的資料中翻譯作品相對而言數量不大,遠遠小於原創作品,而且只涉及賀陽研究的歐化語法成分的幾種而非全部[①]。此外,無論是翻譯小説還是原創小説,過去大多數學者所使用的20世紀的資料很少爲1919年前的,一般都是1919年之後的(即五四運動之後的)。賀陽采用的大量資料中偶爾也能看到清末民初的著作(主要爲摘自所謂"譴責小説"的片段),但總的來説20世紀初、1919年前的資料幾乎没有。因此本研究將主要對20世紀初前三十年的翻譯小説進行較爲系統的分析,同時也將集中關注1919年前這一階段的資料。

如第一部分中所述,當時西方作品的中文譯文主要被刊載在一些文藝雜志上,因此本研究選擇使用當時最有代表性、影響力最大的從事小説刊載的幾個期刊,即:《新小説》(1902—1906年)、《繡像小説》(1903—1906年)、《小説林》(1907—1908年)和《小説月報》(1910—1931年)[②]。收集、摘自這些期刊的資料都是1902年(即《新小説》的創刊年)到1931年(《小説月報》的停刊年)這一時間内發表的,基本上能包括歐化語法的興起及發展最爲重要的階段[③]。

就摘自每種雜志的資料數量而言,本研究選擇分析刊載在《新小説》和《繡像小説》每一號上的翻譯作品以及刊載在《小説林》和《小説月報》每一年

① 就本文研究的兩種歐化語法現象而言,在賀陽所分析的有關動態助詞"着"使用範圍的擴大的資料中似乎没有翻譯作品,而有關"被"字句使用頻率的增加則有之,但與原創作品資料相比要少得多,而且基本上均爲"五四"以來的著作,不包括"五四"之前的20世紀著作。

② 所有這些資料都是本人在北京大學作訪問學者時在北京大學圖書館裏查閲的。就分析的具體資料(即選擇使用哪些期刊和哪些作品)而言,主要以魏紹昌,《中國近代文學大系(史料索引集)》,上海:上海書店,1996;中共中央馬恩列斯著作編譯局研究室編,《五四時期期刊介紹》,北京:人民出版社,1958,等史料索引集及介紹當時文藝期刊有關情況的類似作品爲指導。

③《小説月報》的創刊號出版於1902年11月。當年11、12月號中的翻譯小説都是用文言文寫的。漢語歐化語法現象涉及的主要是白話文而不是文言文,因此在統計數據(即分析歐化成分使用頻率的變化)及舉例方面本研究使用的資料始於1903年第1號。

前兩號的譯文。因此有關20世紀初的資料相對而言比較多,這是因爲如前所述,本研究主要聚焦於以前其他學者基本上有所忽略、但又對考察"現代漢語歐化語法現象的出現是否早於過去有關該現象的研究所推測的那樣"具有很大意義的這一階段。所有這些資料不包括文言文資料,因爲文言文本身是一種極其固定、不容易發生變化(即不容易受到外語影響)的語言,而白話文恰恰相反(尤其是當時尚未規範化的白話文)。每個年份所收集的有關白話文資料數量有所不同,但因爲在觀察歐化成分使用頻率的變化以及統計有關數據方面每個年份的分析資料字數當然必須一樣,所以我們雖然對所有這些資料都進行了儘可能仔細的考察,但在觀察歐化成分使用頻率的變化以及統計有關數據的時候,在所收集的資料中,我們還是以所有年份當中字數最少(即12000個字左右)的年份爲字量指標 。因此,用來作出統計的所有年份有關資料的字數加起來共爲348000個字左右。

就所分析的翻譯作品的原文而言,如第一章中所述,20世紀初翻譯著作中原文爲英文的最多,因此我們所使用的資料中此種翻譯作品也最多,但也有不少原文爲法文或其他歐洲語言的作品。所分析的翻譯資料原文僅限於西方語言(即與嚴格意義上的"歐化"現象有關的語言)不包括日語[1]。

就所分析的翻譯作品的類型而言,本研究使用的主要是小説,但在所收集的有關某個年份翻譯小説資料數量不夠多的情況下,偶爾也使用屬於戲劇類或散文類的作品。

除收集大量翻譯資料外,本研究同時也收集了少量的原創小説,來與翻譯小説中有關歐化成分的語言情況進行比較分析。爲此,我們選擇分析當時最有代表性的五部原創小説的片段。這些資料包括:吴沃堯的《二十年目睹之怪現狀》(1903—1905年間連載於《新小説》上)、李寶嘉的《官場現形記》(1903—1905年間連載於《世界繁華報》上)、魯迅的《阿Q正傳》(1922年出

[1] 日語對漢語語法有可能引起的影響這一問題,可見:沈國威,《現代漢語"歐化語法現象"中的日語因素問題》,《東アジア文化交涉研究》,別册第7,2011年3月,141—150頁。

版）、老舍的《老張的哲學》（1926年）和巴金的《滅亡》（1929年）[①]。與翻譯作品資料相同，原創作品資料中出版在不同年份的每部小説的樣本字量大概也爲12000個字左右，因此用來作出統計的五部作品的字數加起來共爲60000個字左右。在選擇片段時我們採取的是通過電子軟件進行隨機抽樣的方法。

（二）"被"字句使用頻率的增加

1. 已有的研究成果綜述

"被"字句的出現大概可追溯到戰國末期，漢朝的時候基本上已普遍使用。"被"並不是唯一用來引出有標志被動句的介詞，除此之外，還有"見"、"爲"、"叫"（或"教"）、"讓"等介詞，均爲同義詞，有的爲早期的，有的爲晚期的，但到明清時期"被"無疑變成其中最普遍的被動句的標志[②]。儘管如此，漢語中被動句的應用從古代到20世紀初向來不如主動句普遍。換言之，漢語一直以來的一個明顯特點即爲傾向於使用主動形式，而被動形式雖然存在，但其應用僅爲偶然情況，使用頻率遠遠低於西方語言中的被動形式。這一事實到現在尚未改變，但從20世紀初開始——大多數學者認爲主要是1919年之後——漢語中由"被"或類似介詞引出的被動句的使用變得越來越普遍。我們這裏引用賀陽的研究結果來介紹該現象的有關情況。他對明清資料與"五四"以來資料所進行的對比分析的結果顯示，明清小説中"被"字句出現頻率爲4.7例／萬字，而現當代小説中則爲8.9例／萬字，增長將近一倍。此外，除舊白話小説以及現當代小説之外，賀陽的分析資料中還包括部分當代學術著作並發現該種作品中"被"字句的出現頻率高於小説，而且學術譯作中的出現頻率高於原創學術著作 。很明顯，一直以來使用頻率較低的"被"字句卻從20世紀起在與外語的接觸比以前更强的情況下突然開始普遍起來，因此多數學者都認爲這種出現頻率的增加與外語的影響有關[③]。

---

① 如引言已解釋過的，本研究特別注意考察1919年前的階段，因爲大多數學者從未系統地研究過這一段時間，因此本研究的有關資料包括20世紀初（1903—1905年）的兩部著作，即吴沃堯的《二十年目睹之怪現狀》和李寶嘉的《官場現形記》。當然在統計1903—1905年的數據來觀察所分析的歐化成分使用頻率的變化時，我們對兩部著作的有關數據取了平均數。

② 賀陽，《現代漢語歐化語法現象研究》，226頁。

③ 比如：王力，《中國現代語法》，下册，353、354頁；吕叔湘、朱德熙，《語法修辭講話》（1952），《吕叔湘全集》，第4卷，沈陽：遼寧教育出版社，2002，83頁。

上述"被"字句使用頻率的增加主要表現爲兩種情況:1)"被"字句用法上的變化(即"被"字句傳統語義限制的消解);2)有標志被動句的增長及"被"字句贅用現象。

以下我們將會對這兩種現象進行詳細介紹。

(1)"被"字句用法上的變化

前一段落中已提到,漢語裏的"被"字句在漢朝時已普遍起來。然而,從古代到20世紀初"被"這一介詞向來具有某種語義限制,即只能用於表示不幸、不被企望的事情,而且多數時候被用於表示"遭受"、"受損害"或"受支配"的含義。下面是賀陽對此種現象的描述:

> 漢語的"被"字句通常只用於表達消極義,這一語義約束在漢代到五四前長達兩千年左右的時間裏,基本上没有什麽變化。從漢代到唐宋時期,絶大多數"被"字句的語義色彩都是消極義的,對此,王力(1958:430-433)和崔宰榮(2002)都已經予以證明和確認。[①]

如他所説,王力已强調過這一點,其實早在其40年代時編纂的兩部專著中王力已明確指出過"被"字句的這種語義限制:

> 被動句所敘述,若對主語而言,是不如意或不企望的事,如受禍,受欺騙,受損害,或引起不利的結果等等。[②]
>
> 上文説過,"被"字有遭受的意思,因此,被動式所敘述者,對主位而言,必須是不如意或不企望的事。[③]

以下爲王力給出的有關例句[④]:

1. 我們被人欺負了。
2. 老太太被風吹病了。
3. 知是何三被他們打死。

① 賀陽,《現代漢語歐化語法現象研究》,229、230頁。

② 王力,《中國現代語法》,下册,88頁。

③ 王力,《中國語法理論》,128頁。

④ 王力,《中國現代語法》,下册,88頁。

王力和崔宰榮[①]研究的主要是唐宋時期的資料，而賀陽則聚焦於明清時期，來繼續進行此種對“被”以及其他引出被動句的介詞的考察[②]。研究結果顯示，在這一時期的資料中“被”字句大多數時候同樣也用於表示消極意義。更具體地看，表示消極意義的“被”字句佔總量的92.7%，表示中性意義的佔5.6%，而表示積極意義的只佔1.7%。下面是賀陽給出的表示消極意義的“被”字句的有關例句：

1. 當時我父曾被水賊傷生，我母被水賊欺占。(《西遊記》十九回)
2. 這些衙役百姓，一個個被他打得魂飛魄散。(《儒林外史》八回)
3. 我被關了一個多月，悶得慌了。(《二十年目睹之怪現狀》四十三回)

在王力看來，該種語義限制的來源與“被”一詞的詞源有着緊密的關係。在古代漢語裏“被”是動詞，有“遭受”的意思，後來由於某種虚化的過程(或稱語法化過程)則逐漸變成介詞了，在被動句中用來引出施事。雖然如此，介詞“被”在某種程度上一直保留着原來的“消極意義”，因此就産生了上述用法上的約束，即基本上只用於表達不如意的事。吕叔湘、朱德熙等學者對此的看法與王力基本一致：

> 在意義方面，被動式所敘述的行爲對於主語(即被動者)大都是不愉快的：“被他欺負”、“被他騙了”是常見的，“被他寫好”、“被他送來”就不説了。這是因爲“被”字原來的意義是“遭受”，只有對於不愉快的事情我們才説是遭受。[③]

向熹在其1993年的著作中將“被”字句語義限制的消解理解爲虚化過程的結果，此種過程屬於漢語内部演變的歷程，不僅涉及“被”這一動詞，也涉及許多虚化成介詞的其他動詞。在他看來，動詞“被”因其語法化的過程變成介詞後在長達幾百年的時間裏一直保留原來的消極意義，因此基本上只能用於

① 崔宰榮，《唐宋時期被動句的語義色彩》，《語言學論叢》第26輯，北京：商務印書館，2002。

② 其分析資料有：《西遊記》《儒林外史》《紅樓夢》《兒女英雄傳》《二十年目睹之怪現狀》。

③吕叔湘、朱德熙，《語法修辭講話》(1952)，《吕叔湘全集》，第4卷，82、83頁。其他學者的觀點也可見：謝耀基，《現代漢語歐化語法概論》，96頁。

表示不被企望的事,而後來在該種虚化的過程得到進一步發展時介詞“被”也跟着失去了原來的語義限制,開始用於表達中性意義或積極意義的事。换言之,他認爲“被”字句語義限制的消解就是該虚化過程的進一步發展,而與西方語言的影響毫不相關[①]。雖然向熹的看法有一定的道理,但在我們看來似乎不足以解釋“被”字句在20世紀初相對突然的語義限制的消解。當然,如前所述,源語對目標語語法系統的影響再大,一般也不會導致顛覆性或巨大的變化,很難會改變極其穩定的傳統習慣,因此我們在肯定外語語言對漢語語法影響的同時也不排除該過程中同時存在内部因素的可能性。换言之,“被”字句語義限制的消解有可能本來就是屬於漢語自然演變的一種過程(無論它是否與動詞虚化現象有關),但重新激活這種數百年來已經停止的過程、並使其得到新發展的,就應該主要是漢語與没有任何語義限制的西方語言中被動句的接觸。因此,導致“被”字句語義限制的消解很可能同時有兩種因素,一種是内部傾向,一種是外部因素(即西方語言的影響)。也許如果本來没有這種内部的因素,外部的影響就不會起如此大的作用,而正是因爲漢語中原來就有該種内部傾向,外部因素才會有産生較大影響的可能性。比如説,在漢語的歷史上,類似情况還有由單音節詞走向多音節詞的傾向。這種現象的出現既有自然因素(主要與漢語語音系統逐漸簡單化有關),也有外部因素(即漢語與外語的接觸)。賀陽不同意向熹的看法,同時也不提及“内部因素加外部因素”的可能性,而稱“被”字句語義限制的消解這種現象純粹是漢語與西方語言接觸的結果,並爲了進一步地證明這一點,他强調多數其他學者對此都具有同樣的看法:

> 吕叔湘、朱德熙(1951: 83)、趙元任(1968: 587)、丁聲樹(1961: 99)陳建民(1986: 153)以及 Li and Thomson(1981: 99)等也都認爲五四以來“被”字句的這種變化與外語的影響有關。趙元任(1968: 587)還具體分析了這種現象産生的過程,他説“近年來因爲翻譯外文的被動動詞,總把西方語言裏by之類的被動式字眼機械性的翻譯成‘被’字,因此講好處

① 向熹,《簡明漢語史》,下册,501頁:“隨着被動句的發展,‘被’或‘吃’進一步虚化,不再表示對主語不幸或不如意的事實,而僅僅成爲表示被動關係的語法標志。”

或没有處置意義的動詞,也都能加上‘被’字了”。Li and Thomson (1981: 362)還指出,不表示消極義的“被”字句在當代漢語中越來越多,這種變化趨勢顯然是受了印歐語言,尤其是英語的影響。[①]

無論如何,有兩個客觀因素是任何學者都没法否認的: 1)以前漢語中“被”字句一直存在我們所説的語義限制;2)這種語義限制,在西方語言中是不存在的,即被動句可以用來表示不如意、如意或無所謂如意不如意的事情。但到了20世紀初(大多數學者稱是五四運動之後)“被”字在漢語中的傳統語義限制開始消解,用其來表示中性意義或積極意義的情況越來越常見。下面是謝耀基引用的幾個有關例句:[②]

1. 河水被晚霞照得有些微紅……(老舍《駱駝祥子》)

2. 金桂被村裏選成勞動英雄,又選成婦聯會主席,李成又被上級提拔到區上工作。(趙樹理《傳家寶》)

Kubler在巴金小説《家》1931年的原版中也發現一些用來表示中性意義或積極意義的“被”字句。以下分别給出1931年的原版中表示積極意義的“被”字句,以及1957年修訂版中巴金在去掉所有過於歐化的成分時所改爲的主動句[③]:

這些話被周氏和覺心聽得清清楚楚……(1931年版)

周氏和覺心清清楚楚地聽見了這些話……(1957年版)

爲了證明“被”字句語義色彩的變化,賀陽選取了一部分五四運動之後的現代文學和當代文學作品以及一部分當代社科論著,以隨機抽樣的方式對“被”字句語義色彩分布情況進行了考察,並與其曾經分析過的舊白話資料的有關情況進行了比較。考察的數據統計結果顯示,現當代作品中表示消極意

---

① 賀陽,《現代漢語歐化語法現象研究》,236頁。與多數學者的看法不同,程麗麗(《十八世紀中葉到二十世紀末北京話被動式研究》,首都師範大學博士論文,2012年)認爲從18世紀中葉到20世紀90年代北京话中“被”字句的如意、褒義用法,比例没有增加。雖然本研究傾向於同意其他學者的觀點,但其论文无疑值得参考。

② 謝耀基,《現代漢語歐化語法概論》,96、97頁。

③ C. Kubler, *A study of Europeanized Grammar in Modern Written Chinese*, p.94.

義的“被”字句佔總數的53.5%,表示中性意義的佔總數的38.4%,表示積極意義的佔總數的8.1%,由此不難發現,雖然傳統的語義限制尚未完全消解而且“被”字句大多數時候還是用來表示消極意義,但用來表示積極的或中性意義的“被”字句出現頻率大幅上升[①]。如下文將要詳細論述的那樣,本文所采取的資料分析結果同樣顯示,表示積極意義或中性意義的“被”字句使用頻率出現了明顯的增加。如前所述,這種變化與其他語法變化相比,發生得較爲突然,因此這就很可能跟西方語言的影響有關,但與其他學者曾經指出的不同,本研究的結果顯示這種變化的出現應早於1919年,可追溯到20世紀初西方文學著作的翻譯活動已甚爲活躍的時候。

(2) 有標志被動句的增長及“被”字句贅用現象

如前所述,“被”字句從20世紀初開始出現的使用頻率增加的情況,不僅表現爲“被”字句(以及由其他介詞引出的所有被動句)語義限制的消解,而且也表現爲“被”字句贅用現象,即在本可以使用主動句來表示被動意義的情況下卻使用被動句。與西方語言不同,漢語缺乏動詞變位以及嚴格意義上的形態變化,因此其動詞的被動意義不是由動詞本身的詞形變化表達的,而是可以從語境中看出的。在需要强調被動意義時當然可以用“被”字(或其他類似的介詞)來充當被動標記,但作爲孤立語,漢語追求綜合,在某句的被動意義可從語境中看出的情形之下,則傾向於不使用有標志被動句,而使用所謂“無標志被動句”。下面是不同學者爲了説明“無標志被動句”而舉的例子:

1. 你二哥哥的玉丢了。
2. 五兒嚇得哭哭啼啼。(王力)[②]
3. 上身的衣服全打濕了。(Kubler)[③]
4. 摔碎的杯子/修好的電視/扔掉的襪子。(賀陽)[④]

---

① 賀陽,《現代漢語歐化語法現象研究》,234頁。

② 王力,《中國語法理論》,127頁。

③ 該例句Kubler摘自巴金《家》1957年的修訂版,1931年原版中相應的句子中則帶有“被”字:上身的衣服全被打濕了(見:C. Kubler, *A Study of Europeanized Grammar in Modern Written Chinese*, p.95)。

④ 賀陽,《現代漢語歐化語法現象研究》,250頁。

我們這裏先不提供摘自中國近現代文學的例句(有關例句可見下一段落),但順便指出,目前網絡上和口語中像“被嚇得……”/“給嚇得……”或“被丟了”等類似説法經常出現,而根據王力的觀點,這些説法是不需要加上“被”字的 。這進一步地證明了“被”字句使用頻率的不斷增加,而且值得注意的是,這種現象似乎在40年代後——在王力的著作問世之後——仍繼續發展。

此外,主語爲受事語、使用“的”字結構來强調施事語的判斷句,本來也不需要加上“被”字來表示其被動意義:

花瓶是我打破的。(謝耀基)[①]

然而,從20世紀初開始,出現了在以上情況下也使用“被”字句的傾向,即有標志被動句的使用頻率有所增加,開始用在本可以使用無標志被動句的情形下。下面是賀陽提供的幾個有關例句[②]:

1. 現在教育經費都被拉去充作軍事費用用掉了。(巴金《家》,1931年)

2. 宇宙裏的神像全被搗亂了。(馮德英《苦菜花》,1958年)

3. 一個被大家看不上的學生當然不能列在前幾名的。(老舍《牛天賜傳》,1934年)

4. 他們是被關老爺引進文城來的。(老舍《火葬》,1943年)

上述例句均摘自“五四”之後的原創作品。在以下段落中,我們將會講述“五四”前後翻譯作品中的有關情況並與原創作品中的有關情況進行比較來進一步地確定該現象與西方語言影響的關係,同時也將試圖證明其出現早於1919年。

2. 20世紀初西方翻譯小説中的歐化語法成分:“被”字句有關情況

如前所述,過去有關現代漢語歐化語法現象的研究成果顯示,1919年以來由“被”(或類似介詞)引出的被動句使用頻率有所增加。從事這種研究的學者也都在不同程度上分析了該現象的具體表現形式。本研究對使用資料所進行的分析結果進一步地證實了“被”字句使用頻率的增加,此外,也顯示

① 該句子不用説成“花瓶是被我打破的”,可見:謝耀基《現代漢語歐化語法概論》,97頁。

② 賀陽,《現代漢語歐化語法現象研究》,249—251頁。

了翻譯資料中該現象的出現早於原創資料,由此更深一層地證明了其歐化的性質。

翻譯資料中"被"字句的使用頻率早在1905年就出現了明顯的增加(這一年的使用頻率比原創資料中同一年的使用頻率高約三倍),而原創資料中"被"字句使用頻率的增加出現地要晚一些,大概從1922年開始逐漸增加,直到1929年達到最高點 。

就"被"字句在用法上的變化而言,研究的結果顯示出同樣的情況,即在翻譯資料中新用法的最早出現可追溯到1905—1906年,而在原創資料中則要到1926年才開始出現。我們認爲"被"字句的新用法最早出現於翻譯資料中這一事實是極其重要的現象,因爲它似乎能夠進一步證明該現象與西方語言影響的密切關係。

以下兩個表格顯示的是1903—1931年間翻譯作品中有關"被"字句使用頻率的具體情況:

年代	1903	1904	1905	1906	1907	1908	1910	1912	1913	1917
總字數	12000	12000	12000	12000	12000	12000	12000	12000	12000	12000
用例	5	4	15	10	2	5	5	2	4	3
萬字比	4.2	3.3	12.5	8.3	1.7	4.2	4.2	1.7	3.3	2.5

年代	1921	1922	1923	1924	1925	1926	1927	1928	1929	1930	1931
總字數	12000	12000	12000	12000	12000	12000	12000	12000	12000	12000	1200
用例	4	14	10	7	2	6	7	18	7	10	10
萬字比	3.3	11.7	8.3	5.8	1.7	5	5.8	15	5.8	8.3	8.3

以下表格顯示的是1903—1931年間翻譯作品中有關"被"字句用法上變化的具體情況(a式指絶對意義上的消極意義,b式指根據語境可視爲消極意義,c式指中性意義,d式指積極意義):

年代	1903	1904	1905	1906	1907	1908	1910	1912	1913	1917
a式用例	2 (40%)	4 (100%)	6 (40%)	7 (70%)	1 (50%)	2 (40%)	2 (40%)	0 (0%)	3 (75%)	2 (66.7%)
b式用例	3 (60%)	0 (0%)	7 (46.7%)	0 (0%)	1 (50%)	0 (0%)	2 (40%)	2 (100%)	1 (25%)	1 (33.3%)
c式用例	0 (0%)	0 (0%)	2 (13.3%)	1 (10%)	0 (0%)	2 (40%)	1 (20%)	0 (0%)	0 (0%)	0 (0%)
d式用例	0 (0%)	0 (0%)	0 (0%)	2 (20%)	0 (0%)	1 (20%)	0 (0%)	0 (0%)	0 (0%)	0 (0%)
總計	5	4	15	10	2	5	5	2	4	3

年代	1921	1922	1923	1924	1925	1926	1927	1928	1929	1930	1931
a式用例	1 (25%)	8 (57.1%)	6 (60%)	6 (85.7%)	1 (50%)	3 (50%)	5 (71.4%)	9 (50%)	6 (85.7%)	9 (90%)	9 (90%)
b式用例	2 (50%)	4 (28.6%)	0 (0%)	0 (0%)	0 (0%)	0 (0%)	0 (0%)	4 (22.2)	0 (0%)	0 (0%)	0 (0%)
c式用例	1 (25%)	1 (7.1%)	2 (20%)	1 (14.3%)	0 (0%)	3 (50%)	2 (28.6%)	3 (16.7%)	1 (14.3%)	0 (0%)	0 (0%)
d式用例	0 (0%)	1 (7.1%)	2 (20%)	0 (0%)	1 (50%)	0 (0%)	0 (0%)	2 (11.1%)	0 (0%)	1 (10%)	1 (10%)
總計	4	14	10	7	2	6	7	18	7	10	10

下面則是原創作品中有關"被"字句使用頻率以及用法上變化的具體情況(a式指絶對意義上的消極意義,b式指根據語境可視爲消極意義,c式指中性意義,d式指積極意義):

年代	1903—1905	1903—1905	1922	1926	1929
總字數	12000	12000	12000	12000	12000
用例	5	6	9	12	26
萬字比	4.2	5	7.5	10	21.7

年代	1903—1905	1903—1905	1922	1926	1929
a式用例	2 (40%)	1 (16.7%)	7 (77.8%)	3 (25%)	17 (65.4%)
b式用例	3 (60%)	5 (83.3%)	1 (11.1%)	1 (8,3%)	2 (7.7%)
c式用例	0 (0%)	0 (0%)	1 (11.1%)	8 (66,7%)	6 (23%)
d式用例	0 (0%)	0 (0%)	0 (0%)	0 (0%)	1 (3.8%)
總計	5	6	9	12	26

接下來,我們將會提供一些有關"被"字句使用頻率增加的例句。首先需要指出,不管是在翻譯資料中還是原創資料中,多數時候雖然"被"字句表示的是消極意義,但其消極意義不是由句中的動詞本身表達出來的,而是由語境所決定的,也就是説,表達的不是"絶對意義上"的消極之事,比如:

1. 不過先要回去把號衣脱了,不然被上官看見了,不像樣的。(鮑福《毒蛇圈》,《新小説》1905年,第2號,第126頁)[①]

2. 妙兒道:"(……)那個罪犯早晚總要拿到的",顧蘭如搶着問道:"怎麼?還没有拿到嗎?那班警察真是疏忽極了。"瑞福道:"可不是嗎?"妙兒道:"太太,你們可相信,我爹爹自從那天晚上回家之後,從没有專去見過官質問一句。不過當時被那警察長問了幾句就算了。"(鮑福《毒蛇圈》,《新小説》1905年,第1號,第112頁)

3. 她(……)差不多要舉行婚禮了。忽然被我探聽得那少年本是個無賴子弟。(英國韋烈《合歡草》,《小説月報》1910年,第1期,第7頁)[②]

4. "怎麼……父親的屍首葬在草地裏……我怎麼没有知道?"因爲這件事情被坎拿大的同胞知道了要不容你父親睡在地下,所以連你也不知道了。(迦爾威尼《無名氏》,《小説月報》1913年,第3卷,第10號,第26頁)[③]

5. 而且菲基尼愛又已經太大,不能被稱爲"您"了。這是一個障礙,

① 原文:F. Du Boisgobey, "Margot la Balafrée",譯自英文版"In the Serpent's coil"。

② 小説原名和作者不詳。

③ 原文:J. Verne, "Famille sans nom"。

一種相距的力，在他們中間。（佛羅貝爾《簡單的心》，《小説月報》1922年，第13卷，第2號，第16頁）①

6. 説是算被兒子拿去了罷，總還是忽忽不樂；説自己是蟲豸罷，也還是忽忽不樂：他這回才有些感到失敗的苦痛了。（魯迅《阿Q正傳》，1922年）

7. 他回過頭，看見她已經被幾個兵士抬起走了。（巴金《滅亡》，1929年）

以上例句中的"被"都與本身不一定表示消極意義的動詞連用（因此没看上下文的讀者不一定能看出其消極意義）。此外，值得注意的是，按照傳統的習慣，"被"這一介詞並不能與任意動詞連用，而只能與表示"處置性意義"的動詞連用（如：打、殺、撞等動詞），然而，如例句1—5中所示，到了20世紀初"被"字已經開始與不屬於該類的動詞連用，如：看見、聽見、知道、問等動詞。這種新句式的普遍使用雖然不完全違背傳統語法規則，但也可視爲"被"字句歐化現象的初步發展階段。

如前所述，更爲明顯的一種歐化用法是用"被"字句來表示中性意義或積極意義。以下爲在翻譯資料中所觀察到的有關這種用法最有代表性的幾個例句：

1. 此種公墳就由大家公舉了董事經理永遠栽培得花木芬芳。（鮑福《毒蛇圈》，《新月報》1905年，第1號，第113頁）

2. 葛蘭德一個人在那裏唧唧噥噥自言自語卻被站在旁邊的陳家鼐聽了去了，所以也輕輕的問道："麥爾高嗎？爾説的是那一個？姓麥的我認得有六七個呢"。②（鮑福《毒蛇圈》，《新小説》1905年，第2號，第113頁）

3. 在那裏看的人，見我把他們的小孩子丢在地下，就大喊起來彎下腰去拾了泥塊想要扔過來打我，幸虧被紅馬喝住。（司威夫脱《汗漫遊》，

① 原文：G. Flaubert, "Un coeur simple"。

② 該句中的"那一個"意思當然是"哪一個"，但我們這裏還是照抄原文。

《繡像小説》1906年，第66期，第100頁）[①]

4. 我（……）把一件淫襪子掛在竹竿上放在外頭 。被風吹了一夜才晾乾了。（司威夫脱《汗漫遊》，《繡像小説》1906年，第66期，第101頁）

5. 忽然看見了我們的海洋便拼命的囚過來，也是有天幸被他們用力一囚居然抵岸。（司威夫脱《汗漫遊》，《繡像小説》1906年，第66期，第108頁）

6. 但見床上睡着一個美人的背景，身上蓋一條秋香色輕紗單，被我捺住了兩太陽，閉目一想這人便是我的婚妻心上便由不得一震。（英國韋烈《合歡草》，《小説月報》1910年，第1期，第12頁）

7. 我今日的運氣真好，被我選到這一個好地點來停我的飛機。（J. U. Gieiy《紅鴛豔蝶》，《小説月報》1917年，第9卷，第1號，第8頁）[②]

8. 伊向前走，被他的擁抱支持着，他們步子放緩了。（佛羅貝爾《簡單的心》，《小説月報》1922年，第13卷，第1號，第34頁）

9. 伊再不能看出什麽人了，在這被月光染成銀色的洋面上，那個船變成一個漸漸小漸漸小的黑點，終於隱没在晚色的中間去了。（佛羅貝爾《簡單的心》，《小説月報》1922年，第13卷，第2號，第17頁）

10. 有些是像祥柯泰諾一樣的，出身是貴族，卻反抗那些富人底不公平，被窮人底苦惱所感動，而離了他們的世界逃到森林裏面去剪徑，自己造出公平世界來。（維克多·愛甫底眉《緑林好漢包旭》，《小説月報》1923年，第14卷，第2號，第2頁）[③]

11. 大地被裹在青春底芳香裏，吐出温暖的氣息。（維克多·愛甫底眉《緑林好漢包旭》，《小説月報》1923年，第14卷，第2號，第3頁）

12. 他（……）二十二歲被聘爲"Comoedia"劇院舞臺監督（……）。（維克多·愛甫底眉《緑林好漢包旭》，《小説月報》1923年，第14卷，第2號，第5頁）

---

① 原文：J. Swift, "Gulliver's travels"。

② 小説原名和作者不詳，小説月報上寫著的"J. U. Gieiy"似乎爲筆誤，我們估計作者原名爲J. U. Giesy（19世紀末美國作家），但無法證明這一點。

③ 原文：V. Eftimiu, "The Hayducks"。

13. 室内的整理和廚房的事，都由一個老女僕包辦。(莫泊桑《離婚》，《小說月報》1924年，第15卷，第2號，第2頁)①

14. 我知道他入了外交界，人家預料他有個美好的前程。我祝他遇着好時被唤到聖座(教宗)左右。(法郎士《李利特的女兒》，《小說月報》1925年，第16卷，第1號，第5頁)②

15. 曉前，明初開，你的靈魂方在你身中昏睡……猶如湛湛濃霧被旭日慢慢地侵透。(羅曼羅蘭《若望克利司朵夫》，《小說月報》1926年，第17卷，第1號，第2頁)③

16. 玻璃被雨底指頭兒敲動，又當當響了一兩次。(羅曼羅蘭《若望克利司朵夫》，《小說月報》1926年，第14卷，第2號，第9頁)

17. 這初年的光陰在他的腦裏蕩漾著，好似麥浪滚滚，林葉蕭蕭被微風飄動。(羅曼羅蘭《若望克利司朵夫》，《小說月報》1926年，第14卷，第2號，第11頁)

18. 街上一盞油燈迷迷糊糊的照見那些房子都被雪蓋着，那雪紛紛不斷的遠下着。(G. Caldron《小墳屋》，《小說月報》1927年，第18卷，第1號，第1頁)④

19. 從前有一段木頭，躺在一個老木匠的店裏，被一個叫蓋比都的要了，預備做一個木偶。(科洛提《木偶的奇遇》，《小說月報》1927年，第18卷，第3號，第1頁)⑤

20.(……)可是他還是爬着，已經差不多全身被窗裏射出的光所照了。(支魏格《黄昏的故事》，《小說月報》1929年，第20卷，第1號，第177頁)⑥

21.他的請願有時被人笑着容納；有時斷然地或竟嚴厲地被人謝絕

---

① 原文：G. Maupassant, “Un cas de divorce”。

② 原文：A. France, “La fille de Lilitts”。

③ 原文：R. Rolland, “Jean Christophe”。

④ 原文：G. Caldron, “The little stone house”。

⑤ 原文：C. Collodi, “Le avventure di Pinocchio. Storia di un burattino”。

⑥ 原文：S. Zweig, “Untergang eines Herzens”。

了。[①](史帝文生《自殺俱樂部》,《小説月報》1930年,第21卷,第1號,第193頁)[②]

22.“這件事給您猜着了,柏倫德太太”。(巴雷《我們上太太們那兒去嗎》,《小説月報》1931年,第22卷,第1號,第138頁)[③]

上述均爲摘自翻譯資料的例句,接下來提供的則是摘自原創資料的有關“被”字句語義限制消解的例句:

1. 他不放開她的手,她也就没再拒絶而由他握着,握得更緊了一些。(老舍《老張的哲學》,1926年)

2. “你老往死葫蘆裏想,現在維新的事不必認識才有來往! 不管相識不相識,可以被請也可以請人。”(老舍《老張的哲學》,1926年)

3. 龍鳳呆呆的看着水仙花,被風吹的那些花瓣一片一片的顫動,射散着清香。(老舍《老張的哲學》,1926年)

4. 這種妙法被老張學來,於是遇萬不得已之際,也請朋友到家裏吃茶。(老舍《老張的哲學》,1926年)

5. 然後左角上説:“我們舉南飛生!”右角上“……龍樹古!”以次:“張明德”“孫占元”“孫定”“李複才”,大概帶有埋伏的全被舉爲起草委員。主席聽下面喊一聲,他説一聲“通過”。被舉的人們,全向着大衆笑了笑。(老舍《老張的哲學》,1926年)

6. 王秉鈞氣得不做聲,等到杜大心底背影被門關住了以後,他才指着門説:“這是一個反革命! 反革命!”(巴金《滅亡》,1929年)

7. 而他寫的歌詞又爲她所愛讀,愛唱。(巴金《滅亡》,1929年)

8. 他既然同情那個人,那麼他自然希望那人底愛會被李静淑接受,希望他們兩人能過着幸福的愛情生活。(巴金《滅亡》,1929年)

9. 房東母女又向我説起姑娘底種種好處。我從她們底口氣中知道她們很喜歡她。而且也就不知不覺地被房東母女底談話感動了。(巴金

① 該例包含兩個被字句,第一個表示積極意義,第二個表示消極意義。

② 原文:R. L. Stevenson, “The suicide club”。

③ 原文:J. M. Barrie, “Shall we join the ladies”。

《滅亡》,1929 年)

10. 我起先推口説,我是外國人,夾在她們中間會使她們不方便,又故意找了些不大近情理的托辭,但終於半推半就地被房東女兒拉下去了。(巴金《滅亡》,1929 年)

如前一段落中所述,"被"字句使用頻率的增加,不僅表現爲"被"字句語義限制的消解,而且也表現爲被動句有標志形式的增加。這種句式中有時具有强調"被動"之意的實際作用,但有時"被"字的應用也可被理解爲贅用。本文所分析的翻譯資料中這種贅用的情況也較爲常見,比如:

1. 農夫道(……)不是跌傷,是被洋槍打傷的。(法國某《宜春苑》,《新小説》,1903 年,第 6 號,第 2 頁)①

2. 這些緑林好漢,被絞殺的是很多的,在克蘭伏亞,在塔哥雅希底(……)當一班顫怯怯的市衆前面。(維克多愛甫底眉《緑林好漢包旭》,《小説月報》,1923 年,第 14 卷,第 2 號,第 2 頁)

3. 起初我想是(......)伐木的人,可是決不是的,只見被分開了的草的動摇,卻不見人樣的姿態。(《愛犬故事》,《小説月報》,1928 年,第 19 卷,第 1 號,第 52 頁)

4. 近年來(……)他時常認自己是被拋棄的病人。(《黄昏的故事》,《小説月報》,1929 年,第 20 卷,第 1 號,第 170 頁)

下面則是摘自原創資料的有標志被動句使用增加的例句:

1. 第一次是他生下來的第三天,由收生婆把那時候無知無識的他,像小老鼠似的在銅盆裏洗的。(老舍《老張的哲學》,1926 年)

2. 耶穌教是由替天行道的牧師們,不遠萬裏而傳到只信魔鬼不曉得天國的中華。(老舍《老張的哲學》,1926 年)

由此可見,原創資料中"被"字句增加的表現方式與翻譯資料相類似,即一方面表現爲"被"字與本身非絶對意義上的消極或者其消極意義由語境決

① 小説原名和作者不詳。

定的動詞、甚至與表示中性意義和積極意義的動詞進行連用，而另一方面則表現爲其有標志形式的增長，但無論其表現方式如何，“被”字句使用頻率有所增加的情況在原創資料中的出現要晚於在翻譯資料中的出現。

3. 小結

本研究所使用的兩種不同資料，即翻譯資料和原創資料，在所觀察的近三十年時間裏都出現了“被”字句使用頻率增加的情況，而在翻譯資料中，該現象的出現要早於在原創資料中。具體來看，在翻譯資料中這種情況于20世紀初已經开始出現(雖然只涉及個别年份)，而在原創資料中則需要等到1922年這種情況才開始出現。

同樣，兩種資料中都出現了“被”字句在用法上的變化，即“被”字句因傳統語義限制的消解而開始用於表示中性意義或積極意義。在翻譯資料中，該種用法開始擴大的出現也早於原創資料。

我們認爲兩種資料“被”字句的這種使用頻率增加的情況以及用法上的變化出現時間之間的差别，似乎能夠證實該現象與翻譯活動、與西方語言影響的直接關係，因此也能夠進一步地證明其歐化性質。此外，研究結果也顯示，“被”字句的歐化過程不是(如大多數學者曾指出過)1919年之後才開始發生的，而是早在1919年之前就開始發生的。

(三)“着”使用頻率的增加

1. 已有的研究成果綜述

漢語的動態助詞“了”、“着”、“過”原來均爲動詞，後來逐漸語法化而成爲助詞，宋元時代漢語的動態體系基本上已接近完全形成。但到了20世紀初，很可能是由於西方語言的影響，又有了新的發展並得到了進一步的完善。從事研究漢語歐化語法的大多數學者在其著作中都曾提到過西方語言影響下的動態助詞的普遍使用現象①。他們雖然强調使用頻率增加最大的是助詞“着”，但一般還是使用將三個助詞都包括在内的“動態助詞增加”這一種

---

① 謝耀基和賀陽雖然承認“着”爲發展最大的助詞，但明確提到所有動態助詞1919年以後的歐化現象(可見：謝耀基，《現代漢語歐化語法概論》，48頁；賀陽，《現代漢語歐化語法現象研究》，181、182頁)。

泛指性的説法 。然而,本研究的結果顯示,使用頻率惟一有明顯增加的助詞只有“着”,因此在本文中我們將只論述有關這一助詞的情況。接下來,我們將會對過去有關該現象的已有研究成果做一個總結性的介紹,而在下一段落中我們則會詳細介紹本研究的成果。

在西方語言中,基本上任何動詞都可以通過動詞變位來表示進行貌,而漢語則不是這樣。王力是最早强調動態助詞“着”按照傳統習慣的用法僅限於跟在表示具體意義或者動作性頗重的動詞之後的學者。據他説,這可能是因爲“着”原來有“附着”的意思,表示“物質性”意義,而1919年之後由於西方語言的影響這一約束則逐漸消解,“着”開始緊跟在表示抽象意義的動詞(如“希望”、“證明”、“代表”、“包含”等動詞),下面是王力舉的歐化用法的典型例子[①]:

> 關於這問題,還是留待以後再討論,同時希望着别人能有新穎的意見發表。(《今日評論》四卷十二期)

吕叔湘和朱德熙在他們1952年的著作中較爲詳細地列出了所有在理論上不能帶“着”的動詞種類[②]。第一種就是本身不能表示持續性、指動作一下子就完成的動詞,如“進”和“表現”。第二種則是本身已經含有“持續”之意、因此就不必帶上“着”來進一步强調持續性的動詞,如“愛”、“恨”、“擁護”、“需要”等動詞。最後一種不可帶有“着”的動詞爲表示某種動作已經結束之意的動詞,如“説明”等内部結構爲“動詞加上結果補語”的雙音節動詞。兩位學者的看法能夠證明,直到20世紀50年代“着”的傳統用法約束最起碼在理論上還是存在的。但實際上,從20世紀初開始,該約束已經逐漸消解,從而使得“着”開始緊跟在本來按照傳統語法規則不可帶“着”的動詞之後。這種現象主要關涉表示抽象意義和表示持續性的動詞。以下爲賀陽給出的一些例句[③]:

---

① 王力,《中國語法理論》,466頁。

② 吕叔湘、朱德熙,《語法修辭講話》(1952年),《吕叔湘全集》,第4卷,77、78頁。

③ 賀陽,《現代漢語歐化語法現象研究》,182、183頁。

1. 這就證明着我的工作的切實。(魯迅《傷逝》,1925年)

2. 然而因爲這個新興階段的自身内包含着若干矛盾,所以它的鬥爭態度不能堅決。(丙申(即茅盾)《"五四"運動的檢討》,《文學導報》1931年第1卷第2期)

3. 社會主義思想道德集中體現着精神文明建設的性質和方向。(《中共中央關於加强社會主義精神文明建設若幹重要問題的決議》,1996年)

4. 至於文學批評界本身,一直就存在着恪盡職守的批評家。(《人民日報》2000年6月17日)

關於"着"與表示持續性的動詞的連用,Gunn部分地持不同的看法。與其他研究漢語歐化語法的學者不同,這位美國學者在其專著中將"着"的這種用法歸納爲"Indigenous innovations"而不將其歸納爲"Euro-japanese innovations"。他一方面指出,中文譯文著作中"着"新用法的出現不一定與原文中動詞的進行貌有相應關係,另一方面則承認"着"新用法的出現與西方語言語法的影響有關係的可能性,並引用林語堂所舉的中文作品中明顯模仿英語裏表示進行貌的後綴—ing的"着"的例子。因此我們認爲,他傾向於將這種現象理解爲"着"的一種自然演變的同時,也不完全排除其歐化的可能性[①]。

其實,如同所有歐化現象,動態助詞"着"使用頻率的增加也可能本來就是屬於漢語自然演變的一種過程,但後來漢語通過翻譯活動而與西方語言發生的間接接觸在很大程度上也促使並加快了這種過程。Kubler在其專著中引用王力與之相關的看法,並指出"着"的普遍使用有可能同時跟兩個因素有關,一個是外部的,一個是内部的。外部因素當然就是外語的影響,而内部因素就是中國南方方言對普通話的影響(某些南方方言中原來没有"着"這一助詞,母語爲這種方言的人説普通話時不完全掌握"着"的用法,經常用錯,而他們的這種錯用後來對普通話中"着"的用法産生了影響,使其普遍使用,隨便緊跟在任何動詞之後)[②]。

"着"用法的擴大較爲明顯的例子是其與動詞"有"的連用。"有"表示的是

① E. Gunn, *Rewriting Chinese:: Style and Innovation in Twentieth-Century Chinese Prose*, p.194.

② C. Kubler, *A Study of Europeanized Grammar in Modern Written Chinese,* p.64.

“擁有”的意思,“擁有”這一概念本身動作性甚輕,而且含有某種持續性之意,理論上不應屬於可帶“着”的動詞,因此多數學者不贊同“有着”這種不符合傳統語法習慣的説法。最早批評“着”這種用法的學者爲王力,他將其看作一種“變質的歐化”[①]。這是因爲西方語言中表示“有”的動詞也很少用其進行貌,而漢語中“着”的這種濫用很可能是“着”因受了很大程度的歐化過程而出現用法上的擴大之後,自身的進一步發展。Kubler在强調“有着”已普遍使用時引用了鄒國統[②]對這種現象的解釋,説促使“有着”這種説法出現的主要因素之一很可能就是漢語單詞本身由單音節詞變成雙音節詞的傾向。無論“有着”的出現是否也有内部因素,即是否也取決於漢語的自然演變,但在漢語書面語中它已經是很普遍的説法了(後面主要跟抽象意義的賓語,如“有着……意義”、“有着……重要性”等),基本上所有研究漢語歐化語法現象的學者都認爲其普遍使用在某種程度上與外語的影響有關。

前面我們論述的是,動態助詞“着”由於其傳統用法約束之消解,而被普遍使用的情況。如前所述,該現象已導致“着”在用法上的擴大以及不完全符合傳統語法規則、引起多數中國語言學家批評的用法的出現。但導致“着”使用頻率增加的,不僅是其傳統用法約束的部分消解,除此之外,還有助詞“着”本身開始出現的某種“贅用”現象,即在本來没有必要使用“着”的情況下而用之的現象。在該情況下,雖然句中加上“着”按照傳統語法規則不可視爲錯用,但如果去掉“着”,句中的意思并不會改變,句子本來已經很清楚,不用加上“着”。下面是王力提供的有關這種贅用的例句[③]:

胖大漢和巡查都斜了眼研究着老媽子的鉤刀般的鞋尖。(魯迅《示衆》)

上面我們綜合論述了其他學者有關“着”歐化現象的研究成果,在本研究所使用的資料中我們也發現了上面所説的助詞“着”使用頻率的增加。兩種不同資料中有關該過程的具體時間、表現方式和特點請看下一段落。

---

① 王力,《中國語法理論》,466頁。

② 可見:C. Kubler, *A Study of Europeanized Grammar in Modern Written Chinese*, p.68,以及鄒國統,《談談“有着”》,《中國語文》1956年48期,39頁。

③ 王力,《中國語法理論》,356頁。

2. 20世紀初西方翻譯小説中的歐化語法成分:動態助詞"着"有關情況

翻譯資料中動態助詞"着"從1907年起已開始普遍使用,總的來説,除個别使用頻率降低的年份之外,直到20年代末其用法不斷有所增加。以下兩個表格顯示的是1903—1931年間翻譯作品中有關動態助詞"着"使用頻率的具體情況:

年代	1903	1904	1905	1906	1907	1908	1910	1912	1913	1917
總字數	12000	12000	12000	12000	12000	12000	12000	12000	12000	12000
用例	58	28	37	27	104	109	29	103	1	36
萬字比	48.3	23.3	30.8	22.5	86.7	90.8	24.1	85.8	0.8	30

年代	1921	1922	1923	1924	1925	1926	1927	1928	1929	1930	1931
總字數	12000	12000	12000	12000	12000	12000	12000	12000	12000	12000	12000
用例	26	85	64	90	49	81	84	165	263	41	103
萬字比	21.7	70.8	53.3	75	40.8	67.5	70	137.5	219.2	34.2	85.8

如前一段落中所述,同時有兩個原因導致了該使用頻率的增加,其一爲"着"傳統語義限制的消解,其二爲它的一種贅用現象。下面給出的是翻譯資料中所觀察到的有關"着"使用範圍擴大的幾個最有代表性的例子:

1. 花丹聽了看着福來西説道:"將軍休怎地説將軍不來農難道一個人倒獨樂了不成有着將軍這般威武這樣命名同來講講閑話,怎麽不歡迎呢?"(《黑蛇奇談》,《小説林》1907年,第1期,第17頁)

2. 如今富倫克既愛着羅敖一種疫情到死不肯消滅的樣子。(《黑蛇奇談》,《小説林》1908年,第8期,第53頁)

3. 羅敖只説道他已復了本性需要着我哩。(《黑蛇奇談》,《小説林》1908年,第8期,第60頁)

4. 福夫人道:"你成就了便好,農懼着你或者失敗。"(《黑蛇奇談》,《小説林》1908年,第8期,第90頁)

5. 高利佳醫生又道:"都總是自己人,請你陪伴着新郎少坐。"(英國韋烈《合歡草》,《小説月報》1910年,第1期,第11頁)

6. 但是這老人也確説他已經説過這句話了。又繼續着告訴後來的事

情。他以後曾得到了包倫同這個女郎結定婚的一封信。(《忍心》,《小説月報》1921年,第12卷,第1號,第32頁)

7. 因爲要用一種動人的形式去教導孩子們,他送給他們一副雕刻的地理圖。圖中的雕琢表示着世界各地的風物。(佛羅貝爾《簡單的心》,《小説月報》1922年,第13卷,第1號,第36頁)

8."祝你健康,我的勇敢的他索! 但願你的教父長壽可以享受着你的福氣!"(《教父》,《小説月報》1923年,第14卷,第2號,第3頁)

9. 時間之早遲啦、熱啦、冷啦,她一切都關心着。(《伯乃特保姆》,《小説月報》1925年,第16卷,第1號,第3頁)

10. 他放聲大哭。他的母親輕輕地撫摸他。於是苦惱底鋭氣輕減了些。但是他繼續哭着;苦惱還盤踞在他身上。(羅曼羅蘭《若望克利朵夫》,《小説月報》1926年,第17卷,第1號,第9頁)

11. 無處不是迷途。又有些容貌、姿勢、連動、聲音,在他周圍永久地旋環着!(羅曼羅蘭《若望克利朵夫》,《小説月報》1926年,第17卷,第1號,第9頁)

12. 在人相對坐着,默無一聲,互相看望的時候,就是最親近的臉龐在黑暗裏也要覺得成爲老而疏遠的了,並且彷彿從來没有親密的知道過這臉龐,而現在過了若干年後才遠遠裏看見着似的。你既説你不願意静默着,因爲聽那時鐘把時間分切成好幾千塊小碎片是頂難受的,而且在静默裏的呼吸也要變成爲洪大的聲音,和病人的呼吸一般。那末,讓我給你講點什麽,好不好?(支魏格《黄昏的故事》,《小説月報》1929年,第20卷,第1號,第167頁)

13. 在現代的娱樂中還缺少着一種便利即離去這舞臺的一個適當而容易的方法。(史帝文生《自殺俱樂部》,《小説月報》1930年,第21卷,第1號,第198頁)

14. 要是有一位座上客自殺了,小非就一定只會招呼她把屍體搬開,而同時他仍然繼續着獻水果的。(《我們上太太們那兒去嗎》,《小説月報》1931年,第22卷,第1號,第131頁)

15. 他們想着他們的妻,在一陣情愛的激射中想念着她們,雖説他們

結婚已二十年。(《半天玩兒》,《小說月報》1931年,第22卷,第1號,第151頁)

在上述例句中大多數時候"着"緊跟在本身就表示持續性的動詞之後,如"繼續"、"需要"、"缺少"等(見例句3、6、10、13和14)。而且,有時緊跟在本身也含有某種持續性之意的表示心理狀態的動詞,如"愛"(見例句2)、"關心"(例句9)、"懼"(例句4)等。此外,在翻譯資料中"着"有時也緊跟在表達抽象意義、動作性甚輕的動詞後,如例句7中的動詞"表示"。

如前所述,動詞"有"動作性也很輕,按照傳統的語法習慣不可帶"着",但從20世紀開始雖然不少語言學家反對"有着"這一說法,但它卻越來越普遍。例句1就是我們在分析資料中所觀察到的"着"最早用在動詞"有"之後的典型的例子之一。

例句12中動態助詞"着"出現三次,其中第二個用例"看見着"爲錯用,因爲結果補語(如"看見"裏的"見")表示的是完成的動作,因此"着"不能用在其後。這本身也能證明當時它的使用頻率已經開始有很大的增加,因此錯用以及贅用的情況較爲頻繁。

除了上述"着"緊跟在本來不能與其連用的動詞後的例子之外,在翻譯資料中也出現過很多"着"贅用的情況,即"着"緊跟在本來可以與其連用的動詞之後但其在句中的出現是多餘的情況。類似的情形,可見例句11(句中本來已經有表示持續性的"永久地"一狀語,不必加上"着")、15和12[1]。

在本研究所使用的原創資料中,雖然"着"同樣也出現了較爲明顯的使用頻率的增加,但該增加大概始於1926年,遠遠晚於翻譯資料中這種增加的開始。

以下表格顯示的是原創作品中有關動態助詞"着"使用頻率的具體情況:

① 這裏指的是例句12中"着"第三個用例。

年代	1903—1905	1903—1905	1922	1926	1929
總字數	12000	12000	12000	12000	12000
用例	35	63	24	99	69
萬字比	29.2	52.5	20	82.5	57.5

以下爲摘自原創資料的幾個例句：

1. 她固然不能把他底靈魂了解透徹，但這一些日子的觀察使她知道他有着一個高貴的心靈，一顆黄金似的心。(巴金《滅亡》，1929年)

2. 杜大心所夢想着永久的休息果然到了，而且就在幾天以後。(巴金《滅亡》，1929年)

3. 對於那般親眼見着這樣的慘劇而不動心，照常過着奢侈生活的人，我是不能愛的。(巴金《滅亡》，1929年)

由此可見，原創資料中"着"也用在本來不能與其連用的動詞之後，如本身或在某種特殊語境中表示持續性的動詞(例句2和3)和動作性較低的動詞之後(例句1)。

3. 小結

在本研究所使用的兩種不同的資料中，"着"使用頻率都有所增加①。在翻譯資料中這種現象的出現要早於原創資料，而且早於1919年。"着"大概早在1907年就開始普遍起來，後來，除個别使用頻率突然降低的年份之外，一直在保持增長，直到1931年。而原創資料中則必須等到1926年"着"才開始出現使用頻率的明顯增加，我們認爲兩種資料中"着"的這種使用頻率增加的出現時間之間的差别能夠證明該現象很可能是由西方語言的影響所導致的，因此也有助於我們更爲確定地將該現象斷定爲"歐化語法"現象。

結語

通過對以前尚未被系統研究過的20世紀初西方翻譯小説與同一時期的

① 這裏説的"使用頻率的增加"當然也包括前面所論述的用法上的變化。

中國原創小説進行歷時的對比分析,本文試圖以“被”字句使用頻率的增加以及動態助詞“着”使用頻率的增加爲例考察現代漢語“歐化語法”現象的出現和發展。研究的結果顯示兩種歐化語法現象皆起源於翻譯領域,而且它們的出現都早於1919年。

由於時間和眼界所限,本文無法爲歐化語法現象提供全面的介紹,但希望能爲它以及間接語言接觸這兩個仍有待被深入研究的領域提供一些新的見解以供參考。

文獻天地

北京大學圖書館藏中國典籍暨傳統文化研究著作外譯本展覽説明

党寶海
張紅揚
鄒新明

翻譯是不同文化間深入溝通的橋樑，而翻譯家則是橋樑最重要的設計師和建築師。

傳統典籍作爲中國悠久文化的載體，很早就引起周邊國家的濃厚興趣。在古代的漢字文化圈，如朝鮮、日本、越南等國，知識精英能夠直接閱讀漢籍，並用漢文寫作。在這些國家創制了本國文字之後，才逐漸開始翻譯漢籍。在元代，有少量中國史學和醫學書籍在蒙古伊利汗國譯爲波斯文。

隨着16、17世紀大航海時代的來臨，來中國的歐洲傳教士日漸增多。他們成爲漢籍外譯的重要承擔者，漢籍和中國傳統文化的傳播從此進入了一個新的階段。

由於館藏書籍和展覽場地的限制，我們在這裏只能大略按中國傳統四部分類的方法，兼顧文學類著作的完整性，介紹並展示主要外文語種的古籍譯著和一部分中國現當代學者傳統文化研究著作的外譯本。

党寶海單位：北京大學歷史學系
張紅揚、鄒新明單位：北京大學圖書館

一

作爲中國傳統價值倫理的核心載體，儒家經典很早就受到外國譯者的重視。早期傳教士翻譯了十三經中的絶大部分，篳路藍縷，功不可没。

早在16世紀末，意大利耶穌會士羅明堅(Michele Ruggieri，1543—1607)就用拉丁文翻譯了《大學》等著作。其後，意大利傳教士利瑪竇(Matteo Ricci，1552—1610)翻譯了《四書》(未出版)，意大利殷鐸澤(Prosper Intorcetta，1625—1696)和葡萄牙郭納爵(Ignatius da Costa，1599—1666)合譯《大學》《論語》，比利時衛方濟(Franciscus Noël，1651—1729)將《四書》《孝經》《小學》譯爲拉丁文，法國宋君榮(Antoine Gaubil，1689—1759)、馬若瑟(Joseph-Henry H. de Prémare，1666—1735)分别翻譯《尚書》，雷孝思(Jean-Baptiste Régis，1663—1738)譯《易經》，孫璋(Alexandre de La Charme，1695—1767)譯《詩經》，等等。這些儒家經典的譯本，無疑爲當時歐洲的思想界認識東方打開了一扇窗子。更爲重要的是，爲後來歐洲思想啓蒙運動的引領者，如伏爾泰(Voltaire，1694—1778)等人提供了寶貴的思想資源。

19世紀下半葉到20世紀初，中國經部典籍迎來了外譯的黄金時代。除傳教士之外，職業漢學家在漢籍譯介方面日益發揮重要作用。英國的理雅各(James Legge，1815—1897)、法國的顧賽芬(Séraphin Couvreur，1835—1919)是這一時期的代表人物。理雅各翻譯了《四書》《尚書》《竹書紀年》《詩經》《春秋》《左傳》等，以上九種譯著合爲五卷本的《中國經典》(*The Chinese Classics*)刊行。此外，理雅各還譯出《易經》《禮記》《孝經》等。顧賽芬翻譯了《四書》《詩經》《尚書》《儀禮》《禮記》《左傳》等。與這兩位傑出的翻譯家同時，歐洲的其他翻譯家也取得了令人矚目的成就。如法國畢歐(Edouard C. Biot，1774—1862)翻譯了《周禮》，俄國柏百福(П. С. Попов，1842—1913)翻譯了《論語》《孟子》，等等。

在20世紀，儘管一些重要的儒家經典已經有了歐洲主要文字的譯本，仍不斷有新的譯作出現，反映出中國經典受到經久不衰的關注。英國韋利(Arthur Waley，1889—1966)譯《詩經》、《論語》；德國衛禮賢(Richard Wilhelm，1873—1930)譯《論語》《大學》《孟子》《禮記》《孔子家語》；美國華兹生(Burton

Watson)、加拿大安樂哲(Roger T. Ames)分别譯《論語》;日本倉石武四郎等譯《四書》、竹内照夫譯《禮記》《左傳》等。

由於儒家經典涵蓋深廣,其譯本的影響力絶不局限於哲學、倫理學、政治學等領域,富於哲學思辨和東方神秘主義精神的《易經》、古詩集《詩經》等著作在20世紀的影響歷久彌新。古老的《易經》曾令17世紀德國數學家、哲學家萊布尼兹(Gottfried W. Leibniz,1646—1716)着迷,在20世紀仍以其深邃的哲理吸引着樂於思考的人們——衛禮賢翻譯的《易經》長銷不衰,對心理學家榮格(Carl G. Jung,1875—1961)、諾貝爾文學獎得主黑塞(Hermann Hesse,1877—1962)等人均有一定影響。美國詩人龐德(Ezra Pound,1885—1972)翻譯的《詩經》直接推動了美國"意象派"詩歌的發展。

二

子部書籍中的名著很早就引起外國翻譯家的關注。

在18世紀,法國耶穌會士錢德明(Jean-Joseph-Marie Amiot,1718—1793)翻譯了《孫子兵法》《六韜》等。

19世紀後期開始,大量子部經典被譯爲外文,英國理雅各譯《老子》《莊子》,法國戴遂良(Léon Wieger)譯《老子》《莊子》等。德國衛禮賢系統翻譯道家典籍,包括《老子》《莊子》《列子》,此外,他還翻譯了《吕氏春秋》。德國弗爾克(Alfred Forke)翻譯了《論衡》《墨子》,洪濤生(Vincenz Hundhausen)譯《莊子》;法國列維(Jean Lévi)、荷蘭戴聞達(Jan J. L. Duyvendak)、俄國嵇遼拉(Леонард С. Переломов)分别譯《商君書》;英國翟林奈(Lionel Giles)譯《孫子兵法》,韋利譯《老子》;美國德效騫(Homer Dubs)譯《荀子》,華兹生譯《墨子》《荀子》《韓非子》《莊子》,日本金谷治、倉石武四郎、小川環樹等人譯《老子》《莊子》《列子》《吴子》,竹内照夫譯《韓非子》,等等。

另外,新發現的重要古籍被很快譯爲外文,如長沙馬王堆帛書本和郭店楚簡本《道德經》均由美國韓禄伯(Robert Henricks)譯爲英文。

佛教和道教典籍多有外譯本,如印度高僧龍樹《大智度論》鳩摩羅什漢譯本的法譯本、《妙法蓮華經》的英譯本,唐代僧人慧然《臨濟録》的法、日譯本等。北宋僧人克勤《碧巖録》有德、日譯本,法國漢學家沙畹(Édouard

Chavannes, 1865—1918)有《三藏經節譯》,德國衛禮賢譯《太乙金華宗旨》、法國康得謨(Max Kaltenmark)和日本澤田瑞穗分別譯《列仙傳》,荷蘭施舟人(Kristofer M. Schipper)譯《漢武帝内傳》《莊子》《老子》,並長期主持歐洲《道藏》翻譯項目。

歸入子部的古代科學典籍被譯成外文的也頗爲可觀。1856年,法國漢學家儒蓮(Stanislas Julien, 1797—1873)翻譯了清代陶瓷專著《景德鎮陶録》和《本草綱目》的若干篇章。20世紀50—70年代,以藪内清爲首的日本學者翻譯了中國古代科技名著《天工開物》《夢溪筆談》等。

三

歷史、地理類著作按中國傳統的書籍分類,屬於"史"部。相對于儒家經典,史部著作的翻譯只是略晚而已。

18世紀,法國耶穌會士馮秉正(Joseph F. de Moyria de Mailla, 1669—1748)翻譯《通鑒綱目》、《續通鑒綱目》,編成十二卷本的《中國通史》(*Histoire générale de la Chine, ou, Annales de cet empire*)。

19世紀,作爲中外關係史重要史料的古代行記、僧傳得到譯者的青睞,《佛國記》《大唐西域記》《大唐西域求法高僧傳》《大慈恩寺三藏法師傳》《南海寄歸内法傳》《長春真人西遊記》等,譯成英、法、俄等國文字,譯文多出自當時的漢學大家,如法國雷慕沙(Jean-Pierre Abel-Rémusat, 1788—1832)、儒蓮、沙畹、英國理雅各、俄國巴拉第(Петр И. Кафароф, 1817—1878)等。有些著作的譯本還不止一種。

20世紀初,沙畹翻譯了《史記》全書,生前只出版了約三分之一,譯筆精練準確,在學界享有佳譽。此外,有德國夏德(Friedrich Hirth)和美國柔克義(William W. Rockhill)譯《諸番志》,法國伯希和(Paul Pelliot)譯《真臘風土記》,英國韋利譯《長春真人西遊記》,等等。

自20世紀中葉,有更多史地典籍被譯爲外文,包括美國華兹生選譯《史記》《漢書》,美國倪豪士(William H. Nienhauser, Jr.)、蘇聯越特金(P. B. Вяткин)等譯《史記》,美國德效騫譯《漢書》,戴仁柱(Richard Davis)譯《新五代史》,日本增田涉、小川環樹等譯《史記》,小竹武夫譯《漢書》,島田正郎譯《遼史》,内田吟風等人編譯《中國正史西域伝の譯註》《騎馬民族史:正史北狄伝》,英國詹納爾(W. J. F. Jenner)和日本入矢義高分别譯《洛陽伽藍記》,入矢義高和梅原郁譯《東京夢華録》,森鹿三等譯《水經注》,梅原郁譯《宋名臣言行録》,等等。

四

歐洲對中國文學作品的譯介,雖然晚於儒家經典的翻譯,但文學以其優雅的辭藻、生動的情節,容易獲得更廣泛的讀者群。

在18世紀,元雜劇《趙氏孤兒》被傳教士馬若瑟譯介到法國,思想家伏爾泰將其改寫爲《中國孤兒》(*L'orphelin de la Chine*)劇本並公演,引起轟動,劇中所體現的中國文化精神對當時法國文化界有一定影響。令人意想不到的是,小説《好逑傳》成爲18世紀最受譯者重視的中國文學作品,有法、英、德文多個譯本。

19世紀,德國嘎伯冷兹(Hans C. von der Gabelentz,1807—1874)翻譯了《金瓶梅》,未公開出版。法國巴贊(Antoine P. Bazin,1799—1862)譯元代南戲名作《琵琶記》等。通俗小説《玉嬌梨》有多個譯本,以法國儒蓮的翻譯最佳。

20世紀上半葉,德國的庫恩(Franz Kuhn,1884—1961)譯出大量中國古代小説,包括《今古奇觀》《玉蜻蜓》《二度梅》《三國演義》《水滸傳》《金瓶梅》《紅樓夢》等。中國的很多作品通過庫恩的譯筆介紹到更多歐洲國家——不少中國小説的外文譯本是從庫恩的德譯本轉譯的。此外,德國葛禄博(Wilhelm Grube)等譯《封神演義》,德國祁拔兄弟(Otto Kibat、Artur Kibat)和英國埃傑頓(Clement Egerton)分别譯《金瓶梅》,英國翟理斯(Herbert A. Giles)譯《聊齋志異》,韋利譯《西遊記》,諾貝爾文學獎得主、美國作家賽珍珠(Pearl S. Buck)譯《水滸傳》,日本松枝茂夫譯《紅樓夢》,等等。

20世紀下半葉,在中國古代小説的翻譯方面,出現了大量優秀譯著。《紅樓夢》有英國霍克思(David Hawkes)、閔福德(John Minford)英譯本、日本伊藤漱平、井波陵一日譯本,李治華、李雅歌(Jacqueline Alezais)法譯本,史華慈(Rainer Schwarz)、吴漠汀(Martin Woesler)德譯本,帕納秀克(B. A. Панасюк)俄譯本,崔溶澈、高旼喜韓譯本,《金瓶梅》有雷危安(André Lévy)法譯本,小野忍、村上知行、土屋英明多個日譯本,馬努辛(В. С. Манухин)俄譯本,芮效衛(David T. Roy)英譯本等。《西遊記》有雷危安法譯本,余國藩、詹納爾兩個英譯本,松枝茂夫、太田辰夫、中野美代子等多個日譯本,羅高壽(А. П. Рогачёв)俄譯本。《聊齋志異》有立間祥介、增田涉等多個日譯本,吴德明(Yves Hervouet)法譯本,阿理克(В. М. Алексеев)俄譯本。《三國演義》的日譯本頗多,立間祥介、小川環樹、金田純一郎的譯本較佳,俄譯本則出自漢學家帕納秀克。《水滸傳》有譚霞客(Jacques Dars)法譯本,松枝茂夫、村上知行、駒田信二等多個日譯本、羅高壽俄譯本。此外,還應提及松枝茂夫選譯的《三言二拍》,德國鮑吾剛(Wolfgang Bauer)、傅海博(Herbert Franke)主持編譯的中國古代短篇小説選《金匱》,等等。

在戲劇的翻譯方面,德國洪濤生有突出貢獻,他翻譯了元代雜劇經典《西廂記》、南戲代表作《琵琶記》、明代湯顯祖《牡丹亭》等。在日本,有青木正兒、吉川幸次郎研究並翻譯元雜劇。

中國古代的詩詞和散文以其特有的藝術魅力吸引着外國漢學家們。

在漢詩外譯的初期,最受重視的是位列儒家經典的詩歌集《詩經》。18—19世紀,有少量唐宋詩歌被譯出。20世紀上半葉,英國韋利翻譯了大量中國古詩,包括《楚辭》,李白、白居易、袁枚等人的詩歌,以及敦煌曲子詞與變文等,享譽譯林。德國洪濤生編譯了中國古代詩集,霍福民(Alfred Hoffmann)翻譯了南唐後主李煜的詞,編譯了唐宋詞選。美國洛威爾(Amy Lowell)選譯的中國古詩集題爲《松花箋》(*Fir-Flower Tablets*),對當時美國國内的詩歌創作有較大影響。

20世紀後期,在詩歌翻譯方面,美國華兹生和法國戴密微(Paul Demiéville)作出了很大貢獻,前者翻譯了《杜甫詩選》《蘇東坡集》《寒山詩集》《蒙求詩》和多種古詩、古賦選集,後者翻譯了《王梵志詩》《太公家教》,主持編譯了《中國古詩選》。此外,英國霍克思翻譯了《楚辭章句》《杜甫集》,日本松枝茂夫等譯《陶淵明全集》,鈴木虎雄等人譯《玉臺新詠集》《杜詩》,小川環樹等譯《王維詩集》《蘇東坡詩選》《宋詩選》,吉川幸次郎等人選譯唐詩,蘇聯吉托維奇(А. И. Гитович)翻譯了《唐詩三百首》等。

古代駢文、散文的很多經典篇章,其翻譯難度絲毫不遜於詩歌。在19—

20世紀上半葉被譯爲外文的以零散篇章爲主。20世紀後期,有多種重要譯著出版,如美國康達維(David R. Knechtges)譯《文選》,馬瑞志(Richard Mather)譯《世説新語》,艾朗諾(Ronald Egan)譯《歐陽修集》,日本吉川忠夫編譯《魏晉清談集》,意大利蘭珊德(Alessandra C. Lavagnino)譯《文心雕龍》,日本宇野精一、興膳宏、斯波六郎等譯《文選》,等等。本次展覽也收入一些譯成外文的著名散文集,如松枝茂夫譯《陶庵夢憶》《浮生六記》等。

五

中國學者對理解本國典籍和文化有先天的優勢,但作爲優秀的翻譯家,還需要同時極其嫻熟地駕馭别國的書面語言。19世紀到20世紀初期,中國學者主要是協助外國傳教士和學者進行翻譯。典型事例有王韜協助理雅各翻譯儒家經典,勞乃宣幫助衛禮賢譯道家經典,江亢虎和賓納(Witter Bynner)合譯《唐詩三百首》,等等。

到了20世紀,中國學人,尤其是在國外受過良好教育的學者們,開始在漢籍的外譯方面發揮更大作用。20世紀初期,辜鴻銘翻譯了《論語》《中庸》,蔡廷幹編譯了唐詩集,等等。辜鴻銘熱衷向西方人宣傳東方的文化和精神,他的譯著多採用"意譯法",在當時的歐洲有較高評價。

20世紀下半葉,海内外的中國學者在漢籍外譯方面有更大創獲。在國内,楊憲益、戴乃迭(Gladys Yang)夫婦合作翻譯了大量文學作品,如《紅樓夢》《儒林外史》《〈詩經〉選》《離騷》《〈史記〉選》《漢魏六朝詩文選》《漢魏六朝小説選》《陶淵明詩選》《唐傳奇》《唐宋詩文選》《關漢卿雜劇選 》《長生殿》《中國古代寓言》等等。許淵沖在中國古典詩詞、戲曲的英譯、法譯方面著述豐富。在海外,長期執教英國的香港學者劉殿爵翻譯了《論語》《孟子》《老子》《孫臏兵法》等,均享很高聲譽。此外,陳榮捷譯《近思録》《傳習録》《六祖壇經》,編譯《中國哲學原始資料》,秦家懿譯《王陽明書信》《明儒學案》,張鍾元、林振述譯《老子》,劉家槐譯《老子》《莊子》《列子》,方志彤選譯《資治通鑒》,鄧嗣禹譯《顔氏家訓》,王伊同譯《洛陽伽藍記》,李祁譯《徐霞客遊記》,施友忠譯《文心雕龍》,葉維廉譯《漢詩英華》,王際真節譯《紅樓夢》等,爲中國傳統文化的傳播起到了積極作用。

六

20世紀，中國學者在中國傳統文化研究方面有很多成就，不少優秀的研究專著被翻譯成外文。由於北京大學在中國傳統文化研究方面所具有的地位，我們以北京大學學者的著作爲主，展示部分被海外漢學家翻譯的中國學者研究專著，作爲本次展覽的終章。

在哲學方面，有胡適《中國哲學史大綱》、中國禪學史日文譯文集，馮友蘭《中國哲學史》英譯本，梁漱溟《東西文化及其哲學》的法譯本、日譯本，任繼愈《漢唐佛教思想論集》《中國佛教史》日譯本等。

歷史學方面，顧頡剛《〈古史辨〉自序》對中國思想文化史和先秦史的研究有里程碑式的貢獻，很早就由漢學家恒慕義(Arthur W. Hummel)譯爲英文。此外，還展出梁啓超《清代學術概論》日譯本、錢穆《中國歷代政治得失》英譯本、胡適自傳《四十自述》日譯本、陳垣《元西域人華化考》英譯本、袁行霈等主編《中華文明史》英譯本。

文學研究方面，收入袁行霈《中國詩歌藝術研究》日譯本、《陶淵明影像——文學史與繪畫史之交叉研究》韓譯本、褚斌傑《中國古代文體概論》日譯本、駱玉明《簡明中國文學史》英譯本等。

窺豹一斑，通過這些外譯的學術專著，我們不妨换一個視角，嘗試用海外漢學家的觀察角度，審視我國近現代人文學術走過的道路。

北京大學圖書館藏中國典籍暨傳統文化研究著作外譯本展覽目録

經部　11種

易經(圖1)

譯本題名：*Y-king, antiquissimus Sinarum liber quem ex latina interpretatione: P. Regis aliorumque ex Soc. Jesu p.p*（拉丁文）

譯者：Julius Mohl

出版地：Stuttgartiae; Tubingae

出版者：J. G. Cottae

出版年：1834—1839

易經(圖2)

譯本題名：*A Translation of the Confucian [Yih king] : or the "Classic of Change"*（英文）

譯者：Thomas Russell Hillier McClatchie（麥格基）

出版地：Shanghai; London

出版者：American Presbyterian Mission Press; Trubner & Co.

出版年：1876

詩經

譯本題名：*Confucii Chi-king, sive, Liber carminum*（拉丁文）

譯者：Alexandre de Lacharme（1695—1767）(孫璋)

出版地：Stuttgartiae; Tubingae

出版者：Sumptibus J. G. Cottae

出版年：1830

周禮(圖3)

譯本題名：*Le Tcheou-li, ou, Rites des Tcheou*（法文）

譯者：Edouard Constant Biot（畢歐）

出版地：Paris

出版者：Imprimerie nationale

出版年：1851

左傳　左丘明(春秋)
譯本題名: 春秋左氏伝(日文)
譯者: 小倉芳彦
出版地: 東京
出版者: 岩波书店
出版年: 1988—1989

左傳　左丘明(春秋)
譯本題名: 春秋左氏伝(日文)
譯者: 竹内照夫
出版地: 東京
出版者: 平凡社
出版年: 1968

四書
譯本題名: 論語　孟子　大学　中庸(日文)
譯者: 倉石武四郎
出版地: 東京
出版者: 筑摩書房
出版年: 1968

大學　中庸　論語
譯本題名: *Confucius: the Great Digest, the Unwobbling Pivot, and the Analects*(英文)
譯者: Ezra Pound(龐德)
出版地: New York
出版者: New Directions Pub. Corp.
出版年: 1969, 1951

論語　孔子(春秋,前551—479)
譯本題名: *The Analects of Confucius*(英文)
譯者: Arthur Waley(韋利)
出版地: New York
出版者: Random House
出版年: 1938

論語　孔子(春秋,前551—479)
譯本題名: 論語(日文)
譯者: 金谷治
出版地: 東京
出版者: 岩波書店
出版年: 1963

論語　孔子(春秋,前551—479)
譯本題名: Беседы и суждения(俄文)
譯者: П. С. Попов(柏百福)
出版地: Москва
出版者: Артефакт-пресс, Пан пресс
出版年: 2013

史部　21種

史記　司馬遷(西漢,前145—86)
譯本題名: *Les Mémoires historiques de Se-ma Ts'ien*(法文)
譯者: Édouard Chavannes(沙畹)
出版地: Paris
出版者: Ernest Leroux
出版年: 1895—1905

史記　司馬遷(西漢,前145—86)
譯本題名: *The Grand Scribe's Records*(英文)
譯者: Tsai-fa Cheng(鄭再發); Lü Zongli(吕宗力); William H. Nienhauser, Jr.; Robert Reynolds
出版地: Bloomington
出版者: Indiana University Press
出版年: 1994—

史記·司馬相如傳　司馬遷(西漢,前145—86)
譯本題名: *Le chapitre 117 du Che-ki (biographie de Sseu-ma Siang-jou)*(法文)
譯者: Yves Hervouet(吴德明)
出版地: Paris
出版者: Presses Universitaires de France
出版年: 1972

史記　司馬遷(西漢,前145—86)
譯本題名: Исторические записки (Ши цзи)(俄文)
譯者: Вяткин, Рудольф Всеволодович; Всеволод Сергеевич Таскин
出版地: Москва
出版者: Наука
出版年: 1972—

史記·世家　司馬遷(西漢,前145—86)
譯本題名: 史記世家(日文)
譯者: 小川環樹
出版地: 東京
出版者: 岩波書店
出版年: 1980—1991

漢書　班固(東漢,32—92)
譯本題名: *The History of the Former Han Dynasty*(英文)
譯者: Homer H. Dubs(德效騫)
出版地: Baltimore
出版者: Waverly Press
出版年: 1938—1955

後漢書 范晔(南朝宋,398—445)
譯本題名: 後汉书(日文)
譯者: 吉川忠夫
出版地: 東京
出版者: 岩波書店
出版年: 2001

通鑒綱目 朱熹(南宋,1130—1200)
續通鑒綱目 商輅(明,1414—1486)
譯本題名: *Histoire generale de la Chine, ou, Annales de cet empire* (法文)
譯者: Joseph Anne Marie Moyriac de Mailla(馮秉正)
出版地: Paris
出版者: Ph. D. Pierres
出版年: 1777—1785

佛國記 法顯(東晉,334—422)
譯本題名: *Foĕ Kouĕ Ki, ou, Relation des royaumes bouddhiques: voyage dans la Tartarie, dans l' Afghanistan et dans l' Inde, éxecuté, a la fin du IVe siècle* (法文)
譯者: Abel Rémusat(雷慕沙); Julius von Klaproth; Ernest Augustin Xavier Clerc de Landresse
出版地: Paris
出版者: Imprimerie royale
出版年: 1936

佛國記 法顯(東晉,334—422)
宋雲行記 宋雲(北魏)
譯本題名: *Travels of Fah-Hian and Sung-Yun : Buddhist pilgrims, from China to India (400 A.D. and 518 A. D.)* (英文)
譯者: Samuel Beal
出版地: London
出版者: Trubner
出版年: 1869

佛國記 法顯(東晉,334—422)
譯本題名: *The Pilgrimage of Fa Hian* (英文)
譯者: 不詳
出版地: Calcutta
出版者: J. Thomas, Baptist Mission Press
出版年: 1848

洛陽伽藍記 楊衒之(北魏)
譯本題名: *Memories of Loyang: Yang Hsüan-Chih and the Lost Capital (493—534)* (英文)
譯者: W. J. F. Jenner(詹納爾)
出版地: Oxford; New York
出版者: Clarendon Press; Oxford University Press
出版年: 1981

洛陽伽藍記　楊衒之(北魏)
譯本題名: 洛陽伽藍記(日文)
譯者: 入矢義高
出版地: 東京
出版者: 平凡社
出版年: 1990

洛陽伽藍記　楊衒之(北魏)
水經注　酈道元(北魏,470—527)
譯本題名: 洛陽伽藍記 水経注(日文)
譯者: 入矢義高;森鹿三;日比野丈夫
出版地: 東京
出版者: 平凡社
出版年: 1974

大唐西域記　玄奘(唐,596—664)
譯本題名: 大唐西域記(日文)
譯者: 水谷真成
出版地: 東京
出版者: 平凡社
出版年: 1971

大唐西域記　玄奘(唐,596—664)
譯本題名: *Mémoires sur les contrées occidentales* (法文)
譯者: Stanislas Julien(儒蓮)
出版地: Paris
出版者: L'Imprimerie impériale
出版年: 1857—1858

大唐西域求法高僧傳　義净(唐,635—713)
譯本題名: *Mémoire composé à l'époque de la grande dynastie T'ang sur les religieux éminents qui allèrent chercher la loi dans les pays d'Occident* (法文)
譯者: Édouard Chavannes(沙畹)
出版地: Paris
出版者: E. Leroux
出版年: 1894

南海寄歸内法傳　義净(唐,635—713)
譯本題名: *A Record of the Buddhist Religion as Practised in India and the Malay Archipelago (A.D. 671—695)* (英文)
譯者: J. Takakusu(高楠順次郎)
出版地: Oxford
出版者: Clarendon Press
出版年: 1896

東京夢華録　孟元老(北宋)
譯本題名: 東京夢華録(日文)
譯者: 入矢義高;梅原郁
出版地: 東京
出版者: 岩波書店
出版年: 1983

真臘風土記　周達觀(元)
譯本題名: *Mémoires sur les coutumes du Cambodge de Tcheou Ta-kouan* (法文)
譯者: Paul Pelliot(伯希和); Georg Cœdès; Paul Demiéville(戴密微)
出版地: Paris
出版者: Librairie d'Amérique et d'Orient
出版年: 1951

諸蕃志　趙汝适(南宋,1170—1231)
譯本題名: *Chau Ju-Kua: His Work on the Chinese and Arab Trade in the Twelfth and Thirteenth Centuries* (英文)
譯者: Friedrich Hirth(夏德); W. W. Rockhill(柔克義)
出版地: St. Petersburg
出版者: Printing office of the Imperial Academy
出版年: 1912

子部　28種

老子　莊子　列子
譯本題名: *Philosophes taoïstes, tome 1, Lao-tseu, Tchouang-tseu, Lie-tseu* (法文)
譯者: Liou Kia-hway(劉嘉槐); Benedykt Grynpas
出版地: Paris
出版者: Gallimard
出版年: 1980

老子　莊子　孫子　列子　吴子
譯本題名: 老子　莊子　孫子　列子　吴子(日文)
譯者: 金谷治;倉石武四郎;関正郎;福永光司;村山吉廣;金谷治
出版地: 東京
出版者: 平凡社
出版年: 1937

道德經 老子(春秋)
譯本題名：*Das Eine：Als Weltgesetz und Vorbild*（德文）
譯者：Vincenz Hundhausen(洪濤生)
出版地：Peking
出版者：Verlag der Pekinger Pappelinsel
出版年：1942

道德經 老子(春秋)
譯本題名：Дао дэ цзин：Книга о пути и его силе（俄文）
譯者：А. Костенко
出版地：Алматы
出版者：Жеті жарғы
出版年：2012

道德經(郭店楚簡) 老子(春秋)
譯本題名：*Lao Tzu's Tao Te Ching: A Translation of the Startling New Documents Found at Guodian*（英文）
譯者：Robert G. Henricks(韓禄伯)
出版地：New York
出版者：Columbia University Press
出版年：2000

孫子兵法 孫子(春秋)
譯本題名：*Sun-tzu: The Art of Warfare*（英文）
譯者：Roger T. Ames（安樂哲）
出版地：New York
出版者：Ballantine Books
出版年：1993

莊子 莊周(戰國,前369—286)
譯本題名：*Die Weisheit des Dschuang-dse*（德文）
譯者：Vincenz Hundhausen（洪濤生）
出版地：Peking; Leipzig
出版者：Pekinger Verlag
出版年：1926

墨子 墨翟(戰國,前480—420)
譯本題名：*Mê Ti des Sozialethikers und seiner Schüler philosophische Werke*（德文）
譯者：Alfred Forke（弗爾克）
出版地：Berlin
出版者：Kommissionsverlag der Vereinigung wissenschaftlicher verleger
出版年：1922

商君書　商鞅（戰國，前390—338）
譯本題名：*Le livre du prince Shang*（法文）
譯者：Jean Lévi
出版地：Paris
出版者：Flammarion
出版年：1981

荀子　荀況（戰國，前313—238）
譯本題名：*Xun Zi (Siun Tseu)*（法文）
譯者：Ivan P. Kamenarović
出版地：Paris
出版者：Éditions du Cerf
出版年：1987

大智度論　龍樹菩薩造，鳩摩羅什譯
譯本題名：*Le traité de la grande vertu de sagesse de Nāgārjuna*（法文）
譯者：Étienne Lamotte
出版地：Louvain
出版者：Bureaux du Muséon
出版年：1944

荀子　荀況（戰國，前313—238）
譯本題名：*The works of Hsüntze*（英文）
譯者：Homer H. Dubs（德效騫）
出版地：London
出版者：Arthur Probsthain
出版年：1928

淮南子·主術訓　劉安（西漢，前179—122）
譯本題名：*The Art of Rulership: A Study in Ancient Chinese Political Thought*（英文）
譯者：Roger T. Ames（安樂哲）
出版地：Albany
出版者：State University of New York Press
出版年：1994

三藏經節譯
譯本題名：*Cinq cents contes et apologues extraits du Tripiţaka chinois*（法文）
譯者：Édouard Chavannes（沙畹）
出版地：Paris
出版者：E. Leroux
出版年：1910—1934

碧巖録　克勤(北宋,1063—1135)
譯本題名: *Bi-yän-lu: Meister Yüan-wu's Niederschrift von der Smaragdenen Felswand, Verfasst auf dem Djia-schan bei Li in Hunan zwischen 1111 und 1115 im Druck erschienen in Sïtschuan um 1300* (德文)
譯者: Wilhelm Gundert
出版地: München
出版者: C. Hanser
出版年: 1960

碧巖録　克勤(北宋,1063—1135)
譯本題名: 碧巖録(日文)
譯者: 入矢義高
出版地: 東京
出版者: 岩波書店
出版年: 1992—1996

臨濟録　慧然
譯本題名: 臨済録(日文)
譯者: 入矢義高
出版地: 東京
出版者: 岩波書店
出版年: 1989

臨濟録　慧然
譯本題名: *Entretiens de Lin-tsi* (法文)
譯者: Paul Demiéville(戴密微)
出版地: Paris
出版者: Fayard
出版年: 1972

穆天子傳
譯本題名: *Le Mu tianzi zhuan* (法文)
譯者: Rémi Mathieu
出版地: Paris
出版者: Presses universitaires de France
出版年: 1978

列仙傳　劉向(西漢)
譯本題名: 列仙伝(日文)
譯者: 沢田瑞穂
出版地: 東京
出版者: 平凡社
出版年: 1969

列仙傳　劉向(西漢)
譯本題名：*Le Lie-sien tchouan: biographies légendaires des immortels taoïstes de l' antiquité* (法文)
譯者：Max Kaltenmark(康德謨)
出版地：Pékin
出版者：巴黎大學北京漢學研究所
出版年：1953

漢武帝内傳
譯本題名：*L' Empereur Wou des Han dans la légende taoiste = Han Wou-ti Nei-Tchouan* (法文)
譯者：Kristofer Marinus Schipper(施舟人)
出版地：Paris
出版者：École française d'Extrême-Orient
出版年：1965

菜根譚　洪應明(明,1573—1620)
譯本題名：菜根譚(日文)
譯者：今井宇三郎
出版地：東京
出版者：岩波書店
出版年：1975

陶庵夢憶　張岱(明,1597—1676)
譯本題名：陶庵夢憶(日文)
譯者：松枝茂夫
出版地：東京
出版者：岩波書店
出版年：1981

浮生六記　沈復(清,1763—1807)
譯本題名：浮生六記(日文)
譯者：松枝茂夫
出版地：東京
出版者：岩波書店
出版年：1981

千金方　孫思邈(唐,581—682)
譯本題名：*Prescriptions d'acuponcture valant mille onces d'or: traité d'acuponcture de Sun Simiao du VIIe siècle* (法文)
譯者：Catherine Despeux(戴思博)
出版地：Paris
出版者：Guy Trédaniel
出版年：1987

歷代名畫記　張彥遠(唐,815—875)
譯本題名:歷代名畫記(日文)
譯者:小野勝年
出版地:東京
出版者:岩波書店
出版年:1938

景德鎮陶録　藍浦(清)
譯本題名:*Histoire et fabrication de la porcelaine chinoise*(法文)
譯者:Stanislas Julien(儒蓮)
出版地:Pairs
出版者:Mallet-Bachelier
出版年:1856

文學　62種

楚辭章句　劉向(西漢)
譯本題名:*Ch'u Tz'u:The Songs of the South: An Ancient Chinese Anthology*(英文)
譯者:David Hawkes(霍克思)
出版地:Oxford
出版者:Clarendon Press
出版年:1959

文選　蕭統(南朝梁,501—531)
譯本題名:文選(日文)
譯者:斯波六郎;花房英樹
出版地:東京
出版者:筑摩書房
出版年:1963

文選　蕭統(南朝梁,501—531)
譯本題名:*Wen xuan, or, Selections of Refined Literature*(英文)
譯者:David R. Knechtges(康達維)
出版地:Princeton, N.J.
出版者:Princeton University Press
出版年:1982—1996

文選　蕭統(南朝梁,501—531)
譯本題名:文選(日文)
譯者:興膳宏;川合康三
出版地:東京
出版者:角川書店
出版年:1988

文選　蕭統(南朝梁,501—531)
譯本題名: 文選(日文)
譯者: 宇野精一;平岡武夫
出版地: 東京
出版者: 集英社
出版年: 1974—1976

玉臺新詠集　徐陵(南朝梁,507—583)
譯本題名: 玉臺新詠集(日文)
譯者: 鈴木虎雄
出版地: 東京
出版者: 岩波書店
出版年: 1953—1956

陶淵明集　陶淵明(東晉,365—427)
譯本題名: *T'ao Yüan-ming (AD 365—427): His Works and their Meaning* (英文)
譯者: Albert Richard Davis
出版地: Hong Kong
出版者: Hong Kong University Press
出版年: 1983

陶淵明全集　陶淵明(東晉,365—427)
譯本題名: 陶淵明全集(日文)
譯者: 松枝茂夫;和田武司
出版地: 東京
出版者: 岩波書店
出版年: 1991

王梵志詩　太公家教　王梵志(唐,590—660)
譯本題名: *L'œuvre de Wang le zélateur (Wang Fan-tche) ; Suivi des Instructions domestiques de l'aïeul (T'ai-kong kia-kiao) ; Poèmes populaires des T'ang* (法文)
譯者: Paul Demiéville(戴密微)
出版地: Paris
出版者: Collège de France, Institut des hautes etudes chinoises
出版年: 1982

王維詩集　王維(唐,701—761)
譯本題名: 王維詩集(日文)
譯者: 小川環樹;都留春雄;入谷仙介
出版地: 東京
出版者: 岩波書店
出版年: 1972

王維詩選　王維(唐,701—761)
譯本題名：*Poems of Wang Wei*（英文）
譯者：G. W. Robinson
出版地：Harmondsworth, Middlesex, England
出版者：Penguin Books
出版年：1973

杜甫集　杜甫(唐,712—770)
譯本題名：*A Little Primer of Tu Fu*（英文）
譯者：David Hawkes(霍克思)
出版地：Oxford
出版者：Clarendon Press
出版年：1967

杜詩　杜甫(唐,712—770)
譯本題名：杜詩(日文)
譯者：鈴木虎雄;黒川洋一
出版地：東京
出版者：岩波書店
出版年：1963—1966

唐詩
譯本題名：*Poésies de l'époque des Thang*（法文）
譯者：D'Hervey-Saint-Denys
出版地：Paris
出版者：Amyot
出版年：1862

唐詩三百首
譯本題名：Три танских поэта：Ли Бо, Ван Вэй, Ду Фу: Триста стихотворений（俄文）
譯者：Гитович, Александр Ильич
出版地：Москва
出版者：Издательство восточной литературы
出版年：1960

歐陽修集　歐陽修(北宋,1007—1072)
譯本題名：*The Literary Works of Ou-yang Hsiu*（英文）
譯者：Ronald Egan(艾朗諾)
出版地：Cambridge
出版者：Cambridge University Press
出版年：1984

蘇東坡詩選　蘇軾（北宋，1036—1101）
譯本題名：蘇東坡詩選（日文）
譯者：小川環樹；山本和義
出版地：東京
出版者：岩波書店
出版年：1975

春花秋月：木版畫與宋詞
譯本題名：*Frühlingsblüten und Herbstmond : Ein Holzeschnittband mit Liedern aus der Sung-Zeit, 960—1279*（德文）
譯者：Alfred Hoffmann（霍福民）
出版地：Köln
出版者：Greven Verlag
出版年：1951

德譯中國詩歌
譯本題名：*Chinesische Dichter in deutscher Sprache*（德文）
譯者：Vincenz Hundhausen（洪濤生）
出版地：Peking
出版者：Pekinger verlag
出版年：1926

中國古典詩選
譯本題名：*Anthologie de la poésie chinoise classique*（法文）
譯者：Paul Demiéville（戴密微）
出版地：Paris
出版者：Gallimard
出版年：1962

西廂記　王實甫（元，1295—1307）
譯本題名：*Das Westzimmer: ein chinesisches Singspiel aus dem dreizehnten Jahrhundert*（德文）
譯者：Vincenz Hundhausen（洪濤生）
出版地：Leipzig
出版者：Im Erich Roth-Verlag/Eisenach
出版年：1926

琵琶記　高明（元，1306—1359）
譯本題名：*Die Laute*（德文）
譯者：Vincenz Hundhausen（洪濤生）
出版地：Peking
出版者：Pekinger verlag
出版年：1930

琵琶記　高明(元,1306—1359)(圖4)
譯本題名：*Le pi-pa-ki, ou, L'histoire du Luth*（法文）
譯者：Bazin Antoine Pierre Louis(巴贊)
出版地：Paris
出版者：Impr. Royale
出版年：1841

牡丹亭　湯顯祖(明,1550—1616)
譯本題名：*Die Rückkehr der Seele : ein romantisches Drama*（德文）
譯者：Vincenz Hundhausen(洪濤生)
出版地：Zürich; Leipzig
出版者：Rascher Verlag
出版年：1937

牡丹亭　湯顯祖(明,1550—1616)
譯本題名：*Der Blumengarten*（德文）
譯者：Vincenz Hundhausen（洪濤生）
出版地：Peking
出版者：Pekinger verlag
出版年：1933

世説新語　劉義慶(南朝宋,403—444)
譯本題名：*A New Account of Tales of the World*（英文）
譯者：Richard B. Mather(馬瑞志)
出版地：Minneapolis
出版者：University of Minnesota Press
出版年：1976

魏晉清談集
譯本題名：魏晉清談集(日文)
譯者：吉川忠夫
出版地：東京
出版者：講談社
出版年：1986

遊仙窟　張鷟(唐,658?—730?)
譯本題名：遊仙窟(日文)
譯者：今村與志雄
出版地：東京
出版者：岩波書店
出版年：1990

聊齋志異　蒲松齡(清,1640—1715)
譯本題名:聊斎志異(日文)
譯者:立間祥介
出版地:東京
出版者:岩波書店
出版年:1997

聊齋志異　蒲松齡(清,1640—1715)
譯本題名:*Strange Stories from a Chinese Studio*(英文)
譯者:Herbert A. Giles(翟理斯)
出版地:Shanghai
出版者:Kelly & Walsh
出版年:1908

聊齋志異　蒲松齡(清,1640—1715)
譯本題名:*Contes extraordinaires du pavillon du loisir*(法文)
譯者:Yves Hervouet(吴德明)
出版地:Paris
出版者:Gallimard
出版年:1969

聊齋志異　蒲松齡(清,1640—1715)
譯本題名:Рассказы Ляо Чжая о необычайном(俄文)
譯者:Алексеев, Василий Михайлович(阿理克)
出版地:Москва
出版者:Художественная литература
出版年:1983

水滸傳　施耐庵(元,1290—1365)
譯本題名:*Au bord de l'eau*(法文)
譯者:Jacques Dars(譚霞客)
出版地:Paris
出版者:Gallimard
出版年:2005

水滸傳　施耐庵(元,1290—1365)
譯本題名:*All Men Are Brothers*(英文)
譯者:Pearl S. Buck(賽珍珠)
出版地:New York
出版者:The John Day Company
出版年:1933

水滸傳　施耐庵(元,1290—1365)
譯本題名：水滸伝(日文)
譯者：村上知行
出版地：東京
出版者：修道社
出版年：1955

水滸傳　施耐庵(元,1290—1365)
譯本題名：水滸伝(日文)
譯者：松枝茂夫
出版地：東京
出版者：岩波書店
出版年：1959—1960

水滸傳　施耐庵(元,1290—1365)
譯本題名：Речные заводи(俄文)
譯者：A. Рогачев(羅高壽)
出版地：Москва
出版者：Гослитиздат
出版年：1959

金瓶梅詞話　蘭陵笑笑生(明,1526—1593)
譯本題名：*The Plum in the Golden Vase Or Chin P' ing Mei*(英文)
譯者：David Tod Roy(芮效衛)
出版地：Princeton, N. J.
出版者：Princeton University Press
出版年：不詳

金瓶梅詞話　蘭陵笑笑生(明,1526—1593)
譯本題名：*Fleur en fiole d'or*(法文)
譯者：André Lévy(雷威安)
出版地：Paris
出版者：Gallimard
出版年：1985

金瓶梅詞話　蘭陵笑笑生(明,1526—1593)
譯本題名：Цветы сливы в золотой вазе, или Цзинь, Пин, Мэй(俄文)
譯者：Виктор Сергеевич Манухин; С.Хохлова
出版地：Москва
出版者：Художественная литература
出版年：1972—

封神演義　許仲琳(明)
譯本題名: *Die Metamorphosen der Goetter*(德文)
譯者: Wilhelm Grube(顧祿博); Herbert Mueller
出版地: Leiden
出版者: Buchhandlung und Druckerei vormals E. J. Brill
出版年: 1912

三國演義　羅貫中(明,1330—1400)
譯本題名: 三国志 完訳(日文)
譯者: 小川環樹;金田純一郎
出版地: 東京
出版者: 岩波書店
出版年: 1982—1983

三國演義　羅貫中(明,1330—1400)
譯本題名: Троецарствие(俄文)
譯者: Владимир Андреевич Панасюк
出版地: Москва
出版者: Художественная литература
出版年: 1984

西遊記　吴承恩(明,1500—1582)
譯本題名: 西遊記(日文)
譯者: 太田辰夫;鳥居久靖
出版地: 東京
出版者: 平凡社
出版年: 1971—1972

西遊記　吴承恩(明,1500—1582)
譯本題名: *La Pérégrination vers l'Ouest*(法文)
譯者: André Lévy(雷威安)
出版地: Paris
出版者: Gallimard
出版年: 1991

西遊記　吴承恩(明,1500—1582)
譯本題名: *The Journey to the West*(英文)
譯者: Anthony C. Yu(余國藩)
出版地: Chicago
出版者: University of Chicago Press
出版年: 1977—1983

西遊記　吴承恩(明,1500—1582)譯本題名: *Monkey*(英文)
譯者: Arthur Waley(韋利)
出版地: London
出版者: G. Allen & Unwin
出版年: 1942

西遊記　吴承恩(明,1500—1582)譯本題名: *Journey to the West*(英文)
譯者: W. J. F. Jenner(詹納爾)
出版地: Beijing
出版者: Foreign Languages Press
出版年: 1982

西遊記　吴承恩(明,1500—1582)
譯本題名：Сунь Укун —Царь обезьян(俄文)
譯者：А. П. Рогачев(羅高壽)
出版地：Москва
出版者：Художественная литература
出版年：1982

紅樓夢　曹雪芹(清,1717—1763)
譯本題名：*The Story of Stone*(英文)
譯者：David Hawkes(霍克思)
出版地：Harmondsworth
出版者：Penguin
出版年：1973—1986

紅樓夢　曹雪芹(清,1717—1763)
譯本題名：*Le rêve dans le pavillon rouge*(法文)
譯者：Li Tche-Houa(李治華); Jacqueline Alezais(李雅歌)
出版地：Paris
出版者：Gallimard
出版年：1981

紅樓夢　曹雪芹(清,1717—1763)
譯本題名：新訳紅楼夢(日文)
譯者：井波陵一
出版地：東京
出版者：岩波書店
出版年：2013

紅樓夢　曹雪芹(清,1717—1763)
譯本題名：紅樓夢(日文)
譯者：伊藤漱平
出版地：東京
出版者：平凡社
出版年：1958—1960

紅樓夢　曹雪芹(清,1717—1763)
譯本題名：Сон в красном тереме(俄文)
譯者：Владимир Андреевич Панасюк
出版地：Москва
出版者：Художественная литература
出版年：1958

好逑傳　名教中人(清)
譯本題名：*La brise au clair de lune : Le deuxieme livre de génie*（法文）
譯者：Charle George Soulie
出版地：Paris
出版者：B. Grasset
出版年：1925

好逑傳　名教中人(清)
譯本題名：*Hau kiou choaan: histoire chinoise*（法文）
譯者：Marc Eidous
出版地：Lyon
出版者：B. Duplain
出版年：1766

好逑傳　名教中人(清)
譯本題名：*The Fortunate Union: a Romance*（英文）
譯者：Sir John Francis Davis（德庇時）
出版地：London
出版者：Printed from the Oriental Translation Fund
出版年：1829

好逑傳　名教中人(清)
譯本題名：*Hau Kiou choaan: or, The Pleasing History*（英文）
譯者：Thomas Percy; James Wilkinson; Robert Dodsley; James Dodsley
出版地：London
出版者：Printed for R. and J. Dodsley
出版年：1761

玉嬌梨　荻散人(清)
譯本題名：*Les deux cousines*(法文)
譯者：Stanislas Julien(儒蓮)
出版地：Paris
出版者：Didier
出版年：1864

中国小説選
譯本題名：中国小説選(日文)
譯者：金文京
出版地：東京
出版者：角川書店
出版年：1989

中國古典探案小説選譯

譯本題名：*Sept victimes pour un oiseau : et autres histoires policières*（法文）

譯者：André Lévy（雷威安）

出版地：Paris

出版者：Flammarion

出版年：1981

金匱：兩千年中國短篇小説選

譯本題名：*Die goldene Truhe: Chinesische Novellen aus zwei Jahrtausenden*（德文）

譯者：Wolfgang Bauer（鮑吾剛）；Herbert Franke（傅海博）

出版地：München

出版者：C. Hanser

出版年：1988

顧賽芬譯著　6種

詩經

譯本題名：*Cheu King*（法文）

譯者：Séraphin Couvreur（顧賽芬）

出版地：Ho Kien Fou（河間府）

出版者：Imprimerie de la Mission Catholique

出版年：1896

書經

譯本題名：*Chou King*（法文）

譯者：Séraphin Couvreur（顧賽芬）

出版地：Hien Hien（獻縣）

出版者：Imprimerie de la Mission Catholique

出版年：1916

儀禮

譯本題名：*Cérémonial*（法文）

譯者：Séraphin Couvreur（顧賽芬）

出版地：Hsien Hsien（獻縣）

出版者：Imprimerie de la Mission Catholique

出版年：1916

禮記

譯本題名：*Li Ki: memoires sur les bienseances et les ceremonies texte Chinois*（法文）

譯者：Séraphin Couvreur（顧賽芬）

出版地：Ho Kien Fou（河間府）

出版者：Imprimerie de la Mission Catholique

出版年：1913

左傳 左丘明(春秋)

譯本題名: *Tch'ouen ts'iou et Tso tchouan* (法文)

譯者: Séraphin Couvreur(顧賽芬)

出版地: Ho Kien Fou(河間府)

出版者: Imprimerie de la Mission Catholique

出版年: 1914

四書

譯本題名: *Les quatre livres* (法文)

譯者: Séraphin Couvreur (顧賽芬)

出版地: Ho Kien Fou (河間府)

出版者: Imprimerie de la Mission Catholique

出版年: 1895

理雅各譯著 3種

中國經典(圖5)

譯本題名: *The Chinese Classics* (英文)

譯者: James Legge (理雅各)

出版地: Hongkong and London

出版者: Author and Trübner & Co.

出版年: 1861

老子 莊子

譯本題名: *The Sacred Books of China: The Texts of Tâoism* (英文)

譯者: James Legge (理雅各)

出版地: London

出版者: Oxford University Press

出版年: 1927

佛國記 法顯(東晉,334—422)

譯本題名: *A Record of Buddhistic Kingdoms : Being an Account by the Chinese Monk Fa-Hien of His Travels in India and Ceylon (A.D. 399-414) in Search of the Buddhist Books of Discipline* (英文)

譯者: James Legge(理雅各)

出版地: Oxford

出版者: The Clarendon Press

出版年: 1886

衛禮賢譯著　11種

易經

譯本題名：*I ging : das Buch der Wandlungen*（德文）

譯者：Richard Wilhelm（衛禮賢）

出版地：Jena

出版者：E. Diederichs

出版年：1924

大學

譯本題名：*Die höhere Bildung*（德文）

譯者：Richard Wilhelm（衛禮賢）

出版地：Tsingtau

出版者：不詳

出版年：1920

道德經　老子（春秋）

譯本題名：*Lao Tse taote king : das Buch des Alten vom Sinn und Leben*（德文）

譯者：Richard Wilhelm（衛禮賢）

出版地：Jena

出版者：Eugen Diederichs

出版年：1923

論語　孔子（春秋，前551—479）

譯本題名：*Gespräche = Lun Yü / Kungfutse ; aus dem Chinesischen*（德文）

譯者：Richard Wilhelm（衛禮賢）

出版地：Jena

出版者：E. Diederichs

出版年：1923

孟子　孟軻（戰國，前372—289）

譯本題名：*Mong dsi: die Lehrgespräche des Meisters Meng K'o*（德文）

譯者：Richard Wilhelm（衛禮賢）

出版地：Koln

出版者：Eugen Diederichs Verlag

出版年：1982

南華經　莊周（戰國，前369—286）

譯本題名：*Das wahre Buch vom südlichen Blütenland*（德文）

譯者：Richard Wilhelm（衛禮賢）

出版地：Kèoln

出版者：Eugen Diederichs Verlag

出版年：1982

吕氏春秋　吕不韋（戰國，前 292—235）

譯本題名：*Frühling und Herbst des Lü Bu We*（德文）

譯者：Richard Wilhelm（衛禮賢）

出版地：Düsseldorf

出版者：Eugen Diederichs Verlag

出版年：1979

列子　列子（戰國，前 450—375）

譯本題名：*Das wahre Buch vom quellenden Urgrund: (Tschung hü dschen ging) : die Lehren der Philosophen Liä Yü Kou und Yang Dschu*（德文）

譯者：Richard Wilhelm（衛禮賢）

出版地：Jena

出版者：E. Diederichs

出版年：1911

孔子家語　王肅（三國魏，195—256）

譯本題名：*Kungfutse Schulgespräche (Gia Yü)*（德文）

譯者：Richard Wilhelm（衛禮賢）；Hellmut Wilhelm（衛德明）

出版地：Düsseldorf

出版者：Eugen Diederichs

出版年：1961

太乙金華宗旨

譯本題名：*Das Geheimnis der Goldenen Blüte*（德文）

譯者：Richard Wilhelm（衛禮賢）

出版地：München

出版者：Dornverlag

出版年：1929

中國民間童話

譯本題名：*Chinesische Volksmärchen*（德文）

譯者：Richard Wilhelm（衛禮賢）

出版地：Jena

出版者：E. Diederichs

出版年：1927

庫恩譯著　10種

水滸傳　施耐庵(元,1290—1365)
譯本題名：*Die Räuber vom Liang Schan Moor: mit sechzig Holzschnitten*（德文）
譯者：Franz Kuhn（庫恩）
出版地：Frankfurt am Main
出版者：Insel Verlag
出版年：2009

杜十娘怒沉百寶箱　馮夢龍(明，1574—1646)
譯本題名：*Das Juwelenkästchen*（德文）
譯者：Franz Kuhn(庫恩)
出版地：Dresden
出版者：Wilhelm Heyne Verlag
出版年：1937

玉蜻蜓
譯本題名：*Die Jadelibelle: Roman*（德文）
譯者：Franz Kuhn(庫恩)
出版地：Berlin
出版者：Schützen-Verlag
出版年：1936

金瓶梅　蘭陵笑笑生(明，1526—1593)
譯本題名：*Kin Ping Meh: oder, Die abenteuerliche Geschichte von Hsi Men und seinen sechs Frauen*（德文）
譯者：Franz Kuhn（庫恩）
出版地：Frankfurt am Main
出版者：Insel Verlag
出版年：1977

好逑傳　名教中人(清)
譯本題名：*Eisherz und Edeljaspis: oder, die Geschichte einer glücklichen Gattenwahl*（德文）
譯者：Franz Kuhn(庫恩)
出版地：Leipzig
出版者：Insel-Verlag
出版年：1926

蔣興哥重會珍珠衫
譯本題名：*Das Perlenhemd: eine chinesische Liebesgeschichte*（德文）
譯者：Franz Kuhn(庫恩)
出版地：Leipzig
出版者：Insel-Verlag
出版年：不詳

二度梅

譯本題名:*Die Rache des jungen Meh: oder Das Wunder der zweiten Pflaumenblute*(德文)

譯者: Franz Kuhn(庫恩)

出版地: Leipzig

出版者: Insel-Verlag

出版年: 1927

中國古代中短篇小説

譯本題名: *Altchinesische Novellen*(德文)

譯者: Franz Kuhn(庫恩)

出版地: Leipzig

出版者: Insel-Verlag

出版年: 1979

中國中短篇小説經典

譯本題名: *Chinesische meisternovellen*(德文)

譯者: Franz Kuhn(庫恩)

出版地: Leipzig

出版者: Insel-Verlag

出版年: 不詳

《古今圖書集成》"君道"、"治道"選譯

譯本題名: *Chinesische Staatsweisheit*(德文)

譯者: Franz Kuhn(庫恩)

出版地: Darmstadt

出版者: Otto Reichl

出版年: 1923

華兹生譯著　14種

論語　孔子(春秋,前551—479)

譯本題名: *The Analects of Confucius*(英文)

譯者: Burton Watson (華兹生)

出版地: New York

出版者: Columbia University Press

出版年: 2010

荀子　荀況(戰國,前340—245)

譯本題名: *Basic Writings / Hsün Tzu*(英文)

譯者: Burton Watson(華兹生)

出版地: New York

出版者: Columbia University Press

出版年: 1963

莊子　莊周(戰國,前369—286)
譯本題名：*The Complete Works of Chuang Tzu* (英文)
譯者：Burton Watson (華玆生)
出版地：New York
出版者：Columbia University Press
出版年：1968

墨子　墨翟(戰國,前480—420)
荀況　荀況(戰國,前340—245)
韓非　韓非(戰國,前280—233)
譯本題名：*Basic Writings of Mo Tzu, Hsün Tzu, and Han Fei Tzu* (英文)
譯者：Burton Watson(華玆生)
出版地：New York
出版者：Columbia University Press
出版年：1964

韓非子　韓非(戰國,前280—233)
譯本題名：*Han Fei Tzu : Basic Writings* (英文)
譯者：Burton Watson (華玆生)
出版地：New York
出版者：Columbia University Press
出版年：1964

史記　司馬遷(西漢,前145—86)
譯本題名：*Records of the Grand Historian of China* (英文)
譯者：Burton Watson(華玆生)
出版地：New York
出版者：Columbia University Press
出版年：1961

漢書選　班固(東漢,32—92)
譯本題名：*Courtier and Commoner in Ancient China: Selections of the History by Pan Ku* (英文)
譯者：Burton Watson (華玆生)
出版地：New York; London
出版者：Columbia University Press
出版年：1974

漢魏六朝賦選
譯本題名：*Chinese Rhyme-Prose; Poems in the Fu Form from the Han and Six Dynasties Periods* (英文)
譯者：Burton Watson (華玆生)
出版地：New York
出版者：Columbia University Press
出版年：1971

寒山詩集　寒山(唐,627—649)
譯本題名：*Cold Mountain: 100 Poems by the T' ang Poet Han-shan*（英文）
譯者：Burton Watson（華兹生）
出版地：New York
出版者：Columbia University Press
出版年：1970

杜甫詩選　杜甫(唐,712—770)
譯本題名：*The Selected Poems of Du Fu*（英文）
譯者：Burton Watson(華兹生)
出版地：New York
出版者：Columbia University Press
出版年：2002

蘇東坡集　蘇東坡(北宋,1037—1101)
譯本題名：*Su Tung-P'o: Selections from a Sung Dynasty Poet*（英文）
譯者：Burton Watson（華兹生）
出版地：New York
出版者：Columbia University Press
出版年：1965

蒙求詩　李瀚
譯本題名：*Meng ch' iu : Famous Episodes from Chinese History and Legend*（英文）
譯者：Burton Watson(華兹生)
出版地：Tokyo; New York
出版者：Kodansha International
出版年：1979

從早期到十三世紀中國詩歌
譯本題名：*The Columbia Book of Chinese Poetry : From Early Times to the Thirteenth Century*（英文）
譯者：Burton Watson(華兹生)
出版地：New York
出版者：Columbia University Press
出版年：1984

中國二到十二世紀古詩選
譯本題名：*Chinese Lyricism; Shih Poetry from the Second to the Twelfth Century*(英文)
譯者：Burton Watson(華兹生)
出版地：New York
出版者：Columbia University Press
出版年：1971

中國學者譯著　20種

論語　孔子(春秋,前551—479)
譯本題名: *The Analects (Lun yü)*（英文）
譯者: D. C. Lau（劉殿爵）
出版地: Harmondsworth; New York
出版者: Penguin Books
出版年: 1979

孟子　孟軻(戰國,前372—289)
譯本題名: *Mencius*（英文）
譯者: D. C. Lau（劉殿爵）
出版地: Harmondsworth
出版者: Penguin
出版年: 1970

道德經　老子(春秋)
譯本題名: *Tao te ching*（英文）
譯者: D. C. Lau（劉殿爵）
出版地: Hong Kong
出版者: The Chinese University Press
出版年: 1982

孫臏兵法　孫臏(戰國)
譯本題名: *Sun Pin : The Art of Warfare*（英文）
譯者: D. C. Lau（劉殿爵）
出版地: New York
出版者: Ballantine Books
出版年: 1996

論語　孔子(春秋,前551—479)
譯本題名: *The Discourses and Sayings of Confucius*（英文）
譯者: Ku Hung-ming（辜鸿铭）
出版地: Shanghai
出版者: Kelly and Walsh, Ltd.
出版年: 1898

中庸
譯本題名: *The Universal Order, or Conduct of Life*（英文）
譯者: Ku Hung-ming（辜鸿铭）
出版地: Shanghai
出版者: Shanghai Mercury, Ltd.
出版年: 1906

史记　司馬遷(西漢,前145—86)
譯本題名：*Records of the Historian of China*（英文）
譯者：Yang Hsien-yi(楊憲益); Gladys Yang（戴乃迭）
出版地：Hong Kong
出版者：Commercial Press
出版年：1975

離騷　屈原(戰國,前340—278)
譯本題名：*Li Sao : And Other Poems of Chu Yuan*（英文）
譯者：Yang Hsien-yi(楊憲益); Gladys Yang(戴乃迭)
出版地：Peking
出版者：Chinese Literature Press
出版年：1953

漢魏六朝小説選
譯本題名：*The Man Who Sold a Ghost: Chinese Tales of the 3rd-6th Centuries*(英文)
譯者：Yang Hsien-yi(楊憲益); Gladys Yang(戴乃迭)
出版地：Peking
出版者：Foreign Languages Press
出版年：1958

詩經選
譯本題名：*Selections from the "Book of Songs"*（英文）
譯者：Yang Hsien-yi(楊憲益); Gladys Yang(戴乃迭)
出版地：Peking
出版者：Chinese Literature Press
出版年：1983

漢魏六朝詩文選
譯本題名：*Poetry and Prose of the Han, Wei and Six Dynasties*（英文）
譯者：Yang Hsien-yi(楊憲益); Gladys Yang(戴乃迭)
出版地：Peking
出版者：Foreign Languages Press
出版年：2005

陶淵明詩選　陶渊明(東晉,365—427)
譯本題名：*Selected Poems*（英文）
譯者：Yang Hsien-yi(楊憲益); Gladys Yang(戴乃迭)
出版地：Peking
出版者：Chinese Literature Press
出版年：1993

龍女：唐傳奇十種

譯本題名：*The Dragon King's Daughter: Ten Tang Dynasty Stories*（英文）

譯者：Yang Hsien-yi（楊憲益）; Gladys Yang（戴乃迭）

出版地：Peking

出版者：Foreign Languages Press

出版年：1980

關漢卿雜劇選　關漢卿（元，1210—1298）

譯本題名：*Selected Plays of Kuan Han-ching*（英文）

譯者：Yang Hsien-yi（楊憲益）; Gladys Yang（戴乃迭）

出版地：Peking

出版者：Foreign Languages Press

出版年：1958

紅樓夢　曹雪芹（清，1717—1763）

譯本題名：*A Dream of Red Mansions*（英文）

譯者：Yang Hsien-yi（楊憲益）; Gladys Yang（戴乃迭）

出版地：Peking

出版者：Foreign Languages Press

出版年：1978—1980

唐宋詩詞選

譯本題名：*Poetry and Prose of the Tang and Song*（英文）

譯者：Yang Hsien-yi（楊憲益）; Gladys Yang（戴乃迭）

出版地：Peking

出版者：Chinese Literature Press

出版年：1984

杜十娘怒沉百寶箱：中國十到十七世紀小說

譯本題名：*The Courtesan's Jewel Box: Chinese Stories of the Xth-XVIIth Centuries*（英文）

譯者：Yang Hsien-yi（楊憲益）; Gladys Yang（戴乃迭）

出版地：Peking

出版者：Foreign Languages Press

出版年：1957

儒林外史　吳敬梓（清，1701—1754）

譯本題名：*The Scholars*（英文）

譯者：Yang Hsien-yi（楊憲益）; Gladys Yang（戴乃迭）

出版地：Peking

出版者：Foreign Languages Press

出版年：1957

長生殿　洪昇(清,1645—1704)
譯本題名：*The Palace of Eternal Youth*(英文)
譯者：Yang Hsien-yi(楊憲益); Gladys Yang(戴乃迭)
出版地：Peking
出版者：Foreign Languages Press
出版年：1955

中國古代寓言
譯本題名：*Ancient Chinese Fables*(英文)
譯者：Yang Hsien-yi(楊憲益); Gladys Yang(戴乃迭)
出版地：Peking
出版者：Foreign Languages Press
出版年：1957

許淵沖譯著

許淵沖文集　許淵沖(1921—　)
出版地：北京
出版者：海豚出版社
出版年：2013

傳統文化研究著作外譯　20種

清代學術概論　梁啓超(1873—1929)
譯本題名：清代學術概論(日文)
譯者：橋川時雄
出版地：東京
出版者：東華社
出版年：1922

元西域人華化考　陳垣(1880—1971)
譯本題名：*Western and Central Asians in China under the Mongols: Their Transformation into Chinese*(英文)
譯者：Ch'ien Hsing-hai(錢星海); L. Carrington Goodrich(富路特)
出版地：Nettetal
出版者：Steyler Verlag-Wort und Werk
出版年：1989

四十自述　胡適（1891—1962）
譯本題名：四十自述（日文）
譯者：吉川幸次郎
出版地：東京
出版者：創元社
出版年：1940

《胡適文存》、《胡適論學近著》選譯
胡適（1891—1962）
譯本題名：支那禪學之變遷（日文）
譯者：今關天彭
出版地：東京
出版者：東方學藝書院
出版年：1936

中國哲學史大綱（卷上）　胡適（1891—1962）
譯本題名：古代支那思想の新研究（日文）
譯者：楊祥蔭；內田繁隆
出版地：東京
出版者：巖松堂書店
出版年：1925

《古史辨》自序　顧頡剛（1893—1980）
譯本題名：*The Autobiography of a Chinese Historian*（英文）
譯者：Arthur W. Hummel（恒慕義）
出版地：Leyden
出版者：Late E. J. Brill Ltd.
出版年：1931

東西文化及其哲學　梁漱溟（1893—1988）
譯本題名：東西文化とその哲学（日文）
譯者：長谷部茂
出版地：東京
出版者：農山漁村文化協會
出版年：2000

東西文化及其哲學　梁漱溟（1893—1988）
譯本題名：*Les cultures d' Orient et d' Occident et leurs philosophies*（法文）
譯者：Luo Shenyi（羅慎儀）
出版地：Paris
出版者：Presses universitaires de France
出版年：2000

中國歷代政治得失　錢穆（1895—1900）
譯本題名：*Traditional Government in Imperial China, A Critical Analysis*（英文）
譯者：Chün-tu Hsüeh（薛君度）；George O. Totten（陶慕廉）
出版地：Hong Kong
出版者：The Chinese University Press, Hong Kong
出版年：1982

中國哲學史　馮友蘭（1985—1990）
譯本題名：*A History of Chinese Philosophy*（英文）
譯者：Derk Bodde（卜德）
出版地：Peiping
出版者：Henri Vetch
出版年：1937

中國哲學史　馮友蘭（1985—1990）
譯本題名：*A History of Chinese Philosophy*（英文）
譯者：Derk Bodde（卜德）
出版地：Princeton
出版者：Princeton University Press
出版年：1952

老子全譯　任繼愈（1916—2009）
譯本題名：老子訳注（日文）
譯者：坂出祥伸；武田秀夫
出版地：東京
出版者：株式會社東方書店
出版年：1994

漢唐佛教思想論集　任繼愈（1916—2009）
譯本題名：中国仏教思想論集（日文）
譯者：古賀英彦等
出版地：東京
出版者：株式會社東方書店
出版年：1980

中國佛教史　任繼愈（1916—2009）
譯本題名：中国仏教史（日文）
譯者：丘山新；小川隆；河野訓；中條道昭
出版地：東京
出版者：柏書房株式會社
出版年：1992

中華文明史 袁行霈(1936—)
譯本題名: *The History of Chinese Civilization*(英文)
譯者: David R. Knechtges(康達維)
出版地: Cambridge
出版者: Cambridge University Press
出版年: 2012

傳統與現代:人文主義的視界 陳來(1952—)
譯本題名: *Tradition and Modernity*(英文)
譯者: Edmund Ryden
出版地: Leiden; Boston
出版者: Brill
出版年: 2009

簡明中國文學史 駱玉明(1951—)
譯本題名: *A Concise History of Chinese Literature*(英文)
譯者: Ye Yang(葉揚)
出版地: Leiden; Boston
出版者: Brill
出版年: 2011

陶淵明影像:文學史與繪畫史之交叉研究 袁行霈(1936—)
譯本題名: 도연명을 그리다 — 문학과 회화의 경계(韓文)
譯者: 김수연(金秀燕)
出版地: 한국 경기도 파주시
出版者: 태학사
出版年: 2012

中國詩歌藝術研究 袁行霈(1936—)
譯本題名: 詩の芸術性とはなにか(日文)
譯者: 佐竹保子
出版地: 東京
出版者: 汲古書院
出版年: 1993

中國古代文體概論 褚斌杰(1933—2006)
譯本題名: 中国の文章-ジャンルによる文学史(日文)
譯者: 福井佳夫
出版地: 東京
出版者: 汲古書院
出版年: 2004

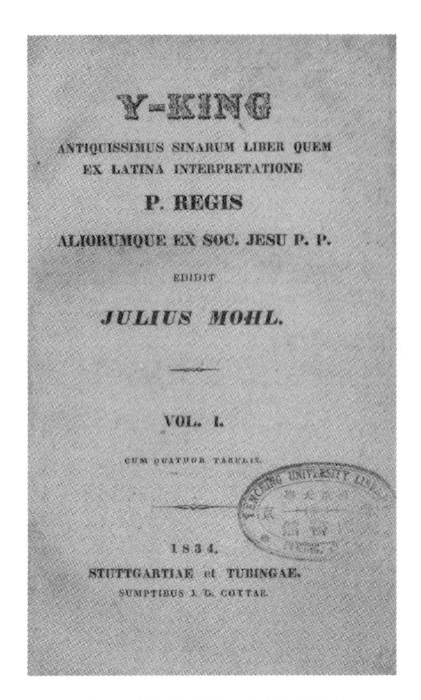

Y-KING
ANTIQUISSIMUS SINARUM LIBER QUEM
EX LATINA INTERPRETATIONE
P. REGIS
ALIORUMQUE EX SOC. JESU P. P.
EDIDIT
JULIUS MOHL.

VOL. I.
CUM QUATUOR TABULIS.

1834.
STUTTGARTIAE et TUBINGAE,
SUMPTIBUS J. G. COTTAE.

圖1 《易經》1834年拉丁譯本

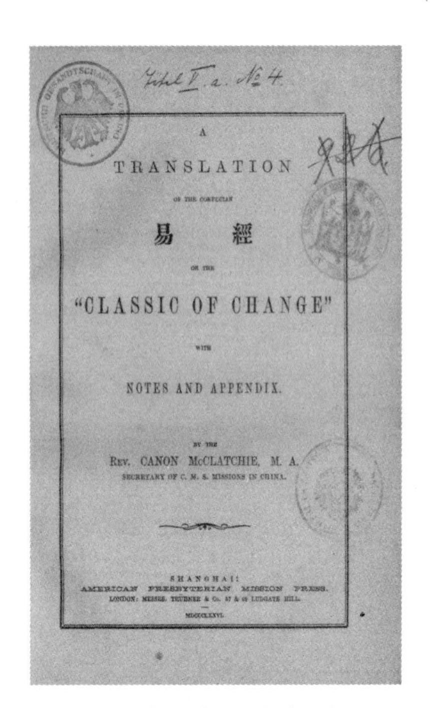

A
TRANSLATION
OF THE CONFUCIAN
易 經
OR THE
"CLASSIC OF CHANGE"
WITH
NOTES AND APPENDIX.
BY THE
REV. CANON McCLATCHIE, M. A.
SECRETARY OF C. M. S. MISSIONS IN CHINA.

SHANGHAI:
AMERICAN PRESBYTERIAN MISSION PRESS.
LONDON: MESSRS. TRÜBNER & Co. 57 & 59 LUDGATE HILL.
MDCCCLXXVI.

圖2 《易經》1876年英譯本

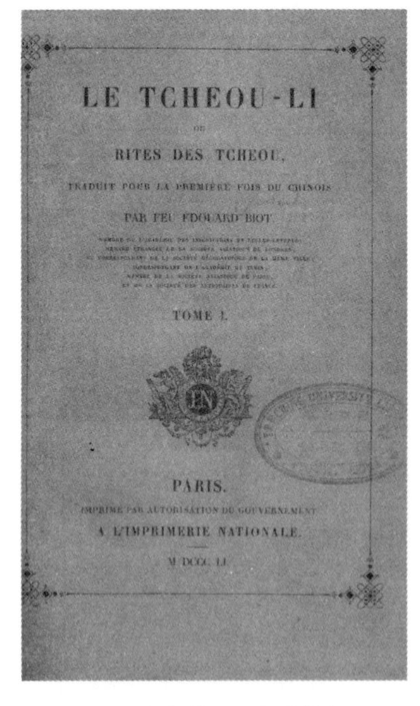

LE TCHEOU-LI
OU
RITES DES TCHEOU,
TRADUIT POUR LA PREMIÈRE FOIS DU CHINOIS
PAR FEU EDOUARD BIOT.

TOME I.

PARIS.
IMPRIMÉ PAR AUTORISATION DU GOUVERNEMENT
A L'IMPRIMERIE NATIONALE.
M DCCC LI.

圖3 《周禮》1851年法譯本

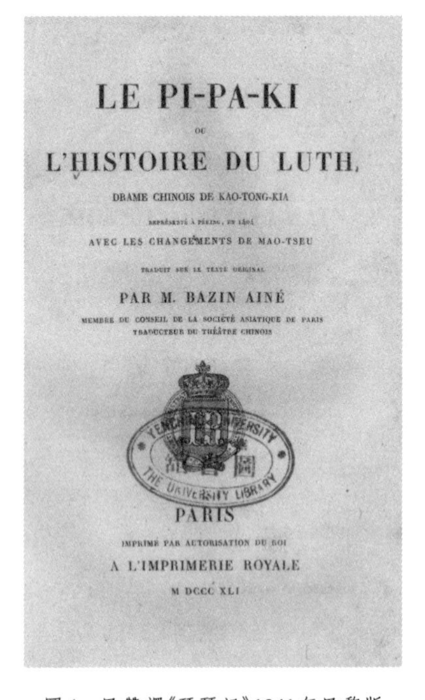

LE PI-PA-KI
OU
L'HISTOIRE DU LUTH,
DRAME CHINOIS DE KAO-TONG-KIA
REPRÉSENTÉ À PÉKING, EN 1404
AVEC LES CHANGEMENTS DE MAO-TSEU
TRADUIT SUR LE TEXTE ORIGINAL
PAR M. BAZIN AINÉ
MEMBRE DU CONSEIL DE LA SOCIÉTÉ ASIATIQUE DE PARIS
TRADUCTEUR DU THÉÂTRE CHINOIS

PARIS
IMPRIMÉ PAR AUTORISATION DU ROI
A L'IMPRIMERIE ROYALE
M DCCC XLI

圖4 巴贊譯《琵琶記》1841年巴黎版

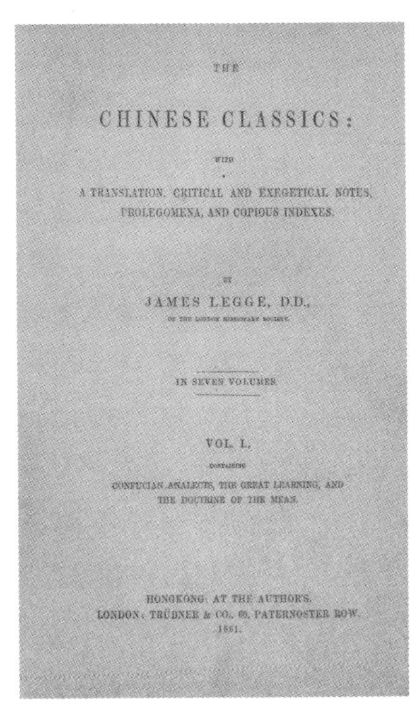

THE

CHINESE CLASSICS:

WITH

A TRANSLATION, CRITICAL AND EXEGETICAL NOTES, PROLEGOMENA, AND COPIOUS INDEXES.

BY

JAMES LEGGE, D.D.,

OF THE LONDON MISSIONARY SOCIETY.

IN SEVEN VOLUMES.

VOL. I.

CONTAINING

CONFUCIAN ANALECTS, THE GREAT LEARNING, AND THE DOCTRINE OF THE MEAN.

HONGKONG: AT THE AUTHOR'S.
LONDON: TRÜBNER & CO., 60, PATERNOSTER ROW.
1861.

圖5　理雅各《中國經典》1861年

圖6　部分展覽圖書

漢學人物

一個被遺忘的晚清大收藏家

——關於景其濬的初步研究

陳　霄

晚清之際,一批具有官宦背景的文人在金石書畫的收藏領域十分活躍,不少成爲頗負時名的收藏家,其中又以顧文彬(字蔚如,號艮庵,1811—1889)的過雲樓最負盛名,"江南收藏甲天下,過雲樓收藏甲江南"。一百多年過去了,不但研究藝術史的學者對過雲樓的故事耳熟能詳,經他收藏的書畫也常出現在各大公私博物館舉辦的展覽中和近年來的一些拍賣會上,爲當代收藏家所追捧。今天,當人們談及晚清的官員收藏家,過雲樓俯瞰群彦,儼然翹楚。然而,生於晚清書畫鑑藏世家的崇彝(字泉孫,號巽庵,1885—1945),在談及當時的書畫收藏史時,卻將一個今天已被人們遺忘的名字"景其濬"和顧文彬相提並論:

> 咸豐、同治年間,……景劍泉閣學(其濬)、顧艮庵(文彬),兩家所收書畫尤精。顧氏所藏有過雲樓書畫記十卷可稽。景氏未有成書,據吾所知,後又歸於張吉人先生(度),又轉售於楊蔭北京卿(壽樞),皆光緒間事也。①

作者單位:美國波士頓大學藝術與建築史系

① 崇彝,《道咸以來朝野雜記》,北京:北京古籍出版社,1982,57—58頁。崇彝乃柏葰(字静濤,? —1859)之孫。

由於景其濬没有詩文集和收藏著録傳世，今天的人們不但對他已極爲陌生，更無從知曉他的收藏曾可以和過雲樓媲美。本文藉助清人著作中的相關文字，尤其是晚清官員的日記、尺牘、書畫著録和其他史料，以及公私收藏中鈐有其收藏印的書畫作品，鉤沉輯佚，對景其濬的生平、交遊和收藏活動作一初步的梳理和研究。

一、生平

景其濬，字劍泉（一作鑑泉），别號師吾儉齋居士、師吾儉齋主人、讀有用書齋主人、句町者卜河漁者[①]，生年未詳，卒於光緒二年（1876）三月。景其濬出生在貴州興義一個有官宦背景的書香世家。曾祖景殿颺（字扶雲）官貴築教諭。祖景震（字乘剛，1742—1811）爲諸生，授錦屏縣教諭[②]。父景壽春（字介菴，1790—1850），嘉慶六年（1801）補府學弟子員，十五年（1810）舉於鄉。道光七至十二年（1826—1832），景壽春相繼任鎮洋縣事、丹徒知縣，賑災有法，治理河道，懲治刁民，百姓得到很多實惠。十四年（1834）調補長洲知縣，十六年（1836）升通州知州，二十二年（1842）署徐州知府，適逢英軍沿海爲寇，直入内地，景壽春團結鄉勇，率先防堵[③]。

景其濬有一兄，名其淦，一弟，名其沅[④]。景其淦爲優貢生，久在江蘇做官，道光二十三年（1843）署新陽知縣，二十四年署靖江知縣，還曾署任金壇知

① 别號“師吾儉齋主人”參見附録一景其濬藏王鑑《壬寅九秋仿古册》。“師吾儉齋主人”“讀有用書齋主人”參見夏穎整理，《景其濬致朱學勤手札》，載上海圖書館歷史文獻研究所編，《歷史文獻》第十輯，上海：上海古籍出版社，2006，76—111頁。“句町者卜河漁者”參見美國波士頓美術館藏傳沈周《江天暮雪卷》畫後景其濬跋。此爲筆者目前發現的三個景其濬的别號。

② 鄒漢勳、朱逢甲纂，《興義府志》卷六三《景震傳》，清咸豐四年刻本，6b—7a。

③《興義府志》卷六一《景壽春傳》，12b—13b頁。

④《景壽春傳》云：“子其淦以優貢生官荆溪知縣，加同知銜，次其濬。”見劉顯世修，楊恩元纂，《貴州通志》“人物志五”，民國三十七年鉛印本，69a。景其淦爲官的時間在道光年間，明顯早於景其濬，加之景其濬有一方“景二”的朱文小印，因此可以肯定，景其淦爲景其濬的兄長。前引《景震傳》云，景壽春還有一子，名其澍（生卒年不詳），曾任貴州黄平州訓導，排在景其濬之前。但不見其他記載説景其濬有此兄。

縣和荆溪知縣[1]。景其沅同治八年(1869)成舉人後,久在四川做官,光緒十七年(1891)署眉山縣令[2],光緒二十四年(1898)署越嶲廳同知,整頓書吏舞弊糧户上米,贏得糧户愛戴並爲他立去思碑[3]。後又任四川酉陽知州[4]。景其澍曾任貴州黄平州訓導[5]。從地方志可知,貴州景氏"其"字輩尚有景其澤(拔貢生)、景其全(舉人)等[6],他們和景其濬的關係有待進一步考證。目前已知景其濬育有兩子,即景方昇(生卒年不詳)和景方昶(1866—?)[7]。景方昶,字旭初,號明久,光緒十五年(1889)進士,五月改翰林院庶吉士,十六年(1890)四月,散館授翰林院編修,曾官至湖南辰州府知府(從四品),賞二品銜,著《東北輿地釋略》[8]。

景氏一家在興義府、乃至整個貴州是極負盛名的。祖孫三代科場甚是順暢,不但景其濬是興義郡北闈中舉的第一人,更爲顯赫的是,興義在明清兩代只出過景其濬和其次子方昶兩個進士[9]。景氏祖孫三代分别在京師、江蘇、四川、湖南等地爲官,這在當時地處偏遠、經濟相對落後、教育資源匱乏的貴州地區實屬罕見。從零散的記述也可看出,他們爲官期間對興義府的關切源源不斷,爲鄉里的文化發展做出了貢獻。如景壽春雖然長年在外省做官,但十分關心鄉里,曾請興義府知府會同士紳代爲購田,歲收租作爲郡士的永久試

---

① 金吴瀾修,朱成熙纂,《崑新兩縣續修合志》卷十六,清光緒六年刊本,28a。又:葉滋森修,褚翔纂,《靖江縣志》卷十,清光緒五年刻本,41b。又:夏宗彝修,汪國鳳纂,《金壇縣志》卷五,清光緒十一年活字本,11b。又:施惠修,吴景牆纂,《宜興荆谿縣新志》卷五,清光緒八年刊本,7b。

② 王銘新修,郭慶琳纂,《眉山縣志》卷九,民國十二年石印本,6a。

③ 越嶲,現越西縣,四川涼山彝族自治州所轄。當時"收米管倉書吏門級巧立敷米、樣米、灰米、掛腳米等名目,糧户上米百般刁難,受害頗深。其沅蒞任,紳糧訴請除弊,並恐該書等阻撓,願於上糧時,每石給錢二百四十文,又向係一米二穀,亦願改爲二米一穀,以資公費。其沅察核屬實,如稟勒石,並將案提存户房使,倉書不得再匿案舞弊,迭稟上憲存案,糧户沾感無暨"。見馬湘纂,孫鏘修,《越嶲廳全志》卷七之一,清光緒三十二年鉛印本,3a。

④《貴州通志》"選舉志三",15b。酉陽,今重慶市東南,清雍正十三年(1735)"改土歸流",酉陽設直隸州,轄酉、秀、黔、彭四縣。

⑤《興義府志》卷六三《景震傳》,6b—7a。

⑥ 張瑛修,鄒漢勳、朱逢甲纂,《興義府志》,清咸豐四年刻本,民國三年貴陽文通書局據刻本鉛印本,9頁。卷首"採訪"對象中提及景其澤、景其全(二人均生卒年不詳)。

⑦《清國史》本傳,引自台灣"中研院"歷史語言研究所"人名權威-人物傳記資料查詢":http://archive.ihp.sinica.edu.tw/html_name/index. php。

⑧ 龐思純,《明清貴州700進士》,貴陽:貴州人民出版社,2005,258頁。

⑨ 龐思純,《明清貴州700進士》,258頁。

費,又爲"郡之應鄉試者,别捐金爲三科卷費"[1]。景其濬爲官期間,亦時時關心家鄉,光緒元年(1875),他將"咸同年間貴州頻遭兵禍,興義府深受其害及官紳士民死難情形"上報朝廷,爲家鄉請蔭襲,又主張將他們的絶産充公[2]。

景其濬在道光二十九年(1849)成爲舉人,咸豐二年(1852)清文宗登極,開恩科,景其濬爲二甲二十三名,朝考選翰林院庶吉士,散館授翰林院編修[3]。咸豐五年(1855)六月十二日,景其濬充浙江鄉試考官[4],七年(1857)充日講起居注官,九月大考二等,以庶子升用,十二月補授右春坊右中允,八年(1858)充順天鄉試同考官[5]。同年十二月廿八日,景其濬奉旨接替翁同龢(字聲甫,號叔平,晚號松禪,1830—1904)選陝甘學政[6]。翁同龢在日記裏也記述了他因病解任回京、由景其濬接任的事:咸豐九年"正月十一日,午刻折差回,開缺折於十二月廿八日呈遞,恩准回京調理,新任放景其濬"[7]。然而就在景其濬準備上任時,因牽涉戊午順天鄉試科場案而臨時罷任,最終由慎毓霖任

---

① 《清國史》本傳,引自台灣"中研院"歷史語言研究所"人名權威-人物傳記資料查詢"。

② 安龍縣史志信息網,"古代人物—景其濬"。http://alsz.gzdaxx.gov.cn/alrw/gdrw/2014-01-02/18646.html。

③ "道光二十九年景其濬應順天鄉試中式,此興郡有北闈舉人之始也。"見《興義府志》卷四九,31b。舉進士及授翰林院編修,見朱寶炯、謝沛霖編,《明清進士題名碑録索引》第三册,上海:上海古籍出版社,1980,2509頁。又:《清國史》本傳,引自台灣"中研院"歷史語言研究所"人名權威-人物傳記資料查詢"。

④ 景其濬抵浙後,薛時雨(字慰農,號澍生,安徽全椒人,1818—1885)曾與之泛舟。薛時雨有《臨江仙·闈後偕景劍泉太史其濬泛湖》一詞記其事:"天上文星湖上落,一湖秋水澄鮮。畫船來往夕陽邊。波光濃似酒,人影淡於煙。珊樹珍奇齊入網,遺珠訪到嬋娟。錦袍紅袖兩翩躚。泥他金縷曲,酬爾玉堂仙。"玉堂仙,翰林學士的雅號。從這首詞的内容可以想象,景其濬咸豐五年(1855)八月秋闈後和薛時雨一同泛湖的場景。景其濬除咸豐五年己卯科浙江鄉試考官和八年充順天鄉試同考官,没有再擔任過考官一職。薛時雨是咸豐三年(1853)的進士,地方志記載薛時雨"咸豐乙卯[1855]來宰嘉興","戊午[1858]由嘉興令調嘉善",可知景其濬監考時,他在浙江做官。見許瑶光修,吴仰賢纂,《嘉興府志》卷八七,清光緒五年刊本,56a。又:江峰青修,顧福仁纂,《重修嘉善縣志》卷十五,清光緒十八年刊本,19b。

⑤ 錢實甫編,《清代職官年表》,北京:中華書局,1980,2970頁。

⑥ 魏秀梅編,《清季職官表附人物録》,台北:近代史研究所,1977,474頁。

⑦ 翁萬戈編,《翁同龢日記》第一卷,上海:中西書局,2011,53頁。

學政[1]。

咸豐十年(1860),景其濬提督河南學政,十一年(1861)擢翰林院侍講學士,留學政任,十二月轉侍讀學士,仍留學政任[2]。《清史本傳》記述了他在河南率領開封全城抵抗捻軍的功績,"會捻逆竄至朱仙鎮,省城戒嚴,其濬會同防守,連挫賊鋒,並捐廉犒賞,衆志愈奮,省城以全。"[3] 景其濬還上疏咸豐整頓治理各省軍務和變革捐輸保舉人員[4]。

同治元年(1862)四月二日,景其濬奏進《歷代君鑑》,慈禧十分高興,諭令:"景其濬所進《歷代君鑑》足資考鏡,著留覽。"[5] 二年(1863)六月,他又上疏推薦河南各州縣的貢生[6]。

同治四年(1865)十一月,景其濬河南學政任滿[7]。次年(1866)四月,充實録館纂修,四月大考獲二等,升用爲詹事府少詹事,十二月編纂《文宗實録》告成,賞三品銜[8]。六年(1867)六月,景其濬上疏彈劾貴州巡撫張亮基(字采臣,號石卿,1809—1871)"玩兵侵餉,縱暴殃民",張亮基後被褫職[9]。七月,景其濬因上奏遵義知府鄭某殉難請卹,卻在朝廷的覆查中,被發現鄭某是陷賊逃歸者,因此被交部議處[10]。不過,從這兩個奏疏可以看出,身爲貴州籍京官的

---

① "九年正月以順天科場獄未定讞詔解任,尋以校閱無私免議。"見《清國史》本傳,引自台灣"中研院"歷史語言研究所"人名權威-人物傳記資料查詢"。又:《清季職官表附人名録》也記載其離職陝甘學政原因爲"科場案罷任"(根據清代起居注册咸豐朝45册,027149頁),474頁。又:翁同龢自編年譜云:"(咸豐九年,三十歲)正月奉批回:准其開缺回京。後任杜瑞聯[棣雲,號鶴田,1831—1991]丁憂,景其濬緣事皆未至。繼之慎毓霖[生卒年不詳]也。"見謝俊美編,《翁同龢集》下册,北京:中華書局,2005,1029頁。

② 錢實甫編,《清代職官年表》,2729—2730頁。

③ 朱仙鎮,位於今河南開封縣西南。1860年(庚申),捻軍爲奪取户部皇倉的糧食,攻掠江蘇北部京杭大運河畔的商業重鎮,駐有南河總督的清江浦(今淮安市主城區),並焚毁清江浦二十里長的街市,以及屬於户部的皇倉和屬於工部的四大船廠。但十五公里外駐有漕運總督的淮安府城(今淮安市淮安區)因爲城牆高大堅固,未能攻下。同年攻打的主要城市還有開封和濟寧。見《清國史》本傳,引自台灣"中研院"歷史語言研究所"人名權威-人物傳記資料查詢"。

④《清國史》本傳,引自台灣"中研院"歷史語言研究所"人名權威-人物傳記資料查詢"。又:《頒發條例(道光元年至同治十三年[1821—1874])》,清同治十三年刻本。

⑤《大清穆宗毅皇帝(同治)實録》卷二一,長春:"大滿洲帝國國務院",1937。

⑥《清國史》本傳,引自台灣"中研院"歷史語言研究所"人名權威-人物傳記資料查詢"。

⑦《清國史》本傳,引自台灣"中研院"歷史語言研究所"人名權威-人物傳記資料查詢"。

⑧《清國史》本傳,引自台灣"中研院"歷史語言研究所"人名權威-人物傳記資料查詢"。

⑨《清國史》本傳,引自台灣"中研院"歷史語言研究所"人名權威-人物傳記資料查詢"。

⑩《翁同龢日記》第二卷,578頁。

景其濬對家鄉的政治參與甚深。

鄭某事件似乎對景其濬的仕途並没有什麽影響，同治七年(1868)五月，景其濬由詹事府少詹遷詹事，八月擢内閣學士兼禮部侍郎銜[①]，達到他官宦生涯的高峰。

同治九年(1870)八月一日，景其濬再次外放，提督安徽學政。他在南下經過天津時，專門去拜訪了直隸總督曾國藩(字伯涵，號滌生，1811—1872)。曾國藩九月十四日日記記載："見客三次，安徽學使景鑑泉談頗久……出門至河干拜景鑑泉。"[②] 同治十一年(1872)五月廿三日，景其濬因牽涉安徽天長縣知縣馮至沂投水自盡一案而交部議處，因案解任[③]。次年(1873)回到北京，充文淵閣直閣事[④]。

光緒元年(1875)歲末，閣學同仁聚餐，景其濬因病重無法赴約[⑤]，次年(1876)三月十六日，他在北京去世[⑥]。

---

① 錢實甫編，《清代職官年表》，1080頁。《清代職官年表》和《清季職官表附人名録》中禮部侍郎一欄並没有官方記載任用景其濬一事，只在别處有所提及(《清季職官表附人名録》，285頁)。又：吴坤修修、何紹基纂，《重修安徽通志》"職名"，清光緒七年刻本，1b，"原任内閣學士兼禮部侍郎御前任提督安徽學政"提及"兼禮部侍郎"銜。

②《曾國藩日記》下册，天津：天津人民出版社，1995，2357頁。九月十四日離景其濬八月一日任命不久，應是景從北京出發沿運河坐船經過天津時拜訪曾國藩，"河干"可能是景所乘的船泊處或是運河邊的住所。

③ "十年，考試泗州，署天長縣知縣馮至沂因辦考差被署盱眙縣知縣路玉階逼勒，投水自盡。十一年五月，諭曰：'前據何璟[字伯玉，號小宋，1817—1888]奏署天長縣知縣馮至沂自盡一案，有牽涉學政情事。景其濬著開安徽學政之缺。俟結案後，再行來京。'旋經兩江總督何璟查明奏聞，諭曰：'此案署盱眙縣知縣路玉階於學政臨考之時，遽行稱疾請假。時馮至沂正在盱眙，景其濬即派巡捕盱眙縣典史逍誠持帖，諭令暫緩回縣，馮至沂疑係路玉階聳留幫辦，並因路玉階禀稱天長考費文内有籌欵多金歸入宦囊之言，以致愁急莫釋，觸發痰疾，遽行投水。景其濬於路玉階請假之時，不令提調官派員辦考，徑遣巡捕官持帖諭留，殊屬措置失宜，著交部議處。'" 見《清國史》本傳，引自台灣"中研院"歷史語言研究所"人名權威-人物傳記資料查詢"。又：《大清穆宗毅皇帝(同治)實録》卷二八〇，"尋議景其濬應比照妄行條奏例降一級調用。得旨：准其抵銷。"

④《清國史》本傳，引自台灣"中研院"歷史語言研究所"人名權威-人物傳記資料查詢"。

⑤ 光緒元年十二月十八日，"閣學備飯請協同批本者，余與叔雨作主，劍泉病不能到也。"見《翁同龢日記》第三卷，1212頁。

⑥ 光緒二年三月十六日，"是日景劍泉物故，家有老母，可傷可傷！"見《翁同龢日記》第三卷，1232頁。

二、交遊

從1852年登進士到1876年在北京去世,景其濬除了一次出任浙江鄉試考官和兩次出任學政外,二十四年的仕宦生涯有十八年是在北京度過的。這十八年中,與之交往密切者,除同鄉著名文人莫友芝(字子偲,號郘亭,1811—1871)外,多爲喜愛收藏金石書畫的京官,代表人物有翁同龢、潘祖蔭(字伯寅,號鄭盦,1830—1890)、朱學勤(字修伯,號結一廬主人,1823—1875)、沈樹鏞(字韻初,號鄭齋,1832—1873)、徐郙(字頌閣,1838—1907)、王懿榮(字正儒,號廉生,1845—1900)等。景其濬收藏金石書畫的活動也和這些友人有着密切關係。

在存世文獻中,翁同龢的日記提供了景其濬和友人交往的資料最多,使筆者能大致重構景其濬與翁同龢等友人的日常互動。翁同龢是咸豐六年(1856)進士,後入翰林,成爲景其濬的同事,又因爲志趣相投,成爲至交。景其濬的名字出現在存世最早的翁同龢日記中:

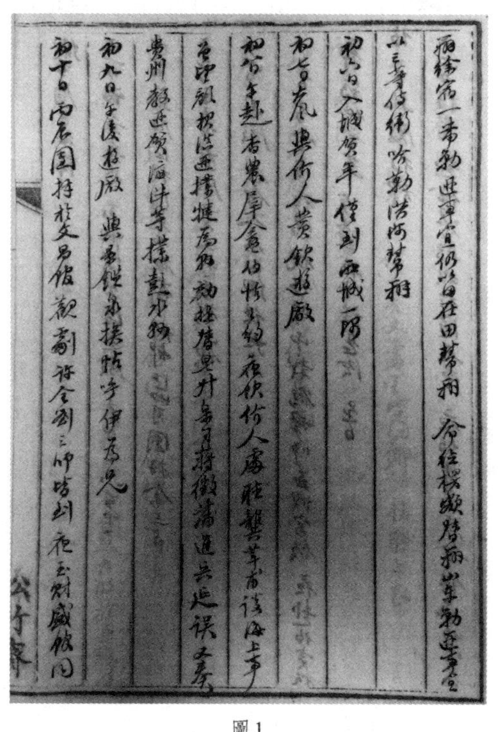

圖1

咸豐八年(1858)八月廿六日:"閱邸抄,知順天考官柏葰、朱鳳標、程庭桂,同考官麟書、……郭夢惠、景其濬、瑞聯……"

咸豐九年己未(1859)正月十一日:"恩准回京調理,新任放景其濬。"

咸豐十年(1860)正月:"初九日,午後遊廠。與景鑑泉换帖,呼伊爲兄。"[①](見圖1)

最後這則日記告訴我們,在1860年初,景其濬和翁同龢换帖成爲結拜兄弟,因景其濬年長,翁稱景爲兄。此後的日記記載了他們更多的交往,如觀劇、宴飲、聊天、贈禮等[②]。聽聞景其濬放河南學政,翁同龢邀約友人爲其賀喜、餞行:

咸豐十年閏三月廿一日:"汪慕杜(汪承元,生卒年不詳)、朱修伯、景鑑泉來。夜瀕石[楊泗孫,1823—1889]來。賀景鑑泉放河南學政。"

四月四日:"五兄與余餞景鑑泉於湖廣館,杏農、辛伯、湛田、瀕石、伯寅作陪。"[③]

景其濬赴河南後,翁同龢仍舊時時關心他的近況,留意他的職官變動[④]。

景其濬赴河南任職前,莫友芝也在北京。咸豐十年(1860),莫友芝目睹内憂外患日深,英法聯軍又已進逼天津塘沽,時局緊張,適逢會試再次落榜,"而截取知縣日期難定,留京生活又無著落,遂決意出京。而所識才俊英彦意紛紛於此期出京,頗有風流雲散之感。"[⑤]不難揣測,時局未靖,而景其濬等知己已经或將陸續離京,莫友芝此時科場失意顯得更爲心酸。打聽到九弟祥芝正在安徽懷寧任職,他決定離京南下,臨行前贈詩衆師友,其中一首即《送景

---

①《翁同龢日記》第一卷,38頁、53頁、71頁。

②"景鑑泉約文昌館觀劇。""夜約杏農[尹耕雲,1814—1877]、犀盦[錢桂森,字辛白,1827—1902]、鑑泉、伯寅、伯恬[周儀暐,生卒年不詳]、星泉[蔣師淵,生卒年不詳]飲。"見《翁同龢日記》第一卷,70—71頁。又:"訪袁聲陔、景鑑泉不遇。"見《翁同龢日記》第一卷,75頁。又:"何白英、景鑑泉來。"見《翁同龢日記》第一卷,77頁。又:"景鑑泉來,贈余筆六枝。"見《翁同龢日記》第一卷,79頁。又:"訪杏農、鑑泉。"見《翁同龢日記》第一卷,80頁。

③《翁同龢日記》第一卷,80—88頁。

④ 咸豐十一年(1861)二月廿三日:"景鑑泉放翰林學士,鄭敦謹放理卿。"見《翁同龢日記》第一卷,132頁。又:八月初五日:"……河南景其濬、四川黄倬……均無庸更换。"見《翁同龢日記》第一卷,161頁。

⑤ 張劍撰,《莫友芝年譜長編》,北京:中華書局,2008,190頁。

鑑泉中允提學河南》,表達了對知己才幹的欣賞和遠大前途的祝福:

儲才本計專提學,重以成均拔萃科。使者非時隆簡授,(非學政例替時,乃以召還前任者,特簡君往。)中州從古擅英多。懸知柱石搜名岳,待挽頹瀾砥濁河。胸次人倫真鑒在,瑰琦從此謝岩阿。[①]

莫友芝離開京城後,八月十五日經臨城到達邯鄲,過梁園、磁州至河南彰德府(府治安陽),又與景其濬重逢[②]。時隔近半年再次見到好友,他又作詩《彰德晤景鑑泉學使試壁留飲》紀念這來之不易的相逢:

星使持衡鎖院開,行車歸馬正虺隤。春風有約留相訪,舊雨仍逢定幾回。笑指群鷗能狎客,促添雙鯽佐銜杯。鄴中名碩曹劉邈,摸索應無失此才。[③]

匆匆一面後,莫友芝便離去,而兩人的書信聯繫卻未曾斷絶[④]。

在景其濬赴河南接替李鴻藻(字季雲,號蘭孫,1820—1897)學政一職前,徐郙也曾想爲其餞行。但景其濬未能赴約,一則可能太忙,二則當年有恩科,景其濬希望徐郙能專心備考。到達河南後,景其濬寫了一信向徐郙致意:

判襼春明,猥承寵餞,衹以匆匆行李,未遑親領盛情,而感荷之私,不啻飽飫郇廚也。辰惟頌閣仁弟世大人履祺豫萃,鼎祉升恒,引跂卿輝,奚如欣抃。兄於清和廿三日抵豫,李蘭翁適已出棚,俟旋署後接篆受事,即日俶裝按試各屬。惟是初持文枋,譾陋自慚,赤水求珠,愧非象網,尚冀南針時錫,藉作韋弦,曷勝盼禱之至。耑泐鳴謝,即頌升祺,諸維愛照,不

---

① 此詩原見於《郘亭遺詩》卷六,九頁。見莫友芝著,龍先緒、符均箋注,《郘亭詩抄箋注》,西安:三秦出版社,2003,570—571頁。

②《莫友芝年譜長編》,203頁。

③《郘亭詩抄箋注》,571—572頁。

④《郘亭日記》:"[咸豐十一年辛酉(1861)五月初一]晴。四月一月中寒雨逾十分之二,至今日夏令乃正。作書复景劍泉提學及其幕客汪芸石。"見《莫友芝年谱长编》,226頁。

宣。愚兄景其濬頓首。①

從"仁弟世大人"的稱呼可推測,景其濬和徐郙的父親徐經(字桓生,號拜庚,1788—1856)認識。徐經是嘉慶二十四年(1819)進士,曾任翰林院編修,久在京城爲官,景其濬應熟識。出身官宦家庭的徐郙,不但很有文才,而且擅長書畫並喜愛收藏,和景其濬有相同的愛好。徐郙早年入國子監學習,咸豐九年(1859)參加順天鄉試,中舉人。因此,景其濬在信中表達了對徐郙高中進士和仕途輝煌的期盼。同治元年(1862),徐郙高中狀元,此後仕途一路順暢。

從河南回京後到同治九年(1870)接安徽學政任前,景其濬在京師和友人頻繁地交往,鑑賞金石書畫成爲他們聚會的一個重要内容。翁同龢的日記有不少記載,爰録數條於下:同治四年(1865)三月初四日:"訪景劍泉,觀所藏《九成宫》二本,一宋拓,一明神廟本也。"同治五年(1866)正月初三日:"晚訪景劍泉,談至暮。"同治六年(1867)二月朔:"夜赴伯寅、若農[李文田,字畲光,號若農,1834—1895]招,在坐者劍泉、辛白、午橋[張丙炎,字午橋,號藥農,生卒年不詳]、海秋[許宗衡,字海秋,1811—1869]、萊山叔侄[孫毓汶,字萊山,1834—1899]、定子[吕耀斗,字庭芷,號定子,1828—1895],暮歸乏極。"同治八年己巳(1869)七月廿三日:"訪晤景劍泉。請陳繼生(惟和)代筆書啓,每月二金,每節二金,託劍泉爲先容。"②

翁同龢在景其濬家觀賞《九成宫醴泉銘》,是當時京師愛好收藏的官員們公餘文化活動的一個縮影,類似的記載屢見不鮮,如翁同龢致景其濬的另兩通信札:

雛鳳本墨光照眼,龢所藏真不足論矣,摩挲一過,敬即奉繳,當於晡時送呈也。劍泉吾師左右。同龢頓首上。初九。

日來感風未能趨造,十六應真墨未識是真曹素功否。(恐非老曹素

① 上海圖書館藏(稿本)。此札或爲幕僚代筆的稿本,因爲用精細的楷書給晚輩寫信並不多見,而且此札没有日期,對於異地之間的往還信札不多見。上海圖書館另有景其濬致長輩潘曾瑩(字申甫,1808—1878)的一通行書信札:"……恭請世伯大人崇安。秋深晚凉,尚希珍攝。世愚姪景其濬頓首。謹啓。匆促間未及恭楷,並乞鑑宥。"景其濬對於他匆促間寫就信札、未能用恭楷完成而對長輩深表歉意。這一定程度説明了他的恭楷多是寫予長輩的。

②《翁同龢日記》第一卷,407頁;第二卷,468頁、546頁、738頁。

功墨。)奉去一錠,乞鑑别,應試者將用以入墨合也。敬上劍泉仁兄老前輩大人。同龢謹啓。十一。[①]

從前文所引翁同龢觀賞景其濬藏《九成宫》拓本和此處對"雛鳳本"的歎賞,到請景其濬爲之鑑定墨,可以看出翁同龢對景其濬家藏和鑑定知識的歆羨。

莫友芝也曾向景其濬借閱碑帖。莫友芝在《吴天發神讖碑鉤本》題識中寫道:"此册後半部'校尉'以下,借於胡魯尊[生卒年不詳]生本,前半墨鉤借景鑑泉本,並舊拓不完,以丁巳[1857]八月在貴陽,己巳[1869]八月在京師摹出。庚申(1860)禮闈後,繩兒又借劉子重[劉銓福,字子重,生卒年不詳]完本鈎補其闕。其本非元石,故朱别之。十月望日,舟行經武昌縣之七磯洪,彙貼成册,題其首,尚有宋人之跋,當别求本摹入。"[②]

景其濬於同治九年(1870)八月一日被任命爲安徽學政,九月以後赴任[③],從北京坐船經運河在閏十月抵達揚州,然後再由淮河入安徽。此時,莫友芝正任揚州書局總校刊,兩位老友在揚州見面後,匆匆揖别。次年(1871)正月十五日,莫友芝致信老友:

劍泉先生年大人閣下:使舟去東經邗江,獲奉數言,頓覺十年客塵,祛除殆盡。匆匆揖别,惟益馳仰。臘杪金陵度歲,奉到賜書,敬悉起居安善,年伯母太夫人又迎養抵皖。自是時月以造士得賢,爲太夫人歡,禄養善養交盡之矣。雖皖中根柢之學,差遜曩昔,首郡士氣,又頗浮動。得吾劍老津津不懈,亟起而震新之,朞月必有可觀,比及三年,一切廢修墜舉可拭目俟也。蒙惠法書四幅,雄偉精詣,蕭條客舍,如睹朱霞,謹當什襲寶藏□□。欲作數紙奉報,氣索不能下筆,俟春和凍解,乃强爲之耳。所示爲親謀外一節,在古昔行之順以易,在今日議之則頗創。聞李見在成

---

① 原件藏常熟市文管會,録文見謝俊美編《翁同龢集》,222頁。謝先生誤將景其濬認作景壽(翁同龢通常稱景壽爲"景額駙",見《翁同龢日記》,第三册,1227頁)。謝先生將信札繫於光緒二年五月也有誤,景劍泉在光緒二年三月去世。從稱謂來看,第一札似是兩者成爲結拜兄弟之前,第二札在此之後。札中所云"雛鳳本",很可能指鄭邸惠園雛鳳樓曾經收藏過的拓本,後歸景其濬。

② 莫友芝著,張劍、陶文鵬、梁光華編,《莫友芝詩文集》下册,北京:人民文學出版社,2008,806頁。

③ 同治九年(1870)九月初五日日記:"邀景劍泉、彭芍亭、夏鷺門、張午橋、潘伯寅飲,江蓉舫、吕定子皆辭未來,客齊已薄暮,散時戌初多。"由此可知,在九月初,景其濬尚未啟程。見《翁同龢日記》第二卷,830頁。

都,京秩小員,亦不許請調,況在大員。必如閣下之意,若以啓札自通,既措辭未便,惟師生面談,差爾無妨。聞明歲有恩科,曷不待録科晤對時一言,爲不著跡乎?友芝萍蓬,習慣里門,還期未知何日,三山客舍即是並州故鄉[①]。方子箴仍有淮南書局之訂,江皋仲春又當遄往。但局中大綱已舉,又遊踪無定也。開正元夕肅復,邀叩侍安,並賀年禧,諸維垂鑒不具。年小弟莫友芝。[②]

莫友芝在信中提到了景其濬幫助他的親友外調職官一事,又爲景其濬出任安徽學政而振興地方學風高興不已。

在揚州會面時,景其濬還給莫友芝帶去一件更爲珍貴的禮物:新出土的漢光禄勳劉曜碑的拓片。此碑"在山東東平州,同治庚午六月新出於州之蘆泉山陽。閏月景鑑泉經邗上贈郘亭"[③]。值得注意的是,此碑在同治九年(1871)六月剛在山東出土,作爲京師的高官,景其濬很快就得到了拓本。深知莫友芝也愛蒐集善本拓片,他以新得拓片相贈。令人惋惜的是,此次相逢成爲兩人的最後一次見面。同治十年秋,莫友芝感冒風寒,醫藥不治,於九月十四日卒於舟中。

潘祖蔭的伯父潘曾瑩(星齋,1808—1878),不但官做得大,而且擅長繪畫。他曾爲景其濬作長卷,景其濬收到畫後,十分欣喜,寫信致謝:

側聞貴體違和,須静攝調元,故未敢摳謁上煩起居。蒙力疾繪長卷,全是徐崇嗣[徐熙孫,北宋畫家]没骨法,謹當什襲,永作家珎,銘感無涯,容再泥謝。[④](見圖2)

同治十二年(1873),景其濬解皖學任後回到京城,三年後去世。從翁同龢的日記和信札中,我們能感受到他去世後家道中落的凄涼。光緒二年

---

① 典出唐代劉皂《長門怨》:"客舍并州數十霜,歸心日夜憶咸陽。無端又渡桑乾水,卻望并州是故鄉。"莫友芝用"三山"喻"并州",意即把異鄉作故鄉。"三山"應當是指南京市西南的三山峰,因在赴揚州任校刊之前,莫友芝曾在金陵任金陵書局總編校。

②《莫友芝年譜長編》,526—527頁。此札係編者據台圖藏《郘亭書函稿》録出,繫年據信函内容及《郘亭日記》。

③ 莫友芝,《宋元舊本書經眼録》附録卷二,清同治刻本。

④ 上海圖書館藏(稿本)。

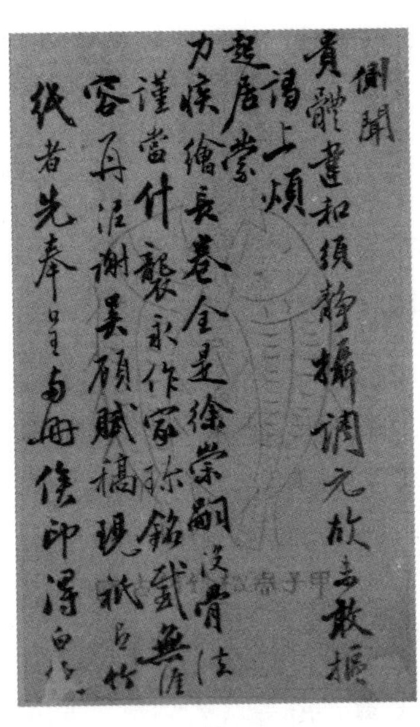

圖2

(1876)景其濬去世後,翁同龢出城“哭景劍泉”[①]。景其濬家有老母和子女,他的去世使家中突然失去主要經濟來源,翁同龢便發信給景其濬的故舊求助。他在光緒二年七月廿二日的日記中寫道:“未出門,爲景劍泉家寫信十函告幫,(外省藩臬道府,己酉、壬子進士。)交子松分寄。”[②] 清代地方官的收入高於京官,所以翁同龢給那些在己酉鄉試、壬子會試與景其濬同時登科並出任布政使(藩)、按察使(臬)、道台、知府的同年們寫信,請他們鼎力相助。

光緒十年(1884),景其濬的母親去世,翁同龢在二月十六日“以三十金助景劍泉之弟官四川者,將侍其母喪歸而無資也。另送二金,其子在此開吊也(派夾槓四員)”[③]。可見景其濬去世後,家道中落,母親過世竟無錢歸葬。

翁同龢在致劉仲良(秉章,字仲良,1826—1905)的信札中,談到拜託時任四川總督的劉秉章照顧景其濬胞弟其沅和長子景方昶一事:

再啓者。景劍泉前輩胞弟其沅,久在蜀中,素承培植,得司馬筴,未

---

①《翁同龢日記》第三卷,1233頁。

②《翁同龢日記》第三卷,1264頁。

③《翁同龢日记》第四卷,1853頁。

綰銅符。其長君[公子]旭林[景方昶]編修捷户讀書，塵中一鶴，令入蜀定省。特附一言，伏望推屋烏之愛。俾刺史得攝方州，則升斗可邀，全家攸賴。專懇。再頌勛祺。諸惟鑒察。不一不一。弟名頓首。①

翁同龢對景其濬家人的持續照顧，足見二人情誼之深重。1901年，已經被罷官回鄉的翁同龢，在家國危亂下回顧他一生的交遊時，寫下這樣一首詩：

平生交遊遍天下，名流韻士如風馳。葉(潤臣)[葉名灃，1807—1859]景(劍泉)沈(仲复)[沈秉成，1823—1895]潘(伯寅)凋謝盡，我篋不受一物遺。森然此例在朋友，九原可鑒千夫知。西風吹雨水洊至，遑恤己病愁民飢。錦鯨還客坐悲嘯，空堂且詠東洲詩。(辛丑五月廿三日。以詩代簡)②

"葉景沈潘凋謝盡"道出了景其濬等友人對於翁同龢的意義。縱使交遊遍及天下，但像景其濬那樣的友人才是他這一生最懷念的。

不同於翁同龢、潘祖蔭等是景其濬的同輩同僚，王懿榮則是一個晚輩。光緒六年(1880)王懿榮中進士時，景其濬已謝世四年，所以兩人最頻繁的交往大致集中在同治十年前後。在《誥封宜人元配蓬萊黄宜人行狀》一文中，王懿榮回憶了他久居北京時與景其濬的交往：

[懿榮]居京師久交遊既廣，每以春秋佳日，與長沙周閣學、吴縣潘侍郎、遵義景閣學、洪洞董研樵檢討、太谷温味秋、儀徵陳六舟、巴陵謝麐伯、餘姚朱肯夫、南皮張香濤、吴縣吴清卿六編修，會稽李蒓客、甘泉秦誼庭、績谿胡荄甫、光山胡石查、遂谿陳逸山五户部，大興劉子重、儀徵陳研香、鄒縣董鳳樵三刑部，元和顧緝廷工部、歙縣鮑子年、長洲許鶴巢兩舍人，遞爲詩酒之會，壺觴幾無虚日。

次年癸酉(1873)秋，中順天副榜第一，九月，宜人陡患傷寒，又誤於醫藥。有江西縣丞丁君者，醫學甚深，景劍泉閣學薦來，使救之，雖就瘉，

① 台北故宫博物院藏。此札書寫時間根據"旭林編修"定在光緒十六年(1890)四月後，因景方昶光緒十六年四月散館，授翰林院編修。

② 見台北故宫博物院藏《清何紹基篆書廣韻》書法册中翁同龢楷書題跋。

而羸弱益甚。[1]

王懿榮所提及在京師與景其濬、潘祖蔭、張之洞(香濤,1837—1909)、吴大澂(清卿,1835—1902)等的交往也間接證明了景其濬和這些人的私交,而薦醫救治王懿榮夫人的細節更可看出了景其濬對這位晚輩的關懷。

王懿榮嗜收藏,凡書籍、字畫、金石文物、印章、錢幣、殘石、瓦當,無不弆藏。北京故宫博物院所藏景其濬致王懿榮的三通信札便展示了二人以收藏爲主題的日常交往:

> 頃所示及河陽[潘祖蔭]所棄一件,暇日當奉上。祈吾兄多打數紙,但求密之,恐尊師[潘祖蔭]我怒也。(中庸[胡義贊]如不知,亦希弗告之。昨所見五色璀璨,真眼福也。許惠鄉味,尚乞分少許。)此請晚安。其濬頓首。廉翁閣下。[2]

王懿榮是潘祖蔭的門人,多年來一直在北京爲其師購買古董奔走[3]。王懿榮向景其濬出示了一件潘祖蔭不打算購買的青銅器,景其濬也不想買,但希望王懿榮能幫他打幾張拓片,並提醒不要讓潘祖蔭和胡義贊知道此事。在另一札中,景其濬也談到了收藏:

> 《畫史彙傳》儻載,至都必奉呈。此時尚未到,遲復爲罪。此請刻安。弟其濬頓首。(角頗佳,而直甚卬。奈何! 雖好之,無力,徒歎息耳。一笑。)王少大老爺。

景其濬的收藏以書畫爲主,信中所説的《畫史彙傳》,又稱《歷朝畫史傳》,清代彭藴燦著。大概是景其濬見到一件此書著録過的畫作,打算等東西到北京後請王懿榮看,徵求他的意見。信中還提到一個青銅酒器角,可能王懿榮向景

---

① 吕偉達編,《王懿榮集》,濟南:齊魯書社,1999,92—94頁。

② 北京故宫博物院藏(稿本)。河陽,即潘祖蔭,此處景其濬用郡望(河陽)代指潘祖蔭,取自曾任河陽縣令而政績累累、名噪一時的西晉文學家潘岳(安仁,247—300),即潘安。中庸,即胡義贊(叔襄,號石查,1831—1902),此處景其濬引洛陽諺語"萬事不理,問伯始;天下中庸,有胡公"(胡公,即胡廣,字伯始,91—172),借胡公指胡義贊。

③ 參見白謙慎,《晚清文物市場和官員收藏活動管窺——以吴大澂及其友人爲中心》,《故宫學術季刊》三十三卷一期(2015年9月)。

其濬推薦此器，但景其濬覺得賣家開價太高，無力購之。但不久景其濬又致信王懿榮，再次談及前信提到的角：

近來廠肆有無新奇可收之物？吾兄有所得否？爲欲廣異聞具示。此請廉兄大人晡安。其濬頓首。（河陽續假屢屢，大有悻悻之意。晤否？昨遣孫姓來，言欲以舊瓷易角。尚未許也。一笑。簠齋打本寄到未？並念。弟處曾訂一分，恐成畫餅矣。奈何！奈何！）王大老爺。

從信的内容可以推測，在寫了前一封信後，經過協商，賣家降價，景其濬將角購入。而此時潘祖蔭也有意於這個角，派古董商孫某到景其濬家，提出用舊瓷换角[①]。景其濬在信中還詢問最近琉璃廠"有無新奇可收之物"。由於王懿榮在北京推銷山東大收藏家陳介祺（簠齋，1813—1884）的拓片，景其濬惦記着遲遲没有寄到的他所訂的拓片。此札的書寫時間大約在1872年左右，此時潘祖蔭開始大肆收藏青銅器，帶動了京師官員收藏青銅器的風氣[②]。雖説景其濬的收藏以書畫和拓片爲主，但在風氣的影響下，也開始收藏青銅器。此時尚未登進士第的王懿榮在户部供職，公餘爲一些喜愛收藏的官員作中介和鑑定。王懿榮中進士後，官至國子監祭酒，成爲晚清著名的金石學家。

江南大收藏家顧文彬也曾在其《過雲樓日記》中記述了同治九年（1870）間他和景其濬在京城的短暫交往[③]：

同治九年五月初四日："往晤胡石查[胡義贊]，見景劍泉托其所銷漢玉各件，皆平常。往拜景劍泉。"[④]

同治九年五月初四日："往晤景劍泉，見其所藏董思翁臨米袖卷、惲

---

① 孫姓古董商應爲琉璃廠古董鋪筠清（又作筠青）齋的主人。吴大澂致王懿榮信札云："昨有事過前門，歸過筠清，亦未見孫四，前卣還百金，渠云百金必不允。兄許以酌加一二十金，渠云日内即送去。鄙意以爲説定，怪某連日不送來，豈孫四昏頭搭腦，竟以鄙言爲游移乎？兄之始意以爲百金以内可得，欲效宋芝山耳。一笑。"見《吉林省圖書館藏名人手札五輯》第二輯，北京：全國圖書館縮微文獻複製中心，2006，195頁。

② 白謙慎，《晚清文物市場與官員收藏活動管窺——以吴大澂及其友人爲中心》。

③ 關於景其濬和顧文彬最早認識的時間暫不清楚。同治九年（1870）閏十月二十日，顧文彬補授浙江寧紹台道員缺，同年九月景其濬赴任安徽學政，此前顧文彬在京城候官時，得以和景其濬來往。此後兩人是否有繼續來往還需要考證。

④ 顧文彬著，蘇州市檔案局、蘇州市過雲樓文化研究會編，《過雲樓日記》（點校本），上海：文匯出版社，2015，23頁。

南田山水軸,皆精。”[①]

同治九年五月十二日:“景劍泉來晤,索觀石田[沈周,字啓南,號石田,1427—1509]《吴山草堂卷》、吴文中[吴彬,字文中,1573—1620]《輞川圖》卷、石谷[王翬,字石谷,1632—1717]《清閟閣圖》卷。”[②]

由此二人短短幾日互相索觀個人收藏的互動不難推知,顧文彬此前很可能與景其濬並無直接來往,但對景其濬的收藏名氣早有耳聞,仰慕已久,兩個大收藏家的此番切磋實在是一段佳話。有趣的是,另幾則日記還從收藏品買賣反映了二人“不打照面”的互動:

同治九年六月廿二日:“德寶齋以舊瓷、印盒、水盂五件售與景劍泉,換其漢玉琴鈎,歸於余,余代換瓷器作價六十兩。此鈎與余昔年得張柳亭漢玉鈎製造如出一手,色澤、分寸若合符節,惟下半鈎所鏤琴軫一凹一凸,似分陰陽,當時必是一對。千百年後,散而復合,洵奇緣也。”[③]

同治九年六月廿七日:“在德寶見白瓷金魚缸,有花紋隱起,據云是宋定窯,未知確否。景劍泉已還價五十金。”[④]

同治九年七月初九日:“在德寶齋見景劍泉托售漢玉五十餘件,内以冕旒一件爲最,然不適於用,未還價。有束髮圈一件(的係舊物不買可惜),還價十二金。”[⑤]

顧文彬記録自己往返德寶齋的活動从側面描繪了二人透過中介討價還價的日常景象。景其濬精於玉器、瓷器鑑賞,收藏甚富,他將漢玉琴鈎最終换給顧文彬更是促成了一對精品的延津之合。在入手新的收藏上,他更是十分敏鋭,動作迅速。

顧景二人在收藏上的切磋很快發展成了深厚的情誼。同治九年八月一日,景其濬放安徽學政任,顧文彬不僅在當天就得知和記述了此事,三天後更

①《過雲樓日記》(點校本),24頁。
②《過雲樓日記》(點校本),26頁。
③《過雲樓日記》(點校本),34—35頁。
④《過雲樓日記》(點校本),35頁。
⑤《過雲樓日記》(點校本),36頁。

專門賀喜[1]。一個月後景其濬便離京赴任,雖然後來兩人並没有再見過面,但短暫的交往在彼此的收藏生涯中絶不失爲一筆濃墨重彩。

在文物收藏方面和景其濬有所交往的,有不少成就與名氣都不如王懿榮和顧文彬者,李玉棻(字真木,號韻湖,生卒年不詳)就是其中的一位。李玉棻的《甌鉢羅室書畫過目考》正是得益於他和景其濬等人的交往、景其濬等人的書畫收藏與平日所見而輯成。他在该書的自序中寫道:

> 余性耽書畫,幼以成癖,每逍遥於古肆,討論於老成,洵知真僞之分全在搜羅之富也,因就正於景劍泉閣學其濬、曾笙巢侍郎協均[曾協均,字笙巢,號韻湖,生卒年不詳],探索兩家秘笈,聆奇瞬美,晷旦不疲,夜分就枕,展轉精思。[2]

曾協均所藏宋元明繪畫甚富,對於古畫精鑑亦有精道之處,非一般人可比。李玉棻向景其濬和曾協均請教,也説明了景其濬的書畫鑑藏功力。延暄(字旭堂或煦堂,生卒年不詳,亦爲當時的收藏家)在《甌鉢羅室書畫過目考》後的題跋中也談到了李玉棻與景其濬、曾協均、皁保(字蔭方,1817—1882)在書畫收藏方面的交往:

> 韻湖司馬自髫齡好古善鑒,咸與當年諸君子遊,如景劍泉、皁蔭方、曾笙巢,均折節訂交,引爲畏友。此卷積三十年眼福,考核精確,每一展軼,儼與名跡相晤對。暄同此癖,幸多砥礪之功,皆屬過眼雲煙,不若君之有心哉。[3]

綜上所述,從初入翰林院到病逝的近二十五年中,和景其濬交往密切的多爲工書畫、精鑑藏的官員。在與這些友人的切磋中,景其濬的鑑賞力不斷提高,收藏品類和規模也逐漸擴大,最終成爲晚清的大收藏家。

---

① 同治九年八月初一日:"景劍泉放安徽。"同治九年八月初四日:"賀洪文卿、汪柳門、景劍泉放差之喜。"見《過雲樓日記》(點校本),40頁。

② 李玉棻,《甌鉢羅室書畫過目考》,盧輔聖編,《中國書畫全書》第十二册,上海:上海書畫出版社,1998,1062頁。李的自序作於清光緒二十年(1894)夏。

③《甌鉢羅室書畫過目考》,1132頁。

三、收藏

景其濬的收藏生涯約始於中進士前後。現藏美國波士頓美術館的傳沈周《江天暮雪卷》上,有一段署名景其濬的題跋,從中可以推知他在咸豐二年(1852)已經開始收藏書畫:

> 庚午[1870]奉使皖江,道出邗上。友人爲言,石田翁江天暮雪卷今藏金瘦仙處。金固江南一大收藏家也,遂爲予和會,得之。廿年購覓,一旦入手,喜不自勝,媵以五截用誌欣幸:廿年京邸典朝衫,名跡何嘗飽眼饞。難得奇觀逢歲晚,蕭蕭寒雪滿松杉。①

景其濬赴安徽學政任經過揚州時,經友人介紹,從江南大收藏家金望喬(字芾亭,號瘦仙,生卒年不詳)處購得沈周的《江天暮雪卷》。購得名作,景其濬不勝欣喜,次年冬天在此卷後作跋並題五首絶句(上引一首)。前兩句自述"廿年京邸典朝衫,名跡何嘗飽眼饞",説明景氏開始收藏活動差不多在二十年前,亦即參加順天鄉試中舉之後。爲官後,他盡其所能購藏金石書畫,以致經常阮囊羞澀②。清代的京官俸禄不高,經常靠地方官以"别敬"和"炭敬"爲名送禮金接濟。景其濬曾兩次外放任學政,有養廉銀、公費、出棚費等額外收入,收入相較京官大增。也正因爲這個原因,當景其濬被任命學政時,他的友人皆爲其慶賀,在上引莫友芝寫給他的信中,才會有"禄養善養交盡之矣"這樣的詞語。但收藏金石書畫、特别是古代名作,畢竟是極爲奢侈的愛好,使得景其濬在經濟上相當拮据,以致他去世後,不但家裏的收藏要變賣,翁同龢還要聯絡各地友人資助。

官員嗜好收藏,有時還會受到友人的訾議和告誡。上海圖書館藏有景其濬致朱學勤的信札多通,其中一通寫道:

> 承示一切,深荷關垂,並誡勿嗜古玩,洞中癥結,敬佩終身。昔年曾有嗜痂癖,到汴以後,惟得汪肖梅《蘭亭》一卷耳,(渠託人攜出求售。)實

① 此卷應爲仿作,署名景其濬的題跋也非真跡,但從題跋的内容分析,其中細節均與歷史情境吻合,應是模仿原本抄録,文字内容可信,故在此予以討論。

② "典朝衫"出自古代典朝服沽酒的典故,此處應泛指傾囊購藏。

無專足至京購覓骨董事。[①]

從景的回信可以看出,人在京師的朱學勤聽説景其濬在河南做學政期間,專門派人到京師購買骨董,因此勸誡他"勿嗜古玩"。所以景其濬在回信中一方面説,朱的勸誡是"洞中癥結,敬佩終身",同時也解釋,自己到河南後,只收了一件汪家的《蘭亭》,而且並非他專門派人到北京去買骨董,而是汪家"託人攜出求售"。

景其濬向汪家購買的是一卷定武《蘭亭》。翁同龢咸豐十一年(1861)七月廿七日日記記載:"夜到辛伯齋觀景鑑泉新購定武《蘭亭》卷,殊不佳。雙鈎本,内如'群'字、'欣於'之'於'字,皆不成點畫,映燈觀之,中粗外細。卷乃休寧汪蓮甫家藏,即藏唐人書《兜沙經》者也,惜此經已質錢於山東富人,不可見矣。"[②] 景其濬覆朱學勤信中提到的汪肖梅,應該就是翁同龢日記提到的休寧汪家。

翁同龢等在京師的賞鑒家對景其濬新收定武《蘭亭》的意見,不久就由朱學勤的信札傳達給了景其濬。景其濬在當年九月初五日覆朱學勤的信中説:"肖梅《蘭亭》氊蠟尚舊,若云定武,談何容易。此帖紛如聚訟,弟亦不能精審,惟收其可藏者而已。"[③]

上引朱學勤的勸誡和景其濬的檢討,讓人想起景其濬的友人吴大澂。白謙慎先生曾以吴大澂爲例,討論晚清喜歡收藏的官員面對"玩物喪志"批評時的焦慮感[④]。晚清是一個充滿内憂外患的時代,在這樣的時代,一個政府高官痴迷收藏容易招致訾議。景其濬出任學政的河南在1860年代備受捻軍竄擾,這一現實,難免使喜愛收藏的景其濬感受到"玩物喪志"的道德壓力。其實,他説自己到了河南只是買了一件書畫,其原因在於河南並非書畫收藏的重鎮,而在北京和江南才更有可能買到書畫精品。

---

①《景其濬致朱學勤手札》,《歷史文獻》第十輯,88頁。

②《翁同龢日記》第一卷,160頁。

③《景其濬致朱學勤手札》,《歷史文獻》第十輯,104頁。

④ 白謙慎, "Antiquarianism in a Time of Crisis: On the Collecting Practices of Late Qing Government Officials, 1861—1911", in Alain Schnapp with Lothar von Falkenhausen, Peter N. Miller, and Tim Murray at el., eds., *Traces, Collections, and Ruins: Towards a Comparative History of Antiquarianism: Comparative Perspective,* ed. Los Angeles: The Getty Research Institute, Los Angeles, 2013, pp. 386-403.

按照《清史本傳》的記載，景其濬在同治四年(1865)十一月河南學政任滿。由於景其濬在十一年(1861)擢翰林院侍講學士，留學政任，十二月轉侍讀學士，仍留學政任，他在任滿前就已經回到北京。翁同龢同治四年三月初四日的日記記載："訪景劍泉，觀所藏《九成宫》二本，一宋拓，一明神廟本也。"初九日"晤價人，遇景劍泉"。閏五月朔，"是日奏派充分教庶吉士(景其濬、徐桐、許其光、許應揆、吴鳳藻、孫毓汶、林天齡)"。七月十五日，"是日庶常館開課，……辰正到館，大教習並諸君皆集矣。大教習：周中堂、全小汀師；小教習：景鑑泉、徐蔭軒……"[①]回北京後，景其濬繼續收藏骨董。此時，太平天國已滅，晚清進入了所謂"同光中興"時期，經濟開始慢慢復甦，收藏活動也日趨活躍。前引景其濬致王懿榮信札顯示，1870年代初，景其濬還加入了京師官員收藏青銅器的活動之中。

景其濬去世後，家中的收藏陸續散出。翁同龢的日記記載：

光緒二年丙子(1876)閏五月廿一日："有持《大觀貼》五册來者，曩嘗欲購未果，爲景劍泉所得，今重對之，不勝悵惘，本有大令兩册，余酷愛之，今逸去矣。(紙墨極佳，宣示表内有臣張□□，張□□一行。)"[②]

光緒五年己卯十月廿九日："得見景劍泉所藏馬文璧畫《中庵圖卷》(皆元跋)。"[③]

光緒六年庚辰十二月十四日："得見董文敏雙畫卷，一仿《煙江疊嶂》(絹本)，一仿北苑(紙本，眉公小輞川詩)。皆景氏物，妙絶，索數百金也。"[④]

光緒七年辛巳三月初六日："博古齋持舊拓《開母廟碑》來，景劍泉物，索百金。"[⑤]

光緒八年壬午四月十二日："含英閣送來景氏藏帖，無極好者。"[⑥]

---

①《翁同龢日記》第一卷，407頁、409頁、429頁、441頁。

②《翁同龢日記》第三卷，1253頁。

③《翁同龢日記》第四卷，1493頁。

④《翁同龢日記》第四卷，1571頁。

⑤《翁同龢日記》第四卷，1692頁。

⑥《翁同龢日記》第四卷，1698頁。"含英閣"，原作"余英閣"，應是釋文錯誤。

景其濬過世不久，就有極好的書畫碑帖流入市場，價格甚高，不少甚至是他在世時衆人歆羡卻無法得到的東西。而數年後，市場所剩的則多是“無極好者”了。

但是，由於景其濬的收藏規模相當大，在1890年代，廠肆中仍有不少他的舊藏。晚清鑑賞家李葆恂（字寶卿，1859—1915）曾這樣寫道：“己丑[1889]入都，獲見孫文恪公、王文敏公、盛意園祭酒、張樵野侍郎、李芝陔太守，皆時所稱賞鑑家也。幸得盡窺所藏。適馮展雲中丞及景劍泉學士故物散在廠肆，去余所居甫一里而近。每治事畢，輒往遊觀，往往流連至日夕始返。”[①] 由於景其濬的收藏没有著録，他收藏的書畫去向變得難以追踪，僅前文崇彝提及者大多相繼歸於張度（字吉人，1830—1904）和楊壽樞（字蔭伯，1863—1944）。

景其濬還是晚清田黄、田白印石的大藏家。關於他收藏田石的正式記録極少，僅陳亮伯（生卒年不詳）在所著《説田石》中簡單地交待了景其濬舊藏田石的去向：

> 余與吴彦復[吴保初，字彦復，1869—1913]初入京，觸眼黄白物，尤氏寶珍、英古兩齋所藏，多山西舊家物。德寶、永寶兩齋，多景劍泉學使其濬家物……景家石悉歸於彦復，今已爲彼佛及吴昌碩[1844—1927]所干没。余所得於英古諸家者，又已不可復見。一時風流雲散，而廠甸爲之減色矣。昔之琳瑯滿目者，今則吉光片羽，罕而彌珍，回首舊遊，恍若隔世。[②]

陳亮伯慨歎的不僅是尤氏和景其濬兩家舊藏的風流雲散，更是對於昔日光景因時代變遷而淪落荒涼的無奈與悵惘。

縱觀景其濬涉獵的收藏門類，書畫應當是最爲精彩的部分。李玉棻《甌鉢羅氏書畫過目考目録》著録的景其濬收藏一共近五十件，其中書法約十五件，餘下三十五件繪畫，大多是明清兩代的山水畫，足見他收藏趣味之廣

---

① 李葆恂，《無益有益齋讀書詩二卷》，翟文選修，王樹枏纂，《奉天通志》卷二二六，《藝文四》，“遺著（清）”，民國二十三年鉛印本。

② 陳亮伯，《説田石》，據民國二十年刊行陳亮伯《説印》（遺稿經崇彝整理而成）。

博[①]。其中一條著録曰:"景劍泉閣學藏有王原祁[字茂京,號麓台,1642—1715]紙本墨山水兩卷,名曰《婁東雙璧》,所謂具有金剛杵者,爲生平佳製。"[②]楊翰(字伯飛,1812—1879)在他的《歸石軒畫談》也提到了兩幅王原祁的手卷,"余己巳[1869]至京,過丁濂甫[丁紹周,字濂甫,1821—1873],出麓臺手卷二,墨筆精絶,知爲景劍泉物,後二君俱歿,不知此畫歸何處。余見司農畫多矣,用筆之濃厚飛翔,則以二卷爲冠。"[③] 此二人所講的正是景其濬藏王原祁最精彩的兩件作品,即《竹溪松嶺》和《九日適成》。

《竹溪松嶺圖卷》是王原祁歷時三年仿董源之作,現藏於北京故宫博物院。據張庚《國朝畫徵録》記載,王原祁繪畫創作過程相當繁複,反復皴染,頗費時日,可知畫家所稱"純任自然,不爲筆便"是求所繪山水之境界和畫風之風格,而非筆墨的施用過程[④]。此畫經景其濬、潘遵祁、龐元濟收藏,畫末景其濬的兩方鑑藏印"劍泉平生癖此"和"景氏子孫寶之"清晰可见,後有他的友人成沂(字子清、芝青,生卒年不詳)題詩曰:

> 昔持千黄金,訪古燕市上。落落司農跡,煙雲空夢想。有時獲一觀,洵非所欣賞。何從覓此圖,古今本同尚。太羹滋味希,羣推朝野望。草色凝緑雨,嵐光摇翠嶂。東領歌雪樵,西洲起漁唱。一片清光來,頓消我俗障。荒率不可攀,允矣推大匠。星使性尚友,寸心千古抗。名姝與駿馬,問直詎堪量。(君以五百金購此兩卷,真麓臺知己也)爲我洗塵容,拭目天懷放。辛未[1871]夏五月上浣興化清道人成沂題於潁川試院。[⑤]

成沂應該是同時見到了《九日適成卷》,所以便在這第二幅畫後再次題詩

---

① 楊峴《遲鴻軒所見書畫録》著録的景其濬藏書畫五十餘件,與李玉棻《甌鉢》著録大致相同,從李玉棻和景其濬的私交看,李玉棻過眼景其濬這些收藏的可能性極大。筆者暫且以李玉棻《甌鉢》爲例對個中著録進行討論,在附録一《景其濬收藏表》中收入了兩本著録。

② 婁東即太倉。以王時敏、王鑑、王原祁爲首的"婁東派"因婁江(即瀏河)東流經過太倉,又稱"太倉派",清代重要畫派之一。

③ 楊翰,《歸石軒畫談》,《中國書畫全書》第十二册,119頁。

④ 張庚,《國朝畫徵録》卷下,上海:江都朱氏藏版萃文書局印行,1862—1908,3頁。又:"純任自然,不爲筆便"見王原祁《竹溪松嶺》畫後長題。見蕭燕翼編,《故宫博物院藏文物珍品全集——四王吴惲繪畫》,香港:香港商務印書館,1996,158—159頁。

⑤ 龐元濟,《虚齋名畫録》卷五,清宣統烏程龐氏上海刻本。又:故宫博物院藏畫集編輯委員會編,《中國歷代繪畫——故宫博物院藏畫集》第七册,北京:人民美術出版社,1991,16—17頁。

讚頌此二美：

雅望海内欽，相逢潁川上。好古如好士，大有濠濮想。瑯琊畫本稀，尺幅恣吟賞。當時重連城，況乃所素尚。兩美期必合，北苑洎公望。春星羅草堂，秋暉吐晴嶂。讀畫勝讀詩，三歎復三唱。一朝快奇觀，鄙懷增妒障。作者洵意造，搜羅亦心匠。秘笈世所無，兩卷鼎能抗。披覽太白浮，百壺未可量。狂夫不敢狂，平生悔奔放。疊前韻並希劍泉先生閣學大人哂政。芝青成沂題。①

成沂在《竹溪松嶺》題詩注中所云"君以五百金購此兩卷，真麓臺知己"，説的正是景其濬曾同時購得王原祁的《竹溪松嶺》和《九日適成》。黄小峰對此五百金作了簡要的解釋：景其濬新任閣學不久，内閣學士雖爲從二品高官，但清朝文職京官收入向來不高，每年俸銀約爲一百五十兩，俸米約七十五石，主要經濟來源是其他一些制度外的收入，儘管如此，五百兩至少佔其年度進項的八分之一②。這也從一個側面印證了景其濬自己所説的"廿年京邸典朝衫，名跡何嘗飽眼饞"，見到名跡不惜花重金購入。

現藏於上海博物館的董源《夏山圖》更是景其濬一件令人歎爲觀止的藏品。董源傳世真跡頗少，此幅是他晚年作品，董其昌（字玄宰，號思白，1555—1636）據《宣和畫譜》的記載定名爲《夏山圖》，從現存收藏印鑑、題跋和著録看，曾經南宋賈似道、元代史崇文以及明代黄琳、袁樞、董其昌遞藏，流傳有序，是描寫江南丘陵風貌最濃烈的一幅作品③。前面提到的在京城見到王原祁"婁東雙璧"的楊翰，在1869年拜訪景其濬時，有幸觀看此畫，慨歎道：

北苑爲山水極則，宋元人皆宗之，真本世難經見。己巳[1869]至京師，過景劍泉學士，收藏甚富，屬余書寶苑齋額。旋出巨幅，蒼渾圓厚，下筆雄偉，有嶄絶巉岩之勢，洵爲後賢所不能到。《夢溪筆談》云，董源多寫江

①《虚齋名畫録》卷五。

② 黄小峰，《"隔世繁華"：清初"四王"繪畫與晚清北京古書畫市場》，中山大學藝術史研究中心編，《藝術史研究》第九輯，2007，166頁。

③ 楊敦堯，《董其昌建構董源〈夏山圖卷〉之初探》，台灣藝大書畫藝術同窗學友會編，《蘇峰男教授服務公職四十年退休紀念書畫藝術論文集》，台北：蕙風堂，2005，463頁。

南真山，不爲奇峭之筆。米芾《畫史》云，董源平淡天真，峰巒出没，雲霧顯晦不裝巧趣，嵐色鬱蒼，枝干勁挺，溪橋漁浦，洲渚掩映，一片江南也。米老此語如爲此幅寫照，余無以名之，於此畫歎觀止矣，上有思翁跋，極爲珍重，恐後人難問津耳。①

現已發現曾經景其濬收藏的繪畫精品還包括南宋宫素然（生卒年不詳）的《明妃出塞圖》、元代錢選的《西旅獻獒圖》、仇英的《募驢圖》，以及清初“四王”、惲壽平、龔賢、石濤等的畫作。在筆者初步統計曾經景其濬收藏的九十餘件書畫中，有十二件“四王”作品，四件惲壽平作品，五件石濤作品。極爲有限的資料使我們無法比較深入地了解景其濬的收藏趣味，同時還需要思考他的收藏活動在多大程度上是他個人趣味的體現，又在多大程度上受到了晚清收藏風尚的影響②。

景其濬收藏過的書法精品包括現藏日本東京國立博物館的米芾《三帖卷》和北京首都博物館藏趙孟頫小楷《大乘妙法蓮花經》卷五等。從上海圖書館藏景其濬致潘曾瑩的信札可以看出，景其濬的行書受到米芾的影響。在景其濬的友人中，潘祖蔭也效法米芾。不論景其濬是早已親近米字還是在收藏這卷書法後逐漸培養出對米芾書風的熱愛，這一藏品至少在陪伴他的歲月裏爲他日常的細緻觀察和用心臨摹提供了最好的範本。

值得注意的是，景其濬舊藏的某些書畫，諸如米芾《三帖卷》和錢選《西旅獻獒圖》，鈐有“乾隆御覽之寶”和“石渠寶笈”等清内府的收藏印，這應是在咸豐十年（1860）九月，英法聯軍侵入北京，圓明園被搶掠焚毁後，流入市場的内府收藏③。同年十月二十日，亦即圓明園被毁一個多月後，翁同龢在潘祖蔭處見到“《茶録》《姜遐碑》二帖，皆淀園散落者，索直甚昂，且留之以待珠還耳”④。此後的數年，圓明園流出的書畫不時在市場上出現，景其濬在京久居，

---

① 楊翰，《歸石軒畫談》，《中國書畫全書》第十二册，72頁。

② 關於這一點，參考黄小峰對景其濬花百金收藏王原祁《竹溪松嶺》《九日適成》一事持有的觀點：高價購藏“四王”畫作並非僅僅是個人愛好，而是當時京城的主流鑑藏風尚。見《“隔世繁華”：清初“四王”繪畫與晚清北京古書畫市場》，《藝術史研究》第九辑，166頁。

③ 白謙慎，《晚清文物市場與官員收藏活動管窺——以吴大澂及其友人爲中心》。

④ 淀園，即圓明園。《翁同龢日記》第一卷，112頁。

自然有很大的機會購買到原内府的收藏。

由於書畫上可以題跋和鈐蓋收藏印，已有著録和新發現景其濬收藏的書畫，常常有他所交好、敬畏、相互砥礪的友人的題跋、觀款和印鑑，反映了景其濬與其他收藏家的互動。當時的大收藏家金望喬和景其濬就先後收藏了沈周的《江天暮雪卷》、王鑑的《仿古山水册》和仇英的《募驢圖》，這一遞藏關係應始於景其濬在揚州從金望喬購得沈周手卷時。在景其濬收藏的畫作上，時能見到同輩摯友沈樹鏞的觀款和鑑藏印，其中包括董源的《夏山圖》、文徵明的《關山積雪圖》和仇英的《募驢圖》等，這有兩種可能：景其濬邀請沈作觀款，或是沈去世後，舊藏被景其濬購得。成沂更是有幸品鑑景其濬諸多收藏，除了在《江天暮雪卷》等書畫上的觀款外，王原祁的《竹溪松嶺》《九日適成》上的兩段題跋直接説明了二人在書畫鑑藏上甚是親密的來往。此外，景其濬舊藏趙孟頫小楷精品《大乘妙法蓮花經》上有王懿榮的觀款，石濤《故城河圖》上有李玉棻的印鑑。

景其濬的碑帖收藏也甚負時名。雖説以往的著録對於碑帖的記録較之於書畫要少，幾件有記録和有景其濬鑑藏印的碑帖頗能證明他的收藏之精。在碑帖收藏上，他保有對待書畫的鑑藏習慣，不一題再題，多在作品的角落鈐上零星兩三枚印，比如"劍泉平生癖此"，"景氏子孫寶之"等，保持原作的光潔完整[①]。

景其濬所藏《宋拓九成宫醴泉銘》和《宋拓皇甫誕碑》都是歐陽詢的名作。其中，現藏上海圖書館的《宋拓皇甫誕碑》，後來成爲吴湖帆的"四歐寶笈"之一，吴湖帆題簽"景劍泉官詹舊藏"，上有景其濬的好友潘祖蔭的觀款，鈐印"吴潘祖蔭章"。

景其濬收藏的碑帖除邀好友作觀款外，亦有通過好友題跋以記述分享鑑藏的樂趣。葉名澧（字潤臣，1811—1859）曾爲景其濬題《宋拓虞永興破邪論序册》云："鑑泉仁兄以所得舊拓虞永興破邪論序見示，謹題一詩奉求正定。

① 有學者指出，景其濬的數方鑑藏印可能是吴熙載（讓之，1799—1870）所刻，其篆法、章法、刀法皆與吴熙載晚年風貌吻合。見劉嘉成，《衆里尋她千百度——吴讓之印拓尋踪》，引自台灣藝術大學美術學院造形藝術研究所網站。李玉棻《甌鉢羅氏書畫過目考目録》亦談到"景劍泉閣學藏有[讓之]三行篆書屏四幀"，可知景其濬曾收藏吴熙載書法。

戊午[1858]十月下瀚弟葉名澧。”[1]

成沂在爲景其濬藏王原祁《婁東雙璧》題跋的同時,亦爲所藏《梁瘞鶴銘拓本》作跋:

> 同治辛未夏六月,於潁川試院獲觀景劍泉學使珍藏翁北平鶴壽本,有高江村長跋,其搨尾有米老石刻觀款,洵佳本也。……秋七月記於皖江舟中。[2]

景其濬有時也應邀爲友人的拓本作題跋,如他曾借觀《宋拓小麻姑仙壇記》,作跋云:

> 顔書《小麻姑壇記》今之詡爲宋搨者,大抵明初覆本,用濃墨精拓,雖黝濃可愛,而精采不存。至原石則又日□刓敝,蹲注内撇之法,無復可尋,此原石舊搨所以難得可貴也。此本的係唐石宋拓,生平僅見,仲衡觀察弆藏雖富,此其最矣。借觀月餘,竊喜年末翰墨有緣,因識以誇眼福。黔南景其濬劍泉甫誌。[3]

現藏台北故宫博物院的景其濬舊藏《漢李孟初神祠碑拓》,其出土、流轉、傳拓,頗具傳奇性。原碑立於南陽,漢代時南陽郡經濟文化發達,又爲光武帝劉秀發迹之地,號稱“帝鄉”,當時立碑樹碣者衆多,但時代更迭,留存至今的漢代碑碣卻是鳳毛麟角[4]。李孟初神祠碑最初立於漢永興二年(154),後在乾隆年間因白河水漲沖出,後又入土。道光初年白河漲水塌岸,此碑又再次出土。由於此碑在乾隆時期有學者曾經著録,莫友芝在咸豐十年(1860)景其濬任河南學政時,向景索拓片,景其濬便向南陽知府金梁索取此碑的拓本,金梁

---

① 葉名澧,《敦夙好齋詩全集》續編卷九,《養痾集》,清光緒十六年葉兆綱刻本,27頁。在此册上,還有晚清著名收藏家裴景福(1854—1926)的題跋:“先師景劍泉閣學,高才博學,尤精賞鑑,此册爲其所藏。甲戌[1874]初入京會[試]一再見,後廿餘年,得此於吴門,簽題藏印俱在,墓門拱矣,不覺出涕。唐初碑刻,歐虞並重,論者以虞爲勝,自應以夫子廟堂碑爲冠。信本小楷寫經外,無煊赫鉅跡,唐一代小楷又因以此論爲冠絶矣。睫庵。”見裴景福,《壯陶閣書畫録》卷二十二,台北:台灣中華書局,1971,6頁。裴景福爲安徽人,應在景其濬任安徽學政時與之相識。

② 原日本三井聽冰閣藏,日本二玄社印,《書跡名品叢刊》,東京:二玄社,1961。

③《宋拓小麻姑仙壇記》,現藏北京故宫博物院。

④ 秦明據黄易《小蓬萊閣金石目》整理《李孟初神祠碑册》,故宫博物院網站,2009。

在南陽縣任伯雨家訪得此碑，移到南陽府衙中，並在石碑漫漶處刻跋，其中便提到了景其濬索拓一事：

今年景劍泉學使書來索此碑，甫訪求得之。因念此碑湮没已久，前四十年爲河流沖出，今又得景學使搜羅之，古蹟幸有若有神鬼呵護。[1]

後任知府傅壽彤爲保護此碑，建亭覆之，並作文記述景其濬等和此碑的因緣：

同里莫子偲知宛有此，索搨於學使景劍泉前輩，時太守金君移置堂下。嗣余守此，更慮剥損，亭以覆之。嗟乎！此碑沈埋沙磧者自漢以來幾二千載，及出，又流落人間，不知寶重。如金君者弆而藏之，誠碑之幸。然非劍泉索搨，碑無由至是，非子偲言劍泉，又無由索搨，然則子偲誠碑之知己歟！[2]

四、餘論

當本文發表時，距景其濬去世差不多正好一百四十年。在景其濬去世後的前五十年，他的名字還不時地出現在清末民初文人的筆記和書畫著録中。隨着歲月的流逝，他漸漸地被淡忘了，以至於今天從事藝術史和收藏史研究的人們很少知道在晚清還曾有過這樣一個大收藏家。本文根據零散而有限的文獻對他的生平、交遊、收藏活動所做的粗略勾勒，或許能幫助人們認識這位曾經名重一時的收藏家。對於古代藝術史研究來説，晚清算是很近的一個時代，但即便如此，我們對晚清的收藏史又有多少瞭解呢？進而推之，除了那些耳熟能詳的大收藏家，我們對更早的收藏史又瞭解多少呢？

本文的附録一，以表格形式列出了筆者從各種文獻和著録、圖録中輯録的景其濬舊藏，計有一百餘件。由於戰爭、動亂和各種天災人禍，加之許多公私收藏中的景氏舊藏未能公開，本文附録所展現的必定只是景氏收藏過的文

---

① 見台北故宫博物院藏《漢李孟初神祠碑墨拓本軸》。

② 潘守廉修，張嘉謀纂，《光緒南陽縣志》卷十，“藝文下·金石”，清光緒三十年刊本。

物中的極小一部分,希望今後有機會不斷補充。

在感歎景其濬這個被遺忘的大收藏家和他不爲人知的舊藏的同時,我們也應該意識到,根據景其濬舊藏進行的造僞活動很早就開始了。波士頓藝術博物館藏傳沈周卷,於1915年入藏該館。20世紀初,日本和歐美的收藏家紛紛開始收藏中國藝術品,大量用來欺騙洋買家的僞作便産生於那時。

北京圖書館於1961年出版的《北京圖書館藏善拓題跋輯録》曾收録了一件傳景其濬舊藏的《明拓北魏張孟龍碑》[①],而曾任職於中國國家圖書館善本部金石組的趙海明先生在《碑帖鑑藏》一書中,判斷此明拓實爲裝潢者陳仰山以清人王瓘藏明拓《張猛龍碑》之珂羅版本作僞而成,並指出此僞本的明顯漏洞在於用紙潔白,字口平滑無明顯凹入痕跡,且填墨痕跡淹死呆板,無傳形拓之拓形與神韻[②]。

再如,目前下落不明的王原祁《九日適成》卷的仿作也出現在近年的文物拍賣市場。不但署名吴大澂的引首題字、畫作本身的筆墨、題跋的書法皆有問題,畫卷上的鑑藏印也有疑問,其中一方朱文長方印"貴陽景氏"也應該是仿刻。在現有真跡諸如趙孟頫《大乘妙法蓮花經》卷五和王鑑《仿古山水册》上,也有一方朱文長方印"貴陽景氏",與《九日適成》卷仿作中的印章對照,可以發現真跡上的印痕刀法流暢、結字章法工穩,左上方有一個明顯的缺口,且位置近乎一樣。然而《九日適成》仿作上的印章邊緣卻完好無損。這些都是我們在重構景其濬的收藏時要予以警惕的。

鳴謝:本文是筆者在白謙慎老師指導下完成的一個研究項目。在收集資料、研究和寫作過程中,白老師都給予了十分具體的教導,並得到了梁穎、秦明、何炎泉、薛龍春諸位先生和黄朋女士的幫助,在此謹表誠摯的謝意。

① 王敏編,《北京圖書館藏善拓題跋輯録》,北京:文物出版社,1961,229頁。。

② 趙海明,《碑帖鑑藏》,天津:天津古籍出版社,2010,215—216頁。

A Forgotten Name: The Prominent Qing Collector Jing Qijun

Abstract: In the history of late Qing dynasty art collecting, several names of important Chinese collectors come easily to mind. Yet, the literati-official Jing Qijun, active in the mid and late nineteenth century, has long remained obscure. But recently, an increasing number of artworks, especially paintings, calligraphies, and rubbings, have been found to bear his names and seals, demonstrating the significance of this prominent collector and connoisseur to Chinese art history. This paper attempts to explore the importance of the collector Jing Qijun through his biography and his social intercourse with other contemporary literati-officials and collectors, and more importantly to conduct historical textural research on his lifelong activity of collecting artworks. It may be inferred that Jing's intimate interaction with other late Qing bureaucratic-amateur collectors was key to the large scale of his collecting and to his recognition as a discerning Qing collector. The story of his collecting activities is a valuable component in reconstructing the mechanisms of late Qing connoisseurship and collection.

附録一　景其濬收藏表[1]

一、書法繪畫

時代	作者	作品	遞藏者	資料出處	現藏地	景其濬鑑藏印
五代	董源	《夏山圖》	賈似道、史崇文、黄琳，袁樞、董其昌、戴熙、徐渭仁、潘遵祁、景其濬等	楊翰，《歸石軒畫談》，《中國書畫全書》（上海：上海書畫出版社，1998）	上海博物館	"景劍泉家藏"
北宋	米芾	《行書三帖卷》（《叔晦帖》《李太師帖》《張季明帖》）	項元汴、乾隆内府、景其濬等	大野修作編，《米芾集》（東京：二玄社，1990）	日本東京國立博物館	"劍泉平生癖此""景氏秘篋墨皇""景氏子孫寶之"
南宋	宫素然	《明妃出塞圖》	梁清標、陸勉、張湯、張寧、景其濬、陸學源、完顔景賢等	《中國書畫名品展》實行委員會編，《大阪市立美術館藏、上海博物館藏中國書畫名品圖録》（東京：二玄社，1994）	日本大阪市立美術館	"劍泉平生癖此""景氏子孫寶之"
元	錢選	《西旅獻獒圖》	乾隆内府、景其濬、裴景福、溥儒等	鈴木敬編，《中國繪畫總合圖録》（東京：東京大學出版社，1998）	私人收藏	"劍泉平生癖此""景氏子孫寶之"

---

① 部分景其濬舊藏因著録不詳，或著録不曾記録所有歷任收藏者，或原作不詳，而缺乏歷任收藏者的完整信息，筆者暫記録爲"景其濬等"（不排除只有景其濬一人舊藏的可能）。部分收藏因著録不曾記録景其濬所有鑑藏印，或原作不詳，或印刷出版物上的作品圖像不清晰，而缺乏其鑑藏印的完整信息，筆者暫記録爲"不詳"。部分收藏的資料出處與現藏地相同，筆者只列出現藏地。在金石碑帖部分，筆者按碑刻的年代順序編排。在根據鑑藏印章列出遞藏者時，有些遞藏者之間的前後關係並不清楚。

續表

時代	作者	作品	遞藏者	資料出處	現藏地	景其濬鑑藏印
元	趙孟頫	《大乘妙法蓮花經·卷五》	張岳崧、張鍾彦、楊壽樞、吴榮光、景其濬、王懿榮等	陸心源,《儀顧堂題跋》(光緒十六年刻本)	北京首都博物館	“貴陽景氏”
元	馬琬	《中庵圖卷》	景其濬等	翁萬戈編,《翁同龢日記》(上海:中西書局,2011)	不詳	不詳
元	不詳	《元人行書詩卷》	景其濬等	上海博物館編,《中國書畫家印鑑款識》(北京:文物出版社,1987)	上海博物館	“劍泉審定”
明	(傳)沈周	《江天暮雪卷》	金望喬、景其濬等		美國波士頓美術館	“劍泉二十年精力所聚”“濬以文籍自娱”“曾訪夷門監者來”“緑梅庵收藏書畫”
明	祝允明	《小楷歌詞二十一首卷》	顧可潛、錢穀、文元發、朱之赤、陸恭、景其濬、關冕鈞、譚敬等	關冕均,《三秋閣書畫録》(蒼梧關氏,1928)	私人收藏	“劍泉平生癖此”“景氏子孫寶之”
明	祝允明,文徵明,王寵	《行書詩合卷》	景其濬等	中國古代書畫鑑定組編,《中國古代書畫圖目》(北京:文物出版社,1986)	上海博物館	“劍泉珍賞”“劍泉真賞”
明	文徵明	《關山積雪圖卷》	董其昌、王永寧、程正揆、景其濬、朱省齋、鍾仁階等	汪砢玉編,《珊瑚綱》,《中國書畫全書》(上海:上海書畫出版社,1998);郁逢慶編,《郁氏書畫題跋記》(辛亥八月順德鄭氏依舊抄本摘印)	私人收藏	“貴陽景氏”

續表

時代	作者	作品	遞藏者	資料出處	現藏地	景其濬鑑藏印
明	仇英	《募驢圖》	朱之蕃、錢天樹、金望喬、景其濬等	端方,《壬寅消夏録》,《續修四庫全書》(上海:上海古籍出版社,1995—1999)	美國華盛頓弗利爾美術館	“景氏”“劍泉平生癖此”
明	丁雲鵬	《白描花卉長卷》	景其濬等	李玉棻,《甌鉢羅室書畫過目考目録》;楊峴,《遲鴻軒所見書畫録》,《中國書畫全書》(上海:上海書畫出版社,1998)	不詳	不詳
明	董其昌	《仿煙江疊嶂圖卷》	景其濬等	《翁同龢日記》	不詳	不詳
明	董其昌	《仿北苑圖卷》	景其濬等	《翁同龢日記》	不詳	不詳
明	董其昌	《別賦》《舞鶴賦》	高士奇、乾隆内府、景其濬等	高士奇,《江邨書畫目》(東方學會,1924)	私人收藏	“景氏”
明	王鐸	《行書詩軸》	景其濬等	《三秋閣書畫録》	不詳	“黔南景劍泉平生欣賞”
明	王猷定	《大行楷五行直幀》	景其濬等	《甌鉢羅室書畫過目考目録》;《遲鴻軒所見書畫録》	不詳	不詳
明	萬壽祺	《采芝女士立幀》	景其濬等	同上	不詳	不詳
清	丁元公	《墨蘭石小幀》	景其濬等	同上	不詳	不詳
清	王鑑	《墨山水紙本長幀》	景其濬等	同上	不詳	不詳

續表

時代	作者	作品	遞藏者	資料出處	現藏地	景其濬鑑藏印
清	王鑑	《壬寅九秋仿古册》(二十幀)	景其濬等	邵松年,《澄蘭室古緣萃録》,《續修四庫全書》(上海:上海古籍出版社,1995—1999)	不詳	“劍泉二十年精力所聚”“師吾儉齋居士”
清	王鑑	《仿古山水水墨册頁》(十幀)	金望喬、景其濬	西泠印社拍賣公司2007秋季藝術品拍賣會	私人收藏	“黔南景劍泉收藏書畫私印”“緑梅村瘦枝閣景氏印”
清	周亮工	《行書跋初拓曹全碑》四頁	景其濬等	《甌鉢羅室書畫過目考目録》;《遲鴻軒所見書畫録》	不詳	不詳
清	湯燕生	《小行楷録自作古文》(五十篇)	景其濬等	同上	不詳	不詳
清	查士標	《二瞻雙絕册》(十二幀)	永瑢、景其濬、彭恭甫	李兵,《二瞻雙絶》(杭州:西泠印社出版社,2008)	不詳	“貴陽景氏”“景劍泉家收藏”
清	龔賢	《江邨圖卷》	景其濬等	《澄蘭室古緣萃録》	上海博物館	不詳
清	(傳)黄媛介	《緑文竹赭石小幀》	景其濬等	《甌鉢羅室書畫過目考目録》;《遲鴻軒所見書畫録》	不詳	不詳
清	張學曾	《設色山水紙本小卷》	景其濬等	同上	不詳	不詳
清	項奎	《墨山水小册》八頁	景其濬等	同上	不詳	不詳
清	八大山人	《書畫册》	景其濬等	王朝聞主編,《八大山人全集》(南昌:江西美術出版社,2000)	上海博物館	“景其濬印”“黔南景氏劍泉收藏”

續表

時代	作者	作品	遞藏者	資料出處	現藏地	景其濬鑑藏印
清	姜宸英	《行書臨王獻之姑比日帖》	景其濬等	《大阪市立美術館藏、上海博物館藏中國書畫名品圖録》	上海博物館	“劍泉平生癖此”
清	吴歷	《秋景山水圖卷》	景其濬、王震等	日本泉屋博古館編,《泉屋博古－中國繪畫》(京都:泉屋博古,1996)	日本泉屋博古館	“劍泉鑑藏”“劍泉”
清	惲壽平	《墨山水小卷》	景其濬等	《甌鉢羅室書畫過目考目録》;《遲鴻軒所見書畫録》	不詳	不詳
清	惲壽平	《仿雲林�石樗散圖卷》	景其濬等	同上	不詳	不詳
清	惲壽平	《臨張中桃花山鳥圖軸》	景其濬等	《中國書畫家印鑑款識》	上海博物館	“黔南景劍泉平生欣賞”
清	惲壽平	《折枝櫻桃紙本立幀》	乾隆内府、景其濬等	《甌鉢羅室書畫過目考目録》;《遲鴻軒所見書畫録》	不詳	不詳
清	方乾	《摹周昉麗人行詩意圖》	梁清標、戴植、侯汝承、景其濬等	郭葆昌,《觶齋書畫録》(台北:漢華文化事業股份有限公司,1971)	私人收藏	不詳
清	王翬	《江山卧遊長卷》	景其濬等	崇彝,《選學齋書畫寓目記續編》,《美術叢書》(台北:藝文印書館,1975)	不詳	“黔南景氏收藏印”
清	王翬	《仿惠崇江南春圖》	景其濬等	《三秋閣書畫録》	不詳	“劍泉”“黔南景劍泉平生欣賞”“黔南景劍泉收藏古書私印”

續表

時代	作者	作品	遞藏者	資料出處	現藏地	景其濬鑑藏印
清	王翬	《漁莊秋霽圖》	景其濬等	《三秋閣書畫録》	不詳	不詳
清	王翬	《仿夏禹玉漁村秋霽圖》	景其濬等	同上	不詳	“劍泉”“黔南景劍泉平生欣賞”“黔南景劍泉收藏古畫私印”
清	王翬	《仿李營邱平林散牧圖》	景其濬等	同上	不詳	“劍泉”“黔南景劍泉平生欣賞”“黔南景劍泉收藏古畫私印”
清	王翬	《臨董羽設色桃源春漲圖》	景其濬等	《甌鉢羅室書畫過目考目録》;《遲鴻軒所見書畫録》	不詳	不詳
清	米漢雯	《淡設色仿房山山水小卷》	屠倬、景其濬等	同上	不詳	不詳
清	石濤	《故城河圖》	景其濬、李玉棻、陳夔麟	陳夔麟編,《寶迂閣書畫録》(民國4年〈1915〉石印本)	私人收藏	“劍泉真賞”
清	石濤	《野色圖册》十二頁	景其濬、張大千	Marilyn Fu and Wen Fong, *The Wilderness Colors of Tao-chi.* (New York: The Metropolitan Museum of Art, 1973)	美國紐約大都會博物館	“景氏珍藏”“劍泉珍賞”
清	石濤	《芝松圖卷》	徐衡、景其濬	澳門藝術博物館編,《至人無法——故宫上博珍藏八大石濤書畫精品》(澳門:澳門藝術博物館,2004)	上海博物館	“景劍泉家藏”“劍泉二十年精力所聚”

續表

時代	作者	作品	遞藏者	資料出處	現藏地	景其濬鑑藏印
清	石濤	《梅竹雙清圖軸》	景其濬、汪士元、何冠五、王季遷	大村西崖編,《中國名畫集》(東京:東京竜文書局,1945)	私人收藏	"劍泉珍賞"
清	石濤	《山水花卉册頁》	景其濬等	王之海編,《石濤書畫全集》(天津:天津人民美術出版社,1995)	上海博物館	"景二""劍泉""景""景劍泉家收藏""劍泉真賞"
清	卞永譽	《墨筆柏石立幀》	景其濬等	《甌鉢羅室書畫過目考目録》;《遲鴻軒所見書畫録》	不詳	不詳
清	楊晉	《與耕煙合作安儀周小照長卷》	陳奕禧、翁方綱、景其濬等	同上	不詳	不詳
清	顧昉	《墨筆仿大癡山橋曳杖圖卷》	景其濬等	同上	不詳	不詳
清	陸道淮	《仿夏禹玉晴江歸棹圖大卷》	景其濬等	同上	不詳	不詳
清	陸道淮	《設色仿大癡冬山圖大幀》	阮元、景其濬等	同上	不詳	不詳
清	曹曰瑛	《小楷絹本册》四十頁	景其濬等	同上	不詳	不詳
清	蔣廷錫	《臨劉寀本菱藻游魚立幀》	景其濬等	同上	不詳	不詳

續表

時代	作者	作品	遞藏者	資料出處	現藏地	景其濬鑑藏印
清	金永熙	《臨范華原夕寒山翠重秋静雁行高小卷》	景其濬等	《甌鉢羅室書畫過目考目録》;《遲鴻軒所見書畫録》	不詳	不詳
清	陳撰	《墨筆竹籬秋影立幀》	景其濬等	同上	不詳	不詳
清	王原祁	《紙本水墨山水卷》	景其濬等	同上	不詳	
清	王原祁	《竹溪松嶺圖卷》	潘遵祁、景其濬、龐元濟等	《歸石軒畫談》;故宫博物院編,《中國歷代繪畫——故宫博物院藏畫集》(北京:人民美術出版社,1991)	北京故宫博物院	"劍泉平生癖此""景氏子孫寶之"
清	王原祁	《九日適成卷》	景其濬、龐元濟等	《甌鉢羅室書畫過目考目録》;《遲鴻軒所見書畫録》	不詳	不詳
清	鄒一桂	《仿管夫人竹石立幀》	景其濬等	同上	不詳	不詳
清	鄒一桂	《仿陳仲美設色薔薇立幀》	景其濬等	同上	不詳	不詳
清	鄒一桂	《仿承旨停車坐愛圖立幀》	景其濬等	同上	不詳	不詳
清	梁巘	《臨虞恭公碑册頁》	奕訢、景其濬	中國嘉德國際拍賣有限公司2014春季拍賣會	私人收藏	"黔南景氏珍藏""劍泉珍賞""黔南景劍泉平生欣賞"

續表

時代	作者	作品	遞藏者	資料出處	現藏地	景其濬鑑藏印
清	顧文渊	《幽篁怪石絹本立幀》	景其濬等	《甌鉢羅室書畫過目考目録》;《遲鴻軒所見書畫録》	不詳	不詳
清	錢大昕	《三行隸書屏》(四幀)	景其濬等	同上	不詳	不詳
清	王文治	《臨晉唐宋明各家行楷册》四十頁	景其濬等	同上	不詳	不詳
清	方婉儀	《墨菊紙本立幀》(羅聘補竹石)	景其濬等	同上	不詳	不詳
清	羅聘	《蒲草圖》	景其濬等	王之海編,《揚州畫派書畫全集·羅聘》(天津:天津人民美術出版社,1999)	廣東省博物館	"景氏劍泉收藏圖記"
清	羅聘	《雜畫卷》	景其濬等	《甌鉢羅室書畫過目考目録》;《遲鴻軒所見書畫録》	不詳	不詳
清	董誥	《仿倪黄山水小册》	景其濬等	同上	不詳	不詳
清	潘恭壽	《設色仿戴文進靈谷春雲圖卷》	景其濬等	同上	不詳	不詳
清	王玖	《咫尺千里山水小册》	景其濬等	《選學齋書畫寓目記續編》	不詳	四印不詳

續表

時代	作者	作品	遞藏者	資料出處	現藏地	景其濬鑑藏印
清	錢坫	《三行篆書屏》(十二幀)	景其濬等	《甌鉢羅室書畫過目考目録》;《遲鴻軒所見書畫録》	不詳	不詳
清	夏翬	《墨竹大屏》(四幀)	景其濬等	同上	不詳	不詳
清	錢元章	《篆書大幀》	景其濬等	同上	不詳	不詳
清	錢元章	《墨松竹大幀》	景其濬等	同上	不詳	不詳
清	邵士燮	《五行隸書大幀》	景其濬等	同上	不詳	不詳
清	陳希濂	《四行隸書大幀》	景其濬等	同上	不詳	不詳
清	陳希濂	《墨花卉立幀》	景其濬等	同上	不詳	不詳
清	黄均	《仿倪迂設色春林山影圖大幀》	景其濬等	同上	不詳	不詳
清	汪士慎	《墨梅長卷》	景其濬等	同上	不詳	不詳
清	汪士慎	《行書絶四首》	景其濬等	同上	不詳	不詳
清	李鱓	《設色花木與鳥大屏》	景其濬等	同上	不詳	不詳

續表

時代	作者	作品	遞藏者	資料出處	現藏地	景其濬鑑藏印
清	韻可	《設色花木大册》	景其濬等	《甌鉢羅室書畫過目考目録》;《遲鴻軒所見書畫録》	不詳	不詳
清	凌瑚	《二喬圖紙本立幀》	景其濬等	同上	不詳	不詳
清	包世臣	《行書屏》(四幀)	景其濬等	同上	不詳	不詳
清	鄧傳密	《五行篆書大幀》	景其濬等	同上	不詳	不詳
清	楊鐸	《設色花鳥屏》(四幀)	景其濬等	同上	不詳	不詳
清	金蕊	《設色山水紙本小册》八頁	景其濬等	同上	不詳	不詳
清	錢璞	《設色花卉紙本小册》	景其濬等	同上	不詳	不詳
清	吴讓之	《三行篆書屏》(四幀)	景其濬等	同上	不詳	不詳
清	吴讓之	《設色花卉屏》(四幀)	景其濬等	同上	不詳	不詳

二、金石碑帖

續表

年代	作品	遞藏者	資料出處	現藏地	景其濬鑑藏印
漢	《漢開母廟碑拓本》	景其濬等	《翁同龢日記》	不詳	不詳
漢	《漢李孟初神祠碑墨拓本》	景其濬等	黄易,《小蓬萊閣金石文字》(台北:新文豐出版公司,1986)	台北故宫博物院	不詳
漢	《漢光禄勳六曜殘碑拓本》	景其濬等	莫友芝,《宋元舊本書經眼録》,《續修四庫全書》(上海:上海古籍出版社,1995—1999)	不詳	不詳
三國魏	《晚明拓魏上尊號碑》	黄易、景其濬等		北京故宫博物院	[景氏劍泉]
三國吴	《吴天發神讖碑墨拓本》	景其濬等	莫友芝,《郘亭所見書略》,張劍撰,《莫友芝年譜長編》(北京:中華書局,2008)	不詳	不詳
南梁	《梁瘞鶴銘拓本》	景其濬等	日本二玄社編,《書跡名品叢刊》(東京:二玄社,1961)	日本三井聽冰閣	[劍泉審定]
北魏	《魏張猛龍太守清頌碑》	成邸、景其濬、張度等	王敏編,《北京圖書館藏善拓題跋輯録》(北京:文物出版社,1990)	北京圖書館	[劍泉珍藏]
唐	《宋拓虞永興破邪論序册》	鄭板橋、景其濬等	葉名澧,《敦夙好齋詩全集》續編《養痾集》(清光緒十六年葉兆綱刻本)	不詳	不詳
唐	《宋拓九成宫醴泉銘》	景其濬等	《翁同龢日記》;日本書道博物館編,《書道博物館圖録》(日本:財團法人台東區藝術文化財團,2000)	日本書道博物館	[劍泉珍賞]

續表

年代	作品	遞藏者	資料出處	現藏地	景其濬鑑藏印
唐	《宋拓皇甫誕碑》	郭尚先、張子駿、景其濬、吴湖帆	上海圖書館編,《上海圖書館藏善本碑帖》(上海:上海古籍出版社,2005)	上海圖書館	[劍泉平生癖此][劍泉收藏寶籍金石印]
唐	《宋拓小麻姑仙壇記》	王時敏、魏之璜、趙有成、齊其家、魏克、洪焯、汪澄海、章載道、姚鼐、景其濬、毛昶熙、袁思鐸、譚澤闓等		北京故宫博物院	[其濬][劍泉眼福][黔南][劍泉氏鑑真]
唐	《清拓顔家廟碑册》	畢際有、李在銑、景其濬等		北京故宫博物院	[貴陽景氏][劍泉珍賞]
北宋	《墨拓大觀帖册》	黄琪、文彭、景其濬、薛華培等		北京故宫博物院	[景氏劍泉收藏圖記]

附録二　景其濬鑑藏印選

1.“景劍泉家收藏”

石濤山水花卉册頁

2.“劍泉珍賞”

宋拓九成宫醴泉銘

3.“景其濬印”

八大山人書畫册頁

4.“景二”

石濤山水花卉册頁

5.“劍泉”

石濤山水花卉册頁

6.“景”

石濤山水花卉册頁

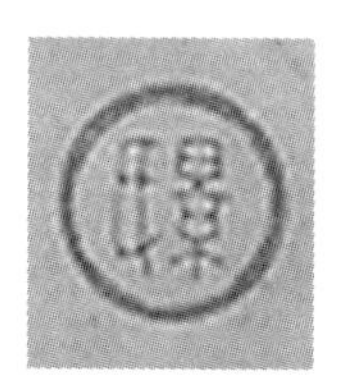

7.“景氏”

仇英募驢圖

8.“劍泉審定”

梁瘞鶴銘拓本

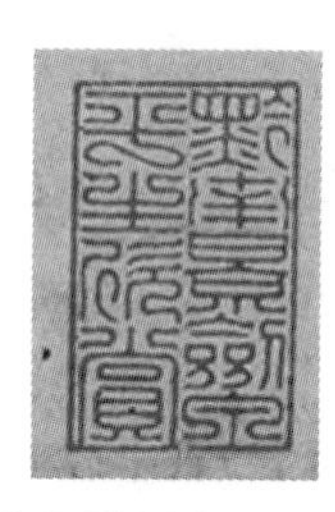

9.“黔南景劍泉平生珍賞”

梁巘臨虞恭公碑册頁

10.“黔南景氏劍泉收藏”

八大山人書畫册頁

11.“貴陽景氏”

趙孟頫大乘妙法蓮花經卷五

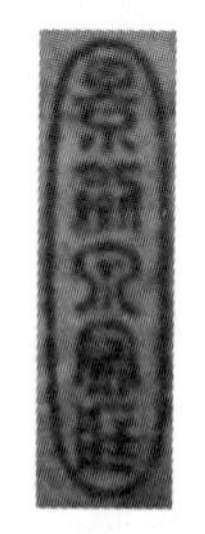

12.“景劍泉家藏”

石濤芝松圖

13.“劍泉真賞”

石濤山水花卉册頁

14.“黔南景氏珍藏”

梁巘臨虞恭公碑册頁

15.“景氏子孫寶之”

米芾三帖卷

16.“劍泉平生癖此”

宋拓皇甫誕碑

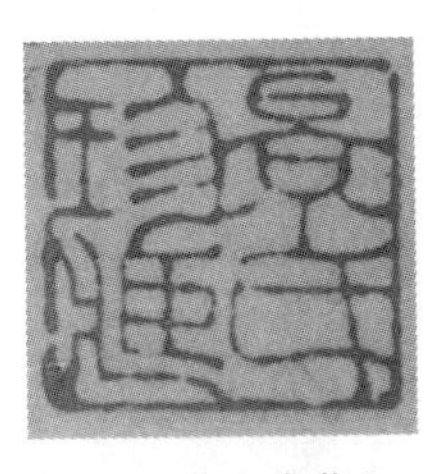

17.“景氏珍藏”

石濤野色圖册

18.“景氏秘篋墨皇”

米芾三帖卷

19.“劍泉收藏寶籍金石印”

宋拓皇甫誕碑

20.“劍泉二十年精力所聚”

石濤芝松圖

命之所繫、魂牽夢縈*

——竹内實心目裏的中國形象

葉楊曦

1923年,正值外國列强瓜分中國之際,在山東的鄉間小鎮"張店",一個日本男孩呱呱墜地。他姓竹内(たけうち),名實(みのる),日後成爲日本國内著名的中國研究學者。2013年,竹内實在京都東北的高野寓所走完了不算平凡的一生,彌留之際家人相伴左右,也算壽終正寢。雖然對於自小體弱多病的竹内實來説,九十辭世當屬高壽,但他的離開仍是日中學術界的損失,對此兩國新聞傳媒皆有報導,學者亦不乏悼念之辭與回憶文章。

2013年7月30日,竹内實在家中就餐時倒下,送醫後不治。8月1日,日本《朝日新聞》《日本經濟新聞》等便刊文記事,强調竹内實"現代中國研究第一人"的身份和關於毛澤東的論述[①]。當日中國主流媒體新華網轉載環球網的消息,又另外加上"曾獲毛澤東接見"的提示[②]。日中傳媒的後續跟進與學

作者單位:山東大學文學與新聞傳播學院

* 本文受山東大學基本科研業務專項資金資助,敬致謝忱。

①《中國研究者の竹内實さん死去　毛沢東論など記す》,《朝日新聞》2013年8月1日,http://www.asahi.com/obituaries/update/0801/OSK201308010161.html。《竹内實氏が死去　現代中國研究の第一人者、京大名譽教授》,《日本經濟新聞》2013年8月1日,http://zh.cn.nikkei.com/gate/big5/www.nikkei.com/article/DGXNASDG0101S_R00C13A8CC0000/。

②《日著名中國研究學者竹内實逝世　曾獲毛澤東接見》,新華網2013年8月1日,http://news.xinhuanet.com/2013-08/01/c_125103523.html。

人的追憶述懷又突出竹内實的“文化大革命”研究、“友好容易理解難”的名言及晚年傾心翻譯的木偶劇《項羽與劉邦》[①]。上述報導與文章涉及的幾個面向都是竹内實學術生涯裏集中用力的地方，他的中國研究興趣廣泛，涉及政治、經濟、軍事、歷史、文學等衆多人文社科領域。

1978年，竹内實的《茶館：中國的風土與世界像》被司馬長風譯爲中文，並以《中國社會史話》爲題由香港文藝書屋出版，是爲中國學界首次譯介其專書[②]。2013年5月，香港天地圖書有限公司刊行程麻編譯之《竹内實的中國觀——第一本中文自選集》，這也是在竹内實有生之年問世的最後一本漢譯論著。頗爲巧合的是，中國學者對竹内實著述的翻譯都與香港有關，甚至從某種意義上講，可以説是以香港爲原點，最終又回到了香港。筆者淺見，香港本地學者有必要而且應該發出聲音，表達自己的意見，追懷、體認並思考竹内實的中國研究。

有鑒於竹内實獨特的生活經歷與豐富的文字著述，本文將在概述其生平與學術的基礎上，從兩本專著入手，對他取鏡中國背後的學術旨歸及最終訴諸的中國觀進行論述，最後和竹内實的中國研究展開對話，體認其治學特點。

一、徘徊與遊走：竹内實的生平和學術

（一）生平

竹内實曾幽默地解釋過自己的名字：“我的名字最好記，人家都説竹子裏面是空的，但我的名字叫‘竹内實’。”[③]竹内實1923年生於中國山東膠濟鐵路沿線的農村，現今已屬淄博市張店區。父母爲普通農民，自日本愛知縣赴華經營日式旅店。5歲時喪父，由母親一人撫養帶大。小學三年級開始學習漢

① 劉燕子，《竹内實與日本“文化大革命”研究》，《開放雜誌》2013年9月，83—87頁。《竹内實，友好容易理解難》，《北京青年報》2013年8月9日，http://bjyouth.ynet.com/3.1/1308/09/8193526.html。《晚年の竹内實さん、中國の劇100話翻訳　項羽と劉邦の物語、原稿2千枚の情熱》，《朝日新聞》2013年8月20日，http://www.asahi.com/shimen/articles/TKY201308190662.html。

② 竹内實，《茶館：中國の風土と世界像》，東京：大修館書店，1974；竹内實著，司馬長風譯，《中國社會史話》，香港：文藝書屋，1978。

③ 徐一平，《我和竹内實先生》，《中華讀書報》2006年11月15日。

語。1934年11歲時，隨母親、弟妹移居僞滿洲國"新京"(今吉林長春)。小學畢業後進入新京商業學校。1942年因考入東京二松學舍專門學校(前身爲漢學塾二松學社，今二松學舍大學)而回到日本。翌年，以"學生上陣"名義於愛知縣豐橋工兵部隊參軍。入伍四月即因病住院，日本戰敗前兩月退伍，未赴前線。1946年考入京都大學文學部，專攻中國文學，師從時任東方文化研究所語言研究室主任的倉石武四郎。大學期間曾擔任冰心的課堂口譯，聽過吉川幸次郎(杜甫)、大手拓次、萩原朔太郎(象徵主義詩歌)、北原白秋、宮澤賢治(新體詩)、貝塚茂樹(中國古代史)、青木正兒(元雜劇)等人的課。1949年，竹内實大學畢業，隨倉石師轉入東京大學大學院(即研究院)。二年級時，參與營救過丸山昇。三年級首次參加示威活動("五·一"事件)，並受其影響，思想開始左傾，加入日本共産黨。大學院修了的同時，竹内實前往東大中國研究所"就職"，後因缺少中國語教員而轉職東京都立大學。1953年夏天，竹内實重返闊别十餘年的中國，以運送俘虜死難者遺骨航船口譯員的身份去過天津，又作爲日中友好協會訪華團口譯員，訪問北京、上海、重慶與瀋陽。此後多次來華，尤其是1960年隨日本文學代表團訪華之際，在上海經陳毅引介，受到毛澤東的點明表揚和接見①，還見到周揚，结识赵树理、老舍，訪問上海電影製片廠，見到趙丹。1962年被日本共産黨除名。1966年擬參加《朝日新聞》"'文化大革命'考察團"，但中國方面邀請函上並無其名字。1970年，因不滿於東京都立大學教職員支持學生"革命造反"，辭去教職。1973年進入京都大學，由助教授直至教授，並出任京都大學人文科學研究所所長。此後曾擔任立命館大學、北京日本學研究中心等多機構的教授。1979年隨日本學術振興會訪問團重回改革開放的中國。1987年從京都大學退休，同年開始擔任日本現代中國研究會會長，直至1994年。1992年獲第三次福岡亞洲文化獎(福岡アジア文化賞)。1999年被日本政府授予勛三等旭日中綬章。2008年生平最後一次來華，赴井岡山參加毛澤東詩詞國際學術研討會，並接受毛新宇採訪。2013年逝世於日本京都，享年九十。

---

① 毛澤東，《美帝國主義是中日兩國人民的共同敵人》(1960年6月21日)，中共中央文獻研究室編，《毛澤東文集》第八卷，北京：人民出版社，1999，200—208頁。附注云："這是毛澤東同日本文學代表團的談話。"

(二) 學術

竹内實是筆耕不輟、勤奮多産的學者。1988年京大人文研主辦的《東方學報》專門刊發過《竹内實教授著作目録》,據筆者目檢,目録包括編著29部,翻譯29部,監修5部,論文更是多達403篇,爲數可觀[①]。一直進行《竹内實著譯年表》編寫工作的程麻根據不完全統計得出,他"一生共出版關於中國的論著、譯著約50種上下,幾乎每月都在報刊上刊登三四篇文章,總數超過千篇以上"[②],字數接近千萬。余項科有更爲詳細的搜集考證,指出從1946年23歲時發表的首篇論文《麵包,即pain》開始直到2014年10月櫻美林大學緊急出版的《轉變的中國 静止的中國》爲止的67年間,竹内實獨著、合著、監修、監譯、合集等書籍不下218種,包括隨筆、譯文、書評等在内的文章則多達1431篇[③]。

竹内實的代表論著包括《毛澤東的詩與人生》(1965,與武田泰淳合著)、《毛澤東與中國共産黨》(1972)、《毛澤東的生涯》(1972)、《魯迅遠景》(1978)、《魯迅周邊》(1981)、《日本人心目裏的中國形象》(1966、1992)、《中國的思想——傳統與現代》(1967、1999)、《現代中國的文學》(1972)、《文化大革命》(1973,編譯)、《茶館——中國的風土與世界像》(1974)、《友好容易理解難》(1980)、《北京——世界都市物語》(1992、1999)、《蟋蟀與革命中國》(2008)、《中國的世界——人、風土、近代》(2009)、《竹内實中國論自選集》(2009)、《漂泊的孔子,復活的論語》(2011)等等。最爲中國學界稱道的,是竹内實監修的二十卷本《毛澤東集》《毛澤東集補卷》,較爲全面地收録了毛澤東在1949年前的著述,體現出京都學派重視文獻資料的傳統學風。另一方面,中國對其作品的譯介除引言部分提及的《中國社會史話》與《竹内實的中國觀》兩書外,程麻主持編譯的十卷本《竹内實文集》無論在數量還是内容上都最爲可觀,涵蓋了竹内實的毛澤東、"文化大革命"、中國傳統文化、中國現代文學、比較文學與文化、中日關係及時事政治等各個研究主題。

---

① 《竹内實教授著作目録》,《東方學報》1988年60期, 733—749頁。

② 程麻,《竹内實:"可以走了"——一位中國學者的哀思與憶念》,《中華讀書報》2013年9月4日,7版。

③ 余項科,《生きている中國—竹内實先生を偲ぶ》,《蒼蒼》2013年57號(竹内實先生の逝去を悼む追悼文および追悼記事),http://www.21ccs.jp/soso/takeutisenseituito/tuitoubun_yo.html。

竹内實著作等身,數量驚人,短期内要想完全瞭解着實不易,但大致可以窺見,他從研究日本人如何看待中國入手,逐漸轉入表達自己的中國觀。這兩點正可以用代表竹内實學術生涯起點與終點的兩部專著來概括,即作爲他山之石的《日本人心目裏的中國形象》及用於反躬自省的《竹内實的中國觀》。

二、他山之石與反躬自省:以兩部專著爲中心

(一)他山之石:《日本人心目裏的中國形象》

《日本人心目裏的中國形象》1966年由春秋社在東京出版,是竹内實獨立完成的最早一部學術專著。日文原名爲《日本人にとっての中國像》,之所以並未使用日本學界流行的"中國觀",竹内實主要是考慮相較於前者的系統,他更關心"中國像"背後"未必能形成體系的印象,甚至是無意識的東西"[①]。作者着力討論"日本人對於中國形象的探求與描述,以及藉助於這種形象對中國和日本問題的思考",他所關心的,是"日本觀念裏與中國有關的那些東西"。書中諸篇"既非'中國論',也不是所謂的'日本論'",而是類似於溝通兩國的橋樑[②]。作者的這些努力似乎可以書中的一句話來概括:想要"把中國作爲'他山之石'"[③]。

此書彙集了作者從20世紀50年代末至60年代初八年間所寫的17篇中國評論文章。竹内實於二戰後開始進入中國研究,此書所收反映了他初登學術舞臺時的思考,從中可以窺見其日後研究的端倪。根據作者原意,這些文章可劃分爲四類:階段性研究評估、當前形勢研究、歷史回顧與整理以及研究方向探討。如結合其中文章的發表或撰寫時間來看,則可較爲清晰地看出作者對本課題的探討從其時之文學現狀入手,進而不斷上溯,推至近代,最後返

---

① 竹内實,《〈日本人心目裏的中國形象〉後記》,程麻譯,《竹内實文集》第一卷《回憶與思考》,北京:中國文聯出版社,2002,141頁。

② 竹内實,《はしがき》,氏著,《日本人にとっての中國像》,東京:岩波書店,1992,276頁。中譯見竹内實,《〈日本人心目裏的中國形象〉序言》,《竹内實文集》第一卷《回憶與思考》,332頁。

③ 竹内實,《戰後文學と中國革命》,竹内實,《日本人にとっての中國像》,東京:春秋社,1966,186頁。中譯見竹内實,《戰後文學與中國革命》,程麻譯,《竹内實文集》第五卷《日中關係研究》,北京:中國文聯出版社,2004,107頁。

回當下,繼續與當代文學對話,並着眼於未來研究。

最早寫成之《昭和文學裏的中國形象》刊佈於1957年,正處於昭和時代(1926—1989)中期,奠定了全書的論述基調。竹内實在前言中就明確表示他對昭和以來文學的考察不滿足於只是勾勒日本人心目中中國形象的變化,更想揭示出這種變化背後折射出的認識主體日本人自身"思想變化、生活變化,以及政治或經濟方面的變化等等",並通過變化軌跡幫助認識到"如今應該怎樣評價中國形象的問題"①。作者結合中國實際而將昭和文學分作戰前(1926—1935)、戰中(1936—1945)和戰後(1946—1955)三個階段。竹内實在具體論述中指出,戰前時期中國形象的基本特點是對無産階級的同情,雖然其抒情至今令人感動,但大部分詩歌、小説、戲劇等在表現革命中國時過於直白,流於公式化與概念化,缺乏感同身受的實際體驗。能夠擺脱這種表面化危險的是反戰文學,由於其中不僅使中國登場,而且以日本士兵爲主體,"通過日本問題反映中國",清醒地意識到作品面對日本讀者。另外,中國的無産階級革命也影響到横光利一在上海的實地經歷,幫助其打開"新感覺派的活路"。竹内實因而稱中國形象在日本文學的微妙轉變中"發揮了即使不是決定性的,也具有相當重要的作用"。戰中十年,文學逐漸被荒廢,所謂"戰争文學"與作爲"滿洲"紀行的"大陸開拓文學"佔據主流,報告文學開始氾濫。其特點是内容空洞,"缺乏真正洞察戰争本質的力量,也没有這方面的熱情"。在日本軍國主義者心目中,中國形象"相當蒼白",這種無知與不理解帶來的淒慘印象不能不唤起竹内實内心"敏感而又沉重的反省"。即使是對戰争有抵抗意識的武田泰淳《司馬遷》與竹内好《魯迅》在"接近正確的中國"上也有段距離。戰後階段方才有日本人深入思考中國問題,無論是武田泰淳的《審判》,還是堀田善衛的《時間》都體現出對中國的贖罪意識。但另一方面,即"積極描繪中國所發生的巨大變化",日本文學中卻表現不夠。此外,在堀田善衛的《喪失祖國》與《歷史》裏的中國遊移擺蕩,邊界模糊,恰恰反映出"輪廓清晰並穩定的日本"並不存在。通過上述分析,竹内實將三個時期文學裏的

① 竹内實,《昭和文學における中國像》,《日本人にとっての中國像》,315頁。中譯見竹内實,《昭和文學裏的中國形象》,《竹内實文集》第五卷《日中關係研究》,2頁。

中國形象歸納成“革命的中國”“空白的中國”與“贖罪的中國”。

爲了探究當代日本文學裏中國形象的源流，竹内實自昭和文學開始向前追溯，從明治、大正之交夏目漱石的《滿韓處處》，到明治末年宇野哲人的《“支那”文明記》，到明治晚期田岡嶺雲的《戰袍餘塵》、安東不二雄的《“支那”漫遊實記》與内藤湖南的《燕山楚水》，再到明治中前期岡千仞的《觀光紀游》，直至明治初年竹添進一郎的《棧雲峽雨日記》，等等。在對上述不同時期中國行紀的詳細分析中，讀者可以看到日本人心目中中國形象的變遷軌跡。這些文學作品的作者大多是漢學者，雖然最初的竹添進一郎與岡千仞皆以漢語行文，但竹内實認爲作爲竹添精神文化母國的中國正是岡千仞所極力批判的。早期對於中國的崇敬思想與精神隨着時間的推移逐漸動摇：截然區分古典中國與現代中國，雖然内藤湖南、宇野哲人、夏目漱石等對儒家聖賢和古代經典依然心嚮往之，但安東不二雄等人對現代文化已毫無興趣，而是更關心經濟、軍事領域的發展，形成“奇妙的二律背反”[①]。更值得注意的是，田岡嶺雲等人否定戰爭的寫作姿態在昭和初年得到繼承，影響到日本無産階級革命文學的創作與發展。

在回顧歷史的同時，竹内實也進一步深化對日本當代文學的研討，而相關主題多是從此前昭和文學的總論中申發出來。戰爭期間，“八紘一宇”“大東亞共榮圈”等口號凸顯出日本人的民族使命感，竹内實分析評論保田與重郎和竹内好等人筆下平民的使命感及其瀰漫與消失。戰後，日本挫敗的國内形勢又使人沮喪，作者結合井上光晴《虚構的吊車》、太宰治《惜别》、田中英光《暈船》、武田泰淳《風媒花》《審判》與堀田善衛的《時間》等討論這種挫折感、屈辱感及由此引發的贖罪意識，讚揚武田與堀田文學作品的責任擔當，尤其是兩人的文學活動在“溝通中國的民族主義與現代日本”[②]之間所發揮的橋樑作用。日本人對於中國革命既有民間的“親近感”，也有官方的“驚恐感”。日本的民族主義必須立足於中國、朝鮮等亞洲民族利益的基礎上，應該結合中國的民主主義與社會主義促發新的日本民族主義，而不是一味“强調日本思

① 竹内實，《明治漢學者の中國紀行》，《日本人にとっての中國像》，231頁。

②《戰後文學と中國革命》，《日本人にとっての中國像》，195頁。中譯見《戰後文學與中國革命》，《竹内實文集》第五卷《日中關係研究》，116頁。

想的自立”[①]。橋本身固然重要,但首先應該夯實地基,重視岸的建設,只有這樣,中日兩國才有可能恢復邦交並加深相互理解。而就在不久的十年之後,竹内實等衆多日本人積極推動和翹首以盼的中日恢復邦交成爲現實。

全書最後,竹内實以同題論文總結既往研究並展望未來。他注意到普通日本人的“舊中國”意識,並抱以懷疑的態度,以便更好地樹立“新中國”觀念。面對“舊”“新”兩個中國,日本人心目中的距離感遵循着“零距離與無限大”[②]的變化模式,而其背後“無非都是多次受到來自中國的刺激所出現的情緒變化”[③]。作者希望今後繼續此項研究時能“對自己以前那些不太自覺的觀念進行反思”。

1992年《日本人心目裏的中國形象》同名再版,内容則做了很大的改動[④]。除了在第二部分“日本人心目裏的中國形象”保留了原書的4篇文章外,另加入第一部分“啊,大東亞共榮圈”,由《難民的思想》《建國的思想》與《安撫的思想》三篇論文組成。在《難民的思想》中,作者結合“滿洲”開拓團活動的記録和重回故地者的描述,探討日本人在“滿洲”的殖民開拓、戰敗時的悲慘境遇與日後反思,指出日本帝國主義侵略在給中國、朝鮮等亞細亞民族帶來災難,引發内戰饑荒的同時,亦造成大批殖民地日僑“難民”的出現。其流毒餘害甚至波及戰後,這些以“大東亞共榮圈”爲信條的“難民”因失去國家保護而出現,其“不定型、無秩序、無價值的生活”[⑤]依然深受“民主主義”的侵擾。《建國的思想》則帶着日本爲何要建立“滿洲國”的疑問,推源溯流,迴歸此概念的歷史脈絡,梳理其建制沿革、地理區劃、開拓活動與住民流動。作者認爲“滿洲國”是“日本對中國侵略下建立的傀儡政權,是對帝國主義行爲的僞裝”,建國思想是“伴隨現實過程發展起來的思想”。最後一篇《安撫的思想》主要論述日軍戰中推行的“宣撫班”策略。“安撫的思想”一方面是“轉向者的

---

①《橋のうえの眼覚め—日中関係の基礎となるもの》,《日本人にとっての中國像》,140頁。中譯見《對所謂“橋樑”關係的思考——談日中關係的基礎》,《日中關係研究》,102頁。

②《日本人にとっての中國像》,《日本人にとっての中國像》,390頁。中譯見《日本人心目裏的中國形象》,《竹内實文集》第一卷《回憶與思考》,243頁。

③《日本人にとっての中國像》,392頁。中譯見《日本人心目裏的中國形象》,245頁。

④ 竹内實本意將此書命名爲《昭和文學裏的中國形象》,出版時的書名是責任編輯改動的結果。

⑤《難民の思想》,《日本人にとっての中國像》,32—33頁。

思想”,另一方面是被稱作“反八路軍”“反共”等“反”的思想。它伴隨軍隊的移動而被宣傳,政權確立後則停用。安撫的最終目的是“必須加深與中國民族的接觸”。“安撫的思想”隨着日本戰敗而被廢止,但“思想”的行爲“安撫”卻並未消除。

上述第一部分最早作於1970年8月15日前後,當時竹内實剛因故辭去東京都立大學教職,出於生計考慮,也爲了紀念這個特殊的日子,他選擇以當時日本歷史學中犯忌的“大東亞共榮圈”爲題。如果説前一個十年他尚能“給人如飢似渴的印象”,那麼20世紀70年代的他則有“被孤立”[①]的感覺。作者少年時代有一半時間在康德紀元的“新京”長春度過,甫及成人,即遭遇所謂“大東亞戰争”,甚至在歸國服役期間試圖回到“滿洲”而未果。他研究“滿洲國”,更多與其莫名其妙、曖昧含糊的身份有關,受童年記憶影響甚深。

“日本人心目裏的中國形象”是竹内實進行中國研究時的重要課題。但他的用意並不止於描述“他山之石”,而是將其帶入自己心目中中國形象的探討之中。從20世紀40年代末初入學術界開始,竹内實在對中國傳統文化與現代社會保有濃厚興趣的同時,又一直以追蹤時政的熱情緊跟時代潮流,從毛澤東到“文化大革命”再到改革開放,不斷就中國國内的熱點問題發表見解。而在這些努力的背後可以看到他借鏡中國以反躬自省的學術姿態。

(二)反躬自省:《竹内實的中國觀》

2009年日本櫻美林大學東北亞總合研究所出版了三卷本的日文《竹内實中國論自選集》,在其卷首,作者有如下説明:“第一卷是與我現在内心仍關注的‘文化大革命’有關的拙文的彙集,第二卷爲‘轉變的中國’,那是我時時注視着的動向,第三卷像是上兩座山脈谷間的平地,集中了影評、魯迅論、風土論與人物論,題爲‘電影與文學’。”

這套書收録了作者討論自己心目中中國形象的相關論述中最具代表性的學術文章。以此爲基礎,結合十卷本中文版《竹内實文集》,中國學者程麻編譯出《竹内實的中國觀》一書,2013年由香港天地圖書有限公司出版。此書

① 《あとがき》,《日本人にとっての中國像》,276頁。中譯見《〈日本人心目裏的中國形象〉後記》,《竹内實文集》第一卷《回憶與思考》,141頁。

有一個副標題，叫"第一本中文自選集"，表明其面對的是中國讀者。

結合日文《竹内實中國論自選集》的編排框架，讀者根據《竹内實的中國觀》的12篇選目可以看出他對於自身的學術定位：自日文本《自選集》第三册《電影與文學》中選入篇目最多，有6篇，其次爲見收於第一册《文化大革命》中的5篇，出自第二册《轉變的中國》中的篇目則爲1篇。

不同於《日本人心目裏的中國形象》，《竹内實的中國觀》按照時間順序編排内中諸文。有學者稱"竹内實來到日本後重新接近中國的切入點是文學"[①]。如果進一步縮小範圍，那麽可以説與戰後成長起來的其他日本第一代現代中國研究者類似，他也是通過魯迅研究進入這一領域。但竹内實也有自己的特點，他涉及魯迅的論述並不僅僅局限在魯迅本身。此書以最早發表之《魯迅和他的弟子們》開篇，作者在文中將重心放在魯迅得意門生蕭軍與胡風在新中國成立後的遭際問題，與當時接連不斷的"批判運動"展開對話。他提出"獨立於政治的'文學'之類觀念確實曾不斷受到'革命'激流的洗禮"[②]的問題，直言不諱地指出當時所謂"向魯迅學習也許只是一個口號"[③]而已，暗含對於魯迅在新中國被誤讀、形象遭受歪曲的不滿。與魯迅直接相關的另有兩篇文章：《阿金考》與《魯迅與孔子》，二者都隱含批判新中國時政的意味。前文的寫作有意模仿"文化大革命"時推動者所作的影射之文。《阿金》是魯迅一篇雜文的題目，主人公是受僱於魯迅對面人家的女傭。作者將文中劃線部分與當時的政治形勢做了細緻的比對，在魯迅其他作品中找出"阿金"的影子，並結合中國文學裏的潑婦形象與現實社會某些"革命者"的投影進行討論，讚揚魯迅"深刻挖掘社會"[④]的努力，將魯迅格言改寫爲"現實之於虚妄，正與文學相同"，指出其文學之特點正在於"以虚構的文學去抨擊虚假的現實"。時隔半個多世紀，竹内實在回憶時明確稱"文章針對的是江青夫人"[⑤]。《魯迅與孔

① 馬場公彦、竹内實著，程麻譯，《竹内實，一身兩棲於日中之間·採訪解説》，《竹内實的中國觀——第一本中文自選集》，340頁。原載馬場公彦，《戰後日本人的中國形象——自日本戰敗到文化大革命、日中恢復邦交》(《戰後日本人の中國像—日本敗戰から文化大革命·日中復交まで》)，東京：新曜社，2010。

②《魯迅和他的弟子們·解題》，《竹内實的中國觀——第一本中文自選集》，11頁。

③《魯迅和他的弟子們》，《竹内實的中國觀——第一本中文自選集》，16頁。

④《阿金考》，《竹内實的中國觀——第一本中文自選集》，149頁。

⑤《阿金考·解題》，《竹内實的中國觀——第一本中文自選集》，126頁。

子》也與現實息息相關。該文本爲作者1981年在京都所作的報告,對象是赴東京參加辛亥革命七十週年學術研討會的中國近代史研究者。"文化大革命"中隨着林彪叛國墜機而開始了"批林批孔"運動,其中曾經反對儒教的魯迅受到頌揚。孔子很早以前在中國便是至聖先師,而魯迅的聖人地位則是1937年毛澤東在延安親口所授:"孔夫子是封建社會的聖人,魯迅則是新中國的聖人。"[①]更加吊詭的是,民國年間,魯迅曾代表官方主持過祭孔大典。作者以祭孔期間的魯迅日記爲主要材料,聯繫其雜文、書信等,分析出魯迅的屈辱感受與複雜心境,認爲他不拒絶擔任祭祀神官出於兩方面的考慮:一是並非完全抵觸陳舊事物,二是迫於生計。但無論如何,他都"没有過於屈服"。

着眼於整個中國現代文學,竹内實總結"中國大陸文化的基礎仍是窑洞文學"[②]。1942年以前"延安壓倒性的主流是城市型文學(戲劇)",而毛澤東《在延安文藝座談會上的講話》則只將"窑洞文學"視作文學。雖然中華人民共和國成立時召開的文藝工作者代表大會期待"城市文學"重回主流,但從延安的"思想改造"運動開始,到反右,再到"文化大革命",經過文學領域重複的批判運動,"城市文學"終被否定。對此,竹内實在"文化大革命"結束後認爲"已經到了打破停滯與惡性循環的時候",必須清除頻繁政治鬥争給文學帶來的嚴重創傷,期待超脱於二者之上的作品,遵循文學自身的運行法則。

1960年訪華期間,竹内實在上海見到了趙丹,當時"不禁湧出了淚水"[③]。這是一種親眼所見印證銀幕接觸時的真情流露。在電影方面,竹内實並非要做專業性的影評,所關注的也不是具體技術的運用,而是"通過電影了解中國動向"[④],更看重影像自身的活力與生機,關注其所傳遞的中國風土人情與文化傳統。

毛澤東與"文化大革命"研究是竹内實中國研究的核心領域,貫穿其整個學術生涯始終。他出版的首部專題性學術論著便是與前輩武田泰淳合作的

① 《魯迅與孔子》,《竹内實的中國觀——第一本中文自選集》,153頁。此句並未收入《毛澤東選集》。

② 《從窑洞文學到城市文學》,《竹内實的中國觀——第一本中文自選集》,214頁。

③ 《電影描繪的風情、風俗與傳統》,《竹内實的中國觀——第一本中文自選集》,215頁。

④ 《電影譜·解題》,《竹内實的中國觀——第一本中文自選集》,255頁。

《毛澤東的詩與人生》，而且除後者所寫《後記》外，正文部分皆由竹内實一人完成[①]。上文提到的1960年中國之行是竹内實一生中最爲難忘的，主要因爲他受到毛澤東的點名表揚。《竹内實的中國觀》之《對話毛澤東——“牛鬼蛇神”及其他》在十卷本中文《竹内實文集》中僅保留了副標題，而此文的特色恰恰如主標題所揭示的：站在一個平等客體的角度與毛澤東展開對話。他對毛澤東的終極定位是“中國最後的革命家和造反派”[②]。“牛鬼蛇神”是唐人杜牧對李賀詩作的評語，毛澤東起初使用該詞時只想表達原意，即“妖魔鬼怪”，但日後詞義轉變。從1963年的等同於“地主、富農、反革命分子”逐步引申，“文化大革命”期間被紅衛兵與造反派利用。不過，竹内實對毛澤東在詞義轉變過程中扮演的角色做了客觀的切分：“不太好説毛澤東應負有多大責任”，但“文化大革命”期間各種運動歸根結底是“黨内政治鬥爭”“權力鬥爭”，客觀上樹立了“毛澤東思想絶對權威”的地位。書中最能代表作者借鏡中國幫助日本人“反躬自省”態度的是《“文化大革命”和日本思考方式》一文。“文化大革命”曾一度在日本廣受贊同，该文藉此以考察“戰後日本一種流行的思維方式”[③]。現實中的“文化大革命”實際與回憶裏的童年記憶給竹内實以倍感熟悉、似曾相識之感，而他又强烈質疑“文化大革命”中的做法。他曾坦率表露過自己對於“文化大革命”的矛盾心理：“既對‘文化大革命’難以共鳴，又對採取那種實際做法的紅衛兵覺得有些熟悉。”[④]新中國成立初期，“群衆”與“潮流”的力量得到肯定，戰後日本人心目中的中國形象是“誇張的現實”。“文化大革命”對他們來説是痛苦的轉變，轉向“令人掃興的現實”，導致兩種喪失：一是“依據中國校正日本喪失方向”，一是“隱藏的天皇制觀念喪失崇拜的對象”，但依舊有日本人對“文化大革命”保持信仰。林彪出逃使“文化大革命”開始受到質疑，而其垮臺前後的兩種形象“生動地反映了中國從‘“文化大革命”中國’走向‘反革命中國’的轉變過程”。日本人看待中國時對“令人掃興的現實”相當討厭，故意視而不見，選擇“誇張的現實”，而後者支撑了日本的

① 武田泰淳、竹内實，《毛澤東 その詩と人生》，東京：文藝春秋社，1965。
②《對話毛澤東——“牛鬼蛇神”及其他》，《竹内實的中國觀——第一本中文自選集》，77頁。
③《“文化大革命”和日本思考方式·解題》，《竹内實的中國觀——第一本中文自選集》，79頁。
④《我心中的紅衛兵·解題》，《竹内實的中國觀——第一本中文自選集》，34頁。

天皇制思想信仰。竹内實大聲疾呼的是希望日本人正視那些"令人掃興的現實",不要試圖忘卻。

本書正文以《根本價值觀:中華思想》結尾。竹内實在文中將"中華思想"當作轉變的中國裏不變的内在價值觀,探討其歷史源流,揭示其本真含義。與此同時,作者不僅發現它在危機意識方面與日本"尊王攘夷"思想的相似性,而且提出現今的政治應該利用"中華思想"等傳統文化分析社會。

"魯迅""電影""毛澤東""文化大革命"與"中華思想"構成了竹内實生平惟一一本中文自選集的關鍵詞,也是他希望呈現給中國讀者的自己心目中中國形象的代表。在某種意義上,這是竹内實生前留給學術界的最後面影:他的研究從文學入手,但不限於此,視野開闊,跨越現代意義上越分越細的學科界限,融匯古今。雖然論述題材有着鮮明的中國特色,但其背後的關懷則是人文日本式的。

三、政治、生活與學術:竹内實的中國研究

(一) 命之所繫、魂牽夢縈

竹内實一生與中國友好,他將個人經歷與生命體驗融入中國研究之中。在戰後日本的中國研究者中,竹内實的情況最爲特殊。他在中國農村出生長大,少年時代在僞"滿洲國"度過,直到成人以後才返回東瀛,因而他對中國的親切感與生俱來。竹内實甚至曾經在1995年寫過題爲《我的故鄉在中國》的隨筆[①]。有學者便認爲:"關於竹内實對中國的認識途徑,印象最深的是那鄉下小鎮的風土人情難以磨滅的烙印,以及對在異國土地上以女人之手撫育了自己的母親的迷戀之情。"[②]竹内實中國觀的根基是其"對故鄉和母親純真依戀的延伸"。

在本文首章,筆者曾花費筆墨詳細介紹竹内實的生平,主要是希望展現其特殊的教育背景與人生經歷,而這兩點的背後有中日雙方的合力在起作

---

① 竹内實,《我的故鄉在中國》,《竹内實文集》第一卷《回憶與思考》,22—24頁。

②《竹内實,一身兩棲於日中之間·採訪解説》,《竹内實的中國觀——第一本中文自選集》,339頁。

用。竹内實在中國度過其成年前的時光,“足足生活了十八年”[①],接受了完整的基礎教育,學會漢語。由於11歲前都生活在山東農村,他對中國的民間文化風俗與中下層民衆的生存狀態有過親身體驗。童年記憶令其印象深刻,雖然他在長春待了7年,也學到不少東西,但“總也忘不了在山東的那段生活”[②],“如今能夠出版這樣的系列文集(筆者注:指《竹内實的中國觀》),原因之一是我出生在山東省”[③]。竹内實涉足中國研究是在1949年新中國成立之後,並將此作爲自己畢生耕耘的學術園地。這固然與其求學期間所受中國語言文學教育、京大與東大兩校不同學風的交匯影響以及京都大學中國研究所的整體氛圍息息相關,更是和他的生命體悟密不可分。在寫作過程中,他“覺得自己好些地方湧生出與中國的天空融合爲一的那種神秘之感”。對其而言,“中國是懷念的故鄉”:

> 或許是那種對中國的眷戀,一種望鄉的情愁,牽引我從事中國的研究。確切地説,除了研究中國,自己已無出路。換句話説,爲了填平無盡的鄉愁,我渴望有關中國的書籍,撰寫有關中國的論文!對我而言,歷來的學説如何論述中國的種種,並不重要,自己如何看待中國,透過各種資料的描述,釐清那些模糊點,並加以援引,使自己的思想得以明確下來,這才是重要的工作。[④]

中國是竹内實命之所繫、魂牽夢縈的地方。在本文重點討論的《竹内實的中國觀》序言中,作者也自稱:“那是我專注於中國研究的産物,與我性命攸關。其中談論的事情,不妨説都和自己的生命有聯繫。”[⑤]

(二)“不當變色龍”

數年前,在日本學者馬場公彦主持的一次訪談中,竹内實稱自己的中國

---

① 竹内實,《自序》,《竹内實的中國觀——第一本中文自選集》,9頁。

② 竹内實,《〈竹内實文集〉自序》,《竹内實文集》第一卷《回憶與思考》,3頁。

③《自序》,《竹内實的中國觀——第一本中文自選集》,10頁。

④ 竹内實,《致中文讀者》,郭興工、黄英哲校訂,《解剖中國的思想——傳統與現代》,台北:前衛出版社,1996,5頁。

⑤《自序》,《竹内實的中國觀——第一本中文自選集》,9頁。

研究最早立足於三點:"一、以自己心中的'中國'爲研究對象;二、進行書桌上的研究;三、有中國人的地方就有'中國'。"[①]

上文提到,竹内實在1960年訪華時曾得到毛澤東的點名表揚,但由於他在此後連續公開批判中國的核試驗及反右、"文化大革命"等運動,因而被中國政府"貼上了'非友好人士'的政治標簽",禁止來華。與此同時,他"在日本被劃爲另一種'政治黑類'"[②]。馬場公彦認爲竹内實的獨特經歷相較他人多了一層"土腥味",這"實際卻成了阻礙其進入研究小團體的主要原因"[③]。但即便如此,竹内實依舊對中國充滿了感情,以上規則正是他在這種孤獨感與親近感籠罩下確立的,並自述"從那以後,我豁然開朗,心謐安寧了。想什麽就寫什麽的,内心無愧,文責自負,不當變色龍"。這種"不當變色龍"的原則使竹内實在當時的中國研究者中獨樹一幟。

他站在自己的立場看待中國問題,既不投合民意,也不逢迎官方,而是進行獨立寫作。他以批判的眼光冷眼旁觀,因而能在日本對"文化大革命"一片贊聲裏洞察其中的隱憂,當日本人將中國共産黨視爲天皇制觀念崇拜的對象時能一語道破他們心目中誇張的中國形象背後的虚幻空洞。因爲"不當變色龍",所以他不會見風使舵,隨波逐流,中日友好運動分裂之際斷絶同所有團體的聯繫,在"學潮"中與學生保持距離,也能因不滿教師支持學生"革命造反"而辭去大學教職。可以説,他的那種孤獨感與主動選擇的自我隔絶、自我邊緣化有關。另一方面,竹内實的"不當變色龍"並不排斥人性的温情,雖然"經常通過以文人爲中心的知識分子批判來看中國",但"文化大革命"期間曾定下戒律:不對自己認識的作家、編輯等中國朋友落井下石。

(三)"友好容易理解難"

上述這段訪談後來被馬場公彦整理爲專章收入《戰後日本人的中國形象》一書,使用了"竹内實:一身兩棲於日中之間"(竹内実　一身で二つの生を生きる)的標題,這較爲準確地反映了竹内實畢生致力的事業:推動中日兩

---

①《竹内實,一身兩棲於日中之間》,《竹内實的中國觀——第一本中文自選集》,338頁。

②《竹内實與日本"文化大革命"研究》,《開放雜誌》2013年9月,86頁。

③《竹内實,一身兩棲於日中之間·採訪解説》,《竹内實的中國觀——第一本中文自選集》,339頁。

國的溝通交流與互相理解。

正式從事中國研究後，他利用一切機會接觸中國：隨日本代表團訪華，接待赴日交流的中國學者，去電影院與銀幕上的中國相遇，等等。不同於從書本獲得的間接文字經驗，現實的中國更能給其以活生生的真實感，從而促進研究的開展。他的中國研究需要一種近似於人類學田野調查的現場感。惟一的例外是從1960—1979年的十九年間，竹內實一直無法重回中國，最大原因便是“文化大革命”。在這種情況下，他給自己的中國研究確立的第二條規則是上文提到的“進行書桌上的研究”。竹內實結合自己對於中國傳統文化的理解，此前來華接觸的中國政治、文學人物與中國社會風土，以及所在教研室訂購的《人民日報》《人民文學》等報刊的閱讀經驗與“文化大革命”展開對話。雖然他能富有預見性地加以批判，提醒日本人不可盲目樂觀，但整個過程中其心情無疑是複雜而壓抑的，所以有學者會説“《“文化大革命”觀察》是先生著作中最‘難産’的一本”[①]。

1980年，竹內實出版了改革開放以來他對於中國時政的看法，題爲《友好容易理解難》[②]。雖然這一表述並非他的獨創，亦見於其他日本學者的論著，但它基本能代表竹內實心目中中日兩國關係從過去到現今的狀況。對於“友好容易理解難”的具體內涵，他曾有過具體解釋：“日本與中國‘友好’或者‘不友好’其實不重要，或者説没那麽重要，兩國之間的歷史、文化、習俗和內在心理，甚至不同的情感、語言表達方式的理解更重要，小圈子，小團體和個人層面上的舒筋活絡，血脈相通更重要。”[③]從以上內容看，他關注的不只是表面上兩國和諧的現象，而更多是民間層面彼此的换位思考與溝通互動。歷史上有很長一段時間，中國和日本都使用相同的書寫體系，雖然口語不通，但能通過“以筆代舌”的筆談形式實現交流。兩國同屬東亞漢文化圈，人民“內心的感受方式、道德觀念、知識結構等，往往是根據某些基本原則而展開”[④]。近代以

① 《竹內實與日本“文化大革命”研究》，《開放雜誌》2013年9月，85頁。

② 竹內實，《友好は易く理解は難し，80年代中國への透視》，東京：サイマル出版会，1980。

③ 《竹內實與日本“文化大革命”研究》，《開放雜誌》2013年9月，86頁。

④ 張伯偉，《從“西方美人”到“東門之女”》，樂黛雲、錢林森等主編，《跨文化對話》28輯，北京：三聯書店，2011，225—226頁。

來，中日間的矛盾與摩擦不斷，關於對方的認識和想像經常不是流於虛幻空泛，就是陷入仇視詆譭，這很大程度上是由於缺乏有效的溝通交流所造成。竹内實一生在中日之間奔波：在中日戰争剛剛完結、兩國外交基本中斷的20世紀50年代，他隨民間團體訪華，爲破冰努力；中國發生"文化大革命"之際，他能冷眼旁觀，未雨綢繆，既反對頭腦過熱，又推動消弭誤解；中國改革開放之後，他追蹤中國現實，促進中國消息在日本民衆間傳播。竹内實雖然做出了"友好容易理解難"的判斷，但也寫過"結冰層層封，雙方不乏遠見人，毅然送春來"[①]這樣的俳句，他對中日關係的前景仍抱有期待。

（四）"中國研究雜家"

上文提到，竹内實在研究日本人心目中的中國形象時，注意聯繫明治、大正兩代，追溯昭和文學的歷史源流。其實不限於此，他的很多研究都是推源溯流，以史爲鑒。竹内實是隨着新中國的成立進入中國研究的，此前並没有現成的日本人關於現代中國的研究範式可供參考。他結合傳統，在古代中國與近代中國裏發現現代中國。通過不斷摸索與反覆實踐，竹内實逐漸走出一條具有自我風格的研究之路，將現代中國研究確立爲一門獨立學科。他在研究中注意保留古典趣味，有些文章如《城牆裏的成熟——對中國歷史和文化的一種視角》[②]及《中華世界的國家與王朝——其如何"統一"?》[③]等，倘若單從標題上看，完全是關於中國傳統的題材，但最後的立足點卻都放在幫助日本人理解現代中國上。

竹内實留下的文字中，有很大一部分篇幅不長，屬於時評社論。雖然與同代人一樣，竹内實也是通過文學進入中國研究，但他另有追蹤新聞時事的熱情，將政治引入學術。《竹内實文集》第七卷《中國改革開放進程追蹤》中收録的文章就是很好的佐證。竹内實的中國研究是將文學、歷史、哲學、政治等結合在一起的綜合研究，而非純粹的文學研究，跨越了現代意義上越分越細

① 于青，《中日名人漢俳聯句》，人民網2009，http://world.people.com.cn/GB/1029/42354/9227938.html。

② 竹内實，《城牆裏的成熟——對中國歷史和文化的一種視角》，程麻譯，《竹内實文集》第九卷《中國歷史與社會評論》，北京：中國文聯出版社，2006，160—172頁。

③《中華世界的國家與王朝——其如何"統一"?》，《中國歷史與社會評論》，293—301頁。

的學科邊界，因而可被稱爲“中國研究雜家”。

竹内實希望作爲思想家被接受：“我想成爲思想家，但現在還只是一名教授”（僕は、思想家になりたかったが、たったの教授でした）[①]。他的文章背後有情懷：一爲相伴始終的鄉愁；二是做中國研究，面對日本問題，介入當下文化建設。從形象學上看，這些文章在潛移默化中起到改變社會集體想像的作用，從而促進日本人與時俱進地形塑自己的中國觀。竹内實爲報章而寫的時評社論受衆廣泛，不僅能引導日本普通民衆對於中國的理解走向正途，而且也在一定程度上幫助日本政府做出有利於中日關係的決策。據此而言，竹内實的中國研究有其思想性的一面，不過相對而言更偏向政治。筆者這樣表述，並非意圖否認竹内實的學術價值，而是想指出學術生産的不同機制。學者可分爲學院派與社會型兩類，竹内實更多屬於第二種類型。竹内實是“中國研究雜家”，如果一定要按現今的學科體系劃分，那麼他的研究更接近於國際政治。學術貢獻並不只是論文與實驗，竹内實的學術觀點對於政府決策的影響本身就是一種學術貢獻。

（五）“竹内中國學”的學術影響

竹内實的中國研究曾被總結爲“竹内中國學”[②]。竹内實本人在日本學術界並非主流人物，不僅由於研究路數異於諸家，不被認同，也與衆多弟子的鼓吹不力有關。不過，“竹内中國學”的學術影響依然有跡可循，例如美國學者傅佛果（Joshua A. Fogel）的中日研究。從嚴格意義上講，傅佛果最多只能算竹内實的私淑弟子。20世紀70年代在哥倫比亞大學攻讀博士學位期間，他曾經獲得獎學金前往京都大學交换一年，當時的指導老師便是竹内實。傅佛果的博士論文是通過遊記文學探討近代到現代日本人中國觀的改變，曾經提及日本旅行者的住宿問題：“早期日本旅行者十分依賴日本使臣或少數日本旅居經商者。但在世紀之交，已經有很多日本旅店、餐館及浴室可供那些深居簡

① 余項科，《生きている中國—竹内實先生を偲ぶ》，《蒼蒼》2013年57號。

② 具體内容可參程麻，《竹内實：“可以走了”——一位中國學者的哀思與憶念》，《中華讀書報》2013年9月4日。

出或因害羞而不准備嘗試住在本地旅店或其他當地設施的日本人選擇。"[①]竹内實便是在傅佛果所描述的環境中出生的,當時他的父母在膠濟鐵路沿線的小鎮經營旅店。另外,上文提到,竹内實曾在《明治漢學者的中國紀行》(《日本人心目裏的中國形象》收録)中以"文化母國"概括明治漢學者心目中的中國,它也被傅佛果譯作"cultural motherland"並在博士論文中襲用,展開議論,並明確表示他"對竹添遊記的閱讀受到竹内實的很大影響"[②]。若就最近的影響而言,竹内實《金印之謎》[③]與傅佛果《一件實物的來世:明治時代有關公元57年金印的辯論》[④]皆由漢光武帝賜予倭國的"漢委奴國王"金印展開,但是討論了不同的主題:竹内實偏於金印的思想史與文化史意義,傅佛果則側重學術史的展開。

四、結語:告别一個學術時代

論述竹内實應注意措辭,因爲謙虚低調是其本人的一貫風格:他本人"在日本,決不是中國研究方面的'第一人',也不是什麽'泰斗'。以前在演講之類的場合經常這樣介紹,我感到很不好意思"[⑤]。在《竹内實文集》的出版慶功會上,他謙稱自己的文章只是"應景之作,不值一文"[⑥]。這種風格也被其家屬秉承,在其去世後,"主要由親人料理先生的後事,不準備接受朋友的弔唁"[⑦]。

竹内實的謙虚源自他對自身狀況的清醒認識,在最後的幾年裏,他常常預感到生命即將終結。2007年前往上海參加毛澤東研究國際研討會時,竹内

---

① Joshua A. Fogel, *The Literature of Travel in the Japanese Rediscovery of China, 1862-1945*, Standford: Standford University Press, 1996, p.66.

② Joshua A. Fogel, *The Literature of Travel in the Japanese Rediscovery of China, 1862-1945*, p. 317 注2。

③ 竹内實,《金印之謎》,程麻譯,《竹内實文集》第八卷《比較文學與文化研究》,北京:中國文聯出版社,2006,286—326頁。

④ 傅佛果著,吴偉明譯,《一件實物的來世,明治時代有關公元57年金印的辯論》,吴偉明編,《在日本尋找中國,現代性及身份認同的中日互動》,香港:中文大學出版社,2013,61—71頁。

⑤《〈竹内實文集〉自序》,《竹内實文集》第一卷《回憶與思考》,3頁。

⑥ 李梓,《日本逐漸淡忘對他的崇敬》,《華人世界》2007年2期, 42頁。

⑦ 程麻,《竹内實:"可以走了"——一位中國學者的哀思與憶念》,《中華讀書報》2013年9月4日。

實曾不無傷感地表示："這可能是我最後一次來到中國了。"[①]2011年撰文紀念魯迅誕辰130週年時，他吃力地表示："去年夏天，倒於酷暑，雖有恢復，卻時好時壞，幾乎不能外出。所謂茫然自失，也正是我眼下的日子。"[②]而就在去世前2天接受程麻家訪的過程中，竹内實突然説出"可以走了"。程麻事後回憶起竹内實的神情語義，感覺"既像送客，也像在自言自语"[③]。筆者則從這短短四個字中讀出了竹内實面對死亡時的那份平静安詳與孤獨蕭索，一方面他覺得自己想要説出的東西已經表達完畢，而另一方面踽踽獨行的竹内實留給世人的最後面影更多了幾分淒涼孤零的感覺。

竹内實有很多重要的學術成果在退休以後方才完成，九十高壽對於延續其學術生命的意義不言而喻。這讓人聯想起一句名言："學術研究不靠拼命靠長命。"類似的表達亦見於竹内實向雙語刊物《藍》題贈的江户儒者詩句："少而學則壯而有爲，壯而學則老而不死，老而學則死而不朽"[④]，它也可以看作是竹内實漫長學術生涯的寫照。

竹内實説過"很希望自己成爲'戰後日本的中國（現當代）研究'這一學術領域中一個被認識與被分析的對象"[⑤]。筆者在初步認識與分析竹内實的中國研究後以爲，他最關心的應該是中日關係向什麼方向發展的問題。對此，中國學者並不樂觀，王曉秋稱"友好不易，理解更難"[⑥]，程麻也認爲"竹内實先生似乎帶走了一個我們曾經熟悉的時代。在他身後到來的新時代，中日關係前景也許將吉凶未卜"[⑦]。竹内實的離開是一個時代的别離，具體而言，就是指戰後日本的第一代現代中國研究者的學術時代。雖然其他同輩學者没有竹内實那種獨特的個人經歷與生命體驗，對於中國的感情也不及竹内實"命之所繫、魂牽夢縈"那樣來得强烈渾厚，但至少大多保持了友好的態度和理解

① 李梓，《竹内實　日本最權威的毛研究專家》，《華人世界》2007年12期，33頁。不過，竹内實最後一次來華是在一年後。

② 竹内實，《近來我的瑣事與魯迅、孔子》，《文藝報》2011年9月16日，11版。

③ 程麻，《竹内實："可以走了"——一位中國學者的哀思與憶念》，《中華讀書報》2013年9月4日。

④ 劉燕子，《竹内實與日本"文化大革命"研究》，《開放雜誌》2013年9月，84頁。

⑤《〈竹内實文集〉自序》，《回憶與思考》，3頁。

⑥ 王曉秋，《友好不易，理解更難——評〈竹内實文集〉第五卷〈日中關係研究〉》，《博覽群書》2006年10期，84—87頁。

⑦ 程麻，《竹内實："可以走了"——一位中國學者的哀思與憶念》，《中華讀書報》2013年9月4日。

的嘗試。隨着時間的推移,與上一輩相比,這種"友好"與"理解"在日本當代學術界日漸式微。2009年竹內實在日文自選集序言中將新中國成立以來的前30年視作"革命的中國",後30年看爲"經濟的中國"。如今,新中國已經走進第三個30年,倘若迴歸"文化的中國",或許有助於推動中日之間的友好,增强彼此的理解。

Life Connected and Dreaming about: China's Image in the Eyes of Takeuchi Minoru

Abstract: As a Japanese scholar of the first generation of China Studies during the post-war period, Takeuchi Minoru was born in China and went back to Japan until he became an adult. He had strong affections with China and tried hard to promote the friendship and understanding between China and Japan. It has already been more than two years since the death of Takeuchi Minoru. The first and last Chinese translations of his works were published in Hong Kong, thus, in some sense Hong Kong is both the start and the end. As a result, Hong Kong local scholars should write something to express their own opinions and commemorate Takeuchi Minoru's achievements in China studies. This article, on the basis of generalizing his academic career, will focus on *Nihonjin ni totte no Chūgokuzō* (En. China in the Hearts of Japanese) and *Zhunei Shi de Zhongguoguan* (En. Takeuchi Minoru's view of China), first and last work of him, and then discuss his academic purpose and view of China. Finally, it will have dialogues with Takeuchi Minoru's China studies and try to summarize its characteristics.

馬可·波羅研究

《馬可·波羅行紀》所記控制天氣法術初探

于　月

馬可·波羅(Marco Polo,1254—1324)早已對控制天氣的法術產生濃厚興趣和特別關注。他在遊記中曾記述蒙古人利用法術呼風喚雨,在戰場上獲勝的奇跡。在講述哈馬底(Camadi)城時馬可·波羅曾談到哈剌兀納思人(Caraunas)[①],他生動描述了哈剌兀納思人製造大霧干擾天氣的過程:

> 當此輩意欲劫掠該省全境時,能倚仗法力,抑或魔術令天色變得如同黑夜,令人無法遠眺,*甚而連己方之同伴亦難以看清*[LT]。彼等可自遠處令天色昏暗,並將*平原上*[L]七日之程的距離盡數覆蓋,*以使本地居民無法自衛*[L]。[②]

伊朗德黑蘭大學歷史系烏蘇吉(Mohammad Bagher Vosooghi)教授在"馬可·波羅讀書班"上這樣解釋道:伊斯蘭文獻中傳説哈剌兀納思人會製造乾霧,乾霧像雲一樣聚集起來,擋住白天的光線。他們製造黑暗,趁機劫掠波斯人。在伊朗南方山區,山上有哨兵負責瞭望。蒙古人實際上利用自然條件製

作者學習單位:北京大學歷史學系

① Paul Pelliot, *Notes on Marco Polo,* Paris: 1959, vol.1, No.122, "Caraunas",pp.183-196. 哈剌兀納思人在蒙古時代專指留在西北印度的韃靼軍人和當地印度婦女所生的後裔。

② A. C. Moule & Paul Pelliot, *Marco Polo: The Description of the World,* London: 1938; New York: AMS Press INC. reprinted, 1976, §36 , "Here he tells of the city of Camadi", pp. 121. 本文譯文遵循原書體例,插入部分用斜體,並注明版本信息。版本縮寫的詳細信息,見該書509—516頁縮寫表。第49節譯文初稿由復旦大學歷史系講師邱軼皓翻譯,後經北京大學"馬可·波羅讀書班"成員集體討論校改。

造霧,使山頂的哨兵看不到山底的敵人,因而不會放煙警告,失去了禦敵的戰機。

蒙古人的天氣法術在戰場上足以出奇制勝,馬可·波羅對此記憶深刻。但是,他在講到上都佛教僧侶的神奇法術時,語氣卻更加充滿震驚:

> 現在你們可能知曉,大汗*每年有三個月在此地的*(FB)宫殿中生活,一旦出現雨、霧或其他糟糕的天氣,他會讓*身邊的*(FB)聰明的占星士、術士爬上*大汗所居宫殿的屋頂*(R),*一旦空中出現暴風雲、雨、霧氣*(FB),*便*(R)憑藉他們的知識和咒語,驅使所有的雲、*雨*(L)及其他惡劣天氣移開,*使它們飄向遠方而不觸及宫殿*(V),如此一來宫殿上空就不會有惡劣的天氣,*甚至不會有一滴雨落在上面*(VB),而其他地方的惡劣天氣則一仍其舊。*雨水、風暴、雷電在此宫殿周遭落下,卻無絲毫及其身*(R)。*而*(V)操作此事的聰明人有兩種,*其一*(VB)唤作吐蕃(Tebet),*另一*(VB)則爲怯失迷兒(Chescemir)。他們是兩個種族,皆爲偶像教徒。他們比其他所有人更通曉妖術、魔法*和控禦魔鬼,在我看來世間之巫師無出其右者*(VB)。①

玉爾(H. Yule)認爲馬可·波羅在這節提到的咒術就是蒙古人盛行的"剳答石"(ǰada)②。蒙古剳答(又作鮓答)之術其實是北方遊牧民族傳統的祈雨術,邵循正認爲剳答之術來自突厥(突厥語作jadamiši),後經乃蠻傳入蒙古③。據《蒙古秘史》《黑韃事略》《南村輟耕録》《山居新語》《元史》《史集》等文獻的記載,這種法術是將名爲"剳答"的石子或牛黄、狗寶等物浸入水中摩挲,

---

① *Marco Polo: The Description of the World*, op cit., §75 , "Here He tells of the City of Ciandu and of the wonderful palace of the Great Kaan", p. 188.第75節譯文初稿由北京大學歷史學系博士生苗潤博翻譯,後經讀書班成員集體討論校改。

② Henry Yule & Henri Cordier, *The Book of Ser Marco Polo*, vol. 1, reprinted edition, New Dehli: Munshiran Manoharlal Publishers Pvt Ltd, 1993, pp. 309-311.

③ 邵循正,《語言與歷史——附論〈馬可波羅遊記〉的史料價值》,《元史論叢》第1輯,北京:中華書局,1982,215頁。

同時口念咒語[1]。成吉思汗征乃蠻之戰，以及拖雷伐金時均曾使用此術[2]。但是，馬可·波羅在"上都"一節關注的其實是佛教密宗僧侶控制天氣的咒術，并非蒙古人傳統的劄答之術。因此，本文將重點探討元代吐蕃和克什米爾密宗僧侶的法術[3]。

一、吐蕃僧侶的法術

元代吐蕃僧人控制天氣的法術特別受到蒙古統治者的青睞，每逢天氣災害便請西僧做佛事禱雨、止風、壓雷甚至鎮海災[4]。藏文史料《賢者喜宴》記載尚·蔡巴(1122—1193)弟子藏巴東庫哇等七人經西夏到蒙古，後被蒙古軍派充牧羊人。適逢天降冰雹暴雨，周圍羊群死傷甚多，他作法祈禱本尊，使所在之處未下冰雹，他的羊隻未受損失。值得注意的是，擁有控制天氣法術的吐蕃僧人，被當時的蒙古人稱爲"能管天的有福德的人"。藏巴東庫哇隨後受到成

① 關於劄答之術的研究，可參考岩井大慧，《遊牧アジア北方民族の禱雨について》，《駒澤史學》1962年第10期，12—19頁；札奇斯欽，《蒙古秘史新譯並註釋》，台北：联經出版公司，1979，180頁；邵循正，《語言與歷史——附論〈馬可波羅遊記〉的史料價值》，《元史論叢》第1輯，215頁；Adam Molnar, *Weather Magic in Inner Asia*, Bloomington: Indiana University, 1994. pp.43-50；Юндэнбатын Болдбаатар, *Монгол Нутаг Дахь Зртний Нуудэлчдийи Задын Шутлэг*, Улаанбаатар, 2011；賓力德巴特爾，《劄答祭祀研究》，烏蘭巴托，2011；馬曉林，《元代國家祭祀研究》第八章《祈雨中的祭祀與社會》，南開大學博士論文，2012，537—538頁。

② 征乃蠻記載見於余大鈞譯注《蒙古秘史》，石家莊：河北人民出版社，2001，188頁。[波斯]拉施特主編，余大鈞、周建奇譯，《史集》，北京：商務印書館，1997，165頁；《元史》卷一《太祖紀》，1976，北京：中華書局，8頁；案，《蒙古秘史》記事未著年代，《元史》《史集》皆繫此事於1202年。拖雷伐金則見於[波斯]志費尼著，何高濟譯，翁獨健校訂，《世界征服者史》，北京：商務印書館，2007，214頁。

③ 目前曾專門就《馬可·波羅行紀》所載法術進行研究的文章有乙坂智子《馬可波羅著作中所描述的藏傳佛教》，《元史論叢》第8輯，南昌：江西教育出版社，2001，62—69頁。筆者在乙坂智子研究的基礎上，搜集相關文獻，擬做進一步的探討。

④《元史》卷五《世祖紀》載："至元元年四月，東平、太原、平陽旱，分遣西僧祈雨。"(96頁)《元史》卷二九《泰定帝紀》載："(泰定元年夏四月)甲戌，命咒師作佛事厭雷。"(646頁)《元史》卷二〇二《釋老傳》載："又嘗造浮屠二百一十有六，實以七寶珠玉，半置海畔，半置水中，以鎮海災。"(4523頁)元代文人對上都的止雨佛事留有不少詩篇："雍容環佩肅千官，空設番僧止雨壇。自是半晴天氣好，螺聲吹起宿雲寒。西番種類不一，每即殊禮燕享大會，則設止雨壇於殿隅，時因所見以發一哂。"(楊允孚《灤京雜詠》卷下，《叢書集成初編》本據《知不足齋叢書》本排印，北京：中華書局，1985，8頁)"實馬珠衣樂事深，只宜晴景不宜陰。西僧解禁連朝雨，清曉傳宣趣賜金。"(宋褧《燕石集》卷九《詐馬宴》，《北京圖書館古籍珍本叢刊》影印清抄本，第92册，北京：書目文獻出版社，1991，193頁)

吉思汗的召見，勸説成吉思汗信仰佛法，後被尊爲告天的長老[①]。密教僧侶控制天氣的法術能夠契合蒙古統治者天命觀的精神崇拜，同時能强烈滿足蒙古遊牧民族對現實功用的需求。密教在元代得以進一步弘揚光大，其原因之一便是密教法術深受蒙古貴族的信賴與推崇。

元代吐蕃僧侶中膽巴（Dam-pa，1230—1303）尤以咒術聞名，他的傳記文獻中不乏大量奇瑞事件[②]。有學者指出，元代的喇嘛僧人非常活躍，但在密教僧侶之中也存在各種不同的類型。八思巴（Ḥphags-pa，1235—1280）活躍於文化、行政方面，沙囉巴（1259—1314）是元代著名的譯經師，而膽巴則是以咒術見長的密教僧侶。元朝初始忽必烈起用八思巴時，旨在從文化方面對密教加以利用，而密教僧侶的咒術、祈禱更好地滿足了蒙古統治者的期待[③]。蒙古統治者對密教的尊崇、對咒術的信仰成爲元代上層社會的風潮，甚至影響到元代民間社會的信仰與習俗。

據史料所載，膽巴曾以祠摩訶葛剌神助元軍攻打南宋，退海都之兵，修佛事使成宗病癒，特别是他控制天氣的法術尤爲靈驗。後世稱他"咒語精密，凡有禱祈，感應之疾，如風馳電卷，不可思議"[④]。《佛祖歷代通載》有云：

> （大德六年，1302）三月二十四日，大駕北巡，命師象輿行駕前。道過雲州龍門，師謂徒衆曰："此地龍物所都，或興風雨，恐驚乘輿，汝等密持神咒以待之。"至暮，雷電果作，四野震怖，獨行殿一境無虞。[⑤]

---

① 轉引自陳慶英《西夏與藏族的歷史、文化、宗教關係試探》，《陳慶英藏學論文集》，北京：中國藏學出版社，2006，159—160頁。

② 膽巴生平的研究，可參閲野上俊静《〈元史·釋老傳〉の研究》，京都：朋友書店，1978，21—27頁。仁慶扎西，《膽巴碑與膽巴》，《仁慶扎西藏學研究文集》，天津：天津古籍出版社，1989，112—124頁。陳得芝，《元代内地藏僧事輯》，原載《中華國學》第1期，香港，1989；收入氏著《蒙元史研究叢稿》，北京：人民出版社，2005，240—245頁。陳慶英、周生文，《元代藏族名僧膽巴國師考》，《中國藏學》1990年第1期，58—67頁。沈衛榮，《元朝國師膽巴非噶瑪巴考》，《元史及北方民族史研究集刊》第12—13合輯，1990，70—74頁。Herbert Franke, "Tan-pa, a Tibetan Lama at the Court of the Great Khans", in *China under Mongol Rule*, Brookfield, Vermont: Variorum, 1994, pp. 157-180.

③ 稻葉正枝，《元のラマ僧膽巴について》，《印度學仏教學研究》第11卷1期，1963年，180—182頁；村岡倫，《元代モンゴル皇族とチベット仏教—成宗テムルの信仰を中心にして》，《仏教史學研究》第39卷第1期，1996年，79—97頁。

④ 釋明河，《補續高僧傳》卷一，《卍新纂續藏經》第77册，日本藏經院，1912，371頁。

⑤ 釋念常，《佛祖歷代通載》卷二二，《大正藏》第49册，台北：新文豐出版公司，1983，726頁。《元史·釋老傳》亦有類似記載。

漢文文獻恰好印證了馬可·波羅的描述，所謂番僧能以咒術驅使風雨遠離大汗宫殿的見聞也絶非無稽之談。除此之外，至元年間懷孟大旱，膽巴曾應世祖之命禱雨[①]。趙孟頫《膽巴碑》稱膽巴在至元七年(1270)到達中國後，"始於五臺山建立道場，行秘密咒法，作諸佛事，祠祭摩訶伽剌。持戒甚嚴，晝夜不懈，屢彰神異，赫然流聞。"[②]憑藉法術的靈驗，膽巴在成宗朝更加受到蒙古統治者的尊崇與信賴。

值得注意的是，在膽巴國師的推動下，五台山、涿州、京畿、杭州等地乃至全國均建立了祭祀摩訶葛剌的神廟[③]。摩訶葛剌神廟成爲密教法術在元代社會流傳的新媒介。除了精通法術的咒師親自唸咒、施法，普通人通過虔誠祭祀也可獲得大黑天神通的護持，以抵禦天災疫病。黑水城文書中的漢文佛教文書，便向我們充分證明大黑天神崇拜在元代曾是何等流行[④]。此外，在黑水城的漢文密教文獻中，也有不少是專門爲祛災、治病、祈雨、防雹、誅殺敵人等特殊用途而念誦的密咒和陀羅尼[⑤]。沈衛榮就此指出：不論是西夏，還是蒙古

---

①《元史·釋老傳》載："中統間，帝師八思巴薦之。時懷孟大旱，世祖命禱之，立雨。又嘗咒食投龍湫，頃之奇花異果上尊湧出波面，取以上進，世祖大悦。"4519頁；陳得芝在《元代内地藏僧事輯》中根據趙孟頫《膽巴碑》考證，《元史·釋老傳》此處當改作"至元間"。參閲《蒙元史研究叢稿》，241頁。

② 趙孟頫，《大元敕賜龍興寺大覺普慈廣照無上帝師之碑》，録文據《全元文》第19册，南京：江蘇古籍出版社，2000，303—305頁。關於《膽巴碑》的研究，可參閲仁慶扎西，《膽巴碑與膽巴》，112—124頁。Hebert Franke, *Chinesischer und tibetischer Buddhismus im China der Yüanzeit Drei Studien*, München, 1996, pp.37-46. 承蒙北京大學歷史學系党寶海老師教示，謹致謝忱！

③《佛祖歷代通載》卷二二載："乙亥(至元十二年，1275)，師具以聞，有旨建神廟於涿之陽。結構横麗，神像威嚴，凡水旱蝗疫，民禱響應。"(726頁)詳見吴世昌《密宗塑像説略》，《羅音室學術論著》第3卷《文史雜著》，北京：中國文藝聯合出版公司，1984，421—456頁；宿白，《元代杭州的藏傳密教及其有關遺跡》，《文物》1990年第10期，55—71頁；王堯，《摩訶葛剌崇拜在北京》，《慶祝王鐘翰先生八十壽辰學術論文集》，沈陽：遼寧大學出版社，1993，441—449頁。

④ 近年，學者在黑水城文獻中發現的西夏、元代時期的密宗抄本，使我們第一次看到求修大黑天的漢文文獻。《黑城出土文書》中見到的佛經鈔本絶大部分是有關念、修大黑天神的咒語和修法儀軌，其中最著名的有《聖觀自在大悲心總持功能依經録》《勝相頂尊總持功能依經録》《佛説金輪佛頂大威德熾盛光如來陀羅尼經》《佛説大傘蓋總持陀羅尼經》和《聖一切如來頂髻中出自傘蓋佛母餘無能亂總持》。《俄藏黑水城文獻》中元代寫本亦有《慈烏大黑要門》和《大黑求修並作法》兩部修習大黑天神的長篇儀軌。參見沈衛榮《序説有關西夏、元朝所傳藏傳密法之漢文文獻——以黑水城所見漢譯藏傳佛教儀軌文書爲中心》，《歐亞學刊》第七輯，2005，169—170頁。

⑤ 沈衛榮，《重構十一至十四世紀的西域佛教史——基於俄藏黑水城漢文佛教文書的探討》，《歷史研究》2006年第5期，28頁。《俄藏黑水城文獻》的密宗抄本文書中有儀軌19件、陀羅尼咒語10件。《黑城出土文書》的佛經抄本還有散見於其他儀軌中的各種祈禱頂尊佛，觀音、文殊等佛，菩薩和護法神的密咒和陀羅尼。參見沈衛榮《序説有關西夏、元朝所傳藏傳密法之漢文文獻》，162頁。

時代,信徒對藏傳佛教的興趣多半與佛教哲學、義理無關,而都集中在有實用價值的密法修行上面①。

二、克什米爾僧侶的法術與宗教活動

如果追溯藏傳佛教密宗法術的淵源,便難以忽略佛法中心克什米爾(Kashmir)②的重要影響。馬可·波羅對克什米爾佛教徒咒術的印象尤其深刻:

> 他們比任何人都熟知邪惡的幻術,這簡直是一個奇跡,能使*聾啞的*[Z]偶像開口說話,*並對人們的請教予以答復*[P]。通過幻術他們*隨心所欲地*[L]改變天氣,將光明或白晝變成黑暗,*將黑暗變成陽光*[VB]。他們通過魔法和智慧行使*其他許多*[L]*神奇*[Z]之事,未曾親見的人難以相信。③

大蒙古國第一位國師就是蒙哥大汗授予的罽賓高僧那摩④。關於這位那摩國師的神奇法術,《大朝國師南無大士重修真定府大龍興寺功德記》(以下簡稱《龍興寺碑》)便有記載:

> 每燒壇持咒,凡有所禱,隨念即應,諸國僧衆咸取則焉……神驗多

---

① 參見沈衛榮《序說有關西夏、元朝所傳藏傳密法之漢文文獻》,166—167頁。

② 在不同時代漢文史料中有罽賓、迦濕彌羅、迦葉彌羅、迦葉彌兒、乞失迷兒、怯失迷兒等譯名。

③ *Marco Polo: The Description of the World,* op cit., § 49, "Here he tells of the province of Chescemir", pp. 139-141.第49節譯文初稿由北京大學歷史學系研究生包曉悦翻譯,後經讀書班成員集體討論校改。

④ 關於那摩生平的相關研究可參閱温玉成《蒙古國的海雲大士與南无國師》,《法音》1985年第4期,收入氏著《中國佛教與考古》,北京:宗教文化出版社,2009,590—594頁;黄春和,《元初那摩國師生平事迹考》,《首都博物館叢刊》第9辑,1994;中村淳,《モンゴル時代の「道仏論争」の実像—クビライの中國支配への道》,《東洋學報》第75卷第3—4期,1994年,229—259頁;中村淳著,陈一鸣譯,《蒙古時代"道佛争論"的真像——忽必烈統治中國之道》,《蒙古學信息》1996年第2期,1—7頁;温玉成,《蒙古國國師克什米爾高僧那摩》,《中國文物報》1995年3月26日,收入氏著《中國佛教與考古》,595—598頁;陈得芝,《元代内地藏僧事輯》,《蒙元史研究叢稿》,234—236頁。馬曉林,《〈大朝國師南無大士重修真定府大龍興寺功德記〉劄記——兼論〈馬可·波羅行紀〉的相關記述》,《國際漢學研究通訊》第6期,北京:北京大學出版社,2013,252—257頁。

端,不能一一縷陳其細。[①]

陳得芝先生指出:“那摩以持咒見長,爲藏傳佛教的得寵於蒙古人,起了過渡作用。”[②]此外,克什米爾佛教的另一特徵便是苦修頭陀行。馬可·波羅在講到克什米爾時,詳細描述了當地佛教徒隱居深山或洞窟中閉關修行,節制飲食,潔身自好[③]。《龍興寺碑》也記載了那摩隱居雪山,苦修頭陀長達十三年,“日中一饗,止宿塚間樹下,慎不再矣”[④]。

那摩國師之後,還有兩位來自克什米爾的高僧被授予“灌頂國師”之號,克什米爾密教僧侣依然活躍於元朝宫廷[⑤]。1305—1308年間,灌頂國師、罽賓國公毗奈耶室利班的荅向愛育黎拔力八達推薦了一本由白蓮派和尚編著的《廬山蓮宗寶鑒》。從“罽賓國公”的稱號可以推測他是克什米爾人,在文獻中這位罽賓國公的命令被稱爲法旨,在武宗朝具有很高的聲望,可惜目前尚未發現詳細記載其生平的文獻[⑥]。

繼毗奈耶室利班的荅之後,罽賓高僧阿麻剌室利板的荅以其高深的佛法修養名揚海外,深受元代皇帝尊崇。他在1324年後受封爲灌頂國師,佩三珠虎符,總制西域僧侣部族。《大元敕賜西天灌頂國師阿麻啦室利板的達建寺公德之碑》詳細記録了其早年經歷:

> 維阿麻啦室利板的達,罽賓國人也,初名阿訶迦。年十一歲,其國王特裏□迦寬出瓦蘇吉,龍王之裔。時建大觀自在寺,有主持曰微麻剌室利板的荅,通三藏,顯、密二乘,天竺諸宗論。聞阿訶迦宿具根性,求於其

---

① 録文見《常山貞石志》卷一五,《石刻史料新編》第2輯第18册,台北:新文豐出版公司,1979。《全元文》第9册,14—17頁。專文研究可參閱張雲《〈重修大龍興功德記〉及其史料價值》,《西藏研究》1994年3期;劉友恒,《一通記録那摩國師行狀的重要佛教碑刻》,《文物春秋》2010年第3期。

②《元代内地藏僧事輯》,《蒙元史研究叢稿》,236頁。

③ *Marco Polo: The Description of the World,* op cit., §49 , “Here he tells of the province of Chescemir”, p. 139-141.

④ 録文見《常山貞石志》卷一五;《全元文》第9册,14—17頁。

⑤ 關於灌頂國師的研究,可參閲松井太著,楊富學、劉宏梅譯,《東西察合台系諸王族與回鶻藏傳佛教徒——再論敦煌出土察合台汗國蒙古文令旨》,《甘肅民族研究》2011年第3期,46—63頁。原文發表在《内陸アジア史研究》2008年第23卷,25—48頁。

⑥ 楊訥編,《元代白蓮教資料彙編》,北京:中華書局,1989,3、6、7、186—187頁;《元史》卷二四《仁宗紀》,543頁。

父波特摩長者，以爲法嗣。初授聲明韻法，次受具足禁戒，乃更令名。未幾，微麻剌室利板的答示跡，以付其大弟子把剌室利板的答，使其卒業……學既精進，爰辭本國，往中印度，禮金剛寶座及菩提樹。行次雅積國，其王曰阿地特牙麻剌，聞師至□，與其國臣延致演説，禮爲國師。仁廟在宥，遂聞師名，遣使往召。雅積國王不敢留行，護送東邁。道出西番，過撒思加瓦地，所至敬奉，如彼雅積。延祐庚申六月，始至上都，英皇初正宸極，搠思斡濟兒國師方被隆遇，即大安殿建内道場。適師來會，集衆舉問，師隨其所請，如響斯答，莫不厭服。大契聖衷，賞賚蕃庶。泰定初元，命往五臺山崇建勝會，乃於金界寺二時講説大菩提行等論。①

我們根據碑文所述，可以發現阿麻啦室利板的達雖出身克什米爾，但他在雅積國與元廷備受推崇很可能緣於他高深的佛法。雖然碑文後文講到他曾做預知之夢，卜地建寺，密祝加持一事，但是，這位罽賓國師很可能並不擅長控制天氣的法術，而是一位以學術見長的密教僧侶。由此也提示了我們，苦修頭陀與擅長法術雖然是克什米爾佛教徒的兩大特徵，但很可能並非每一位克什米爾僧侶都如那摩般擅長法術，可以隨心所欲控制天氣。克什米爾密教僧侶内部也應存在不同的類型，這一點值得我們關注與重視②。

三、印度僧侶的法術與宗教活動

除此之外，那些並非出身罽賓國的印度僧人中多數曾在克什米爾修行，這些梵僧的佛法也具有濃厚的克什米爾佛教背景，擅長法術的天竺僧侶屢屢

---

① 碑文拓片現藏日本京都大學人文科學研究所，編號GEN0213X，拓片的數碼照片可在京都大學人文科學研究所網站上查看，http://kanji.zinbun.kyoto-u.ac.jp/db-machine/imgsrv/takuhon/type_a/html/gen0213x.html。録文見日比野丈夫《五臺山の元碑二通》，《藤原弘道先生古稀紀念史學仏教學論集》第1卷，1973，652—653頁。何啓龍在研究雅濟國的文章中，也曾引用這篇碑文，參見何啓龍《喜馬拉雅山的雅濟王朝之王統與外交——雅濟國（Ya-tshe）跟西藏與元朝及德里·禿魯黑王朝之關係》，《"元代多元文化與社會生活"學術研討會會議手册》上册，2014，178—192頁。

② 北宋克什米爾高僧天息災到達中國後，因通曉梵、漢等語言，奉太宗之命與施護、法天等人設立譯經場，先後翻譯《大方廣菩薩藏文殊師利根本儀軌經》《分别善惡報應經》《聖佛母小字般若經》《觀想佛母般若經》《大乘莊嚴寶王經》《菩提行經》等多部密教經典。天息災精通多種語言，儘管他出身克什米爾，但他仍是一位專職的譯經僧。

見諸各種文獻。西竺僧人蘇陀室利因久慕五臺山文殊菩薩，與弟子七人航海六年抵達金國，深受金代皇帝禮遇。史載蘇陀室利"善閑咒術能通利，神異頗多……在閔宗朝，連陰不霽，特詔登壇，咒龍落地"[①]，可見他是擅長法術之輩。另一位自北印度長途跋涉抵達五臺山的金代密僧吽哈囉悉利，也有神奇的法術，號稱無所不能，"誦諸佛密語，有大神力，能袪疾病，伏猛虎，呼召風雨輒應"[②]。

元代印度名僧指空雖非出身罽賓，但幼年求法於南印度楞伽國，其後遊歷南印度、北印度諸國等地苦修。指空在元貞元年(1295)進入雲南，駐錫雲南長達十六年。居留雲南時期，他掌握了雲南當地方言，聚衆授業，傳播佛法。指空甚至在安寧州(治今雲南安寧市)還傳習了燃頂、焚臂等自殘供養之術，"燃頂焚臂，官民皆然"[③]。然而，在宋代燃頂焚臂等自殘苦行曾被朝廷命令禁止。控制天氣的咒術也是指空擅長之法。他在的哩囉兒國曾用祈雨之術，"燒香一祝，大雨三日"；在雲南祖變寺曾用止雨術，"坐桐樹下，是夜雨，既明，衣不濡。赴其省祈晴，立應"[④]。至治元年(1321)，指空進入湖南，因洞庭湖風雨頗多，指空誦咒息災，平息風浪[⑤]。他還精通多種語言，在元代重新翻譯了《如意咒》《大悲咒》《尊勝咒》《梵語心經》《施食真言》六種藏經密咒。

指空精通密宗佛法、咒術，掌握諸國語言，是一位具備綜合實力的印度高僧。在他抵達大都后，也曾一度受到元朝皇帝的重視。至治三年(1323)，指空北上大都，爲英宗召入禁中講法，名動京師。泰定二年(1325)，他在大都覲見泰定帝，論法稱旨，並於次年受朝廷委派前往高麗傳法。天曆二年(1329)，指空受召返回大都。然而，元文宗身邊僧人衆多，爲得聖寵眷顧，相互之間競爭激烈，不惜構陷傾軋。指空受到衆僧的排擠與陷害，甚至在文宗朝被褫奪僧衣，終不被文宗器重[⑥]。危素曾寫道："天曆皇帝詔諸僧講法禁中，而有媢嫉

①《佛祖歷代通載》卷二〇，685頁。《補續高僧傳》卷一《蘇陀室利傳》記載大致相同。

②《補續高僧傳》卷一《吽哈囉悉利傳》，370頁。《佛祖歷代通載》卷二〇記載大致相同。

③ 李穡，《牧隱稿》卷一二《西天提納薄陀尊者浮屠銘并序》，杜宏剛、邱瑞中、崔昌源輯《韓國文集中的蒙元史料》(下)，桂林：廣西師範大學出版社，2004，496頁下欄。

④《西天提納薄陀尊者浮屠銘并序》，495頁上欄、496頁上欄。

⑤《西天提納薄陀尊者浮屠銘并序》，496頁上欄。

⑥《西天提納薄陀尊者浮屠銘并序》，493頁下欄、496頁下欄。

之者,窘辱不遺餘力。師能安常處順,湛然自晦。居無何,諸僧陷於罪罟。”[①]指空雖具備雄厚的實力,但他性格孤傲,不屑曲意迎奉,與其他西僧發生矛盾,這或許是他在文宗宫廷傳法遭遇挫折的原因。直到順帝即位,指空才重新受到元朝皇帝的禮遇與優待[②]。

活躍於元末明初的尼泊爾僧人具生吉祥(梵名撒哈咱失里或薩曷拶室哩),幼年曾在克什米爾求法,隱居雪山苦修十餘年。因其佛法高深,在來到中國之前,他便已名揚海内外。具生吉祥於至正二十四年(1364)抵達甘肅,元順帝聽聞其聲名,召至大都法雲寺。史載:

> 師諱具生吉祥,别稱板的達,生與釋迦同國,姓刹帝利氏……後從沙門出家,於迦濕彌羅國蘇囉薩寺上座部中,禮速拶那室利爲師。薙落受具,習通五明經律論之學,辯析邪正,雖國之老宿,莫或過焉。然自以言説非究竟法,乃復精修禪定,不出山者十餘年。時慧學沙門迦嘛囉室哩,爲國人所尊,師往謁之,即蒙印可。時諸鄰國争迎供奉,師皆弗赴。嘗慕東震旦國有五台清涼山,是文殊菩薩應現之處。吾[願欲]瞻禮。遂發足,從信度河至突厥,遍歷屈支、高昌諸國。其國王臣喜師至者,無不稟受法戒。凡四閲寒暑,始達甘肅,實元之至正甲辰歲。元主聞師道行,召至燕京,館於大吉祥法雲禪寺。[③]

---

① 危素,《危太樸文集》卷十《文殊師利菩薩無生戒經序》,《元人文集珍本叢刊》,台北:新文豐出版公司,1984,476頁。另外可參考段玉明《指空——最後一位來華的印度高僧》,成都:巴蜀書社,2007,222頁。《文殊師利菩薩最上乘無生戒經》簡稱《無生戒經》,指空譯,至正十三年(1353)初刊於大都,洪武十九年(1386)重刊於高麗。段玉明轉録自韓國學者許興植《印度的燈傳與高麗——指空禪賢》附録。

②《西天提納薄陀尊者浮屠銘并序》載:“天曆初,詔與所幸諸僧講法内庭,天子親臨聽焉。諸僧恃恩,頡頏作氣勢,惡其軋己,沮不得行。未幾,諸僧或誅或斥,而師之名震暴中外……天曆以後,不食不言者十餘年。既言,時時自稱‘我是天下主’,又斥后妃曰:‘皆吾侍也。’聞者怪之,不敢問所以。久而聞於上,上曰:‘彼是法中王,宜其自負如此。何與我家事耶?’”493頁下欄。

③ 釋來復,《西天善世大禪師板的達公設利塔銘》,《石刻史料新編》第3輯第31册,311—314頁。釋來復所作《西天善世禪師塔銘》收入其文集《蒲庵集》,見《蒲庵集》卷六,《禪門逸書初編》第7册,據台北“中央”圖書館藏明洪武刊本影印,91—92頁。引文中方括號文字據《西天善世禪師塔銘》補入。《西天善世禪師塔銘》後收入《全元文》第57册,249—251頁。

具生吉祥不僅精通五明經律論之學,其法術也頗具特色:

師嘗禪餘,則普施法食。仍模佛塔形像,持花香等呪願供獻,散置水陸。時年季秋,河流結凍,四民汲水,抉冰取水,隨冰大小,皆具佛塔形像。耆年碩德,皆議歎訝,莫究端倪。傳聞内庭,密詔詢釋,善其吉凶。師嘗答言:"國家當以金兵興焉,非我靈異於時。"復就内花園結壇,受灌頂净戒,賜衣設供,恩禮稠洽。[①]

然而時值元末動亂之世,元順帝疏於政事,僅一味沉溺修習密教大喜樂等雙修術。具生吉祥雖精通佛法和護佑法術,但已無法滿足順帝對獨特秘密戒法的生理和心理需求。因此雖受禮遇,卻終究難以得寵於御前。"元君間問以事,或對或否,禮接雖隆,而機語不契。"[②]在法雲寺講經授業幾年後,具生吉祥便離開大都前往五臺山。洪武七年(1374),具生吉祥移居金陵(今南京)崇禧寺,明太祖朱元璋下旨召見,特賜銀印,加"西天善世大禪師"之號,統制天下諸山[③]。綜觀其生平,具生吉祥一生在大都、恒山、金陵等地聚徒講學,弘揚佛法,不逐名利,淡泊一生。明太祖極爲推崇、仰慕他的修爲與品行,多次贈詩加以褒獎。我們或許可以推斷,法術很可能並非密教高僧具生吉祥最擅長的方面。

四、密教法術對元代漢傳佛教僧侶及民間社會的影響

元代西僧法術的奥妙甚至吸引了漢地佛僧的興趣,元中期華嚴宗名僧慧印便是一例。五臺山南山寺原名大萬聖祐國寺,始建於元成宗元貞元年(1295)四月,完工於大德元年(1297)三月。寺内現存重要元代文物《印公碑》,是關於元代佛教史的重要史料[④]。慧印,俗姓張,陝西人,幼年從北禪永

---

①《西天善世大禪師板的達公設利塔銘》,311—314頁。《西天善世禪師塔銘》内容源自《西天提納薄陀尊者浮屠銘并序》,文字略有删減,但具生吉祥施行法術這段記載就不見於《西天善世禪師塔銘》。

② 葛寅亮,《金陵梵刹志》卷三七《西天班的答禪師誌略》,天津:天津人民出版社,2007,569頁。

③《西天善世大禪師板的達公設利塔銘》,311—314頁。

④《印公碑》原名《故榮祿大夫大司徒大承天護聖寺住持前五臺大萬聖祐國寺住持寶雲普門宗主廣慧妙辯樹宗弘教大師印公碑銘》。南山寺在臺懷鎮南五里南山坡的最上面,中間的寺名爲善德堂,最下面的寺叫極樂寺。

昌寺真慧大師爲僧,之後遍歷漢地名寺,曾隨華嚴宗大師龍川行育等多位高僧研習佛法。皇慶初年,承答吉太后旨,至京師於安國寺講《華嚴義疏》。皇慶二年(1313),奉旨爲大萬聖祐國寺長講法王。延祐元年(1314),奉旨住持此寺。此寺是元帝室於五臺山所建的第一座寺院。在慧印之前歷届主持分别是華嚴宗高僧仲華文才、幻堂寶嚴和寶嚴之弟寶金,可見元代的五臺山祐國寺是以傳華嚴爲特色的寺院[①]。慧印在延祐六年(1319)受秘密之法於帝師公哥羅古羅思監藏班藏卜(1299—1327)[②],又從上士僧吉學六支秘要。泰定三年(1326),又跟隨上士管加學時輪六支秘要之法。慧印修習密宗的文化意義及待遇規格非比尋常,值得我們深入探究其中深意。

五臺山爲中國四大佛教名山之一,傳説爲文殊道場,漢藏兩系佛教徒對之倍加尊崇。元朝皇室推崇藏傳佛教,歷代皇帝都從物力、人力等方面積極推動藏傳佛教在五臺山的傳播。元仁宗令華嚴宗高僧慧印拜入帝師門下,修習薩迦派教法,便是從思想方面促進藏傳佛教與漢傳佛教融合的最好方法。作爲漢藏佛教文殊信仰聖地的五臺山,也成爲推動二者融合、發展的絶佳平台。

所謂的"秘密之法"很可能是薩迦派密教的核心教法——道果教授。道果法起源於印度,但13世紀初在印度泯滅,唯有藏傳佛教薩迦派保留和發展了該法。道果之法必須由薩迦派師徒秘密傳授。修習道果密法的第一步是擇師,即根據十個條件選擇自己修密法的上師,並由上師考察其是否具有修密的"根器",之後由上師作一次入密門灌頂儀式,方可入密宗道修行的步驟

---

① 温玉成,《五臺山與蒙元時代的佛教》,《五臺山研究》1987年第5期,10頁。

② 參閲稻葉正就《元の帝師に関する研究—系統と年次を中心として》,《大谷大學研究年報》第17卷,1966年,136—156頁。據《元史·釋老傳》所載,公哥羅古羅斯監藏班藏卜在延祐三年(1316)出任帝師,至治三年(1323)去世。在他死後,旺出兒監藏繼任帝師,在任時間是1323—1325年。但是,稻葉正就根據《歷代佛祖通載》《釋氏稽古略續集》等文獻,確定公哥羅古羅斯監藏班藏卜卒年是泰定四年(1327),因此公哥羅古羅斯監藏班藏卜就任帝師時間是1315—1327年。旺出兒監藏僅見於《元史·釋老傳》,别無其他資料記載,稻葉正就據此否定了旺出兒監藏的帝師身份,證明《元史·釋老傳》記載有誤。野上俊静在《元史釋老傳研究》中採納了稻葉正就的觀點,原著發表於1978年,後由余大鈞翻譯題爲《〈元史·釋老傳〉箋注》,收入《北方民族史與蒙古史譯文集》,昆明:雲南人民出版社,2003,554—590頁。

和次第[①]。

“六支秘要”應是六支瑜伽法諸部經典著作的統稱，包括《六支瑜伽教授秘要》《六支瑜伽要訣》《時輪六支瑜伽釋開目論》《六支瑜伽秘密教授》《六支瑜伽》《六支瑜伽秘要》等。六支瑜伽是無上瑜伽部的重要密法之一。六支瑜伽又叫“六加行法”，是時輪金剛乘法修煉氣息的六種方法，指收攝、禪定、運氣、持風、隨念、三摩地。每一支都有一定的修煉特點。在修行層次上，六支瑜伽已經超越觀想修，提高到專修氣息，使氣息不通過意念控制，而能自行截斷的高度，堪稱最高法門[②]。慧印竟然可以拜帝師爲上師，修習薩迦派核心教法與最高法門，一方面在於慧印本人無論氣質與技能均具備修習的條件，即所謂“根器”；從另一方面也顯示出文宗對他格外器重與優待。元代漢傳佛教僧人修習藏傳佛教的社會趨勢值得我們進一步關注與討論。

在修習薩迦派教法的期間，慧印可能也修習了密宗法術，法術的靈驗使其名聲更盛。《印公碑》云：

> 英宗皇帝幸臺山，從上歷諸寺，所至承顧問，訪以至道。至南臺上，以陰雲晦翳，不睹光瑞，命公於文殊像前，致上之誠禱焉。倏睹慶雲流彩，身光煥景，文殊之像，依稀如在明鏡之中。[③]

我們目前無法完全確定慧印曾經修習密宗法術，因爲漢地佛僧修習祈雨祈晴之術的傳統由來已久，自唐代便已非常流行。元廷官方頒布的汉传佛教寺院基本法规《敕修百丈清規》，就是以唐代《百丈清規》爲基礎修訂而成，其中的《報恩章·祈禱》規定祈雨祈晴需諷誦《大悲咒》《消災咒》《大雲咒》[④]。然而不可否認的是，這些經咒仍與密教元素密切相關。元代的密教僧侶廣泛參與官方與民間的祈雨儀式，密僧熟練使用密教咒法，持誦密教經文，無疑强化

---

① 李冀誠，《佛教密宗儀禮窺密》，大連：大連出版社，1991，71—73頁。索南才讓，《西藏密教史》，北京：中國社會科學出版社，1998，372頁。

② 克珠群佩主編，《西藏佛教史》，408頁。

③ 録文據日比野丈夫《五臺山元碑の二通》，《藤原弘道先生古稀紀念史學仏教學論集》第1卷，650—651頁。此外，温玉成在文章中也有《印公碑》録文，但日比野丈夫的録文較好，參閱温玉成《五臺山佛教劄記》，《五臺山研究》1991年第3期，21—24頁。

④ 釋德輝，《敕修百丈清規》卷一《報恩章·祈禱》，《大正藏》第48册，1115頁。此外，唐代不空翻譯的《大雲輪請雨經》就是一部專門記載求雨儀軌的密教經典。

了密教因素對民間祈雨風俗的影響[1]。然而,元代密僧控制天氣的法術對民間祈雨祈晴習俗是否存在新的影響?元代民間祈雨習俗中的密教因素與宋代相比有何延續又有何變化?根據夏廣興的研究,宋代祈雨風俗盛行燃頂、煉臂、斷指、刺血等密宗苦行供養之術,而官方屢次下令加以禁止[2]。筆者目力所及,尚未在元代文獻中找到類似禁令,或許可證明密教自殘供養習俗在元代已不甚流行,但並不意味着消失。上文提及指空在雲南安寧州傳習佛法,也包括燃頂、煉臂自殘之術。這種自殘供養在元代還開始與孝行有關,這些都是元代祈雨習俗的變化之處。《元史·吴希曾傳》云:"父卒,葬之日大雨,希曾跪柩前,炷艾燃腕,火熾,雨止。"[3]元代孝子梅應發爲救母疾,也曾自殘供養。"立北面稽首,以名香然頂、灼臂,叩天乞減己年,以益母壽。是夕,天將雨,陰雲四合,俄剨焉而開,若啓半扉。盡見北斗之六星,惟一星及輔星不見。頃之,雲復合,而合天深黑。及還,至母所,則母已擁衾坐床上。"[4]嚴耀中指出:"宗教存在的歷史就是其不斷世俗化的過程。密教以自殘爲供養之風在佛教内流行了一陣子以後走向世俗,與傳統文化中的忠孝觀念相結合,變成了一種孝行。"[5]吴希曾和梅應發的舉動可能與此有關。上文所談僅僅是一個方面,若要深入探討這些具體問題,仍有待我們進一步發掘更豐富的史料。

五、結語

密宗法術所具備的功能性,充分滿足了蒙古人生產、生活乃至軍事活動的需要。蒙古統治者以"護國祐民"的現實功能爲核心,將西僧施展法術的宗教儀式納入國家祭祀體系之中,成爲其治國安邦的重要輔助。趙改萍指出:這些佛事的舉行,一方面表明當時自然災害的頻繁,以及人們戰勝自然的能力還較爲低下;另一方面也説明佛教賴以生存的根本,在於人們無法抗拒超

---

① 關於宋代祈雨習俗與密教因素的關係,可參閱夏廣興《密教傳持與宋代民俗風情——以宋代祈雨習俗爲中心》,《民俗研究》2015年第1期,104—111頁。

②《宋史》卷二〇《徽宗紀》載:"(大觀四年,1110)二月庚午朔,禁然頂、煉臂、刺血、斷指。"(383頁)《宣和遺事》亦載此事。

③《元史》卷一九七《孝友傳·吴希曾傳》,4460頁。

④ 黄溍,《金華黄先生文集》卷四三《梅孝子傳》,《四部叢刊》初編影元寫本,葉17a。

⑤ 嚴耀中,《漢傳密教》,上海:學林出版社,1999,153頁。

自然、超現實的外部力量，從而心生對佛的信仰，蒙古統治者也不例外[①]。同時，黑水城漢文佛教文書也可證明，密教法術的實用性與功能性，也是其吸引元代信徒修習密教儀軌的核心原因。

根據上文的分析，我們可以確定這樣一個事實：不論是吐蕃喇嘛僧還是克什米爾或者印度僧侶，密僧内部之中存在細緻的類型劃分。諸如以佛法學術見長的講經僧，以語言見長的譯經僧，以法術見長的咒師，甚至還有具備各方面實力的綜合型僧侶。馬可·波羅想向我們傳達的正是元朝皇帝將密教僧侶中精通法術的咒師集中在身邊，爲他進行消災避難的專門化祈禱服務。因此，不論是來自吐蕃、印度、漢地的佛教僧侶還是道教的道士，甚至是基督教徒與穆斯林，如若法術靈驗，掌握呼風喚雨、消災避難的特殊技藝，便有機會服務於大汗身邊。當然，在這些人之中以吐蕃和克什米爾僧侶的法術最爲高明靈驗。基督教士整體而言並不擅長法術，因此被忽必烈評價爲"蠢無所知，庸碌無用"。大汗身邊的咒師集團人數衆多，内部之間互相傾軋，存在着激烈的競争。因此密教僧侶想要在集團中脱穎而出，不僅需要高明的法術，往往還需要精明世故、善於逢迎的處事手段。西僧的法術使其形象在基督徒馬可·波羅心目中遭到醜化與妖魔化；而漢地民間社會雖然敬畏他們的法術，但對以謀私邀寵爲競争目的的西僧也懷有厭惡情緒，對其一直加以詬病[②]。如張昱《輦下曲》有云："西天咒師首蜷髮，不澡不頮身亦殷。裙口何有披紅罽，出入宫闈無靦顔。"[③]

---

① 趙改萍，《元明時期藏傳佛教在内地的發展及影響》，北京：社會科學出版社，2009，86頁。

② 關於元代文人心目中的西僧形象，參閲沈衛榮《神通、妖術和賊髡：論元代文人筆下的番僧形象》，原載《漢學研究》2003年第2期，219—247；收入氏著《西藏歷史和佛教的語文學研究》，上海：上海古籍出版社，2010，514—518頁。

③ 張昱，《張光弼詩集》卷三《輦下曲》，《四部叢刊》續編影印明鈔本，葉18a。

讀《馬可·波羅行紀》“上都城”札記

范佳楠

1275年，長途跋涉的馬可·波羅（Marco Polo，1254—1324）一行終於在上都覲見了忽必烈汗。《馬可·波羅行紀》對這座夏季都城的記載内容豐富，包括城内宫苑、大汗的狩獵宴飲、上都的宗教生活多個方面。本文結合元上都的考古發現，談談對《馬可·波羅行紀》“上都城”中三處史跡的理解。

一、“大理石宫殿”與“第二道牆”的所指

馬可·波羅言及上都風物之時，首先注意到的便是城内氣勢宏偉的“大理石宫殿”（a vest palace of marble）：

> 在此城中，忽必烈汗命人用*精雕細琢的*[VB]大理石和*其他美*[R]石修建了一座宏大的宫殿，*這座宫殿一端的邊牆處於城市中央，另一端邊牆則與城牆相重合*[R]。殿堂、房室和回廊[VB]均塗金且精妙地描繪[VB]，*在内部繪有獸、鳥、樹、花及多種事物的圖畫和形象，如此美妙和精巧，觀之令人愉悦稱奇*[FB]。它是如此的精美絶倫、巧奪天工。此*宫殿的相對方向*[R]建有*第二道*[R]牆，*一端終於宫殿一側的城牆，另一端終於宫殿另一側的城牆*[R]，*圈佔並*[R]圍繞達16*里周長*[P]的*平*[R]地，*以此法確保人們除了從宫殿出發以外絶不能踏入這個城圈*[R]；*它被防禦得如同一座城*

作者學習單位：北京大學考古文博學院

堡[V];*城牆*[VA]内有噴泉、*活水*[VB]河渠以及足夠多的*極美的*[R]草地*和樹林*[VB]。大可汗在那飼養了各種*能叫得上名字*[VB]且*不兇猛的野*[FB]獸,*數量極多*[VB],如牡鹿(hart)、公羊(buck)和麅子(roe-deer),這些都用來餵養海東青和鷹隼,他把它們養在那裏的鷹舍中,*不算鷹隼*[FB],光海東青就*超*[FB]過兩百只。①

R本中"大理石宫殿"位置的信息②,初看令人頗爲費解。學者們對其所指衆説紛紜。石田幹之助認爲"大理石宫殿"是大安閣③。早期到過元上都考察的易恩培(Lawrence Impey)認爲此殿是宫城北牆正中帶雙闕的宫殿基址(穆清閣)④。或説"大理石宫殿"指宫城,"第二道牆"乃皇城⑤。大安閣是上都主殿,有言"大安御閣勢岧亭,華闕中天壯上京"⑥,的確很引人注目。但其位於宫城北部的三條大街交匯之處⑦,距離宫城城牆尚遠,其邊牆無法與城牆相接。且

---

① A. C. Moule & Paul Pelliot, *Marco Polo the Description of the World,* vol.1, London, 1938; New York: AMS Press INC. reprinted, 1976, §75, "Here he tells of the city of Ciandu and of the wonderful palace of the great kaan", p.185. 譯文中,底本文字(F本)用仿宋表示,以仿宋斜體標示其他版本增加的記載,方括號内注明版本信息。版本縮寫信息見該書509—516頁。譯文由筆者譯出初稿,並經由北京大學"馬可·波羅研究班"共同研讀修訂,下引文同此,謹致謝忱!

② 近年意大利學者對R本(16世紀地理學家剌木學本)的研究表明,R本與Z本同屬《馬可·波羅行紀》的B系統版本,包含A系統各抄本(F本、TA本、VA本和VB本等)所不具備的許多細節,這些獨有的信息應有更古老的、接近於馬可原書的來源,是A系統的有效補充。參閱 Eugenio Burgio ed., Giovanni Battista, *Ramusio"editor" del Milione: Trattamento del testo e manipolazione dei modeli,* Roma-Padova: Antenore, 2011。因此R本中獨有的关于上都佈局的記載值得我們重視。

③ 石田幹之助,《元の上都に就いて》,最初連載於《考古學雜誌》第二十八卷第二、八、一二期,1938;修改稿載於《日本大學創立70週年論文集》第1卷,人文科學編,東京:日本大學,1960;此據包國慶譯《關於元上都》,收入葉新民、齊木道爾吉編著《元上都研究文集》,北京:中央民族大學出版社,2003,10—11頁。

④ Lawrence Impey, "Shangdu, Summer Capital of Kublai Khan", *Geography Review,* XV (1924), p.602.

⑤ 李晴,《元上都和中外文化交流》,北京大學碩士論文,2013,7、63頁。

⑥ 周伯琦,《次韻王師魯待制史院題壁二首》,氏著《近光集》卷一,景印文淵閣《四庫全書》第1214册,台北:台灣商務印書館,509頁上。

⑦ 李逸友,《大安閣址考》,《内蒙古文物考古》2001年第2期,8頁;魏堅,《元上都》上册,北京:中國大百科全書出版社,2008,52頁。

大安閣乃元軍遷建北宋熙春閣而成[①],縱然新增了石雕角柱[②],仍是一座以木結構爲主的漢式宫殿。馬可·波羅應該不會把這樣一座漢式宫殿冠以"大理石宫殿"的稱呼。同理,穆清閣也不應是"大理石宫殿"。此外,把"大理石宫殿"認爲是宫城也不合理,因爲宫城和"第二道牆"皇城城垣間隔很近,其間根本無法如馬可·波羅所言佈置苑囿、飼養珍禽。那麽,"大理石宫殿"所指究竟爲何?

若以中國古代城市宫殿作參照,的確很難想像有哪座宫殿像"大理石宫殿"那樣,佔據半城規模。但如果把對"大理石宫殿"的理解不局限於單體或成組的宫殿建築,則會發現元上都皇城恰好符合馬可·波羅眼中"大理石宫殿"的特徵。

元上都位於内蒙古錫林郭勒盟正藍旗上都河鎮東北20公里處,北依龍崗,南臨灤河,由三重城垣構成,城垣由内向外依次是宫城、皇城和外城(圖1)。皇城位於外城東南部,圍繞宫城四周,與宫城呈"回"字形結構。皇城南牆與外城重合,北牆位於外城中央。可見,皇城的位置與文中信息契合無誤。其次,皇城城牆牆基寬12米,現存高度多約六七米,中間爲黄土夯築,内外兩側以自然石塊包砌[③]。漢族士人形象地稱爲"石城"[④]。約英宗至治年間至泰定帝泰定年間(1322—1328)來到中國的鄂多立克(Ordoric,1286—1331)記敘大都汗八里如下:"大汗在這裏有他的駐地,並有一座大宫殿,城牆周長約四英里。其中尚有很多其他的壯麗宫殿。〔因爲在大宫殿的牆内,有第二層

---

① 周伯琦,《扈從上京宫學紀事絶句》:"右二首詠大安閣,故宋汴熙春閣也。遷建上京。"《近光集》卷一,景印文淵閣《四庫全書》第1214册,518頁上;劉祁《歸潛志》卷七:"(金哀宗)正大末……官盡毁之……所存者獨熙春一閣耳。蓋其閣皆桫木壁飾,上下無土泥,雖欲毁之,不能。世豈復有此良匠也。"北京:中華書局,1983,69頁。馮恩學在復原北宋熙春閣形制的基礎上考證大安閣爲七層高閣建築,參閲馮恩學《北宋熙春閣與元上都大安閣形制考》,收入吉林大學邊疆考古研究中心編《邊疆考古研究》第7輯,北京:科學出版社,2008,292—302頁。

② 在清理推測是大安閣遺址的1號宫殿時,基址下層前端東西兩角發現有漢白玉石雕角柱。魏堅,《元上都》上册,321—323頁。石雕表現出典型的元代風格,很可能是元初的增設之物。

③ 魏堅,《元上都》上册,19—20、24頁。

④ 王士熙,《上京次李學士韻》:"山擁石城月上遲,大安閣前避暑時。"《元風雅》卷一二,宛委别藏本,南京:江蘇古籍出版社,1988,368頁;陳旅,《蘇伯修往上京,王君實以高麗笠贈之,且有詩,伯修徵和章,因述往歲追從之悰與今兹睽攜之歎云爾》:"往年飲馬灤河秋,灤水斜抱石城流。"《安雅堂集》卷三,《元代珍本文集彙刊》,台北:"中央圖書館"編印,1970,134頁。

圖1　元上都平面圖

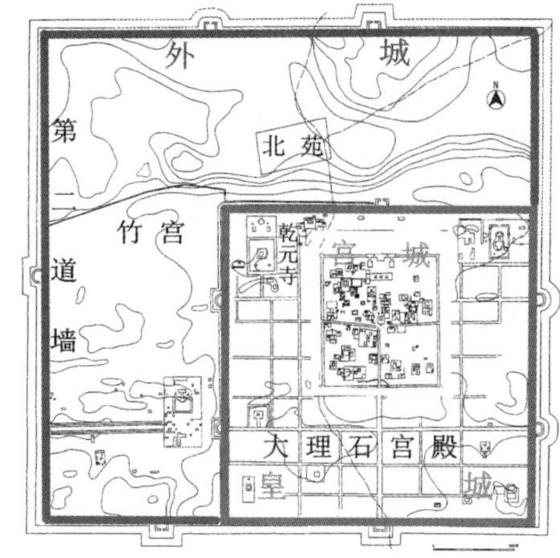

圖2　馬可·波羅眼中的元上都

圍牆……]"[①]鄂多立克是與馬可·波羅文化背景相近的意大利旅行者,從他的記述中可以看出,他筆下的"大宮殿"其實是元大都的皇城,並非皇城内某一具體宫殿。綜上,腦海中没有單座宫殿概念的馬可·波羅完全有可能記住了石牆高聳的皇城外觀,將皇城敘述爲"大理石宫殿"(圖2)。

馬可·波羅隨即談到,"大理石宫殿"的相對方向建有"第二道牆"(a second wall)。這第二道牆具備三項特徵:兩端各自與宫殿兩側的城牆相接、是一個圈佔16里[②]範圍的相對閉合的圍圈、防禦極其嚴密。皇城之外,是在其周邊擴建而成的外城。元上都皇城南側、東側因地形所限無法擴展,只好將皇城南牆、東牆延長築爲外城牆,並新築外城北牆及西牆。這樣一來,平面圖上外城北牆東端與西牆南端分别與皇城東牆及南牆的延長線相接,外城恰好在皇城的西北外圍形成一座閉合城圈。雖然外城不設馬面和角樓,但城外的

---

① [意大利]鄂多立克著,何高濟譯,《鄂多立克東遊録》,北京:中華書局,1981,73頁。

② 此處"里"譯自miles,TA、LT、VA、P本均作15里。哈剌和林和元大都的營造尺約31.6釐米(參閲白石典之《日蒙合作調查蒙古國哈剌和林遺址的收穫》,《考古》1999年第8期,88頁),因此上都也極有可能採此尺度營造。《元典章》記"按式,度地五尺爲步,則是官尺"(卷三十《禮部三·葬禮》,陳高華、張帆、劉曉、党寶海點校本,天津:天津古籍出版社;北京:中華書局,2011,1066頁),故一步約爲158釐米。又《南村輟耕録》卷二一《宫闕制度》:"城方六十里,里二百四十步,分十一門。"(北京:中華書局,2004,250頁)可知元代一里約爲379米。皇城之外延伸的外城東牆長815米,南牆長820米;新築之北牆和西牆均長2220米,外城在皇城基礎上擴展的牆體長度共計6075米,相當於元代的16里。恰與馬可·波羅記載的"第二道牆"圈佔的長度相對應。或許馬可·波羅曾聽説過外城建設的長度里數並加以記録。

北關設軍營[1],外城中部建有隔牆。在此情形下,對一般民衆來説,外城尤其是外城北半部已是防衛嚴密的禁區。的確除去從"大理石宫殿"——皇城出發之外,自由進出並不容易。因此,若"大理石宫殿"是皇城這一推測無誤,"第二道牆"極有可能指的是外城(圖2)。

外城修建年代史料無載。考古工作證實,外城建設晚於宫城和皇城。元憲宗六年(1256),劉秉忠奉命營建開平,初步建設用時三年[2]。此時只是按照藩府規格營建開平,並不是按都城的規格來設計的[3]。外城的構築方式與宫城、皇城明顯不同,它的建設打破了西關街道,説明開平的最初設計可能只包括了宫城和皇城兩重城垣,外城是在皇城和西關關廂形成之後加築的,其應該是在開平升爲上都之後逐步形成[4]。開平升爲上都是在中統四年(1263)五月。中統五年八月頒佈《建國都詔》:"中書省奏'開平府闕庭所在,加號上都。外,燕京修營宫室,分立省部,四方會同,乞亦正名'事。准奏,可稱中都路,其府號大興。"[5]宣佈燕京改名爲中都,兩都制被確立。開平升爲上都之後,元朝政府必會將其完善至都城規模。

都城建制的最佳範本便是三重城垣的宋金開封城。至元四年(1267)"以正月丁未之吉,始城大都"[6],劉秉忠仍舊是主要設計者。至元十三年大都建成[7],其規劃從宋金開封城和《周禮·考工記》中吸取了營養[8]。

在此背景的催化下,上都外城的建設很可能在兩都制形成之後不久便已

---

① 魏堅,《元上都》上册,41—42頁。

②《元史》卷四《世祖紀》:"(憲宗六年)歲丙辰,春三月,命僧子聰卜地於桓州東、灤水北,城開平府,經營宫室。"北京:中華書局,1976,60頁。《元史》卷一五七《劉秉忠傳》:"初,帝命秉忠相地於桓州東灤水北,建城郭于龍岡,三年而畢,名曰開平。"3693頁。

③ 李逸友,《内蒙古古代城址所見城市制度》,收入《中國考古學會第五次年會論文集》,北京:文物出版社,1985,145頁。

④ 魏堅,《元上都》上册,24頁。

⑤《元典章》卷一《詔令卷之一》,6頁。

⑥ 虞集,《大都城隍廟碑》,《道園學古録》卷二三,王雲五主編,《萬有文庫》第二集七百種,上海:商務印書館,1937,387頁。

⑦《元史》卷一四七《張柔傳附張弘略傳》:"(至元)十三年,(大都)城成,賜内帑金扣、玳瑁卮,授中奉大夫、淮東道宣慰史。"3477頁。

⑧ 關於元大都規劃思想的研究,參閱趙正之《元大都平面規劃復原的研究》,《科技史文集》(二)《建築史專輯》,上海:上海科學技術出版社,1979,14—27頁;黄建軍、于希賢,《〈周禮·考工記〉與元大都規劃》,《文博》2002年第3期,41—46頁。

開展。況且至元十一年農曆元旦，忽必烈已在上都舉行大朝會[①]，昭示大元之風儀。並且，馬可·波羅於至元十七年出使哈剌章(雲南)，此前的五年他在大都和上都生活[②]。1275—1280年間，他有充分的機會熟悉上都。馬可·波羅離開中國是在1291年[③]。上都外城或完成於忽必烈統治的晚期[④]。因此，馬可·波羅完全有親眼目見皇城外擴建而成的外城，並將其形容爲"第二道牆"的可能。

二、"竹宫"之形貌與位置

介紹完大理石宮殿和第二道牆，馬可·波羅描繪了大汗飼養珍禽異獸的苑囿和大汗行豹獵的場景。接下來，馬可·波羅提到被"第二道牆"環繞的苑囿的中央有一座全竹材的殿堂——"竹宫"。他對"竹宫"的外形尤其是建造方法描述十分詳細：

> 再者，你或許知道在*如此*[V]被一道牆環繞的那座苑囿的中央，*就在有一片最美樹林的地方*[R]，大可汗修建了一座全竹材(cane)的宏偉的殿宇或*涼廊*(loggia)[L]*作爲他的居所*[VB]，*其上有美觀的塗金和施漆的柱子*，[R]*每根柱之上端都盤踞着一條巨龍*[L]，*通體塗金*，*龍尾繞柱*，*龍頭撐頂棚*，*臂向外伸出*，*一爪向右以支撐頂棚*，*另一爪以同樣的方式向左*[R]；此外，殿之內*外*[P]被通體塗金，飾以或*繪以*[VA]十分精巧的鳥獸。*此殿*[TA]的頂部也全爲竹材，*塗金且*[R]施漆得如此之好、如此濃厚，雨水便無

---

① 《元史》卷五《世祖紀》："(至元十一年)二月癸酉，車駕幸上都。詔諸路總管史權等二十三人赴上都大朝會。"96頁。

② 馬可·波羅出使雲南的時間和路線，參閱陳得芝《馬可波羅在中國的旅程及其年代》，初刊《元史及北方民族史研究集刊》第10期，1986，此據氏著《蒙元史研究叢稿》，北京：人民出版社，2005，430—438頁。

③ 楊志玖，《關於馬可波羅離華的一段漢文記載》，初刊《文史雜誌》1卷12期，1941，重刊於《南開大學學報》1979年第3期，此據氏著《元史三論》，89—96頁；此外，伯希和及波義耳的研究成果同樣支持馬可·波羅離華於1291年。對三位學者研究的評述參閱 F. W. Cleaves, "A Chinese Source Bearing on Marco Polo's Departure from China and a Persian Source on His Arrival in Persia", *Harvard Journal of Asiatic Studies*, vol.46,1976, pp.181-203。

④ 魏堅老師2015年5月29日在北京大學中古史中心舉辦講座《元上都：擁抱文明的廢墟》時，口頭提出此觀點。

法[32c]侵害,*繪畫也絶不會被沖洗掉*;[P]*對從未親眼目睹此景的人來説,此爲世上最奇妙費解之事*[VB];我將告知你它是如何以竹材建成的。你或許真切地知道*建造這些房子所用的*[VB]那些竹子粗逾三*或四*[LT]掌,*圓形*[VB],長十到十五步不等。人們*在竹節處*[TA]將竹材*横*[FB]截爲兩段,一個接着另一個竹節進行,再*縱向地*[LT]*從中間將它們劈開*[L],如此一片竹瓦(tile)便製成了;*每劈一次製成兩片竹瓦*[P]。這些粗竹材如此*之大可製成柱、梁和隔板*(partitions)[P],*它們可用作他途,因爲他們用這些竹材完成許多其他工事*[FB],[因此]人們可自始至終用它們蓋起一整*座*[P]屋頂。——*這座位于苑囿正中的大可汗的*[VA]這座宫殿,我之前告訴過你,全賴竹材*建造*[L]。*但爲了防風,每片竹瓦必以釘固定*[R]。*他們把那些竹材組裝、連接得如此緊密,故*[VB]*它們保護房屋免受雨水侵害,並能將雨水排下*[P]。而且,大可汗將之安排得十分精妙,無論何時,只要他願意,他就能命人*輕易地*[L]搬運*或搭建*[L]、*組裝或拆卸它,卻無任何損失*[P],*當其被立起、組裝好時*[P],有逾兩百根*非常堅固*[R]的絲繩*環繞一周*[R]將其立住,*如同拉帳篷般*[P],*不然由於竹材質輕,會被風吹倒在地*[R]。[①]

在馬可·波羅的記載中,這座"竹宫"全以竹材製成,擁有裝飾堂皇的金漆柱子,並具備可移動性。上述特徵與大蒙古國時期漠北汗庭的失剌斡耳朵(金帳)十分相近。《史集》記載窩闊台命人在離哈剌和林一天行程處建立一處宫殿,那裹"搭起了一座大帳,其中可容千人,這座大帳從來也不拆卸收起。它的掛鉤是黄金做的,帳内覆有織物,被稱爲昔剌斡耳朵"[②]。《世界征服者史》也記載了窩闊台在和林修建的"契丹帳殿","它的牆是用格子木製成,而它的頂蓬用的是織金料子,同時它整個覆以白氈:這個地方叫做昔剌斡耳朵"[③]。《黑韃事略》將窩闊台的金帳記載如下:"凡韃主獵帳所在,皆曰窩裏陀,其金帳(注:柱以金製,故名),……霆至草地時立金帳,想是以本朝皇帝親遣使臣

① *Marco Polo: the Description of the World,* vol.1,pp.186-187.

② [波斯]拉施特主編,余大鈞、周建奇譯,《史集》第二卷,北京:商務印書館,1983,70頁。

③ [伊朗]志費尼著,J.A.波伊勒英譯,何高濟譯,《世界征服者史》上册,北京:商務印書館,2011,262頁。

來，故立之以示壯觀……其制即是草地中大氈帳，上下用氈爲衣，中間用柳編爲窗眼，透明，用千餘條索拽住一門閾與柱，皆以金裹，故名，中可容數百人，韃主帳中所坐胡床如禪寺講座，亦飾以金。"①貴由汗的金帳被奉教皇之命出使蒙古的傳教士普蘭諾·加賓尼(Plano Carpini，1182—1252)記録下來，它是貴由汗舉行繼位典禮的地方："我們離開那裏，同大家一起，騎馬來到三、四裏格外的另一處地方，在那裏，在群山之間的一條河附近的一片美麗的草原之上，已經樹立了另一座帳幕，這座帳幕，他們稱之爲金斡耳朵。原來定於聖母升天節(feast of the Assumption of Our Lady)〔8月15日〕在這裏爲貴由舉行登極典禮，……這座帳幕的帳柱貼以金箔，帳柱與其他木梁連結處，以金釘釘之，在帳幕裏面，帳頂與四壁覆以織錦，不過，帳幕外面則覆以其他材料。"②

法國國家圖書館藏有一副14世紀的波斯細密畫(圖3)，其中繪有蒙古大汗坐在移動宫帳内的形象，帳内以織金錦覆蓋。另一幅藏于德國國家圖書館的創作於1307年的細密畫《讀古蘭經的蒙古王子》(圖4)，則以在帳頂、帳壁繪格線的方式體現出宫帳之木骨。

圖3　波斯細密畫所見移動宫帳

圖4　波斯細密畫《讀古蘭經的蒙古王子》所見宫帳之木骨

① (宋)彭大雅撰，(宋)徐霆疏，王國維箋證，《黑韃事略箋證》，收入《王國維先生全集初編》第7册，台北：大通書局，1976，2822—2821頁。

② [英]道森編，吕浦譯，周良霄注，《出使蒙古記》，北京：中國社會科學出版社，1983，61—62頁。

上都失剌斡耳朵之形貌留存於不少文人的詩句中。柳貫《觀失剌斡耳朵御宴回》:"毳幕承空拄繡楣,彩繩亘地掣文霓。辰旗忽動祀光下,甲帳徐開殿影齊。芍藥名花圍簇坐,葡萄法酒拆封泥。御前賜酺千官醉,恩覺中天雨露低。"詩後有注:"車駕駐蹕,即賜近臣灑馬奶子御筵,設氈殿失剌斡耳朵,深廣可容數千人。"[①]説明失剌斡耳朵又名氈殿,以毛氈覆頂,並以彩繩固定於地。詩文中常與御宴、詐馬宴聯繫起來的地點還有棕毛殿,該殿應該就是失剌斡耳朵,只不過棕毛殿的名稱出現較晚[②]。

元末鄭泳《詐馬賦》:"皇上清暑上京,歲以季夏六月大會親王,宴于棕毛之殿三日。百官五品之上,賜只孫之衣,皆乘詐馬入宴,……甍無五色之琉璃兮,覆以栟櫚之皮;恐颶風之動摇兮,絙以黄絨之施。上鴟吻而門垂兮,下系鐵杙而莫移。"[③]此處所記棕毛殿外觀與上述柳貫詩文中的氈殿恰好可以相互對應。可見,上都的失剌斡耳朵或棕毛殿,承襲自大蒙古國時期的傳統,是一座大型穹廬。

將馬可·波羅對"竹宫"的記録與上述文獻和圖像材料對比可知,"竹宫"是失剌斡耳朵(棕毛殿)無疑。其以木材爲骨、木材間以釘連接、以繩索拉起固定、帳柱塗金的特徵可與其他記載相互印證。

英國學者約翰·曼(John Man)曾在他關於上都的書中以《馬可·波羅行記》爲本繪出竹宫的模擬形象[④]。失剌斡耳朵的内部遍佈花紋秀麗的織金錦,這便是馬可·波羅所説"殿之内外被通體塗金,飾以或繪以十分精巧的鳥獸"的來源。其次,馬可·波羅對"竹宫"的描寫也不乏誇大及失實之處。失剌斡耳朵依靠施漆完好的頂蓬防雨不一定屬實,實際上起到隔水作用的是覆蓋於帳頂的"栟櫚之皮"。上述記載提到製作可容千人的金帳——失剌斡耳朵骨架部分的材料是木頭,並未像馬可·波羅一樣强調是竹材。挺拔粗壯的竹子

---

① 柳貫,《柳待制文集》卷五,《四部叢刊》縮印本,葉2b。

② 棕毛殿名稱最早出現於成宗大德年間。參閲陳高華、史衛民,《元上都》,吉林:吉林教育出版社,1988,123—124頁。

③ 李修生主編,《全元文》第57册,卷一七六一,南京:鳳凰出版社,2004,869—871頁。原作"宴于棕王之殿三日",此"棕王"應是"棕毛"之誤,故改。

④ John Man, *Xandu: Marco Polo and Europe' s Discovery of the East*, London: Transworld Publishers, 2010, p.261.

是南方的物産,如果上都的失剌斡耳朵以竹建成,其材必運自南方。

關於上都失剌斡耳朵的位置,前賢主要有三種觀點。陳高華、史衛民認爲,上都的失剌斡耳朵在西内,而西内是城外的離宫[①]。按照前文對《馬可·波羅行紀》"上都城"的翻譯及馬可·波羅眼中"第二道牆"的分析,上都的失剌斡耳朵——"竹宫"必定位於外城牆圈起的範圍之内。1937年日本東亞考古學會調查上都,他們認爲外城北部(即該報告中的外苑城)遺跡較少,是瑞林院和失剌斡耳朵宫帳所在地[②]。王大方也認爲皇城北部復仁門外的高崗上是失剌斡耳朵的所在[③]。然而外城北部東西向的高崗是全城地勢至高之處,此處並非設宫帳的最佳地點。據《黑韃事略》,失剌斡耳朵所在地需具備"其地卷阿負坡阜以殺風勢"[④]的地形特點。將失剌斡耳朵設在地勢最高的北苑高崗上並不利於防風。魏堅認爲西内在元上都外城西部,"竹宫"在西内。上都皇城西、北兩側的外城之間,有一道東西向隔牆,將外城分隔爲互不相通的西、北兩部分。距皇城西北角樓西南220米處,有一直徑約140米的略微凸起的高地,上述隔牆在此修築爲弧形,可能作拱衛圓形宫帳之用,這一處高地極有可能就是棕毛殿的所在[⑤](圖2)。筆者贊同此觀點,以下從前賢關注不多的與失剌斡耳朵活動有關的宗教人物入手,進一步論證"竹宫"位於此地。

衆所周知,失剌斡耳朵不僅是大汗的行在,還是詐馬宴及各種御宴舉辦的地點。爲大汗服務的藏傳佛教僧侣需奉命作法,以確保詐馬宴期間風和日麗。

宋褧《詐馬宴》:"寶馬珠衣樂事深,只宜晴景不宜陰。西僧解禁連朝雨,清曉傳宣趣賜金。"[⑥]有趣的是,此番藏傳佛教僧侣行控禦天氣法術的場面在《馬可·波羅行紀》中也被大幅渲染,説明此景很可能在上都真實發生過。不僅如此,馬可·波羅還提到,大汗每年8月28日離開上都返回大都,此日要在

---

① 陳高華、史衛民,《元上都》,120—124頁。

② 東亞考古學會,《上都—蒙古ドロンノヘルに於けゐ元代都址の调查》,1941,28頁。

③《元上都遺址考古有新發現,棕毛殿皇家糧倉位置得到確認》,《光明日報》2014年6月17日,09版。

④《黑韃事略箋證》,2821頁。

⑤ 魏堅,《元上都》上册,69頁。

⑥《燕石集》卷九,《北京圖書館古籍珍本叢刊》第92册,北京:書目文獻出版社,1998,193頁。

竹宫所在的苑囿中舉行灑馬奶酒祭祀,馬奶酒祭祀的指導者是占星術士和偶像教徒(藏傳佛教僧侣)[①]。楊允孚在《灤京雜詠》中有詩一首:"内宴重開馬湩澆,嚴程有旨出丹霄。羽林衛士桓桓集,太僕龍車款款調。"詩下自注:"馬湩,馬妳子也。每年八月開馬妳子宴,始奏起程。"[②]楊允孚詩證實了馬可·波羅所記灑馬奶之後啓程返回大都的真實性。占星術士和偶像教徒們侍奉在大汗身旁,向大汗要求舉辦勝大的節日[③]。可見,占星術士和偶像教徒,尤其是藏傳佛教僧侣與大汗關係十分密切。因此,他們必須經常來往於大汗在上都的兩處居所——"大理石宫殿"和"竹宫"之間。

因宫城北牆無門,若失剌斡耳朵位於北苑,從宫城去該地路徑迂回,需從宫城東門東華門出城,繞行至皇城北門復仁門,由此進入外苑城。若要從北苑返回大都,必須再次折回到宫城區。倘若失剌斡耳朵位於西内,大汗的車駕只需從宫城出發向西行,穿過皇城西門便可到達。返回上都之日,在西内失剌斡耳朵行馬奶祭祀後,可經由外城西門直接出上都。並且,皇城西北角的方形院落是藏傳佛教寺院乾元寺,該寺建於至元十一年(1274),即馬可·波羅來上都之前夕,規模仿照大都的大護國仁王寺[④]。乾元寺的位置及建制表明其乃上都藏傳佛教寺院中地位最崇者,與大汗親近之僧衆當承事於此。這樣一來,乾元寺之僧侣便可方便穿行於宫城和失剌斡耳朵之間,主持祭祀、法術等活動。

三、結論

綜上,從形貌、位置來看,馬可·波羅口中的"大理石宫殿"可對應爲元上都皇城。相應地,"第二道牆"指的是在皇城之外擴建而成的外城。伴隨着兩都制的確立,原先開平城的兩重城規劃已不符合都城的建制。元上都的外城很可能於開平升爲上都後不久開始建設,完工於忽必烈在位的晚期。馬可·波羅在中國期間,有親見外城的可能。而"竹宫"是大汗常居的金色大帳失剌

① *Marco Polo: the Description of the World,* vol.1, p.187.

②《灤京雜詠》卷下,《知不足齋叢書》本,1b—2a葉。

③ *Marco Polo: the Description of the World,* vol.1, pp.189-190.

④ 袁桷,《華嚴寺碑》,《清容居士集》卷二五,《四部叢刊》縮印本,18a—21b葉。

斡耳朵，馬可·波羅對其形貌特徵的描寫有準確之處，也不乏誇大的成分。從設帳之地形及參與失剌斡耳朵宴飲、祭祀活動的人群行動便利性的角度考慮，這座帳殿更有可能位於上都西内。

將《馬可·波羅行記》之記載與元上都考古成果及其他文獻加以對照便可發現，此節中不少記録有據可依。其史料價值應該被肯定。文中對上都的描寫，充斥着宴遊、享樂、奇幻的因素，産生此現象的緣由，除了騎士小説家文字加工的影響，主要還是基於上都這座夏季都城的實際功能。正是暗藏於避暑生活之中的民族融合與文明彙聚支撑着龐大帝國的運轉。

附記：本文的一些認識曾在北京大學"馬可·波羅研究班"進行匯報討論，受到了研究班諸位老師、同學的啓發和幫助。對"第二道牆"的看法曾受到党寶海、李鳴飛、羅帥三位老師及求芝蓉等同學的匡正和提醒。謹此致謝！

順風相送:從《馬可·波羅行紀》中記載的祈風儀式談起

李鳴飛

13世紀後期,意大利旅行家馬可·波羅同他的父親和叔父經過西亞和中亞來到中國,留居十七年後又經海路至伊朗回到意大利。他的旅行記《馬可·波羅行紀》中記載了往返途中和旅居中國所見到的方方面面的情況。這部行記自面世後,各種抄本和譯本層出不窮,在作者生前已有過多次改訂,到上世紀初,學者調查清理出的各種版本已有143種之多[①],這些版本可以大致分爲三個系統,即F本系統、R本系統和Z本系統。

雖然學界公認F本是目前所知最早的也是最原始的版本,現存絶大多數抄本和譯本都屬F系統,但學者也指出F本並不完整,其内容已距原本相去甚遠[②]。R本即賴麥錫(G. B. Ramusio)刊本,此刊本的意義不下於任何一個抄本,因爲其中有五分之一的内容不見於F系列中的任何一個抄本。以至於Z本被發現之前,R本的可靠性一直受到學者的質疑。Z本直到19世紀初才被

作者單位:中國社會科學院歷史研究所

① A. C. Moule & Paul Pelliot, *Marco Polo: the Description of the World*, vol.I, pp. 509-516提供了143種抄本的列表,London: George Routledge & Sons Limited, 1938.

② Edited from the Elizabethan translation of John Frampton ; with introduction, notes and appendixes by N.M. Penzer, *The Most Noble and Famous Travels of Marco Polo: Together with the Travels of Nicolò de Conti*, London : Adam and Charles Black, 1937, pp. xviii-xxxi. Translated from the text of L.F. Benedetto by Professor Aldo Ricci, with an introduction and index by Sir E. Denison Ross, *The Travels of Marco Polo*, London: G. Routledge & Sons, 1931, pp.x-xi.

發現，此抄本並不完整，其中三分之一內容被節略，然而保留下來的部分與F本非常一致。此外Z本有200段F系列抄本中没有的內容，其中五分之三見於R本。也就是説，Z本不但證明了R本的可靠性，同時還提供了大約80段全新的內容[①]。

現在認爲Z本是一個比已知抄本都要好的拉丁文譯本。不幸的是該本在開頭進行了大量删節，幸運的是這個抄本越往後越完整，貝內代托（L. F. Benedetto）教授曾提出一個設想，即此本的抄寫者本來打算根據自己的興趣編寫一個原本的縮寫本，但隨着他越來越喜歡這本書，以至於漸漸的一個詞都捨不得删了[②]。正因如此，在關於海路部分的第三卷中有大量只見於Z本的內容，之前一直未被我國學者所注意。

目前國內最常見的《馬可·波羅行記》譯本分别是馮承鈞譯本[③]和陳開俊等譯本[④]，其中馮承鈞本譯自沙海昂（A. J. H. Charignon）譯注的頗節（M. G. Pauthier）校本，該本選用的底本是三個宫廷法語修訂本，內容與F本大同小異，馮承鈞在翻譯過程中參考亨利·裕爾的合校本，加入了一些他認爲有價值的R本內容。陳開俊本則譯自科姆羅夫（Manuel Komroff）編訂的普及本，該本來自馬爾斯登（William Marsden）英譯的R本。其他中譯本也都大同小異，不出F本和R本內容[⑤]。惟有1936年出版的張星烺譯《馬哥孛羅遊記》[⑥]不同。這個譯本來自里奇（Aldo Ricci）英譯的貝內帶托1932年意大利文譯本。貝內帶托譯本是以F本爲底本，加入他調查所得各種版本中有價值的內容，其

---

① 關於《馬可·波羅行記》一書的版本系統，參見任榮康《〈馬可·波羅行記〉版本史簡述》，《中國文化研究集刊》第五輯，上海：復旦大學出版社，1987，219—245頁；李鳴飛，《馬可·波羅前往中國之路——〈馬可·波羅遊記〉伊朗部分研究》，第一部分"《遊記》版本及研究概况"，博士後出站報告（未發表）。

② Translated from the text of L.F. Benedetto by Professor Aldo Ricci, with an introduction and index by Sir E. Denison Ross, *The Travels of Marco Polo*, pp.x-xi.

③ 馮承鈞譯，《馬可波羅行紀》，上海：商務印書館，1936年第一版，1947年第三版，北京：中華書局，1954；上海：上海書店出版社，2001。

④ 陳開俊等譯，《馬可波羅遊記》，福州：福建科學技術出版社，1981。

⑤ 如魏易的《元代客卿馬哥博羅遊記》（北京：正蒙書局，1913）譯自馬爾斯登本。張星烺的《張譯馬哥孛羅遊記》（未完）（北美印刷局印刷，燕京大學圖書館發行，1929）譯自亨利·裕爾合校本，該本以F本爲底本，加入R本內容。李季的《馬可波羅遊記》（上海：亞東圖書館，1936）和梁生智的《馬可·波羅遊記》（北京：中國文史出版社，1998）均譯自科姆羅夫本。

⑥ 張星烺譯，《馬哥孛羅遊記》，上海：商務印書館，《萬有文庫》，1936。

中最重要的是他新發現的Z本複製本，因此貝内帶托本中包含了F本、R本和Z本三大系統的所有内容，張星烺所譯的這個版本也是目前内容最爲完整的譯本，可惜這個譯本印量很少，雖經兩次重印[①]，現在仍很難找到，因此該本的學術價值並未被學者所注意。

幸而現在有了更好的《行記》合校本——穆勒(A. C. Moule)、伯希和(Paul Pelliot)校譯的 *The Description of the World*(以下稱爲《世界寰宇記》)[②]。與貝内帶托本相比，這個合校本所參考的抄本更多，標出了各段内容的版本來源，更重要的是，此本所使用的是在貝内帶托本出版後才找到的Z本原本，因此所加入的Z本内容更爲準確。這個合校本中有很多值得我們注意的内容，例如本文將要討論的，馬可·波羅所記航海前的占卜活動。

《世界寰宇記》第158節對應馮承鈞譯本第一五七章《首志印度述所見之異物並及人民之風俗》[③]和陳開俊等譯本第三卷第一章《印度　它的居民的禮儀、風俗和特別著名的奇聞軼事 航海用的船舶種類》[④]，但有一段内容在後兩個譯本中均無，是關於出海之前的占卜活動：

> 且我們要告訴你，當任何一艘船要起航，他們怎樣確認在旅途中生意會順利還是不順。船上的人確實會有一個籠子，是用柳條製作的栅欄，籠子的每個角和面都綁着繩子，因此共有八條繩子，在另一端它們都繫着長索。他們再找一個蠢貨或醉鬼並把他綁在籠子上，這樣就不會有聰明人或正直的人使自己冒這種風險。當颳大風的時候做這件事。他們確實會把籠子迎着風放置，大風會把籠子吹起來升入空中，那個人就被長索拽着。若在空中籠子向風的方向傾斜，他們會朝他們稍微拉一下長索，這樣籠子就正了，然後他們再放開長索，於是籠子繼續升高。如果它又向下傾斜，他們就繼續這樣拉長索，直到籠子被擺正上升，然後他們放

① 張星烺譯，《馬哥孛羅遊記》，上海：商務印書館，《漢譯世界名著》，1937；台北：台灣商務印書館，《人人文庫》，1972。

② A. C. Moule & Paul Pelliot, *Marco Polo: the Description of the World,* London: George Routledge & Sons Limited, 1938.

③ 馮承鈞譯，《馬可波羅行紀》，第385—386頁。

④ 陳開俊等譯，《馬可波羅遊記》，第197—198頁。

開一些長索，就這樣它會升得極高，高不可見，只要長索夠長。這種試驗就這樣進行，即如果籠子直直升上天，他們就説這場試驗所針對的船將會有一次既快又興旺的旅程，並且所有的商人爲了遠航蜂擁而來坐這艘船。如果籠子没能升上去，那麽没有商人會願意進入試驗所針對的這條船，因爲他們説她將無法完成航行，會被不祥所困。因此那艘船就要在港口呆一年。[①]

這段記載可以説是獨一無二，筆者尚未在中文史料見到類似把人裝在籠子裏，像放風箏一樣放起來，以預測出航是否順利的記載。從祭祀的内容來看，無非是祈求出航之後有風力相助，且風向和順，使船隻順利遠行，如入高空而不可見。

雖然没有見過類似内容的占卜記載，但古代航海，船隻的動力以風力爲主，因此祈求風力相助的祭祀活動是非常常見的，中國古代的海外貿易在宋元時期最爲繁盛，所以宋元時代所記録的祈風活動也最多最詳細。在北宋時期已有官方舉行的祈風儀式。宋代泉州地區爲海外貿易的中心，市舶税收在當地財政中地位重要，因此地方官員對於海路暢通、順水順風非常重視，每年要在九日山舉行祈風祭典。九日山位於泉州西郊7公里處的晉江下游北岸，山高大約80多米，山下留下大量石刻，其中有10通與宋代官方的祈風祭典有關[②]。

南宋著名理學家真德秀曾爲泉州知州，在任期間大力發展海外貿易，史書記載：當時“番舶畏苛征，至者歲不三四，德秀首寬之，至者驟增至三十六艘”[③]。他每年兩次舉行官方的祈風祭典，其文集中保留了爲這種祭祀所做的《祈風文》，文中把祈風的目的説得很清楚：“惟泉爲州，所恃以足公私之用者，蕃舶也。舶之至時與不時者，風也。而能使風之從律而不愆者，神也。是以國有典祀，俾守土之臣一歲而再禱焉。嗚呼，郡計之殫至此極矣，民力之耗亦既甚矣，引領南望，日需其至，以寬倒垂之急者，唯此而矣。神其大彰厥靈，俾

① A. C. Moule & Paul Pelliot, *Marco Polo: the Description of the World*, pp.356-357.

② 參見莊景輝，《泉州宋代祈風石刻考釋》，《江西文物》1989年2期，58、89—94頁。

③《宋史》卷四三七《真德秀傳》，北京：中華書局，1977，12960頁。

波濤晏清，舳艫安行，順風揚帆，一日千里，畢至而無梗焉，是則吏與民之大願也。謹頓首以請。”[①]

主持祈風的除了泉州長官外也有市舶司官員，南宋初年提舉閩舶的林之奇有一部《拙齋文集》，其中也收録了兩篇“祈風文”。他在《祈風舶司祭文》中還説明了真德秀所謂的“一歲而再禱”，分爲“祈”和“報”兩次不同的祭祀活動：“夫祭有祈焉，有報焉。祈也者，所以先神而致其禱；報也者，所以後神而答其賜。祈不可以爲報，而報不可以爲祈，自古然也。而舶事之歲舉事，祀典於神則異乎？是於夏之祈有冬之報，於冬之祈有夏之報，風之舒慘，每以時應，則祠之踈數，必以時舉，如循環之不窮……”[②]所謂一年兩祭，自然是因爲冬季的東北季風送船南下出航，夏季的西南季風則助船返回，因此祭祀祈求這一年兩次的季風和順。

中國傳統的祈風祭祀並没有特別强調登高的行爲，多爲傳統的儒家祭祀，以撰寫祭文、貢獻豬羊肴酒爲主。在明代的《使琉球録》中有不少對這類祭祀活動的詳細記録，例如陳侃就記録了出使琉球前造船時祭祀，“福州府備豕二羊二”，船造好出塢、竪桅、浮水、治繚時“皆有祭，行祭禮皆如初”[③]。使者一行平安歸來後再次祭祀，由“翰林院撰文一道，行令福建政司備辦祭物香帛，仍委本司堂上官致祭……令福建布政司於廣石海神廟備二壇，一舉於啓行之時而爲之祈，一舉於回還之日而爲之報”[④]。這裏也延續了一祈一報的祭祀傳統。

在海上交通非常頻繁的宋元時期，史料中也零星記載了海外船舶和水手的祭祀活動。元代陳秀明的《東坡詩話録》中提到：“吴中每暑月則東南風數日，甚者至逾旬而止，吴人名之曰舶趠風，海外舶船禱於神而得之。”[⑤]從這條記録來看，海外船舶的祭祀活動在當時是頗爲普遍的。

南宋方信孺曾在番禺擔任縣尉[⑥]，他在詩集中提到當地有一番塔，“始于

---

① 真德秀，《真西山文集》卷五〇《祈風文》，《四部叢刊》影明正德刊本。

② 林之奇，《拙齋文集》卷十九《祈風文》《祈風舶司祭文》，文淵閣《四庫全書》本。

③ 陈侃，《使琉球録·使事記略》，明嘉靖刻本，二葉、六葉。

④ 謝傑，《使琉球録》卷下，明嘉靖刻本，二十一葉。

⑤ 陳秀明，《東坡詩話録》，《四庫全書存目叢書》集四一六，濟南：齊魯書社，1997。

⑥《宋史》卷三九五《方信孺傳》，12059頁。

唐時,曰懷聖塔,輪囷直上,凡六百十五丈,絶無等級。其顙標一金雞,隨風南北。每歲五六月,夸人率以五鼓登其絶頂,叫佛號以祈風……”他爲此賦詩一首:“半天縹緲認飛翬,一柱輪囷幾十圍。絶頂五更鈴共語,金雞風轉片帆歸。”[①]“夸”即“夷”字,每年五六月份在蕃塔上祈風之夷人,很可能是來自印度洋或東南亞地區的水手,祈禱的目的自然希望夏季的西南季風和順,保祐來自印度洋和東南亞的船隻順利航行。登上高塔,大聲禱祝祈風,這一風俗與馬可·波羅的記載就更爲接近了,因此《世界寰宇記》中記録的很可能是來自穆斯林水手的祭祀習慣。

明代以鄭和下西洋爲主題的演義小説《三寶太監下西洋記》中也提到以登高爲特征的活動。《三寶太監下西洋記》寫於明代萬曆年間,内容多是神怪故事,其中記載寶船入江和入海的祭祀,亦爲傳統的“殺豬殺羊,備辦香燭紙馬”[②]或“備辦祭品……開讀祭文”[③]。然而在描寫造船時,提到一個火頭登高的故事。文中説出海之前製造寶船時有神明相助,因此不到八個月就大功告成。永樂皇帝不相信有天神相助使得造船如此迅捷,要“三日齋、七日戒,親至寶船廠内,要九張桌子單層起來,果是天神飛身而上,此心才信”。齋戒之後,永樂帝“排了御駕,文武百官扈從,徑往寶船廠來。廠裏已是單層了九張金漆桌子,御駕親臨,即時要個天神出現……未久之間,只見廚下一個燒鍋的火頭,蓬頭跣足,走將出來,對衆匠人説道:‘我在這裏無功食禄,過了七個月,今日替衆人出這一力吧。只是你們都要吆喝着一聲天神出現,助我之興,我才得像果真的。’衆人吆喝一聲道:‘天神出現哩!’倒是個好火頭,翻身就在九張桌子上去了……萬歲爺心上十分快活,今日天神助力,明日西洋有功可知”[④]。這裏描寫一位蓬頭跣足的燒鍋火頭,翻身上了九層桌子,永樂帝因此相信造船確有天神相助,預示下西洋有功可知。雖然描寫的事件發生在造船時,與馬可·波羅記載的航行前的占卜在時間和場合上並不相同。然而兩者卻都包含“蠢貨”和“登高”這兩個相似因素。不排除這種可能性,即明代前期

① 方信孺,《南海百詠》,宛委别藏本,江蘇古籍出版社,1988,20頁。
② 羅懋登,《三寶太監下西洋記》(上),西安:三秦出版社,1996,207頁。
③ 羅懋登,《三寶太監下西洋記》(上),223頁。
④ 羅懋登,《三寶太監下西洋記》(上),182頁。

仍有這種來自外藩的航行前令"蠢貨"以某種形式登高的儀式,在明代海禁之後逐漸消失,後來變成傳説故事,改頭换面被寫進了小説之中。

《馬可·波羅行紀》中記載的這段出海前的祭祀活動,很可能來自印度洋地區和東南亞水手。在海路交往繁榮的宋元時期,這類祭祀活動在中文史料中也可見跡象,在當時大概頗爲普遍,爲人熟知。然而到明代中後期,航海活動中記載的祭祀就不再出現類似因素,大概與明代的禁海政策有關,阿拉伯航船不再像宋元時期那樣頻繁出現在中國海域,中國水手的祭祀也回歸到傳統的儒家祭祀方式。馬可·波羅的記載則爲我們留下了寶貴的資料,展現了當時中西海洋文明交匯的一個剪影。

從額里合牙到額里折兀勒

——馬可·波羅的唐兀大省考察日記

北京大學"馬可·波羅讀書班"

考察日期:2014年8月11日至28日

考察成員:

隊長:榮新江、党寶海

副隊長:馬曉林、羅帥

隊員:孟嗣徽、段晴、李肖、姚崇新、畢波、文欣、包曉悦、陳春曉、范佳楠、馮鶴昌、付馬、胡曉丹、李昀、苗潤博、求芝蓉、沈琛、田衛衛、于月、張曉慧、鄭燕燕

2014年8月11日　星期一　北京　晴

19:10,馬可·波羅考察隊在北京西站會合,登上前往銀川的T277次列車。隊長榮新江和党寶海因參加"粟特人在中國:考古發現與出土文獻的新印證"學術研討會,不與考察隊同行,各自乘坐飛機前往銀川。銀川在馬可·波羅的筆下被稱作"額里合牙"(Egrigaia),他是這樣描述的:

當離開額里折兀勒(Ergiuul,即涼州),向東方騎行八日程,會發現一個被稱爲"額里合牙"的很好的大區,在它治下,有很多城市和村莊,它也隸名於唐兀(Tangut)大省。又,它境内的首府被稱作"哈拉善"(Calacian)。那裏定居的人們絶大部分是偶像教徒,那裏有不少薩拉森

人，那裏還有三座非常美麗的基督教堂，所遵循的教法爲聶思脱里派。又，此外，他們都隸屬於偉大的韃靼可汗。[①]

8月12日　星期二　銀川　晴

7:30到達銀川火車站，8:45至市區如家酒店住宿。安置完畢後，大多數隊員在寧夏考古研究所的安排下，於9:40乘車前往西夏王陵參觀，參加"粟特人在中國"研討會的芮柯（Christiane Reck）和西安考古研究院的辛龍一同前往。西夏王陵位於銀川市西南約30公里處，地處賀蘭山東麓中段的洪積扇地帶。考察隊車輛於10:18到達陵區門口。陵區大門的門樓上書四個鎏金西夏文大字，譯成漢語即"大白高國"。之所以叫"白高國"，據説是因爲懷念党項羌族最早的生活地區白河左近的緣故。此白河就是黄河的上游，因爲水色不黄，而有白河之號。10:30，我們持寧夏文物考古所介紹信免費進入陵區參觀。

西夏王陵陵區南北長10公里，東西寬5公里，現存帝陵9座，陪葬墓150餘座[②]，包括西夏博物館、西夏史話藝術館、西夏碑林、3號帝陵遺址等四個部分。我們首先進入陵區内的西夏王陵博物館參觀。該博物館接近大漠黄沙，爲米白色建築，共兩層，主要展覽王陵、黑水城以及寧夏周邊地區出土的西夏時期文物（多爲複製品），包括多件西夏文文書（《瓜州審案記録》《借貸契約》等）、印刷雕版、印章（正德、天慶、雍寧等時期的多方銅印）、雕塑、碑刻、繪畫、建築構件（滴水、鴟吻），以及百字咒經幢、木筆架、木桌、陶模、銅錢、銅鏡、牛皮鞋、繡花鞋、馬牌等[③]。雕塑中，可以見到多件人面鳥身的迦陵頻伽雕像，這是西夏藝術與信仰中非常有特色的形象；還可看到大石雕刻而成的力士志文支座。館藏碑刻中，有帶有"破醜氏先封……州列二十餘"等字樣的漢文殘碑，1972年出自西夏陵區161號陪葬墓；帶有"食邑三……[皇]太后也[仁]……當齠亂之年……長曰志和鎮國"等字樣的漢文殘碑[④]，1974年出自西夏陵區6號陵，等等。在博物館二樓的牆壁上，我們看到了多件絹本繪畫，包括《上樂金

① A. C. Moule & P. Pelliot, *Marco Polo: The Description of the World*, I, London: Carter Lane, 1938, p. 181. 漢譯文爲馬可·波羅讀書班集體翻譯成果，下引文亦同。

② 許成、杜玉冰，《西夏陵——中國田野考古報告》，北京：東方出版社，1995。

③ 參王志平、王昌豐、王爽編著，《西夏博物館》，銀川：寧夏人民出版社，2006。

④ 參李範文編釋，《西夏陵墓出土殘碑粹編》，北京：文物出版社，1984，圖版陸玖：M8CHB:4。

剛圖》《護法神像圖》《玄武大帝圖》《熾盛光十一曜圖》《月孛圖》等。此外，館中還有圖文介紹了108塔、拜寺口雙塔、康濟禪寺塔、拜寺溝方塔、宏佛塔、承天寺塔等一些西夏佛塔的情況，並展出了部分相關文物。

圖1　西夏王陵陵塔

由於時間緊促，我們參觀完西夏王陵博物館後，放棄了藝術博物館，直接前往陵區東南位置的3號王陵①。3號陵爲李元昊陵墓，是陵區最大的王陵，現存獻殿、角闕、月城等遺跡，南門爲陵園正門，由門道和門闕組成，周圍曾出土大量建築構件。作爲陵園主體建築的陵塔（圖1），原爲圓形密簷塔，塔身黄土夯築，外部木構建築支撑而成七級浮屠，基礎直徑約36米，現存高度約24米，體現出西夏的佛教信仰。西夏陵在規模和保存狀況方面遠不及唐宋帝陵，地下没有地宫，地面只能看到一系列土堆遺存。據指示牌介紹，地面應有闕臺、碑亭、月城、内城、獻殿、墓道封土、陵臺和角臺等遺跡。其中，闕臺周圍曾清理出土迦陵頻伽、摩羯、海獅、塔刹等建築裝飾構件，碑亭臺基上曾出土大量西夏文碑銘殘塊，以及石人、石馬等石像生殘塊。有意思的是，墓道直至陵

① 參寧夏文物考古研究所等，《西夏三號陵——地面遺跡發掘報告》，北京：科學出版社，2007；銀川西夏陵區管理處編，《西夏陵突出普遍價值研究》，北京：科學出版社，2013，144—157頁。

臺,未在中軸線上,而是略爲偏移。墓道本身爲階梯式坡道,呈前窄後寬的梯形,長46米,目測可見其填土隆起,呈魚脊狀。在墓道的末端、陵臺前面有一個巨大的凹坑,此爲一盜坑,直徑達20米,甚爲醒目。陵園圍牆和巨大的墳丘、雙闕依舊清晰可見。南闕外側,鐵網鐵架上放置着數萬件板瓦、筒瓦、滴水等建築構件殘片,這些殘留物令人得以想像當年西夏帝陵地面建築的建築風格與宏偉規模。此陵西北方的1、2、4號陵尚未開放,遠遠望去,白色的陵丘映襯着遠方青黑色的賀蘭山脈和蔚藍的天空,顯得格外迷人。

12:30,大家乘車離開西夏王陵,13:15左右返回"粟特人在中國"研討會的舉辦地點柏悦酒店。14:08,在隊長榮新江的帶領下乘原車前往寧夏自治區博物館參觀,朱安耐(Annette Juliano)、樂仲迪(Judith Lerner)、芮柯三位學者一同前往。大約14:25到達博物館。寧夏博物館一共三層:一層爲岩畫陳列,二層爲通史陳列,三層是回族文化展覽。我們主要參觀了二層的通史陳列部分,特别是其中的"粟特與絲綢之路"陳列館。該陳列館展示了大批固原、靈州等地出土的反映中西文化交流的文物,如拜占庭金幣、粟特鎏金銀瓶、李賢夫婦墓陶俑等[①]。16:45離開寧夏博物館。晚上自由活動。

8月13日　星期三　銀川　晴

上午,"粟特人在中國:考古發現與出土文獻的新印證"國際學術研討會開幕。考察隊全體隊員參會聽講,並協助會務。

8月14日　星期四　銀川　晴

上午和下午聽會。中午休會期間,13:00,党寶海、朱玉麒、李肖、姚崇新、馮鶴昌、鄭燕燕、付馬、于月、范佳楠、沈琛、李昀、胡曉丹等隊員參觀了寧夏文物考古研究所,一同參觀的還有參加會議的屈濤等人;榮新江老師則帶領羅帥、陳春曉、田衛衛、包曉悦四位同學,會同寧夏文物考古研究所的羅豐所長、沙燕會計共同商討會務和考察安排等事宜。14:00結束參觀,返回會場聽會。

18:30,"粟特人在中國"會議閉幕。晚餐後考察隊召開簡短會議,强調了

① 參李進增、陳永耘,《朔色長天——寧夏博物館藏歷史文物集萃》,北京:文物出版社,2013。

接下來在考察過程中的注意事項。

8月15日　星期五　銀川—額濟納旗　晴

考察隊今日離開銀川,前往内蒙古西部的阿拉善盟額濟納旗。在寧夏文物考古所的幫助下,考察隊在銀川租用了一輛大巴車,這輛車及兩位司機師傅將陪伴我們度過接下来的全部考察行程。甘肅簡牘博物館館長、研究員張德芳先生應邀作爲專家顧問,將一路同行考察,直至酒泉。部分考察隊員隨"粟特人在中國"會議參會人員前往固原,參觀新發現的粟特人墓葬及須彌山石窟,之後將飛抵敦煌,與考察隊其他成員會合。

圖2　阿拉善博物館藏東漢石刻殘件

7:50,自柏悦酒店出發。9:50,到達阿拉善盟政府駐地巴彦浩特。10:10,到達阿拉善博物館,李晉賀館長接待。博物館建築宏大,大家先後參觀了歷史陳列館與石刻陳列館。歷史陳列館以西夏和元代器物居多,亦有幾件唐代的捐贈品,如鎏金佛龕和銀箔小篆經書。石刻陳列館收藏了數十件形態各異的草原石人,如雙手合十石人、雙頭石人、羊身人首石人、石羅漢、成雙的石羊,遺憾的是,這些雕刻多未標明出土地與年代信息。在石刻陳列館,榮新江老師發現了標記爲殘功德碑的石刻,並立即告知了從事相關研究的朱玉麒教授。陳列室中藏有一件東漢石刻殘件(圖2),採集自阿拉善左旗騰格里蘇木通湖山，高129、寬138、厚11.2釐米，沙質岩，於自然石面上鑿刻，隸書，陰刻，内容記載了漢武帝列四郡至居延、北地，以及銘文鐫刻期間漢帝國的河西邊疆情況，其文略曰：

漢武威郡本\漢武帝排逐匈奴北置朔方\西置張掖磧\列鄣塞西界張掖居延□\匈奴遭王莽之亂\北地郡拆塞□更於郡\□□山沙之外吉□置蓬火先\□民無警□遠耳目\□□永初□處造作\□西北虜□耳目□也\□□太守□漢武時\□□四年□徒小吉吏\□功……①

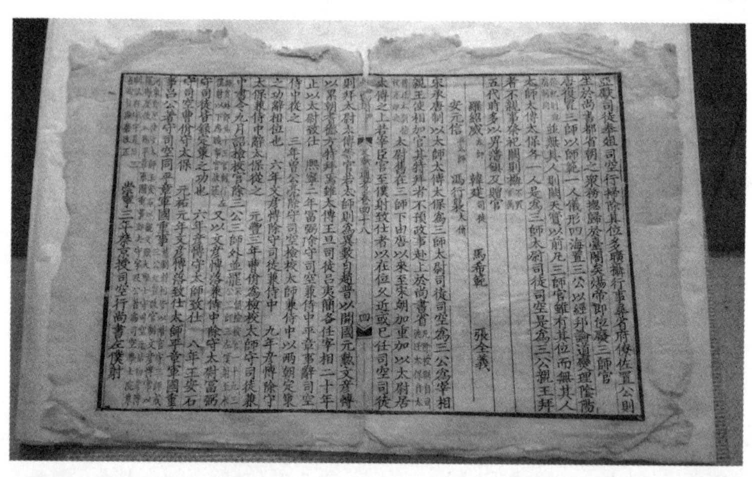

圖3　額濟納旗出土的元代杭州版刻《文獻通考》零葉

① 録文參考了阿拉善博物館編著《阿拉善博物館》，北京：文物出版社，2013，28頁。

另外，該館還藏有額濟納旗出土的元代杭州版刻《文獻通考》零葉（圖3）。這也是一件十分重要的文物，它反映了元代杭州印刷業的發達，以及當時的杭州與西北内地之間的文化聯繫。榮新江教授曾指出宋元時期杭州與河西具有很多聯繫，敦煌發現的元代杭州印刷的西夏文佛典，敦煌莫高窟北區發現的回鶻文寫本193+194號文書所記qingsai tavar"行在緞子"，吐魯番發現的大都（北京）、杭州印刷的漢文、回鶻文佛典以及元代杭州信實徐鋪和泰和樓大街南某家製作的金箔，德國所藏吐魯番寫本秦觀《海康書事十首》第三首習字（Ch.3800和Ch.3801）、宋版《新唐書·石雄傳》（Ch 2132v、Ch 2286v、Ch 3623v、Ch 3761、Ch 3903v五殘片綴合），都表明當時在敦煌、吐魯番、大都與杭州之間存在着一個貿易圈、貿易網絡，值得我們關注和研究。結束參觀後，每位考察隊員獲贈兩册博物館編制的小册子[①]，對於瞭解館藏情況多有助益。

11:00，大家合影留念後啓程奔赴額濟納旗。15:30到達烏力吉，在路邊小店吴忠馬三餐廳稍事午飯之後繼續前行。22:45，考察隊到達額濟納旗，入住金洋酒店。

8月16日　星期六　額濟納旗　晴

8:15，出發赴黑城遺址，一路南行，起先水草肥美，往南則漸爲荒漠。8:50到達黑城。黑城，即《馬可·波羅行紀》所載之"亦集乃城"：

> 從前述甘州城出發騎行十二日，人們可於這十二日路程結束時看到一座名爲亦集乃的城市。城在北方沙漠的起始處，屬唐兀省的一部分。且言所有該省之居民均是偶像教徒。他們擁有非常多的駱駝及各色別畜。此地盛産馴養的蒼鷺及隼，且品種優良。此地果樹與動物豐富。他們以農産品及牲畜爲生，亦不經商。行人在此城攜帶路上所需之四十日糧，蓋離此亦集乃城後，向北騎行四十日穿越一片沙漠，既無民居，亦無客棧，惟夏季可見人煙，緣冬季酷寒。在該沙漠的山、谷之中，人們確實可找到許多水源，其間多有梭子魚、大魚及小魚、野獸，野驢尤多。蓋因

① 阿拉善博物館編著，《阿拉善博物館》；巴爾那、許小燕、景學義編著，《阿拉善蒙古族傳統節慶禮俗》，銀川：陽光出版社，2011。

該沙漠邊緣間或有小松林。①

馬可·波羅在甘州(今張掖)逗留期間,曾聽聞了亦集乃城的一些事情,因此他簡要記載了該城的宗教、物産、交通等情況。科兹洛夫(P. K. Kozlov)、斯坦因(M. A. Stein)曾經對黑城遺址進行過發掘,獲得了大批文書。中華人民共和國成立後,内蒙古文物考古研究所也進行了大量的考古發掘工作,亦發現不少文書。三者都繪有黑城遺址平面圖,後兩者發表了詳細的考古報告②,科兹洛夫則僅留下探險記《蒙古、安多和死城哈喇浩特》③。

圖4 黑城遺址西北角城牆外佛塔

① A. C. Moule & P. Pelliot, *Marco Polo: The Description of the World*, I, pp. 160-161.

② M. A. Stein, *Innermost Asia: Detailed Report of Explorations in Central Asia, Kansu and Eastern Iran*, Oxford: Clarendon Press, 1928, pp. 429-506; 内蒙古文物考古研究所等,《内蒙古黑城考古發掘紀要》,《文物》1987年7期,1—23頁。

③ 彼·庫·柯兹洛夫著,王希隆、丁淑琴譯,《蒙古、安多和死城哈喇浩特》(完整版),蘭州:蘭州大學出版社,2011。

我們在張德芳老師和額濟納旗文物所的杜慶軍老師的帶領下，對黑城遺址進行了細緻的踏查。城中東北角與西門附近風沙堆積嚴重，科兹洛夫所看到的一些遺跡已被掩埋，内蒙古考古所發掘的總管府也難覓蹤跡。黑城中佛教遺跡清晰可見，最顯著的是西北角城牆上的五座藏傳佛教佛塔（圖4），爲具有粗相輪特徵之噶當覺頓式塔[①]。城内外還有多處寺廟遺址。馬可·波羅説“所有該省之居民均是偶像教徒”[②]，從現今所留存的諸多佛教遺跡，可想見馬可·波羅時代此地佛教信仰之盛況。除佛教外，黑城還存留下了其他宗教的遺跡，如西南城牆外的元代伊斯蘭教建築遺跡（圖5）等。城内地面上散落大量碎陶片和碎瓷片。出城後，在榮老師的帶領下，往西走約5分鐘，到達一處佛塔遺址，即科兹洛夫、斯坦因所獲文書的主要來源地，塔身已不存，惟餘塔基。黑城文書是與敦煌文書齊名的重要出土文獻，是研究西夏、金、元歷史的原始資料。黑城文書包含了多種語言文字，涉及政治制度、經濟生活、宗教信仰等方方面面内容。其中關於當地物産和社會生活的内容，無疑是研究馬可·波羅關於亦集乃記載的必不可少的資料。

圖5　黑城遺址西南角伊斯蘭教建築

① 宿白，《張掖河流域13~14世紀的藏傳佛教遺跡》，原載《北京大學學報》1993年2期，收入氏著《藏傳佛教寺院考古》，北京：文物出版社，1996，251—255頁。

② A. C. Moule & P. Pelliot, *Marco Polo: The Description of the World*, I, p. 160.

11:20離開黑水城，乘車東赴唐代大同城遺址，11:30到達。大同城爲回字形方城，分内城與外郭，至今四面仍殘存高大的城牆，高五六米，寬三四米。大同城始建於漢代，隋代時稱作"大同城鎮"，唐代改置同城鎮。同城在唐代是隸屬甘州的軍事要塞，垂拱(685—688)初，安北都護府一度設在這裏，後來遷出。再後來這裏成立同城守捉。天寶二載(743)，升級爲寧寇軍[①]。我們所見到的就是唐代遺留的同城遺址。大家圍繞内城城牆對其進行了迅速的踏查，11:45乘車離開，12:30到達賓館附近的一家餐廳用餐。

14:30出發赴額濟納旗博物館，8分鐘後到達。當地博物館領導接待。博物館藏品豐富，以漢代、西夏、元代藏品最多。其中漢代烽燧木簡、印章、緑城遺址出土文物，捐贈品西涼"涼造新泉"錢幣，西夏紅廟遺址出土雕塑和壁畫，以及元代蒙漢藏文文書、印章、版刻《文獻通考》零葉等最爲重要。16:20，大家離開博物館，乘車北赴中蒙邊境的策克口岸，17:30到達，到中蒙界碑處拍照留念。18:20離開口岸。晚上自由活動。

8月17日　星期日　嘉峪關市　晴

圖6　肩水金關故址

8:10出發向酒泉行進，一路西南行。11:05到達地灣城遺址，考察隊在張德芳老師的指引下在此停留考察。地灣城遺址爲漢代肩水候官所在，由三塢和一鄣組成，現僅存一鄣(圖6)。鄣呈方形，21.6×21.6米，建於一夯土臺基上；鄣牆基寬6米，頂寬2.8~3.5米，殘高7.9~8.2米，係夯土板築，夯層0.2米，在北牆離地3米和東南牆角離地2米處，有兩排小圓木洞。西南兩面的院牆基址仍然可以辨認。候官位於邊牆線上，邊牆由南向北延伸，沿邊牆往北走約500米至肩水金關，僅餘低矮的土丘，

① 孟憲實，《從同城鎮到寧寇軍》，收入沈衛榮、中尾正義、史金波主編《黑水城人文與環境研究》，北京：中國人民大學出版社，2007，396—405頁。

但仍可辨其輪廓[1]。1930年代,中瑞西北科學考察團在該遺址發現萬餘枚漢代木簡[2];1970年代,我國考古工作者又對此地進行了系統的發掘,出土上萬枚木簡[3]。12:30參觀完畢,大家乘車沿314省道南行。馬可·波羅記載的從酒泉到亦集乃的道路,基本上應當是傍漢代長城烽燧線而行的道路。

途經張掖河、金塔縣。17:10到達酒泉市。酒泉即《馬可·波羅行紀》所載之肅州,他記載道:

> 這州裏有衆多的城市和村莊,並且首府本身也被稱作肅州。這裏也有一些聶斯脱里派基督教徒,這裏的居民主要是偶像教徒。他們屬於大汗治下,如前述其他州一般。肅州這一州和這兩個其他的我較早之前告訴你們的州,哈密力和欣斤塔剌思,所在的整個的大省,被稱作唐兀。在這幾個州的所有的山中,能找到很多最上乘的大黄,商人們購買它然後帶它到全世界各處,對這裏出産的其他商品則不會如此。這是真的,經過這裏的旅人不帶着當地的牲畜就不敢進那些山,因爲山裏長着某種毒草,如果牲畜吃了它就會失去蹄子,但是當地出生的動物能辨别那種草而在進食時避開它。這裏有大量的畜群和農作物。他們以從地裏獲取的農産品和他們的畜群爲生,但是他們完全不做貿易。全州人都是健康的,居民皮膚棕色。[4]

抵達酒泉後,我們先參觀了酒泉市肅州區博物館,王館長接待。館内文物以兩漢魏晉時期文物居多,最著名者爲北涼程段兒造像石塔,爲該館鎮館之寶。其餘文物以酒泉地區出土的魏晉南北朝時期墓葬壁畫磚最有特色,陳列數十塊,多出土於高閘溝遺址。

17:40離開肅州區博物館,乘車西行15分鐘,到達丁家閘5號壁畫墓。丁

---

① 參看甘肅居延考古隊,《居延漢代遺址的發掘和新出土的簡册文物》,《文物》1978年1期,1—25頁;吴礽驤,《河西漢塞調查與研究》,北京:文物出版社,2005,160—165頁。

② 勞榦,《居延漢簡·考釋之部》,台北:"中研院"歷史語言研究所,1960(初版於1943—44年);勞榦,《居延漢簡·圖版之部》,台北:"中研院"歷史語言研究所,1957;中國科學院考古研究所,《居延漢簡甲編》,北京:科學出版社,1959;中國社會科學院考古研究所,《居延漢簡甲乙編》,北京:中華書局,1980;謝桂華、李均明、朱國炤,《居延漢簡釋文合校》,北京:文物出版社,1987。

③ 甘肅簡牘保護研究中心等編,《肩水金關漢簡》,上海:中西書局,2013。

④ A. C. Moule & P. Pelliot, *Marco Polo: The Description of the World*, I, p. 158.

家閘5號墓因其繪於前墓室的精美壁畫而著名,是十六國時期的貴族墓葬①。墓室位於地下12米處,墓室、墓道均十分狹窄,因此大家分兩組排隊進入。墓室經一千六百年仍未坍塌,四壁壁畫雖經水泡卻依然色彩繽紛。

18:40離開5號墓,乘車西行約8分鐘到西涼王陵。西涼王陵位於地下20米處,約有6層樓的深度,墓道寬闊,墓道兩側有6個耳室,爲魏晉南北朝時期墓葬所罕見。墓室分前後兩室,外加左右兩個耳室,前室與後室之間有一段甬道,甬道盡頭爲主室石門,用一整塊花崗岩製成。墓室至少被三次盜掘,因此石門被打碎,堆於右側耳室内。2001年4月,由於墓葬盜洞上方土層塌陷,考古人員進入墓室内部進行搶救性清理②。其時墓内已盜掘一空,唯餘一具身首異處的人骨架,首在主室,而身在右耳室。前室甬道兩側的牆壁上存留兩塊繪有執笏的大臣形象的彩繪墓磚,考古人員因此認定該墓屬於在酒泉立國的西涼王李暠之墓。19:10,大家乘車返回市内,吃完晚飯趕到嘉峪關市。

8月18日　星期一　嘉峪關—酒泉　晴

8:20,出發前往嘉峪關關城和長城博物館,約8:50到達。嘉峪關城保存較好,城牆、衙門都較爲完整地保存下來,但由於近些年過度的商業化開發和肆意修繕,使得"天下雄關"的氣勢蕩然無存,出現了如此景象:明代關城上懸掛着清代的黄龍旗,而關城内部又擺設了蒙古軍帳。

爲了節省開支,榮老師、段老師和曾經考察過嘉峪關的同學留在關城外等待。鑒於嘉峪關遇到的門票昂貴問題,榮老師聯繫甘肅省文物局,得到馬玉萍局長的親切關懷,文物局給甘肅省各相關文物單位發文,大力支持本次考察,接待北京大學馬可·波羅考察隊一行。

10:10,大家離開嘉峪關關城,乘車東南行,前往考察文殊山石窟。文殊山石窟位於酒泉市以南15公里的嘉峪關山中,行政上歸張掖市肅南裕固族自治縣管轄。石窟始建於北魏,從蒙元時期轉變爲藏傳佛教石窟寺,至今仍然香火不斷。在石窟寺群正中的文殊寺中,原立有漢文—回鶻文合璧的《重修文

① 甘肅省文物考古研究所編,《酒泉十六國墓壁畫》,北京:文物出版社,1989。
② 肅州區博物館,《酒泉小土山墓葬清理簡報》,《隴右文博》2004年2期,17—20頁。

殊寺碑》,現藏於肅南縣民族博物館。此碑由蒙古喃答失太子立於公元1326年,記載了豳王家族重修文殊寺的過程,具有豐富的史料價值。耿世民、張寶璽二位學者曾對此碑作了譯釋[①],它對研究河西的察合台系後裔、佛教信仰狀況、文殊寺修建歷史等方面具有重要價值,尤其是碑文中將出伯家族列爲察合台系繼承人這一罕見記載,是考量察合台家族歷史的獨特史料[②]。在文殊山山谷中,以前分佈着大量寺觀建築,但很多在清朝同治年間(1862—1874)因戰亂而毁滅[③]。

11:10,到達文殊山下的祁豐藏族鄉。在當地的文殊山文管所索老師的帶領下,乘車5分鐘抵達文殊山下,下車後沿山道登上文殊山前山第3窟,進入窟内參觀。該窟僅存主室,根據中心塔柱的形式可以斷爲北魏石窟,石窟四壁壁畫年代比較複雜,壁畫底層爲表現世俗生活的壁畫,並且存有大量的回鶻文、藏文、漢文題記[④]。姚崇新老師認爲這些壁畫時代應該在回鶻、西夏時期,漢文題記則更晚至明、清、民國皆有。第3窟以北數米爲第4窟,爲北魏時期所造窟,僅存主室,窟内壁畫有兩壁已殘,其餘壁畫比較完整,上層天宫伎樂圖,以下繪千佛,再下依次繪説法圖、供養人像、卷草、垂帳,然已無題記留存。第4窟北側尚有殘窟,未編號,有殘壁畫,内容不可辨識[⑤]。12:00前往後山禪窟,左右兩側分佈禪窟數十個,遠處可見寺院殘址,大家進入右側禪窟參觀,窟内空無一物,片刻即歸。12:35,乘車返回嘉峪關市内用餐。14:15出發前往敦煌,19:43到達敦煌市内的陽關酒店。

---

① 耿世民、張寶璽,《元回鶻文〈重修文殊寺碑〉初釋》,《考古學報》1986年2期,253—264頁。

② 杉山正明,《豳王チュベイとその系譜—元明史料と『ムーイッズル·アンサリ』の比較を通じて》,杉山正明《モンゴル帝國と大元ウルス》,京都:京都大學學術出版會,2004,242—287頁;張寶璽,《喃答失太子〈有元重修文殊寺碑銘〉再考》,仲高主編《龜兹學研究》第3輯,烏魯木齊:新疆大學出版社,2008,201—210頁;張海娟、楊富學,《蒙古豳王家族與河西西域佛教》,《敦煌學輯刊》2011年4期,94頁。

③ 史岩,《酒泉文殊山的石窟寺院遺跡》,《文物參考資料》1956年7期,53—59頁。

④ 參看伊斯拉菲爾·玉蘇甫、張寶璽,《文殊山萬佛洞回鶻文題記》,新疆吐魯番學研究院編《語言背後的歷史——西域古典語言學高峰論壇論文集》,上海:上海古籍出版社,2012,94—106頁。

⑤ 參看甘肅省文物工作隊,《馬蹄寺、文殊山、昌馬諸石窟調查簡報》,《文物》1965年3期,18—19頁。

8月19日　星期二　敦煌　雨

敦煌在漢唐時期一直是絲綢之路的咽喉，在漢代是河西四郡之一，唐代稱爲沙州。馬可·波羅到訪過這裏，仍稱之爲“沙州”，説它位於唐兀省境内，並描述了此地的經濟、宗教信仰，特别是與佛教有關的風俗見聞[①]。他在《行紀》裏提到沙州有許多佛教的殿堂與僧院，但没有明確提及莫高窟。

考察隊自今天開始調查敦煌附近的遺跡。另外，英國倫敦大學亞非學院尼古拉斯·辛姆斯–威廉姆斯（Nicholas Sims-Williams）教授、厄修拉·辛姆斯—威廉姆斯（Ursula Sims-Williams）資深館員、德國柏林科學院吐魯番研究所德金（Desmond Durkin-Meisterernst）所長等學者隨隊考察，敦煌研究院考古研究所陳菊霞研究員接待並做學術指導。

圖7　大方盤城遺址

上午考察敦煌西線遺跡。7:55出發，9:25經過小方盤城，9:40到達大方盤城。大方盤城（圖7），斯坦因編號爲T. XVIII，並認爲它是漢唐時期的河倉城[②]，但李正宇先生根據實地測量和文獻對比，認爲此遺址實爲西漢時期的昌

① A. C. Moule & P. Pelliot, *Marco Polo: The Description of the World,* I, pp. 150-154.

② M. A. Stein, *Serindia: Detailed Report of Archaeological Explorations in Central Asia and Western-most China*, Oxford: Clarendon Press, 1921, pp. 712-717.

安倉[①]。由於時間緊張，大家用了十餘分鐘的時間冒雨對該城址進行了踏查。9:50離開，10:12返回小方盤城。小方盤城（圖8），斯坦因編號爲T. XIV，即漢代玉門關[②]，大家利用半個小時的時間對其進行了踏查。10:40考察隊離開玉門關赴西千佛洞，11:37到達。由於人數較多，隊員分作2組，主要參觀了第7（敦研院編號005）、9（敦研院編號007）、12（敦研院編號10）、16（敦研院編號13）、18窟（敦研院編號15）等窟[③]。在回鶻可汗畫像前，榮新江老師和大家討論了有關西州回鶻與沙州關係問題。

圖8　小方盤城遺址

13:30離開西千佛洞重返敦煌研究院，14:30到達。到達後先參觀了研究院的圖書資料室。15:00開始聽辛姆斯–威廉姆斯教授的講演"粟特語的研究史"，講座梳理了粟特語的發現和研究簡史，並對哈薩克庫爾托貝（Kultobe）遺址的粟特文泥版銘文進行了介紹[④]。講座持續了兩個小時，由張元林先生主持，付馬擔任翻譯，彭金章、沙武田等人也參加了講座。講座結束後返回酒店，晚上自由行動。

---

① 李正宇，《敦煌大方盤城及河倉城新考》，《敦煌研究》1991年4期，72—79頁。關於河倉城的可能位置，參看李岩雲《敦煌河倉城址考》，《敦煌研究》2013年6期，86—92頁。

② M. A. Stein, *Serindia*, pp. 683-688；向達，《兩關雜考》，氏著《唐代長安與西域文明》，石家莊：河北教育出版社，2001，328—383頁。

③ 參看霍熙亮整理，《敦煌西千佛洞内容總録》，收入敦煌研究院編《敦煌石窟内容總録》，北京：文物出版社，1996，195—203頁；霍熙亮編，《榆林窟、西千佛洞内容總録》，收入敦煌研究院編《中國石窟·安西榆林窟》，北京：文物出版社，2012，254—269頁。

④ 有關庫爾托貝粟特語銘文的介紹與研究，可參看N. Sims-Williams and F. Grenet, "The Sogdian inscriptions of Kultobe." *Shygys* 1, 2006, pp. 95-111；N. Sims-Williams, F. Grenet and A. Podushkin, "Les plus anciens monuments de la langue sogdienne: les inscriptions de Kultobe au Kazakhstan." In: *Académie des Inscriptions et Belles-Lettres. Comptes rendus des séances de l' année* 2007 [2009], pp. 1005-1034；畢波，《考古新發現所見康居與粟特》，張德芳主編《甘肅省第二屆簡牘學國際學術研討會論文集》，上海：上海古籍出版社，2012，99—109頁。

8月20日　星期三　敦煌　多雲轉雨

今天的主要行程是參觀莫高窟。8:05自酒店出發,8:45到達莫高窟南區。在陳菊霞老師的帶領下,我們根據所擬定的洞窟號進行了細緻的考察。上午主要參觀南區石窟,依次參觀了61(含62)、71、85、96、100、220、148、237、246、249、259、427、428、420、423、17、16等有代表性的石窟,這些石窟大體涵蓋了各個時段所開洞窟與所繪壁畫①。對莫高窟南區的參觀一直持續到13:50,考察的最後,榮老師還領着考察隊員參觀了王圓籙修建的三清觀,其中陳列了有關20世紀莫高窟遭到西方列强劫掠的一些歷史照片。參觀完後,大家在莫高窟旁邊的一家餐廳用餐。

圖9　莫高窟北區石窟

15:00,大家乘車到附近的敦煌研究院院部聽彭金章研究員的講座,彭先生用半個多小時談了莫高窟北區考古的大概情況,之後又詳細介紹了敦煌北區特窟465窟的情況。16:30講座結束,彭先生帶領考察隊參觀北區石窟。16:50到達北區石窟(圖9)。主要參觀了465、B77窟以及幾個底層的禪窟和僧房窟②。465窟及其附近是莫高窟中規模最大的一處藏傳佛教密教窟室,其開鑿年代在13世紀後半③,正是馬可·波羅經過沙州的時期。馬可·波羅也曾詳細地描述沙州佛教的興盛情況:

> 他們有許多大寺院和普通廟宇,其中佈滿各種神像。信徒們對這些神像奉獻很多祭品並備加崇敬,還非常敬畏並虔誠禮拜。你可能知道該州有一個風俗,所有有小孩的人,一旦小孩出生,就要每年養一頭羊以感

① 參看敦煌研究院編,《敦煌莫高窟供養人題記》,北京:文物出版社,1986;敦煌研究院編,《敦煌石窟内容總録》,1996。

② 參看彭金章、王建軍,《敦煌莫高窟北區石窟》(3卷本),北京:文物出版社,2000—2004;彭金章主編,《敦煌莫高窟北區石窟研究》,蘭州:甘肅教育出版社,2011。

③ 宿白,《敦煌莫高窟密教遺跡札記》,原載《文物》1989年10期,修改後收入氏著《中國石窟寺研究》,北京:文物出版社,1996,303—308頁。

念神像。在年初或是他們的神像節日的時候，那些將羊養肥了的父親們會牽羊攜子來到神像之前，在那裏致以他極高的敬意並設宴以感念神像，除了大人們，小孩也要參與。做完這些之後，他們將整隻肥羊煮熟，然後極虔誠地將其再一次帶到神像之前，直到僧人他們説完其事由和祈禱之後才離去。……

你可能知道這個省的另一個風俗，世界上所有的偶像崇拜者，無論男女，當他們將要死的時候，都要遺命其他人焚燒其屍體。①

18:20參觀結束，大家合影留念。晚上自由活動。

8月21日　星期四　敦煌　晴

9:00出發，六七分鐘後到達敦煌市博物館，博物館藏品豐富，其中尤其以兩件北涼石塔最有價值②。北涼石塔是中國現存最早的一批石塔，對於研究佛教造像藝術具有重要意義。這兩件北涼石塔與酒泉市博物館所藏程段兒造像石塔同爲國寶級文物。除去藏品外，博物館的展覽設計和牆壁上的解説圖表都十分專業。

10:35再赴莫高窟，此次主要參觀窟外的莫高窟陳列館。陳列館分兩部分，一部分是復原石窟，其壁畫均爲手繪原狀復原，雕塑也是原比例復原，主要有217、275、276、285等窟，幾位老師都作了詳細的講解和分析。尤其是285窟因其壁畫具有濃郁的嚈噠風格而受到大家的矚目，該窟始建於西魏，在元代時改建爲藏傳密教窟室③。陳列館的另一部分爲北區石窟文物特展，陳列了北區出土的文物和文書，其中尤以敍利亞文景教文書（圖10）和銅十字架（圖11）最爲珍貴④。馬可·波羅描述沙州的宗教狀況時説："這個州的居民基

---

① A. C. Moule & P. Pelliot, *Marco Polo: The Description of the World*, I, pp. 57-58.

② 參暨遠志，《北涼石塔所反映的佛教史問題》，顏廷亮、王亨通主編《炳靈寺石窟學術研討會》，蘭州：甘肅人民出版社，2003，275—290頁。

③ 宿白，《參觀敦煌莫高窟第285號窟札記》，原載《文物參考資料》1956年2期，修改後收入氏著《中國石窟寺研究》，206—213頁。

④ 參段晴，《敦煌新出土敍利亞文書釋讀報告（續篇）》，《敦煌研究》2000年4期，120—127頁；姜伯勤，《敦煌莫高窟北區新發現的景教藝術》，《藝術史研究》第6輯，廣州：中山大學出版社，2004，337—352頁。

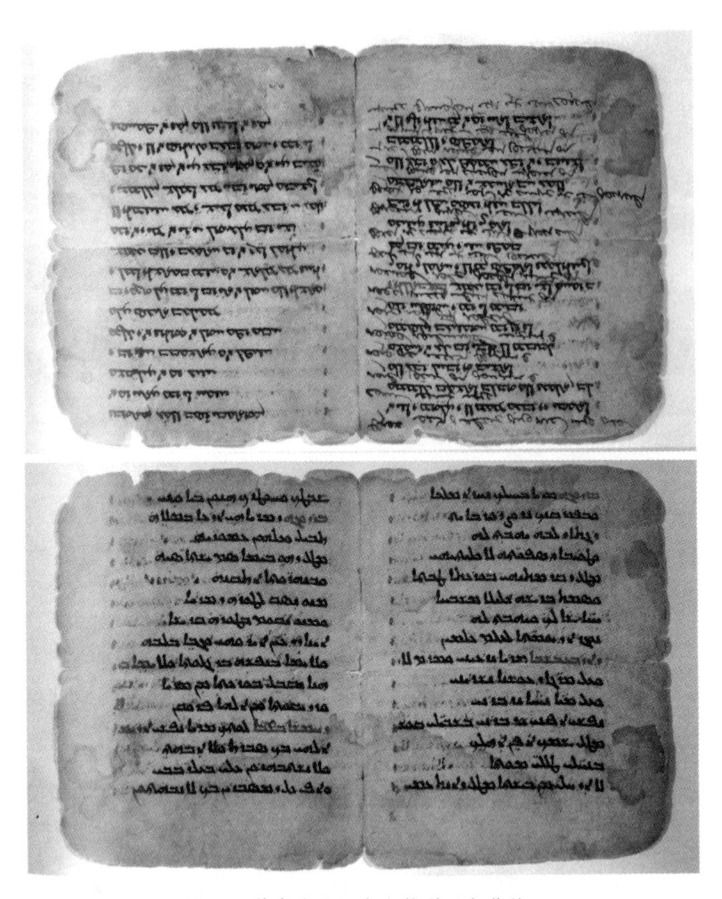

圖10　莫高窟北區出土敘利亞文舊約

圖11　莫高窟陳列館藏銅十字架

本都是偶像崇拜者，雖然的確有一些信奉景教的突厥人和一些薩拉森人。”①景教文書和銅十字架的發現，印證了他的説法。

11:50參觀結束後用餐，14:00結束。飯後考察隊又參觀了李其瓊紀念館，其中陳列了已故敦煌研究院壁畫臨摹藝術家李其瓊的作品。15:00，大家轉赴敦煌研究院召開前期考察總結會議。會議由段晴老師主持，大家各自講了自己的考察

① A. C. Moule & P. Pelliot, *Marco Polo: The Description of the World*, I, pp. 57-58.

收穫,又對下一階段的考察進行了安排,榮新江老師做了最後總結。會議持續一小時,之後散會。晚上自由活動。

8月22日　星期五　玉門市　雨

8:45出發赴榆林窟,10:20到瓜州境,後轉向東南行。10:45過破城子遺址,即漢之廣至、唐之懸泉,城牆保存完好,矗立於道左,考察隊員下車參觀片刻。10:57自大道轉向小道,穿越河床中的道路,11:03到達鎖陽城鎮之後,轉向西南,11:20入水硤口,道路右側爲渾濁的榆林河(又名踏實河)水,出山谷後轉向東南行,11:30到達榆林窟。敦煌研究院的宋子貞所長接待,恰逢敦煌研究院的一些老師正在榆林窟中考察,因此考察隊先到榆林窟的食堂用餐,12:50吃完,隨後在宋老師的帶領下參觀榆林窟。

圖12　榆林窟

榆林窟分佈於榆林河東西兩岸崖壁上(圖12)[①],我們先到西崖參觀31、32、33、35等窟,這些窟皆具有明顯的于闐色彩。其中31窟因繪有于闐王李聖

① 參敦煌研究院編,《中國石窟·安西榆林窟》,文物出版社,2012。

天及于闐王后曹氏的供養像而被學者比定爲敦煌文書中的“天子窟”[1]。敦煌石窟中明確繪有于闐王供養像的洞窟有莫高窟454、98窟以及該窟,該窟年代晚於98窟,但不會晚到宋代。32窟亦爲五代石窟,主室東壁門北畫普賢變,背景爲牛角山以及毗沙門天王決海圖,與東壁門南的以五臺山爲背景的文殊變可能存在某種對應關係。主室南壁存有一排瑞像圖與八大守護神圖,八大守護神居中,這種排列方式比較少見。33窟存曹元忠父子的供養像,主室南壁的毗沙門天王決海圖與瑞像圖同樣引起了大家的注意。35窟主室甬道南北壁各存曹延禄與于闐公主供養人像。討論正酣,其時雨勢稍大,河上小橋有淹没之勢,那樣大家將被困在西岸。於是考察隊在宋老師的帶領下迅速渡河到東岸,繼續參觀了21、25兩窟,兩窟中存在多處藏、蒙古、回鶻、西夏文題記[2]。有學者認爲,25窟是爲了紀念821—822年唐蕃會盟而建,這在敦煌藏文文獻P.T.16和IOL TIB J 751 I中有詳細記載[3]。15:20,離開榆林窟赴瓜州。

16:45到達瓜州博物館參觀,博物館王館長接待。瓜州博物館自1998年以後未曾新進文物,鎮館之寶唐代象牙塔也被徵調至國家博物館,因此館内並未見到特别的文物。17:10參觀結束,榮新江、段晴、孟嗣徽、姚崇新、李肖等學者由於要繼續在莫高窟作考察,而乘車返回敦煌,其餘的隊員在党寶海的帶領下繼續沿着河西走廊東行考察。19:00到達玉門,入住玉門賓館。

8月23日　星期六　張掖　晴

上午8:30從玉門市出發,一路東行,10:10到達嘉峪關市,10:46到達全國重點文物保護單位新城—果園魏晉墓群。該墓群位於嘉峪關市東北20公里

---

① 沙武田,《敦煌石窟于闐國王畫像研究》,《新疆師範大學學報》2006年4期,22—30頁。

② 史金波、白濱,《莫高窟榆林窟西夏文題記研究》,《考古學報》1982年3期,367—386頁;敦煌研究院考古研究所等,《敦煌石窟回鶻蒙文題記考察報告》,《敦煌研究》1990年4期,1—19頁;哈密頓、楊富學、牛汝極,《榆林窟回鶻文題記譯釋》,《敦煌研究》1998年2期,39—55頁;J. Hamilton and Niu Ruji, Inscriptions ouïgoures des grottes bouddhiques de Yulin, *Journal Asiatique* 286, 1998, pp. 127-210;謝繼勝、黄維忠,《榆林窟第25窟壁畫藏文題記釋讀》,《文物》2007年4期,70—78頁;Dai Matsui, Revising the Uigur Inscriptions of the Yulin Caves,《内陸アジア言語の研究》23號, 2008, pp. 17—33;陸離,《關於榆林窟第25窟壁畫藏文題記釋讀的兩個問題》,《西北民族大學學報》2010年4期,54—57頁。

③ 卡普斯坦,《德噶玉采的會盟寺:比定和圖像闡釋》,霍巍、李永憲主編《西藏考古與藝術國際學術研討會論文集》,成都:四川人民出版社,2004,98—127頁。

處的戈壁灘上，處於嘉峪關市新城鄉和酒泉市果園鄉的交界地帶。經普查統計，有1700多座磚墓，多屬魏晉十六國時期，目前經考古清理的有十餘座，其中編號爲1、3、4、5、6、7、12、13號墓的墓室内均有彩繪壁畫①。不過，對外開放的僅有6號墓。

6號墓爲魏晉時期貴族的壁畫墓，墓室位於十幾米深的地下，墓道狹窄，僅容一人通過。該墓爲夫妻合葬墓，墓室分爲前室、中室與後室，墓室已遭盜掘，幾乎没有文物出土，僅餘牆壁上的壁畫磚。前室壁畫磚主要爲畜牧、耕種、出行圖，中室壁畫主要爲生活起居圖，宴飲場景栩栩如生。後室主要是書册、妝盒和粉餅等畫像，整個壁畫比較全面地反映了墓主人的現實生活，因此具有極高的學術價值②。從6號墓出來，考察隊還參觀了墓區入口處的魏晉壁畫藝術陳列展，包括5、6、7號墓的一些磚畫、隨葬品和棺板畫的實物和圖片。11:45參觀結束，12:20到達酒泉路邊的一家麵館吃午飯，13:30繼續向東奔赴張掖。

16:27到達張掖市區的大佛寺(圖13)。大佛寺始建於西夏永安元年(1098)，原名迦葉如來寺，爲全國重點文物保護單位。今天的大佛寺爲張掖市博物館所在地，寺内無僧人。建築群以大佛殿爲中心，殿内有長達34.5米的木胎泥塑卧佛一尊，是中國現存最大的室内卧佛像。馬可·波羅記述"甘州"時，生動描述了大佛寺内大小佛像環立的局面：

圖13 張掖大佛寺

> 於彼之城，偶像教徒據其傳統與風俗建有許多至爲華麗之僧院及衆多殿堂。僧院之内，造像數量及種類極多。此外，我要告訴你，其中一些

① 甘肅省文物隊等，《嘉峪關壁畫墓發掘報告》，北京：文物出版社，1985。

② 參張寶璽編，《嘉峪關酒泉魏晉十六國墓壁畫》，蘭州：甘肅人民出版社，2001，151—214頁。

長有十步上下。像有木刻者，有泥塑或陶燒者，有石雕者，兼有青銅鑄造者；諸像貼金，製作精良，美妙絶倫。此類造像，或稍大，或矮小。這些大像作平躺卧佛式，其餘諸小像環立四周，勢若信徒謙恭禮拜。大像較諸小像尊崇甚矣。[①]

今日一見，竟與馬可·波羅之描述一般無二。寺内建築及大佛經歷代修葺。1956年，大佛内曾流散出一批明代佛教經卷，《文物》1978年第10期曾發表了一組以“張掖大佛寺明永樂佛曲”爲主題的文章對它們進行了介紹[②]。這些經卷中的一部分在大佛寺文物陳列室裏展出。1971年，大佛寺内出土了5枚薩珊銀幣，其中庫思老二世(Khosrau Ⅱ，591—628年在位)時期1枚，卑路斯(Peroz，459—484年在位)時期3枚，另一枚不詳；伴出的尚有唐朝開元通寶銅錢以及明朝正統元寶金片1枚，不過這些錢幣的埋藏時間尚不清楚[③]。我們在大佛寺文物陳列室裏見到了這批薩珊銀幣中的兩枚。

圖14　黑水國遺址

① A. C. Moule & P. Pelliot, *Marco Polo: The Description of the World,* I, pp. 158-160.

② 劉觀民，《佛曲遇見記》，《文物》1978年10期，54—55頁；張掖市博物館，《張掖大佛寺明永樂佛曲目録》，《文物》1978年10期，55—61頁；劉觀民輯，《張掖大佛寺明永樂佛曲所見南北曲名輯録》，《文物》1978年10期，62—63頁。

③ 康柳碩，《中國境内出土發現的波斯薩珊銀幣》，《新疆錢幣》2004年3期，65頁。

圖15　黑水國遺址殘留陶片

18:00離開大佛寺,沿312國道奔赴張掖城西的黑水國遺址,19:00到達。黑水國遺址區包括南城和北城兩座城址、漢墓群以及近年發掘的西城驛遺址等遺跡單位。西城驛遺址曾入選中國社會科學院考古學論壇"2013年中國考古6大新發現",其年代距今4100—3600年,歷經馬廠晚期、西城驛二期、四壩早期三個時期,是一處新石器時代晚期至青銅時代的重要遺址①。然而,由於遺址區多爲農田和廠房,尋訪頗爲不易,最終只踏查了南城。南城呈長方形,東西長約258米、南北寬約226米(圖14)。此城始建於唐代,西夏以後修復沿用,明代後廢棄②。根據史料記載,唐代的鞏笓驛③、元代的八剌哈孫驛(即西城驛)④、明代的小沙河驛均在這裏。現在殘存的是明代小沙河驛的遺址,仍保留着較完整的高達七八米的城牆,城内建築遺跡依稀可辨,地表散落着一些黑釉和緑釉陶片(圖15)。考察隊19:30離開南城遺址,返回市内。

8月24日　星期日　張掖　晴轉雨

8:20自酒店出發前往馬蹄寺,9:30到達。馬蹄寺石窟群位於張掖市區以南65公里,在肅南裕固族自治縣境内,由勝果寺、普光寺、千佛洞、金塔寺、上中下觀音洞七處石窟組成,共有70多處窟龕。各處石窟之間相距2至10公里不等,綿延分佈於馬蹄山山前。馬蹄山古名臨松山,前涼張天錫(363—376年在位)於臨松山下置臨松郡⑤。《晉書》記載,十六國時期河西的儒學大家郭瑀

① 陳國科等,《甘肅張掖市西城驛遺址》,《考古》2014年7期,3—17頁。

② 王北辰,《甘肅黑水國古城考》,原載《西北史地》1990年2期,收入氏著《王北辰西北歷史地理論文集》,北京:學苑出版社,2000,117—129頁;吴正科,《絲路古城黑水國》,蘭州:甘肅人民出版社,2008,85—89頁。

③《新唐書》卷四〇《地理志》,中華書局標點本,1975,1042頁。

④《元史》卷二八《英宗紀》,中華書局標點本,1976,621頁。

⑤《晉書》卷一四《地理志》,中華書局標點本,1974,434頁。

"隱於臨松薤谷,鑿石窟而居"①。因此,馬蹄寺石窟的始鑿被認爲是郭瑀,清乾隆《甘州府志》即將馬蹄寺石窟稱作"薤谷石室",並説"石窟始於郭瑀及其弟子,而後人擴之,加以佛像"②。可見,馬蹄寺是來自漢文化的儒學和來自印度文化的佛教在河西走廊交融碰撞的一個生動實例。

圖16 馬蹄寺北寺石窟

由於時間有限,我們主要考察了北寺石窟(圖16)和千佛洞③。首先依次參觀了7號窟(站佛殿)、3號窟(三十三天)、8號窟(藏佛殿)。這些窟現在爲格魯派藏傳佛教洞窟④,其中明清以前塑像、壁畫幾乎無存。隨後考察了千佛洞,這部分石窟開鑿於峭壁之上,分爲東區、中區、西區三部分,東區主要爲浮雕舍利塔,現未開放,因此主要參觀了中區與西區。其中一窟中存涼州瑞像一軀,時代較早,最爲珍貴⑤。

15:30離開馬蹄寺返回張掖市,17:07到達市内木塔寺。木塔寺始建於唐代,現存木塔爲1926年重修,塔後有藏經樓,爲清代建築,兩者形成一個建築

①《晉書》卷九四《隱逸傳》,2454頁。

② 鍾賡起著,張志純等校注,《甘州府志校注》,蘭州:甘肅文化出版社,2008,128—129頁。

③ 參看甘肅省文物工作隊,《馬蹄寺、文殊山、昌馬諸石窟調查簡報》,《文物》1965年3期,13—18頁;姚桂蘭、格桑美卓,《張掖馬蹄寺石窟群内容總録》,《敦煌學輯刊》1995年2期,75—81頁。

④ 參看宿白,《張掖河流域13~14世紀的藏傳佛教遺跡》,氏著《藏傳佛教寺院考古》,255—259頁;張掖市文物保護研究所,《祁連山北麓馬蹄寺石窟群浮雕舍利塔考古調查簡報》,《華夏考古》2014年4期,39—49頁。

⑤ 張善慶,《甘肅張掖市馬蹄寺千佛洞涼州瑞像再考》,《四川文物》2009年3期,80—84頁。

群。其後考察隊步行幾分鐘到達附近的西來寺參觀。西來寺現爲國家級文物保護單位。因時間緊迫,大家匆匆查看,於18:00離開。晚上自由行動。

8月25日　星期一　武威　晴

8:30出發,驅車再赴西來寺,在寺門口與前來接待的張掖學者任積泉見面,在任先生的引領下再次參觀西來寺。此次詳細地考察了西偏殿拱頂的壁畫,壁畫與拱頂的建築年代約在同時,應爲明代遺存,並未發現唐代的遺物。之後,步行約1.5公里到達附近的清代總兵府,爲清代一位總兵的私宅。其後又步行至附近的馬可·波羅街,馬可·波羅街今稱歐式街,街中央有2003年所立的馬可·波羅塑像(圖17)。9:45離開後,乘車八分鐘到一處現存的明清糧倉,爲省級文物保護單位,現在正在修繕。其後步行數公里到尚未開放的張掖鎮遠樓,在任先生的介紹下得以進入參觀。該樓爲明代建築,四面有八塊匾額,一角存唐代銅鐘,氣勢雄偉,有萬國來賓之氣勢。

圖17　張掖馬可·波羅街塑像

11:00離開張掖,驅車向東赴永昌縣。13:20到達永昌縣,在路邊一家麵館匆匆吃完午飯後北行前往北御山(虎頭山)中的聖容寺,15:55到達。聖容寺在永昌縣城西北約10公里處,爲涼州瑞像的發源地,寺中佛殿内現存一處軀體與山石合爲一體的無首石佛(圖18),即爲北周出現的涼州瑞像。1981年,武威城北出土了《涼州御山石佛瑞像因緣記》殘碑,爲唐天寶元年(742)楊播所撰,此碑記載了有關涼州瑞像的故事:北魏太延元年(435),劉薩訶行至涼州番和

圖18　聖容寺涼州瑞像

圖19　聖容寺北塔

縣(故城在今永昌縣西10公里)時,面北頂禮,預言御山崖上當有像出,世亂則像缺首,世平像乃完全。至北魏正光三年(522),佛像果現而無首;至北周時,有石佛首在涼州城東七里出現,迎戴於石佛肩上而絲毫不差。北周皇帝聞之,遂於保定元年(561)下詔在御山山前建瑞相寺,隋煬帝大業五年(609),改名爲感通寺。唐貞觀十八年(644),玄奘自印度東還,途中歇住感通寺[①]。劉薩訶關於涼州瑞像之預言廣泛流傳,敦煌莫高窟231窟壁畫中存有一幅聖容瑞像,題作"盤和都督府御谷山番和縣北聖容瑞像"。寺院南北兩側的山頂上存兩座佛塔,形制與唐代佛塔類似,相傳爲唐塔,北側佛塔較大(圖19),現正在修繕,外壁仍殘存少量壁畫。南側山頂佛塔較小(圖20),並無壁畫,考察隊員依次登上北南兩塔參觀。在前往南塔的河邊崖壁上,可以看見一處用梵文、藏文、八思巴文、畏兀文、西夏文和漢文六種文字書寫的六字真言摩崖石刻[②]。17:30離開聖容寺,前往武威市,19:10到達。晚上自由活動。

圖20　聖容寺南塔

① 孫修身、党壽山,《〈涼州御山石佛瑞像因緣記〉考釋》,《敦煌研究》1983年創刊號,102—107頁;党壽山,《永昌聖容寺的歷史變遷探賾》,《敦煌研究》2014年4期,101—108頁。

② 參張思温,《甘肅省永昌縣後大寺(聖容寺)六體文字石刻》,《西北民族研究》1989年2期,210—212頁。

8月26日　星期二　武威　晴

武威，漢武帝所設河西四郡之一，粟特文古信札記作“姑臧”，十六國時期先後作爲前涼、後涼、南涼和北涼的都城，西夏時爲陪都，馬可·波羅曾途徑此地，稱這裏爲“額里折兀勒”（Ergiuul）：

> 當離開甘州這個我已經告訴你的上述大區，騎行經過一個地區，足有五日程，……在朝日出方向的那五日程的終點，可以看到一個被稱作“額里折兀勒”的王國。它屬於大汗的統治區，也是上述非常大的唐兀大省的一部分，該王國的確包括幾個别的好的王國。……首府城市被稱作“額里折兀勒”。①

8:30考察武威文廟。武威文廟是全國三大孔廟建築群之一，素有“隴右學宫之冠”之美譽，現爲武威市博物館所在地。由於時值文廟大修繕，博物館通史陳列文物已入庫房，僅有石刻陳列室可參觀。此陳列室中展出了十多通武威地區出土的重要石刻，包括《亦都護高昌王世勳碑》、涼州區徵集的《北涼石造像塔》、《康阿達墓誌》、《唐弘化公主墓誌》、《唐吐谷渾王慕容忠家族墓誌》、《元敏公禪師碑》等。其中《亦都護高昌王世勳碑》是最爲重要的元代石刻資料（圖21）。此碑正面爲漢文，背面是回鶻文。碑文詳細記載了火州（吐魯番）畏兀兒亦都護高昌王入仕蒙古後，受海都、都哇侵佔而遷居永昌，及與蒙元皇室聯姻的史實。此碑的漢文碑文最早於1964年由黄文弼先生刊佈②，80年代党壽山先生又撰寫了《亦都護高昌王世勳碑考》一文，糾正了黄先生的若

圖21　亦都護高昌王世勳碑

① A. C. Moule & P. Pelliot, *Marco Polo: The Description of the World*, I, pp. 178-179.

② 黄文弼，《亦都護高昌王世勳碑復原並校記》，原載《文物》1964年2期，收入黄烈編《黄文弼歷史考古論集》，北京：文物出版社，1989，184—193頁。

干錯誤，使碑文内容更加準確[①]。此碑回鶻文部分的研究開始得較晚，直至1980年才由耿世民先生刊佈，他對碑文全文進行了轉寫、翻譯和注釋，並附詞彙表[②]；之後卡哈爾·巴拉提和劉迎勝合撰了《亦都護高昌王世勳碑回鶻文碑文之校勘與研究》，在耿世民研究的基礎上增添了160多個詞彙，確定了一些原先不能肯定的字句[③]。《亦都護高昌王世勳碑》具有重要的史料價值，爲研究元代畏兀兒地區民族、社會及蒙古人在當地的軍事活動提供了依據[④]。

10:00離開文廟，進入旁邊的武威西夏博物館參觀。西夏博物館是一座以河西地區所出西夏文物爲收藏對象的專題博物館，其中最著名的是出自武威大雲寺的西夏碑《涼州重修護國寺感通塔碑》，正面漢文，背面西夏文，碑中記載了武威在西夏時期的地位："大夏開國，奄有西土，涼爲輔郡，亦已百載。"其他重要文物還包括七八件西夏時期的漢文及西夏文文書、3件西夏文符牌、3件藏文衆神祈願木牘。

10:40離開西夏博物館驅車赴大雲寺。大雲寺爲武威最早的寺院，前涼張天錫（363—376年在位）時期在涼州舍殿建寺，名爲宏藏寺，武周時改爲大雲寺，西夏時稱爲護國寺，元末毀於兵燹。明洪武十六年（1383）日本浄土宗僧人志滿來涼州修復此寺。現大雲寺經過1927年地震後僅鐘樓爲明代建築，上有銅鐘一口，不知其年代[⑤]。11:20離開大雲寺，驅車赴海藏寺，11:45到達。海藏寺最早爲薩迦班智達所建涼州四寺之一，但現在的寺院中已無明確的蒙元遺跡，並且轉變爲漢式寺院。唯獨寺内天王殿的"古靈鈞臺"據宿白先生推測應爲薩班時代建築[⑥]，靈鈞臺臺基仍可見時代不早的夯土層。寺内有多通清

---

① 党壽山，《亦都護高昌王世勳碑考》，《考古與文物》，1983年1期，96—101頁。

② 耿世民，《回鶻文〈亦都護高昌王世勳碑〉研究》，原載《考古學報》，1980年4期，515—530頁，收入氏著《耿世民新疆文史論集》，北京：中央民族大學出版社，2001，400—434頁；Geng Shimin & J. Hamilton, L' inscription ouigoure de la Stele Commemorative des Iduq qut de Qoco, *Turcica*, XIII, 1981, pp. 10-54。

③ 卡哈爾·巴拉提、劉迎勝，《亦都護高昌王世勳碑回鶻文碑文之校勘與研究》，《元史及北方民族史研究集刊》第8輯，1984，57—106頁。

④ 參党寶海，《元代火州之戰年代辨正》，《歐亞學刊》第3輯，2002，217—229頁；阿爾丁夫，《關於〈亦都護高昌王世勳碑〉所記"高昌回鶻"方位問題》，《西域研究》2003年3期，65—69頁等。

⑤ 黎大祥，《涼州大雲寺的歷史變遷》，《隴右文博》2008年2期，30—44頁。

⑥ 宿白，《武威蒙元時期的藏傳佛教遺跡》，原載《文物天地》1992年3期，修改後收入氏著《藏傳佛教寺院考古》，266頁。

代碑記，對海藏寺的歷史記述尤詳，寺内最早的建築是最後的大殿——無量殿，殿前左側立《海藏寺藏經閣碑》，右立孫倩撰修葺碑，據碑文可知海藏寺改爲漢式寺院約發生在明成化（1465—1487）時期。

12:05離開海藏寺，驅車赴武威城北永昌鎮石碑村探訪著名的西寧王碑。12：30到達石碑村，訪於路人得知西寧王碑即在道路一側，現爲鐵栅欄圍住，無人看守，可通過栅欄上的孔洞鑽入，考察隊員一一進入，栅欄之内有一塔亭，高大的西寧王碑即樹立其中（圖22）。西寧王碑全稱《大元敕賜追封西寧王忻都公神道碑》，正面漢文，背面畏兀體蒙文，立於元至正二十二年（1362），記述了元朝畏兀兒忻都公家族在河西的事蹟，是現存元代最晚、最完整的畏兀體蒙文石碑，也是河西地區最大的漢文—蒙文合璧石碑。早在1949年，美國學者柯立甫（Francis Woodman Cleaves）就發表了《1362年所立蒙漢合璧忻都公碑》，刊佈了此碑的漢、蒙文拓片[1]，學者多將此碑與前述《亦都護高昌王世勳碑》共同利用，研究畏兀兒民族變遷、元代西北地方軍事活動等問題[2]。西寧王碑是研究元史的重要石刻資料，今日得見原碑，大飽眼福。

圖22　西寧王忻都公神道碑

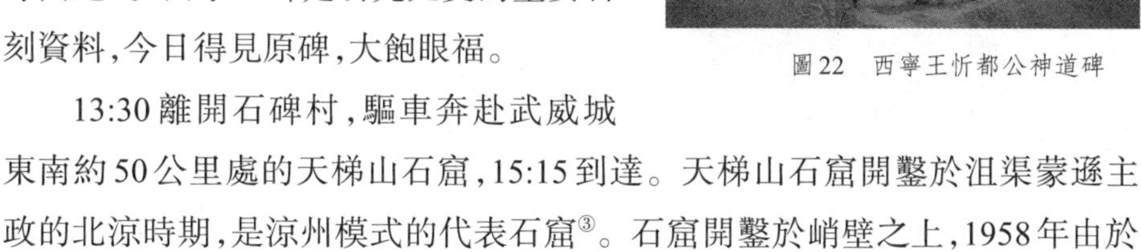

13:30離開石碑村，驅車奔赴武威城東南約50公里處的天梯山石窟，15:15到達。天梯山石窟開鑿於沮渠蒙遜主政的北涼時期，是涼州模式的代表石窟[3]。石窟開鑿於峭壁之上，1958年由於

① 參F. W. Cleaves, “The Sino-Mongolian Inscription of 1362 in Memory of Prince Hindu,” *Harvard Journal of Asiatic Studies* 12/1-2, 1949, pp. 1-133；黎大祥，《元代永昌路故城及碑刻》，《隴右文博》2001年1期，26—28頁。

② 參劉迎勝，《察合台汗國史研究》，上海：上海古籍出版社，2011，259—277頁；賈叢江，《元代畏兀兒遷居永昌事輯》，殷晴編《吐魯番學新論》，烏魯木齊：新疆人民出版社，2006，728—737頁。

③ 宿白，《涼州石窟遺跡和“涼州模式”》，《考古學報》1986年4期，修改後收入氏著《中國石窟寺研究》，39—51頁。

修建黄羊河水庫，曾將整個石窟群的許多塑像與壁畫搬遷至甘肅省博物館，而大佛窟的下半部也被水庫淹沒，直到近些年才在大佛窟周圍修建了一道防滲堤，並對大佛進行了修復，使大佛得以重見天日。除大佛之外，其餘佛窟並未開放，一些雕塑與壁畫仍存於甘肅省博物館[①]。因此，考察隊僅就大佛窟進行了考察。大佛兩側存6身塑像，分别是兩大弟子阿難、迦葉，兩菩薩文殊、普賢，兩天王多聞、廣目，佛頭兩側石壁上殘存壁畫，爲明清重繪。

16:20離開天梯山石窟，前往武威城東南約20公里處的白塔寺，17:07到達。白塔寺是藏傳佛教涼州四部寺之一[②]，因其内建有供奉西藏薩迦派第四任法王薩迦班智達靈骨的白塔而得名。薩迦班智達（簡稱薩班），是元代重要的藏傳佛教大師，1230年代，蒙古軍隊征伐西藏之初，薩班代表西藏各派勢力，攜10歲的八思巴和6歲的恰那多吉，前來河西覲見鎮守涼州的蒙古宗王闊端。在涼州期間，他與闊端建立了良好的友誼，並使闊端皈依了佛教。1247年初，闊端與薩班在這裏進行了"涼州會談"，就西藏歸屬問題達成協議，這是有關西藏歸屬中央統治的重大歷史事件。此後薩班一直駐於此寺，直至1251年圓寂。薩班圓寂後，闊端按照藏式佛塔形式，修建了這座靈骨塔。1999年，中國社會科學院考古研究所與甘肅省文物考古研究所組成聯合考古隊，重點對薩班靈骨塔塔基進行了發掘[③]，對寺内其他遺跡進行了小面積的試掘，將部分地下建築遺跡用玻璃板覆蓋，以供參觀。靈骨塔塔基爲直徑七八米的土坯塔基，頗爲宏大（圖23）。靈骨塔臺基出土文物現陳列於寺内的涼

圖23　薩班靈骨塔

① 參敦煌研究院等編著，《武威天梯山石窟》，北京：文物出版社，2000。

② 參宿白，《武威蒙元時期的藏傳佛教遺跡》，《藏傳佛教寺院考古》，264—266頁。

③ 中國社會科學院考古研究所等，《甘肅武威市白塔寺遺址1999年的發掘》，《考古》2003年6期，52—69頁。

州會談紀念館，包括元代磚雕、龍頭、瓦當、滴水等建築構件，藏文六字真言石刻，黑釉、黄釉瓷器，以及重修塔碑等。白塔寺内的建築在元末毁於兵燹，明以後屢修屢毁，現存塔林均爲2000年之後爲旅遊目的而新修。16:00離開白塔寺，驅車返回武威。晚上自由活動。

8月27日　星期三　蘭州　晴

8:40，考察隊驅車離開武威，沿連霍高速向蘭州進發，12:10到達蘭州市區，但因堵車直到13:30才到達永安賓館。考察隊員辦理入住後，到附近飯館各自吃午飯。陳春曉、胡曉丹因私先行乘機離開，退出最後兩天的考察。除苗潤博隻身前往甘肅省圖書館查閲古籍外，其餘考察隊員皆於14:40乘車赴甘肅省博物館，15:20到達。館内正舉行"絲綢之路——西北遺珍"全國巡迴展覽，包含了陝西歷史博物館、青海博物館、新疆博物館、寧夏博物館、固原博物館、西安市博物館等多家單位的珍貴文物，件件皆珍品，由此可以飽覽各省絲綢之路出土文物的精華。可惜博物館16:40即開始清場閉館，因此未得遍睹，抱憾而歸。晚上自由活動。

8月28日　星期四　蘭州　晴

8:40吃完早飯退房，赴蘭州火車站寄存行李，隨後在甘肅省文物考古研究所李老師的帶領下赴考古所。10:20到考古所，張德芳老師在考古所門口迎接。考察隊員先後參觀了考古所文物修復室、文物保護實驗室，在馬家塬墓地車馬陳列室内觀摩了甘肅張家川馬家塬戰國墓地出土的文物，該墓地的發掘始於2006年，目前仍在進行[①]。該墓地的主人乃戰國時期的戎人[②]，裝點華麗的車馬器及與西方有關的釉陶杯頗爲引人注目[③]。甘肅省考古所對該墓地

---

① 甘肅省文物考古研究所等，《2006年度甘肅張家川回族自治縣馬家塬戰國墓地發掘簡報》，《文物》2008年9期，4—28頁；早期秦文化聯合考古隊等，《張家川馬家塬戰國墓地2007—2008年發掘簡報》，《文物》2009年10期，25—51頁；又《張家川馬家塬戰國墓地2008—2009年發掘簡報》，《文物》2010年10期，4—26頁；又《張家川馬家塬戰國墓地2010—2011年發掘簡報》，《文物》2012年8期，4—26頁。

② 王輝，《張家川馬家塬墓地相關問題初探》，《文物》2009年10期，70—77頁。

③ 馬家塬墓地出土文物可參看甘肅省文物考古研究所編著《西戎遺珍——馬家塬戰國墓地出土文物》，北京：文物出版社，2014。

圖24 考察隊合影

内車馬器和裝飾品採取了整體打包再回室内清理的先進作業方法，清理前先用X射線探傷儀預判，不僅能實現邊發掘邊保護，還可最大程度地保存、觀察出土信息，從而爲精確復原提供依據[①]。

其後在張老師的帶領下參觀了收藏居延漢簡的地下庫房，張老師讓工作人員取出十幾盒木簡供考察隊參觀，他本人則爲大家講解居延漢簡的發掘、整理和出版情况，並就大家關於漢簡的提問一一作答，令大家受益匪淺。

12:00，張老師請大家到馬子禄牛肉麵館吃面。飯後大家同張老師告别，然後乘公車再赴省博參觀，14:30到達。除繼續參觀“絲綢之路”巡迴展外，大家還觀看了通史陳列廳以及佛教藝術展廳的佛教造像，造像中有3件北涼石塔尤爲珍貴[②]。16:40離開省博，各自吃完晚餐，陸續前往火車站，於當晚乘火車離開蘭州，翌日抵京。

馬可·波羅讀書班河西三省區考察之旅至此圓滿結束（圖24）。

附記：本文由沈琛整理考察日記形成底本，由考察隊員傳遞補充，並經羅帥、陳春曉、党寶海、榮新江陸續訂補整理而成。

① 參看韓飛、王輝、馬燕如，《甘肅張家川馬家塬出土車廂側板的實驗室考古清理》，《文物》2014年6期，39—43頁。

② 參看宿白《涼州石窟遺跡和“涼州模式”》，《中國石窟寺研究》，42—46頁。

杭州天目山考察日記

田衛衛

2015年1月24日上午八點，我們一些從北京赴杭參加“馬可·波羅與杭州”會議的師生，如期在入住酒店的門口與陳高華先生等其他從北京、南京趕來的學者，以及數位杭州本地學者集合，約十分鐘後，在杭州文史研究會的組織和安排下，坐上大巴專車，開赴臨安城。此行的目的，是到天目山考察元代的佛教遺址。

説到天目山中的元代佛教遺址，最爲知名的當爲中峰塔院。中峰塔院目前有院牆和中峰墓塔等遺跡，是元代名僧中峰明本的長眠之所。據《八十八祖道影傳贊》《西天目祖山志》等文獻記載，明本禪師號幻住，又號中峰，法號智覺，曾任西天目山住持。明本自幼喜佛事，稍通文墨即誦經，而且常常誦讀至深夜。九歲時其母去世，十五歲時決心出家，元世祖至元二十四年（1287）24歲時在天目山受剃染于禪宗寺高峰和尚。在天目山修行期間，明本常年不懈地堅持白天勞作，夜晚誦經，苦習佛法。而且，他“既得妙法于高峰妙公，唯恐人知而挽其出世，深自韜晦”[①]。元成宗大德元年（1297）至仁宗延祐五年（1318），明本數次出山，四方遊歷，曾先後結庵、船居於蘇浙皖，廣傳佛法，名動當時，王公大臣多致禮從師者。至大元年（1308），賜號法慧禪師。延祐五年，賜號佛慈圓照廣慧禪師，至治三年（1323）坐化，又賜謚普應國師。明本禪

作者學習單位：北京大學歷史學系

① 宋濂，《姑蘇幻住庵記》。

師不僅佛法精通，而且書法頗佳，又能曲善詩，在文學上很有造詣，與當時著名的散曲家馮子振、書法家趙孟頫等名家均交往頗深。明代陳繼儒在《書畫史》中稱明本"書類柳葉，雖未入格，亦自是一家"。屈廣燕《高麗忠宣王與中峰和尚的交往》、林正秋《元代浙江與日本的佛教文化交流史》等研究文章[①]，都記述了明本與高麗、日本來元人士之間的交流與交往。他的不少真跡當時由日本留學僧帶回日本，珍藏至今。此次我們前去考察的中峰塔院，無論是院牆還是墓塔本身，均爲元代所建，對元代佛教史的研究很有價值。

經過大約一個半小時的行駛，汽車到達了天目山景區的大門入口。這裏空氣清新，眼界開闊，令人精神爲之一振。在工作人員的引領下，我們持團體通票魚貫而入，依次通過檢票閘機，步行進入上山小巴候車月臺。因爲上山的路較窄，且據説這段不短的路程上有上百個急轉彎兒，非極其熟悉地形的熟練司機，不能保證安全駕駛通過，故不得不在此處换乘此地所提供的上山專用小型客車。因爲我們到達時其他遊客不多，且又是團體預約，已提前聯絡安排妥當，故集中乘坐了兩台上山專用車，繼續向山頂進發。車行山間，一路上風景無限，雙目應接不暇。或者一個轉彎後看到了蒼茫山谷中展向遠方的煙火村莊，或者又一個轉彎後欣賞了碧波粼粼的高位水庫；或者是晴空碧藍、一眺萬里的山外遠景，或者是白雪皚皚、緑樹成蔭的車前山路。剛開始發車的時候，筆者還默默數記着拐到了第幾個彎兒，可是後來就逐漸目力、心力不濟了，只忙着看風景以及應對連續轉彎兒带来的眩暈感了。

10:20，我們再次到達了一處小型停車場，據稱，此停車場唤作"龍鳳尖"，海拔1100米，是距離我們要考察的開山老殿、中峰和尚塔院等元代佛教遺址最近的停車場了。從此處出發，我們正式開始了徒步進山、踏尋元代遺跡的旅程。

一路按照路徑圖指示的路線前進，進入大樹王景區檢票口大門，沿着右側山路前行，不久即看到了一灣清水，據池畔的石刻記載，這裏叫做半月池（圖1），工作人員介紹説，這個池子後來重新用水泥修葺過，根據當地代代流傳的傳説，元代時這裏曾常有放生的活動，只是後來隨着禪院的走出山間、步

① 《東疆學刊》第21卷4期，2004年；《杭州師範學院院報》2002年第1期。

圖1

入山下紅塵，這裏的放生池也就隨之人煙稀冷，不復當年了。

繼續前行，大約五分鐘後，我們到達了幻住庵。

幻住庵，其前身本是元代高僧中峰和尚的宴坐草廬。中峰和尚在雲遊歸天目之後，爲了達到“誓不歷職”、遠避朝廷打擾的想法，便自稱爲“幻子”，退居此地，食果木之實，飲山澗之水，閉門苦志修行。凡是他所居住之處，均命之爲“幻住”，這裏就是其中的一處。本來久已廢棄，幾十年前重建，稱作“新茅蓬”，現在則是一家高山土菜館（圖2），令我們一行人唏噓不已。

圖2

出幻住庵之後繼續前行，大約十分鐘後，我們到達了開山老殿。開山老殿位於獅子岩東，寺名曰“獅子正宗禪寺”，始建於元至正十六年（1356），壘砌石室，結茅爲廬。在唐朝末年，這裏曾建有寶智、明空等禪院，但直到獅子正宗禪寺建成，天目佛事方才大盛，且名僧輩出，最終成爲臨濟宗的中興之地。元末明初時期，此殿曾先後兩次毀於兵火。清初，禪源寺建成。之後，香火遷至山下，舊址被稱爲開山老殿。

隨人群徐徐而行登上臺階，站在不大的院門口，一眼可見院中建築物的門上高懸着“獅子正宗禪寺”的匾額。在步過不深的庭院之後，邁上三兩級入殿的臺階，可以在殿門口的右側看到長長的玻璃罩式几案，罩内擺放着中國美術學院出版社出版的葉水林主編《趙孟頫書奉天目中峰和尚心經》印本。

進入殿門，首先映入眼簾的，是端坐於大殿正中間高臺上的中峰等三位名僧的塑像，在主位中峰的身後，是上通屋頂、下接台基式樣的一扇大屏風，上面黄底黑字書寫着“禪宗臨濟中興祖庭”及記事，中峰像的頭頂則高懸“江南古佛”四個大字的横匾。與匾額相配的，是三位和尚身前兩側的柱子和對聯：“有幾分證據説幾分話，做一天和尚撞一天鐘”，據稱，此對聯爲1935年胡適親筆所書。從塑像右側繞行長方形的中心塑像壇一周，可看到屏風背面印有大幅清嘉慶十一年(1806)所繪製的天目山地圖，以及散落設展於大殿内的一些文物，例如逐一罩在玻璃罩内的數尊佛頭，據介紹爲唐物，造型有髻有冠，均精美細緻，且下設四方形豎長條細高座，使佛像大致與人肩部齊高，頗利信徒和學者繞行三周，虔心膜拜或觀覽揣摩。此外，還有天目尺八、天目木

圖3

葉盞、袈裟、銅缽等展品，也都各自有其典故和來歷。

步下開山老殿院落大門的臺階，回首門前高懸着的"開山老殿"匾額（葉淺予1984年手書，圖3），此地在元一代鼎盛時期香火旺盛的情景令人遐想。

繼續前行，經過葉水林書"真跡碑亭"碑、冠軍樹，即到達中峰塔院。此地皆爲元代建築，頗顯滄桑。沿着古舊而時見殷緑苔蘚的長條石板拾階而上，行至臺階盡頭，中峰墓塔赫然立於眼前。整個塔院外圓内方，均爲石塊壘砌而成，約2米有餘的石塊壘砌的圍牆，圈出了目測直徑約20米左右的一片圓形的開闊空間。在圓形空間中心處、高約八十釐米的長方形石塊壘砌的台基上，立着一座一面開有小龕的帶屋頂式石壘長方形墓塔（圖4），墓塔包括台基在内高2.5米有餘。小小的開口小龕内還擺放着紅黄相間的人工所製供花。台基前豎立一塊方形石碑，上面題寫有"中峰墓"三個大字，以及數行介紹其事蹟的小字。故宫博物院的孟嗣徽老師、北大的党寶海老師針對幾位同學的提問，對中峰塔院進行了相關知識的現場教學，使同學們得以將一些專業詞彙和現場實物結合，印象深刻地感受到了書本知識的形象化。值得一提的是，墓塔的右前方，即入口一側的院内，堆有一兩堆沙土，據稱是當地準備修葺中峰塔院、以備旅遊開發之用的。這一消息令師生們頗爲修葺活動是否會破壞原址遺建，以及修葺前後的建築風格是否能完美融合一致而擔心了一番。

圖4

離開中峰塔院，行至此路線距離入口的最遠處——大樹王所在地，仰望着高高聳立的樹冠，觸摸着兩三人難以合抱的樹幹，大家感慨不已。熟悉環境的當地學者在附近還給我們指引介紹了一棵高約三十餘米、軀幹筆直的参天雲杉，並表達了希望在場的碩博士研究生們"生活中堅韌正直、治學上勇往直前，最終成爲像這棵杉樹一樣挺拔正直的好學者、大專家"的美好祝願。之後，一行人原路返回至大巴車停車處。

經此天目山一行，不僅眼見了元代佛教建築遺跡，使得大家對書本上的描述有了從抽象到具象的轉變，而且在熱情的當地學者的介紹下，一路認識了國家級保護植物夏蠟梅等不少物種，更從老師們對酸雨造成的天目山植被的變化、以及進一步對一個地區一個民族的影響的討論中，感受到以前自己關心不多的環境史研究的重要。時間雖然短暫，但師生們在天然氧吧、歷史滄桑的自然與人文環境中，不知不覺、潛移默化地受到了薰陶和感染。天目山之行使我們把古代佛教的書本知識和身歷眼見相結合，受益匪淺。

"絲綢之路上的杭州"

——馬可·波羅與杭州第二次研討會紀要

包曉悦　羅　帥　整理

2015年1月25日,"絲綢之路上的杭州"——"馬可·波羅與杭州"第二次研討會在杭州召開。本次會議由杭州文史研究會與北京大學國際漢學家研修基地"馬可·波羅項目組"聯合舉辦,來自中國社會科學院、南京大學、浙江大學、杭州文史研究會的學者以及北京大學"馬可·波羅項目組"的部分師生參加了會議。

會議開幕式由杭州文史研究會常務副會長宋傳水主持,杭州市政府與北京大學歷史學系領導分别致辭。杭州市政協副主席趙光育首先對各位學者的到來表示歡迎,認爲近年來國家提出的"一帶一路"戰略將重現杭州的城市地位,使其深厚文化底蘊再度焕發光彩,希望諸位學者能跨越地域界限,深入地方史研究。北大歷史系主任張帆教授指出,馬可·波羅研究在史學領域具有永久的價值;作爲隋唐宋元時期高度繁榮的城市,杭州同樣具有高度的研究價值;與第一届"馬可·波羅與杭州"研討會相比,本次會議將有更多青年學者發表研究成果,預示着關於杭州城市與馬可·波羅的研究將長期進行下去。杭州市副市長張建庭介紹了杭州正在進行的"尋找當代的馬可·波羅"項目,指出本次會議凸顯了杭州的歷史地位,具有很大的現實意義,並爲杭州的

包曉悦學習單位：北京大學歷史學系
羅帥單位：北京大學國際漢學家研修基地

開放與發展提供歷史借鑒。

會議的學術討論由張帆教授主持。"馬可·波羅項目組"的6位老師在會上作了報告。北京大學國際漢學家研修基地博士後羅帥首先進行了題爲《沉船瓷器與唐宋海上絲綢之路》的報告,梳理了南海與東南亞海域打撈的唐宋沉船的情況,在此基礎上以唐代後期的"黑石號"和南宋的"南海一號"沉船出水瓷器的窯系爲例,探討了唐宋時期遠洋貿易物資的匯聚地、始發港和貿易網絡的變化,以及外銷瓷的若干特點。南開大學歷史學院講師馬曉林的報告《宋元漢藏文化交流之一例——以真武與大黑天爲中心》,通過分析真武與大黑天故事的三個不同文本,探討了元代大黑天信仰的起源與發展,進而考察了宋元時期杭州的真武、大黑天遺跡。北京大學歷史學系党寶海副教授《元代杭州與絲綢之路上的漢文化》考察了兩件元代文獻實物:內蒙古額濟納旗黑水城遺址出土的元代杭州西湖書院泰定元年(1324)刻本《文獻通考》,和新疆吐魯番出土的由杭州漢人刻工刊刻印刷的回鶻文《佛説天地八陽神咒經》,指出元代杭州是文化商品的重要生產地和輸出地。多年研治波斯語言文化的北京大學外國語學院王一丹教授發表了題爲《波斯語和阿拉伯語文獻中的杭州物産》的報告,主要介紹了十三四世紀波斯語、阿拉伯語文獻中有明確記載的出自杭州(行在)的三種物産:楊梅、花緞(kimkhā)和竹器(藤器),重點討論了行在的楊梅。這項研究不僅表明杭州在宋元東西方文化交流中具有重要地位,也反映了波斯語在當時絲綢之路交往中起到了不可替代的溝通作用。故宫博物院研究員孟嗣徽的報告《中峰明本與入元日僧》指出,明州港(今寧波)是佛教東傳日本的要津,尤其在宋元時期,明州港成爲中國禪宗直接東傳的一條要道,並考察了元代先後往天目山隨中峰明本禪師參學的二十一位禪僧,以及他們與日本"幻住派"的密切聯繫和相互影響。北京師範大學歷史學院王東平教授以《西域少數民族語言中的"絲綢"》爲題發言,介紹了《突厥語大詞典》《華夷譯語》《回回館譯語》《回回館雜字》《高昌館雜字》等文獻中收録的與絲織品有關的詞彙,這些字書中的絲織品詞彙反映了西域少數民族對中原絲織品的認知情況。

北京大學"馬可·波羅項目組"的部分研究生也作了發言。北京大學歷史學系碩士研究生張曉慧的報告《〈馬可·波羅行紀〉中關於杭州城市建設的幾個細節》,結合穆勒《"行在":〈馬可·波羅行紀〉注》中對《馬可·波羅行紀》"行在"一節的研究,介紹了書中描述的行在的道路、市場、河渠、浴室等幾個有關城市建設的細節。北京大學歷史學系碩士研究生包曉悦作了題爲《慕阿德與杭州》的報告,介紹了研究馬可·波羅的重要學者慕阿德(A. C. Moule,又譯作莫爾、穆勒等)與杭州的不解之緣。慕阿德早年在杭州的生活經歷深刻影響到他日後的學術興趣,他的不少論著與杭州相關,在一些文章中,慕阿德還會深情回憶起幼年時所見的杭州風物,體現出他對杭州的深厚感情。北京大學歷史學系博士研究生苗潤博的報告《中華再造善本所收杭世駿〈金史補闕〉辨僞》指出,新近出版的《中華再造善本》所收抄本《金史補闕》題清代杭州著名文人杭世駿撰,但實際是一部僞書,從内容到目録皆抄襲《大金國志》。北京大學歷史學系博士研究生求芝蓉報告《徐琰與元初杭州詩壇》,徐琰作爲"東平四傑"之一,繼承了元好問的中州文獻傳統,在至元(1264—1294)後期宦遊江南期間,將中原詞章之學帶到了江南,對江南文壇産生了衝擊,促進了南北文風融合。

除了"馬可·波羅項目組"的成員外,還有其他學者宣讀了報告。南京大學歷史學系高榮盛教授的報告《馬可·波羅所記元代的胡椒與胡椒貿易》,結合馬可·波羅對杭州胡椒消耗量的記載,探討了中國古代的胡椒消費與貿易,認爲元代胡椒消費可能已經平民化,並指出宋代東西方貿易的三大集散地爲印尼群島的蘇門答臘、馬來半島的吉達以及印度半島的古里。中國社會科學院歷史研究所劉曉研究員以《絲路雜論》爲題,介紹了元代杭州市舶司的發展與演變,並根據《元典章》《經世大典》等文獻關於泉州至杭州一線設置水站的記載,探討了泉州貨物北上杭州之路。浙江大學歷史學系王永傑博士的報告《〈馬可·波羅行紀〉對西方中國地理觀的影響》,指出《馬可·波羅行紀》中所記的地名對於西方中世紀地圖的繪製産生了深遠影響,推動了蒙元時期的西方地理學對東方的認識的進步,許多十五六世紀繪製的地圖上都大量引用《馬可·波羅行紀》中的地名。

此外,杭州文史研究會的郭萬年介紹了他們編譯的《日本學者杭州研究譯文集》。該書有六七十萬字,其中收録有關於馬可·波羅研究的文章,以及20世紀二三十年代桑原騭藏的書信。繼浙江大學歷史學系盧向前教授的簡短發言之後，中國社會科學院歷史研究所陳高華教授也發言參與討論，主要介紹了"南海一號"沉船考古的最新進展與成果，他根據近期的實地考察指出，"南海一號"可能是宋末元初的沉船，可與《馬可·波羅行紀》中有關元代海船的記載進行比較研究。陳教授認爲，"南海一號"會對古代海上貿易的研究帶來巨大衝擊，掀起新一輪的海上絲綢之路研究熱潮。

會議最後，由北京大學中國古代史研究中心榮新江教授和中國人民大學歷史學院包偉民教授進行學術總結。榮新江教授首先回憶了泉州沉船出土時的轟動效應，强調了沉船考古的重要性，然後指出有關杭州與海上絲綢之路研究中幾個值得注意的問題：一是海上絲綢之路的研究與内河的研究是分不開的；二是明代海禁對杭州等港口與城市的影響；三是學術研究一定要尊重史實，不能過分拔高五代宋元之前杭州港口的地位。包偉民教授首先談論了海貿轉内河銷售的輻射問題，然後分析了地方史研究對於塑造城市的文化性和城市"城格"的影響。在二位精彩的總結發言之後，研討會圓滿落下帷幕。

研究综览

韓國文學研究者翻譯中國文學專著之經驗

——《陶淵明影像》的翻譯:“按脈式”的閱讀和譯者的“自新”

金秀燕

一、韓國文學研究者翻譯中國文學專著的契機

韓國文學研究者翻譯中國文學專著,很不容易,特别是翻譯古典學術名著。因爲不僅需要熟悉現代漢語,而且需要掌握古代漢語以及中國歷史文化的背景知識。我的專業是韓國古典文學,具體研究領域是朝鮮時代(1392—1910)的小説文學。韓國古典文學作品大多以漢字撰成,所以我平時十分關注與漢學相關的内容。每次訪問中國時,我都會去看看中國古典文學方面的研究專著。2009年,我偶然發現了《古代小説書目簡論》(潘建國,山西人民出版社,2005)一書。雖然是本薄薄的册子,但在韓國古典文學研究領域裏,屬於難得一見的“目録學”著作。在文學研究當中,目録學是需要投入相當功力的研究領域之一,當時一看到這本書,我就對該書取得的具體研究成果肅然起敬。第一次讀這本書的時候,就爲書中所涵蓋的龐大資訊而驚訝,而且覺得應該重新把這本書再讀一遍,於是,我採用了並非“用眼閱讀”而是“以手閱讀”的讀書方式。

作者單位:韓國梨花女子大學校國文科

對於一本好書，我都是“以手閲讀”的，即在閲讀的同時用電腦進行文字輸入。因爲，比起“用眼閲讀”，“以手閲讀”的時候，能夠更好地細細品味作者所寫的字字句句。像《古代小説書目簡論》這樣的外國書籍，我就會一邊閲讀一邊用韓國語輸入到電腦中。就這樣，“以手閲讀”完之後，我得到了譯爲韓語的初稿。我將這文稿介紹給韓國文史學著名學者林熒澤先生。他看後説“這是一部能夠開闊研究者視野的好書”。于是，我策劃將此稿件在韓國出版。後來，這本書以《중국 고소설 목록학 원론》(中國古小説目録學原論)爲題於2010年出版，並在2011年被韓國文化體育觀光部評選为“優秀學術圖書”。

翻譯是最好的精讀方式，因爲翻譯一部著作的時候，字裏行間都要細細查看。而“以手閲讀”的方式，就是精讀的方法之一。“以手閲讀”方法也並非没有淵源。它與17世紀韓國的官僚文人、同時也是小説《九雲夢》的作者西浦金萬重(1637—1692)所提出的閲讀方法相關。金萬重在《西浦漫筆》中，對當代學人的讀書習慣進行了批判。他説“(當代學者)如學醫者，唯讀脈訣，而不按自己三部”，就是説，讀書人好像學習醫術的人一樣，只讀和“脈”相關的書，而連自己的寸關尺三脈都不去把一把。接着他提出“按脈讀書”的方法，其實就是指，要帶着自己的觀點和視角，去好好把握原著的脈絡及意義。

這一點，在將古典資料翻譯成現代文字、或是將外國圖書翻譯成本國文字的時候，尤爲重要。翻譯，是將古書和外國書籍所構成的知識結構，通過另一種語言表達，複製到現代和本國的知識結構的一個過程。因此，翻譯工作必須慎重而準確。接觸了金萬重的“按脈讀書法”之後，我想到了翻譯者需要“按脈”的兩層含義:一是，真如用手切脈一樣，一邊讀，一邊用手去敲打鍵盤;另一個是，去切作者和讀者的心脈，能將兩者結合起來地閲讀。當然，用後者來衡量，難免會有不盡人意的時候。不過，我還是盡力一個字一個字地讀出作者的真意，再重新一個字、一個字如實地傳遞給讀者，就這樣，一點一點進行着自己人文學術著作翻譯的旅程。通過這樣的過程，中國人文學術著作《陶淵明影像》的韓文版誕生了。

二、《陶淵明影像》的翻譯："按脈式"的閱讀與譯者的"自新"

(1) 連接作者和讀者的兩個"按脈"階段

《陶淵明影像》(袁行霈,中華書局,2009)的翻譯工作主要完成於2011年夏天。作爲"國際漢學家訪華研究計劃"的訪問學者,我在北京大學逗留了兩個月之久。在這段時間,我利用第一個月完成了這本書的韓文初稿翻譯。翻譯之前,我先制定了一個爲期20餘日的計劃,規定每天必須完成的頁數,然後動手翻譯。在自己能力所及的情況下,將完成草稿的時限縮至最短,這樣做的目的是爲了掌握原著的文體和表達方面的特徵。在將古書翻譯成現代文字的時候,也是大同小異的。起初,我們對作者的文風和特別一些的表達會感到很陌生,但是通過第一次閱讀(或者草稿翻譯)就會漸漸熟悉起來,也會重新認識之前不甚理解的地方。如果草稿翻譯拖至一個月以上,即使之後重新再讀,仍然很難擺脱對作者的陌生感。如果譯者對原著的作者都感到陌生,則譯著的讀者與原著作者之間的距離會更加遥遠。使讀者和作者接近是譯者的重要任務。

在進行二次翻譯的時候,我投入了大量的時間。雖説是"二次翻譯",裏面包括了多次反復修改的過程。這一階段的工作,在原著所要傳達的意義範圍之内,充分考慮韓國的文化背景和韓國讀者的思維方式,去選擇合適的語句和表達,以此完成對草稿的修改和潤色。必要的時候,更要添加一些説明,來幫助韓國讀者理解相對陌生的中國歷史文化。在二次翻譯的過程中,我還會把原著中的慣用語翻譯成與其意義相似的韓國語慣用表達。有時,也會爲了已經習慣韓國語文章結構的讀者,將原著的語序進行調整;或是爲了方便讀者,而添加一些修辭。這樣做的目的,都是爲了把原著的内容儘量準確、如實地傳達給讀者。不過,我認爲最重要的,還是要讓讀者能夠體會到作者的本意、真心。

舉個例子,韓文版《陶淵明影像》的作者序文,我採用了與正文完全不同的風格。正文採用揭示學術深度和人文情感的客觀謂語;在作者序文中,韓國讀者會感到面前彷彿一位學術界老前輩,態度親切和藹,以敬語語重心長地對年輕學者講述。我認爲這樣翻譯,可以讓韓國讀者更加真切地感受到原著作者的儒雅之氣。

另一個例子是韓文版的書名。"陶淵明影像",如果用韓國語直譯,就變成"陶淵明的印象(도연명의 이미지)"、"陶淵明的形象(도연명의 형상)"之類。這樣的直譯,對耳濡目染於漢字文化中的韓國讀者來説,都是比較熟悉的語言表達。但是,我必須考慮,如何通過韓語的美感來呈現原著的主題思想,怎樣才能將原著所包涵的"對陶淵明的情感",以富有文學性、詩性、繪畫性的表達傳遞給韓國讀者。想了很長時間,最後,我決定將書名翻譯成"도연명(陶淵明)을 그리다[geurida]"。韓語中的"그리다"是一个多義詞,既有"想念某人"的意義,也有"繪畫、畫畫"的含義。韓國人看到這個詞語,馬上就會産生兩種觀念:"繪畫"和"想念"。繪畫,其實是心之所念的。人能畫出來的,其實就是心之所想的。《도연명을 그리다》講的就是畫陶淵明,想陶淵明。因此,我認爲,在韓語"그리다"一詞中,韓國讀者將會在思想感情上領會到原著作者的本意。事實似乎印證了我的這種想法:在韓國,學術書籍出版業的現實情況其實是非常不景氣的,一般以700本的印數進行印刷出版,即使是這樣,多數書籍也仍然會留下相當的庫存。可是,韓文版《陶淵明影像》初版的1000部,竟在十天之内就銷售一空了,並在2013年被韓國文化體育觀光部評選爲"優秀學術圖書"。

(2) 中韓文學傳統的相遇之處

《陶淵明影像》是關於中國東晉詩人陶淵明(365—427)的著作。如果説至今爲止,關於陶淵明的著作,都是將其限制在文學史的大舞臺上進行研究的,那可以説,這本書則將陶淵明置於中國繪畫史的滔滔大浪之中了。從5世紀南朝宋的陸探微所畫《歸去來辭圖》,到清朝黄慎(1687—?)的《桃花源圖》,從大概五十幅畫作中,可以見到各種各樣的"陶淵明"。這些畫作大體分爲三類。一類取材於陶淵明的《歸去來辭》或者《桃花源記》;另一類取材於"採菊""漉酒""虎溪三笑"等與陶淵明相關的逸事;還有一類,則是陶淵明的肖像畫。

當然,這本書並不是單純地展示陶淵明相關畫作。爲了幫助讀者理解畫作,通過引用與畫作相關的陶淵明詩文、或者當代人以陶淵明爲題材所作的詩文,將文學史與繪畫史的交點呈現出來。换言之,這本書是從文學和繪畫相融合的角度,來闡釋陶淵明的人生。他清貧的生活、崇高的人格、卓越的文學世界，不同時代的人對他的詮釋有着怎樣的差異，這一問題在《陶淵明影像》中都得到了盡善盡美的解答。

這本書引用了大量詩文。和繪畫作品一樣，書中的詩文同樣出自古人之手。將中國古典詩詞翻譯成韓國語的過程中，不僅要準確傳達每一個"詩語"的涵義，向讀者傳遞"詩"這種體裁所具有的美感，同樣十分重要。"詩"是一種抒情體裁，具有不同於一般散文的格律。在通過美好的詩語準確傳達詩意的同時，還要考慮到語言的音樂性。這是翻譯漢詩時，最要仔細琢磨、推敲的地方。

爲此，我特意將韓國古典詩歌的格律"移植"到漢詩翻譯中去。詩，不同於散文，所以譯文也要有詩的韻味。古代漢詩大體分爲五言、七言，並講究平仄。韓國的古代詩歌以"音步律"爲詩歌的基本韻律單位，通過"音數律"和"對句"的反復，組成詩歌的節奏。音步律，是將一句詩分爲幾段（即幾個音步），詩中的每一句都通過這樣幾段（幾個音步）的反復，形成詩歌的節奏。韓國的古典詩歌主要以3音步和4音步爲主。也就是説，將一句詩分爲3段或者4段來吟誦。每一段（即每一音步）裏通常都會有3到4個韓國字，有時也會有2個或者5個字。特別是3音步的詩句，多以"3(4)-4(3)-5"爲基本形式。例如下文：

（3音步實例）

狗吠深巷中　후미진 / 마을에서 / 개들이 짖고 //

雞鳴桑樹顛　뽕나무 / 꼭대기에 / 닭이 우누나 //.

——節選自陶淵明《歸園田居》

靖節固昭曠　도연명은 / 진실로 / 밝고 넓어서 //

歸來侶蓬蒿　돌아와 / 쑥대와 / 짝하며 사네 //.

新霜著疏柳　늘어진 / 버들에는 / 서리 내리고 //

大風起江濤　거센 바람 / 불어와 / 강물결 이네 //.

東籬理黄華　동쪽의 / 울타리에 / 국화 심으나 //

意不在芳醪　국화주를 / 염두에 / 둔 것 아니네 //.

白衣挈壺至　흰 옷 입은 / 사람이 / 술 가져오니 //

徑醉還遊遨　어느새 / 취하여 / 다시금 노네 //.

悠然見南山　그윽하게 / 남산을 / 바라보노니 //
意與秋氣高　가을날 / 기상처럼 / 그 뜻 높아라 //

——蘇軾《題李伯時淵明東籬圖》

（4音步实例）
小邑弦歌始數旬　작은 마을 / 풍악 울려 / 벼슬한 지 / 수십 일에 //
迷途才覺便歸身　길 헤매다 / 깨닫고는 / 곧장 몸을 / 돌이켰네 //.
欲從典午完高節　도연명의 / 절개를 / 따르려 / 하기에 //
聊與無懷作外臣　지방의 / 고을살이 / 할 마음이 / 없어라 //.

——葛勝仲《跋陶淵明歸去來圖》

邀陶淵明把酒碗　도연명을 / 청해서는 / 술잔을 / 높이 들고 //
送陆修静過虎溪　육수정을 / 보낼 때는 / 호계를 / 건넜네 //.
胸次九流清似鏡　마음속의 / 구류가 / 거울같이 / 맑으니 //
人間萬事醉如泥　인간세상 / 만사에 / 흠뻑 취해 / 보노라 //.

——黄庭堅《戲效禪月作遠公詠》

3音步的詩句給人輕快之感，而4音步則相對優雅、莊重。3音步和4音步，始于新羅時代（57—935）的“鄉歌”和高麗時代（918—1392）的“高麗歌謠”，直到朝鮮王朝，成爲當代典型詩歌體裁“時調”“歌辭”“民謠”的主要格律。在人們熟悉的調子裏，填上詩句，則會更富有親切感，進而更加容易産生情感上的共鳴。因此，直至今日，韓國的童謠和教科書上的歌辭，大都選用3音步或者4音步這兩種形式。由此可知，音步律對韓國人來説，是非常親切熟悉的傳統詩歌格律，所以翻譯漢詩的時候，把這種格律引用進去，則會讓讀者感受到傳統詩歌的美感。當然，不應該機械地把所有的詩句都套用進3音步或者4音步中去，不過，在意義傳達所允許的範圍内，我儘量在譯文中讓讀者感覺到這種格律的存在。

另外，正文的解説部分，也盡力做到“不散”。有時，如同翻譯漢詩，將敘述内容以句子爲單位，在句子中也採用3音步或者4音步。讀起來流暢的文章，才是好文章。流暢之文，有一共同特點，那就是，即便是散文，也具有歌曲般的節奏,讀起來朗朗上口。韓國語雖然没有聲調,但是通過音步和對句、子音和母音的相互協調,同樣可以組成節奏感很强的句子。爲了確認句子是否讀起來自然而流暢,在進行最終校正的時候,我總是出聲念一遍譯文。

(3) 研究者的自新過程

翻譯是一種精讀的過程。精讀就要做到"善讀"。所謂善讀,並不僅僅指不誤讀作者的本意,準確地理解、把握原著;讀書要在成爲讀者自新的過程之時,才稱得上是善讀。《陶淵明影像》的翻譯,對作爲韓國文學研究者的我來説,是一次拓展研究領域的契機。在翻譯這本書的過程中,我從作者所提出的文學史和繪畫史的交點這一命題中得到啓示,對當前研究的封閉性有了一定的反省。5世紀的詩人陶淵明成爲國際性的文化符號,這引發了我對如下問題的思考:在要求重新認識古典價值的今天,古典文學研究者究竟應該具備怎樣的研究視角。

事實上,很久以前,陶淵明就是韓國人的"朋友"了。新羅元勝王四年(788),在審核官吏考試中有一科叫做"讀書三品科",而收録陶淵明作品的《文選》正是這科考試的内容。新羅的文人崔致遠所作《冬日遊山寺》中有"曾接陶公詩酒興,世途名利已忘機"一句。崔致遠之後,更有無數文人吟誦陶淵明或者寫次韻詩相和。僅僅是對其《歸去來辭》進行和韻的作品就有150餘篇之多。陶淵明的"歸去來"成爲無數文人憧憬的對象,而陶淵明所描繪的"桃花源"更成爲他們所期望的最終歸依之所。因此,在韓國文學研究中,陶淵明也是一個常談常新的話題。但是,在韓國學術界,一直以來,陶淵明幾乎被局限在文學研究範圍内。

《陶淵明影像》的末尾提到韓國畫家安堅(1418—1452)的《夢遊桃源圖》。當時,我想韓國的"陶淵明影像"也一定不少。於是,我決定開始尋找韓國先人所留下的陶淵明具體形象。其結果是,可以收集到從15世紀到20世紀,朝鮮王朝期間所留下的陶淵明相關畫作大約20餘幅。下面就是這些畫作的目録:

- 15世紀

  安堅(1418—1452)

  《夢遊桃源圖》(作於1447年),絹本,彩色,38.6cm×106.0cm,日本天理大學圖書館。

● 17世紀

金明國(1600—?)

《虎溪三笑圖》,絹本,水墨,17.1cm×10.7cm,韓國國立中央博物館。

● 18世紀

1. 鄭敾(1676—1759)

《東籬採菊》(扇畫),紙本,淡彩,22.7cm×59.7cm,韓國國立中央博物館。

《悠然見南山》(扇畫),紙本,淡彩, 22.8cm×62.7cm,韓國國立中央博物館。

《歸去來圖》(屏風畫3幅),絹本,水墨,24.5cm×22.5cm,謙齋鄭敾紀念館。

2. 李夏坤(1677—1724)

《桃源問津》,絹本,彩色,澗松美術館。

3. 咸世輝(1680—?)

《高士策杖》,紙本,水墨,23.9cm×15.3cm,澗松美術館。

4. 尹德熙(1685—1776)

《三笑圖》(作於1732年),紙本,淡彩,29.1cm×18.2cm,韓國國立中央博物館。

5. 李匡師(1705—1777)

《桃源圖》,紙本,淡彩,29.4cm×43.6cm,個人收藏。

6. 崔北(1712—1786)

《虎溪三笑》,紙本,淡彩,29.7cm×21.0cm,澗松美術館。

《高士逍遥》,絹本,淡彩,31.6cm×17.6cm,澗松美術館。

7. 元命維(1740—1774)

《桃園春色》,絹本,淡彩,27.1cm×19.8cm,澗松美術館。

8. 金弘道(1745—?)

《五柳歸莊》,紙本,淡彩,111.0cm×52.6cm,三星美術館(LeeUm)。

● 19世紀

1. 田琦(1825—1854)

《歸去來圖》(作於1853年),紙本,淡彩,34.0cm×109.0cm,三星美術館(LeeUm)。

2. 無名氏

《桃源行圖》,紙本,淡彩,166cm×58.8cm,梨花女子大學校博物館。

3. 金秀哲(1820?—1888?)

《武陵春色》,紙本,淡彩,45.6cm×150.5cm,澗松美術館。

● 20世紀

1. 池雲英(1852—1935)

《撫松觀山》,絹本,淡彩,30.5cm×26.5cm,澗松美術館。

2. 趙錫晉(1853—1920)

《歸去來圖》,紙本,水墨淡彩,38.5cm×114cm,個人收藏。

3. 安中植(1861—1919)

《桃園問津》(作於1913年),絹本,彩色,165.2cm×70.3cm,三星美術館(LeeUm)。

《桃源行舟》(作於1915年),絹本,淡彩,143.5cm×50.7cm,韓國國立中央博物館。

4. 卞寬植(1899—1976)

《武陵桃源》(作於1961年),紙本,水墨淡彩,52.0cm×66.0cm,高麗大學校博物館。

正如袁行霈教授所述,在韓國的陶淵明相關畫作中,《夢遊桃源圖》可以稱得上是首屈一指的。畫家安堅聽了安平大君(1418—1453)的説夢之後,畫了這幅《夢遊桃源圖》。雖是依夢説而畫,但是畫中卻淋漓盡致地展現了安平大君心中的理想世界,也表達出他的雄心抱負。這幅畫作於1447年,時逢世宗大王“治世”的太平時代，那是創制韓文後文化成果倍出的時代，正是那個時代留下了《東國正韻》和《月印千江之曲》等重要文物。在建設世宗時代文化偉業的過程中，當時的世子文宗、首陽大君、安平大君等王子們都發揮了相當的作用。尤其是安平大君作爲當時卓越的藝術家，在詩、書、畫等方面都造詣頗深，連集賢殿的學者們都對他讚歎不已。他在藝術方面的才華還成了古典小説《雲英傳》的主要題材之一。

另外，1438年左右在咸京道邊境討伐野人一事也豐富了安平大君的經歷。像這樣一個具有非凡文化品位的王子，不僅參與朝政，還同當代的精英人才申叔舟、朴彭年、崔恒等一同編纂了《東國正韻》。而陪同自己夢遊武陵的人正是申叔舟、朴彭年和崔恒！這是非同尋常的。安平大君的夢其實正是自己那遠大抱負的具體表現。此畫完成後不久，1450年，孱弱多病的端宗一即位，安平大君就掌控了名爲“黄票政事（王子薦舉人事的權利）”的人事權，作爲朝政中强大的實力派發揮着重要影響力。正是由於安堅深諳安平大君的心思和抱負，才會突破從右向左展開的一般畫卷形式，將左下方作爲原點，視線朝上，在右上方呈現出一座奇異而高聳的山峰。另外，現實世界被濃縮在畫卷下方的入口處，而畫卷正中的大部分空間則展現了王子夢想中恢弘的世界。

圖1
佚名
《桃源行圖》

圖2
安中植
《桃源问津》

《夢遊桃源圖》中的“桃源”既非《桃花源記》中素樸的山野村落,也非文徵明《桃源别境圖》中平静祥和的“世外别境”。和中國歷代畫家所作的“桃源圖”相比,不難想像安平大君夢中的“桃源”是怎樣一番情景。除了安堅的《夢遊桃源圖》以外,李匡師的《桃源圖》、佚名的《桃源行圖》(圖1)、安中植的《桃源問津》(圖2)等都是描繪作者夢中“桃源”的作品。

除了“桃源之夢”以外,陶淵明和他的其他作品也不斷被前人言及並通過繪畫表現出來。其中《歸去來兮辭》和《飲酒》是最受歡迎的次韻原詩,與這兩首詩相關的繪畫作品數量也相對較多。18世紀的鄭敾是我們非常熟悉的著名田園畫畫家,我們幾乎從少年時代就開始學習他創作的真景山水畫的意義。而鄭敾也曾在扇子和屏風上展現過陶淵明的形象。他在扇子上畫的《東籬採菊》(圖3)和《悠然見南山》(圖4)取意於名詩《飲酒》中的名句。以“采菊東籬下的陶淵明”和“悠然見南山的陶淵明” 掀起一陣文化熱潮——吟詩一首,歡飲一杯,世上怎會有比這更瀟灑的生活呢! 鄭敾將《歸去來兮辭》的詩意表現在屏風上的《歸去來圖》,目前僅存《雲無心以出岫》《撫孤松而盤桓》《問征夫以前路》三幅。

圖3　郑敾《東籬採菊》

圖4　鄭敾《悠然見南山》

“歸去來”既是遠離塵世的理想,也彷彿是對現實生活的一種安慰,這一點對畫家來説也不例外。他們不斷地通過畫卷來表現“歸去來”,這其實反應了他們對“歸去來”的一種渴望。

雖然朝鮮時期的代表畫家金弘道以風俗畫成名,但是,他怎不會偶爾有扔掉手中“書堂”和“摔跤場”(金弘道代表作的内容),重新回到故鄉之家的衝動呢? 他的《五柳歸莊》(圖5)也許就是爲表達這樣的情感而作。年輕的畫家田琦畫了一幅《歸去來圖》,據説是受御醫石經李基福(1791—?)之託而創作的,不過,此畫作於他病逝前不久,據此,我們可以想像此圖中包涵着永遠的

"歸去來"之意。在田琦的作品中,我特別喜歡《梅花書屋圖》:在雪白的梅花悄然盛開的山中,有一間格外雅致的書屋,從屋中彷彿傳出磨墨運筆的聲音;另外,有一位友人正步行于小溪的木橋之上,那樣子,好像是在邁過現實和仙境的界限一般。此圖的畫風與桃花盛開的桃源圖相似。在這幅畫中,田琦主要表達了"回歸理想之鄉"的願望,這願望就是"歸去來"和"桃源"。這樣的"歸去來"之夢,一直延續到20世紀初,朝鮮時代最後一位畫家趙錫晉的《歸去來圖》就是最好的證明。現代繪畫藝術家們似乎也没有完全放棄那個"桃源夢"。

除上述作品以外,在韓世徽的《高士策杖》(圖6)、崔北的《高士逍遥》(圖7)、池雲英的《撫松觀山》(圖8)等畫中,都能發現戴著葛巾,拄着拐杖,悠然自在、逍遥一身輕的處士陶淵明。雖然在這些畫作中,並没有指明畫中人是誰,但是從帶着葛巾、穿着道服,或是眺望遠方的深邃眼神,還有平静地撫摸松樹的樣子中就可以毫不猶豫地判斷,那就是陶淵明。在這些畫作中,處士陶淵明有時是一個人,有時則會和朋友們一起登場,《虎溪三笑圖》屬於後者。在金明國的《虎溪三笑圖》、尹德熙的《三笑圖》、崔北的《虎溪三笑》(圖9)中,畫着東晉高僧慧遠、道士陸静修,還有處士陶淵明。"虎溪三笑"的典故是通過慧遠的破戒來闡釋儒、佛、道的匯通,但是比起數萬字寫成的深奥書籍,一幅畫中包含的豪放笑聲,彷彿傳達了更多的故事。

以上圖畫目録、説明及畫作收在韓語版《陶淵明影像》的"譯者後記"中,

圖5 金弘道
《五柳歸庄》

圖6 韓世徽
《高士策杖》

圖7 崔北
《高士逍遥》

圖8 池雲英
《撫松觀山》

圖9 崔北
《虎溪三笑》

這樣一來,韓國讀者就可以和原著正文中所言及的中國陶淵明作品進行比較。在這個過程中,一直拘泥于文學研究者身份的我,得到了拓寬研究視角的機會,認識到以文學爲基礎、兼具對文化和文藝的鑒別能力之必要性。作爲韓國文學研究者,翻譯這本書,是如同把脈一樣的精讀行爲,而在其中則包含了開闊自身眼界的自新過程。

三、翻譯人文學術,傳播人文精神

東西方的所有學術著作翻譯,主導着文明的傳播,是開啓新時代的動力之一。歐洲通過大量翻譯伊斯蘭文獻進而促成文藝復興;中國通過翻譯佛教經文揭開了一個新的時代;韓國通過翻譯儒家經典和新儒學教理書,確立了朝鮮王朝600年的學術根基。學術翻譯是各國學術相互碰撞、相互理解、相互溝通並開啓新時代的出發點。但是,人文類學術著作的翻譯,往往由於價值觀方面存在差異,不能夠快速地進行語言上的轉换並流通。因此,出版過程常常出現一拖再拖的現象。2014年10月在韓國出版的《古代小説作家簡論》(歐陽健,山西人民出版社,2005;韓文版書名爲《中國古小説作家考證學原論》,韓國昭明出版社),其實早在2010年就已經譯完了。當時簽約的出版社是專門出版專業書籍的地方,所以資金方面常常是捉襟見肘,結果,由於出版社的經營狀況不佳,這本書最終没能如期出版。2014年1月,我完成了《西遊記漫話》(林庚,北京出版社,2011)的初稿翻譯工作,但是簽約出版一事仍然比預期晚了許多。

另外,學術界對於翻譯工作的評價也不是很高。根據《關於人文學領域學術成果和評價的標準範式》(韓國研究財團,2014)所載調查結果,參與問卷調查的192名韓國人文學領域學者中,有90%的人主張,在評價研究成績的時候,應該將優秀的譯著也作爲研究成果來進行評價。這反應了到目前爲止,翻譯學術書籍並没有被看做是研究成果中的一部分的現實。這使得有些學者不願意參與需要花費很長時間的翻譯工作。現在應重視翻譯的價值。

世界級的人文學者們,經過長期的鑽研,寫出相關專著,並通過其專著提出促進人類世界發展的進步的人文精神。可以説,翻譯世界級人文學者的專著,這是傳播人文精神的重要工作。比起資本的流通,這是更爲重要的事情。

全真語言的嘉年華會

——余國藩英譯本《西遊記》修訂版問世小識

李奭學

20世紀80年代初,余國藩以英文譯畢《西遊記》,皇皇四巨册,近二千二百頁。此書最大的特點有三:第一,余國藩緊扣黃肅秋在1592年百回本(金陵世德堂本)的基礎上編定的《西遊記》英譯。原文中没有一個字句,没有一首詩,没有一首詞,也没有一首賦,他略過不譯,乃此前所有英譯本中,唯一可稱"真正"的足本《西遊記》英譯,和亞瑟·衛理(Arthur Waley, 1889—1966)諸氏所譯大異其趣[①]。第二,衛理等人乃爲一般讀者而譯,而余譯本卻著眼於學界,是爲學者而譯。所以譯本中《導論》與"箋注"等所謂"副文本"(paratext)並出,充分反映余國藩的學術見解。我們明乎此,再剔除某些余國藩爲保留原文旨趣而做的異化(foreignization)翻譯的句式,則——第三——這套《西遊記》英文之典雅,當世英語譯者中罕見其匹,故而也是内化(domestication)翻譯的典型[②]。上述三個大特點中,尤以《導論》之精與"箋注"之詳,向爲學界稱道。

歲月不居,如今物换星移,三十載歲月匆匆已過,而余譯本《西遊記》的修

作者單位:台灣中研院中國文哲研究所

① 參見李奭學,《兩腳踏東西文化·一心評宇宙文章——〈余國藩西遊記論集〉編譯序》,載余國藩著,李奭學譯,《余國藩西遊記論集》,台北:聯經出版公司,1989,3—5頁。

② Anthony C. Yu, trans. and ed., *The Journey to the West*, revised ed., 4 vols. (Chicago: University of Chicago Press, 2012).此書以下簡稱"2 *JW*"。簡稱之後的第一個數字指卷數,第二個指頁數。以下他書的類似注法,同。

訂版也在2012年秋冬之交出版面世。首版的英譯,余國藩耗時十三載,始克有成。新版也非順手推出:2004年,余國藩爲完成新版,毅然從執教近四十載的芝加哥大學引退,在書房中又窮七年之力,完成這四册二千頁的修訂大業。方之舊版,修訂版英譯本《西遊記》同樣特色顯然。此其間,他還應芝大出版社之請,爲一般讀者整理出一本節本《西遊記》,依舊是特色顯然①。余國藩所修訂者,非唯小説中人物之名的羅馬音譯,由昔日通行的威翟氏系統易爲今日"政治正確"的中文拼音,某些當時爲符合中文語境而不得不生硬譯來的語句,如今也隨着余國藩筆端益發成熟而改得更爲精緻。話説回來,羅馬拼音乃皮相之易,而余國藩的英文頗受古典拉丁文風影響,原本就是當世一流,從用字遣詞到句法皆稱祭酒,不在話下。新版的余譯本《西遊記》最大的特色,我以爲"詮釋"二字,盡得風流②。

有文明以來,我們對於宇宙萬象的看法,幾唯"詮釋"而已。這點中西皆然,但是余國藩以英文詮釋《西遊記》,根本言之,觀念應該衍自歐洲傳統,荷馬詮釋學或費羅(Philo, 約前20—50)到聖奥斯定(St. Augustine of Hippo, 354—430)一脈的《聖經》解經學乃他繼承的西方遺澤。百回本《西遊記》不是童稚遊戲之書,蓋從1592年首見以來,多數傳統批家評者都以"證道書"視之。所證者何道? 這才是癥結,是翻譯時余國藩必需面對的"詮釋"問題。由此延伸者,乃西方人所謂的"寓言"(allegory)或中國傳統文論所稱的"托喻"。余譯本《西遊記》首版的《導論》中,余國藩再三强調是書所述係儒、釋、道三教一家的傳統之見,而這明顯可見的小説家言,自是首回須菩提祖師仙駕所在的"靈台方寸山,斜月三星洞"的象徵意涵③。修訂版的余譯,這方面的强調當在首版之上,這可由版式上特地凸顯(indented)上引兩句話而得窺一斑,蓋版式也是翻譯策略之一。余譯如下:

---

① Anthony C. Yu, *The Monkey and the Monk* (Chicago: University of Chicago Press, 2006).雖然是爲一般讀者節譯,此書仍保留章回小説的回目與插詩等特色,毫不含糊。

② "翻譯"的本質,高達美也有類似之見,詳 Hans-Georg Gadamer, *Truth and Method* (London: Sheed and Ward, 1989), p. 384。

③ 吴承恩,《西遊記》,台北:華正書局,1982,頁10。另見 Anthony C. Yu, "The Formation of Fiction in *The Journey to the West*," *Asia Major* 3rd series, 21.1 (2008): 30-33。華正版的《西遊記》即1954年北京作家出版社黄肅秋編訂本的台灣重印本;此書下文簡稱《西》,頁碼隨文夾註。

The Mountain of Mind and Heart;

The Cave of Slanting Moon and Three Stars.

(2 *JW*, 1: 113)[①]

"靈台"與"方寸"固然都指"心",但是不譯字而譯音,這裹也不能算錯。既然譯之以字,就表示余國藩開顯了山名的隱喻,開始以寓言詮釋《西遊記》。"斜月三星洞"是字謎,乃譯家最難過的關卡,非賴說明不爲功。字謎一旦破解了,"心"的重要性再增,就像小說中須菩提祖師繼之由"猢"字拆到"猻"字,終於決定讓猴王姓"孫"名"悟空"一般。祖師問悟空之"姓",而悟空答以無"性"(《西》,頁11)。在這裹,"姓"與"性"諧音雙關,兩者又變成整本小說衆多的宗教文字遊戲之一,乃寓言的大關目,涉及《西遊記》隨後的人物刻畫。總而言之,《西遊記》首回悟空所遊的山名與洞名象"心",已挑明全書大旨爲何,所寓也不僅是小說接下來屢屢以"心猿"稱之的小說要角孫悟空。時迄於明,中國宗教傳統——無論是釋是道是儒——早已累積出一套以"修心"爲慎獨的重要工夫,而在陽明學派的心學上發展到了巔峰。"心"字代表三教,《西遊記》以此開書,深意饒富。此一托喻式的稗官章法,譯家若老練,筆底必然見真章。

話雖如此,英譯本《西遊記》修訂之後,余國藩對全書的解讀恐怕已把原先所以爲的三教合一的意圖稍稍給降低了。第七十回中,朱紫國王因妖道賽太歲的先鋒吵擾内庭,懷疑悟空是否能轉危爲安,逼得一向好面子的孫大聖賦詩自證,像須菩提祖師一樣,提到他能令"三家會在金丹路"(《西》,頁796)。而這裹的"家",不完全是儒家、道家或佛家那個"家"字,"派别"的内涵强過其他。余國藩原來直譯爲"Three Houses"(1 *JW*, 3: 325),語意太强,修訂版改爲"Three Parties"(2 *JW*, 3: 287),就比較符合此處的文化語境,也表示小說的詮釋生變。講得清楚一點,"Three Parties"可以回譯爲"三個志同道合的團體",比原先所譯更近中文原意。此外,此一名詞的注文也大變,特别費辭要我們回想第二回寫須菩提祖師的往事。祖師登壇"妙演三乘教",他"說一會道,講一會禪,三家配合本如然"(《西》,頁13)。祖師話雖如此,所演此刻顯

① 1983年完成的首版余譯本《西遊記》,本文以下簡稱"1 *JW*"。

遭變質，蓋悟空硬把我們的回想轉成了“金丹之會”，亦即上述“三家”已從儒、釋、道轉爲丹家特重的“精”“氣”“神”(2 *JW*, 3: 508n2 and 4; also see 1 *JW*, 3: 450n6)。“三家”既然“會在金丹路”，所謂“三教合一”，似乎也就變成了小説家的幌子，所寓實則是胡適(1891—1962)寫《西遊記》考證時並不以爲意的“金丹大道”[1]。

在佛教史上，須菩提祖師絶非小角色。他本是佛祖座下十大弟子之一，《維摩詰經·弟子品》特謂之“解空第一”[2]。此所以他爲猴王命名爲“悟空”。話説回來，佛門解空的概念，宋人石泰(1022—1158)的《還原篇》卻以“不識神仙術”而譏之爲“釋門打頑空”(《秘功》，頁300)。或因此故，《西遊記》的敍述者在第一回才説要“打破頑空須悟空”(《西》，頁12)，從而把佛、道結合爲一了——或許還得添加個孔門的訓示！小説所寫的須菩提既有“西方妙相”，也外加了個仙家“祖師”的封號，居然還形容他演的是“三家”的妙義，儼然又具儒門的神韻(《西》，頁10及13)。有趣的是這位“須菩提祖師”教會了悟空神變之道後，隨即尋了個藉口，將猴王逐出師門，還嚴令他在外不得自報家門。此後，“須菩提祖師”就從《西遊記》消失，説來詭異，係小説結構中罕見的出格之舉[3]。我想這點應和他身爲釋迦座下聖僧的地位有關[4]。余國藩或應在這上面做點文章：他師承新亞里士多德學派中人[5]，小説結構是閲讀訓練上的重點。

再談余譯本的詮釋中最重要的丹道。我們細瞧《西遊記》第一回：須菩提祖師命猴王的姓氏時，把“孫”字拆爲“子系”，在隨後的丹道寓言中，馬上又説

---

① 胡適，《〈西遊記〉考證》，見《胡適文存》，4集，台北：遠東圖書公司，1953，2: 390。另見Yu, “The Formation of Fiction in *The Journey to the West*,” pp. 36-37。

② (後秦)僧肇撰，《注維摩詰經·弟子品》，見高楠順次郎、渡邊海旭主編，《大正新脩大藏經》，東京：大正一切經刊行會，1924，38: 349。這一點，陸揚也談過，不過他是從《金剛經》立論，見所著《中國佛教文學中祖形象的演變——以道安、惠能和孫悟空爲中心》，見田浩(Hoyt Tillman)編，《文化與歷史的追蹤：余英時教授八秩壽慶論文集》，台北：聯經出版公司，2009，655頁。

③ 第十七回在一首自報家門的排律中，悟空提到他曾在“靈台山上採藥苗”，而那山上“有個老仙長，……老孫拜他爲師父，指我長生路一條”(《西》，頁192)。但全詩並未托出須菩提之名，悟空算是謹守師命了。

④ 李奭學，《欲望小説·小説欲望：論〈紅樓夢〉與〈西遊記〉》，《國際漢學研究通訊》第7期，2013年6月，67—68頁。此文以下簡稱“李著”。

⑤ 參見李奭學，《兩腳踏東西文化·一心評宇宙文章——〈余國藩西遊記論集〉編譯序》，載余國藩著，李奭學譯，《余國藩西遊記論集》，2頁。

出一句關係全書的關鍵語:"正合嬰兒之本論。"(《西》,頁11)此地的"嬰兒",指的當係對應於"姹女"的道教内丹術語。余譯本首版中,余國藩已十分看重《西遊記》中的丹術,尤其是煉氣還虚的内丹功,且曾另行撰文再三强調之[①]。修訂版裏他濃墨再抹,這方面的强調遠遠超乎我們的想像。余國藩翻修《西遊記》的英譯本,自此幾可謂從丹道寓言出發了。此中和柳存仁(1917—2009)的契合頗有趣。柳氏發表在《明報月刊》上的《全真教與小説〈西遊記〉》[②],把前人著作如《西遊記龍門心傳》中發明未盡之處重予發微,舉證歷歷,而一句"撰寫者是全真教中人"(2 *JW*, 1: 31),已足以令余國藩重探卷帙浩繁的《道藏》,把王重陽(1113—1170)、全真七子及金元兩代全真道的著作再予覆案,爲英譯本從詮釋到注釋增添風華。據我粗估,除了《鳴鶴餘音》等初版中即已詳究的著作外,修訂版中余國藩另行詳探了《洞玄金玉集》《洞淵集》《水雲集》《磻溪集》與《漸悟集》等不下十餘部早期的全真教詩集(2 *JW*, 1: 43—51),而且大有發明。所以修訂後的英譯《西遊記》,再非曩昔魯迅單純所以爲的"神魔小説",而是如斯賓塞(Edmund Spenser,約1552—1599)《仙后》(*The Faerie Queene*)一般寓意多重的"托喻"之作。新版便在此一概念下修正譯文,重新詮釋,而要如此修訂,余國藩當然得採"稠密翻譯"(thick translation)的策略,再用《導論》,用"章節附註"等等"副文本"强化自己的翻譯詮釋學。

瘦辭隱語之外,道教早期如《太平經》——甚至包括魏伯陽(fl. 147—167)的《周易參同契》——中的語言,大多師法漢末開始流行的駢儷文風,騷體的味道强,而用韻的現象更是顯著。不過這些道經用語古拙,語彙神秘,仍然可

① 參見 Anthony C. Yu, "Two Literary Examples of Religious Journey: *The Commedia and The Journey to the West*"; "Religion and Literature in China: The 'Obscure Way' of *The Journey to the West*"; and "The Real Tripitaka Revisited: International Religion and National Politics" 三文,俱見 Yu, *Comparative Journeys: Essays on Literature and Religion East and West* (New York: Columbia University, 2009), pp. 129-203。

②《全真教與小説〈西遊記〉》原載香港《明報月刊》第233—237期(1985),依序爲頁55—62、59—64、85—94、85—90,以及70—74;現已收入柳存仁,《和風堂文集》3册,上海:上海古籍出版社,1991,3:1319—1391。

見文學之美,如今業已研究者衆[1]。全真教雖隸道教,但遲至金代才發軔,宋元則已分有南北兩宗。北宗早期的人物如王重陽與馬丹陽(1123—1183)等人,乃一批苦修傾向頗重的文人或半文人。他們如上所舉的著作,其實長於道情而拙於技術性的論述。舉例言之,《西遊記》第五十回的回前詩《南柯子》,正是全真七子中的馬當陽所填,不過《西遊記》稍有更動,藉以發顯全書的全真氣息:[2]

心地頻頻掃,塵情細細除,莫教坑塹陷毘盧。本體常清净,方可論元初。

性燭須挑剔,曹溪任呼吸,勿令猿馬氣聲粗。晝夜綿綿息,方顯是工夫。

(《西》,頁575)

這首詞除了强調"心"的重要外,表面上没有一語提及修行内丹的方法,也就是只講"性",不講"命",乃典型的金代全真教的文學代表。余譯本《西遊記》初版,也只當是道情之作而英譯之。"曹溪"是禪宗六祖慧能(638—713)弘法之處,而《南柯子》打一開頭實則就在回應惠能擊敗神秀(601—675)的那首著名的佛偈,以故像《西遊記》結合佛、道一樣,又讓全真的堅忍搭上了佛教的禪心。北宗早期的文學,《南柯子》當可窺一斑而見全豹,唯《易經》以還的丹術談得確欠火候。

余國藩的英譯本《西遊記》初版時,我想並未意識到《南柯子》乃馬丹陽所填,所以譯罷之後,僅在"曹溪"上加了一條小注(1 *JW*, 2: 438n1),要求讀者參看《導論》中另一條注釋(1 *JW*, 1: 513n8),而這條所注也不過如上我之所述,並未將全詩那早期全真的況味盡情道出。三十年後,余國藩對《西遊記》認識

---

① 劉祖國,《漢語學界道經語言研究的回顧與展望》,《漢學研究通訊》第32卷第3期,台北:漢學研究中心,2013年8月,8—16頁。有關《周易參同契》所用之韻,參見蕭聖中,《周易參同契的韻轍與平仄》,載《宗教實踐與文學創作暨〈中國宗教文學史〉編撰國際學術研討會論文集》,3卷,高雄:佛光山,2014,1: 427—433。

② 這首詞見於《正統道藏》787《漸悟集》卷下,台北:新文豐,1985,1—2頁。相關探討及其改動處可參見柳存仁,《和風堂文集》,3: 1344。但柳氏以爲《南柯子》無關第五十回,這點我覺得仍有商榷的餘地。

大增,《南柯子》的絶妙處他已可擇其精華而譯出。所謂“工夫”,他也了解係全真借禪修所求的内丹之道(“internal and meditative alchemy,” 2 *JW*, 2: 392n1),而緊隨此一體認而來的,當然是重修舊譯,而且幾乎是全盤改譯。余國藩所耗功夫之大,可以想見。

全真北宗始自王重陽的努力,包括信仰與修持等等,借一句吴光正的話,乃“多以口傳密授和文學書寫的方式實施和呈現”,不但示範了苦修道士苦行的工夫,也刷新了整部道教文學史①。然而這些口傳密授和文學書寫必定有其目的,没有撰者不想登上最高的修煉境界而成仙了道。談到這一點,修煉者豈可没有丹道口訣等技術性的論述,豈能不探討修“性”與修“命”的進境?包括馬當陽——甚至是王重陽——在内的金元全真道,確實不擅類此的論述。所幸他們仍能在中國道教文獻中自尋“師承”,而就《西遊記》言之,這類修“命”的技術文本多從《易經》出發,我以爲最稱典型者有三:一爲東漢末年魏伯陽的《周易參同契》;二爲宋人張伯端(987—1082)的《悟真篇》;第三是余國藩英譯《西遊記》時可能未及參看的一本副文本頗衆的專著,亦即上舉石泰的《還原篇》及其元明從者的箋注。

我們猶記得《西遊記》首回,須菩提祖師擬爲悟空覓一“姓”氏。他因悟空爲猴,故而從“猢猻”替他取“姓”爲“孫”,又把“孫”字拆爲“子系”,其後隨口便評道:“正合嬰兒之本論。”余譯本譯“子”爲“boy”,把“系”解爲“baby”,其實都是在爲那“嬰兒之本論”(“fundamental Doctrine of the Baby Boy”)預奠理論基礎(2 *JW*, 1: 115),以便接下來可借翻譯申論内丹之道。上引括弧中的“Baby Boy”,是認識“嬰兒”真諦之後的直譯,英文讀來應有慣用法之異,就好像把“效犬馬之勞”(《西》,頁40)直譯爲“to serve one as dogs or as horses” (2 *JW*, 1: 150) 一樣生硬,講英語的人士讀來或感“洋腔洋調”,而是的,學術翻譯的譯者常見類此的直譯策略,我們無足爲異,何況由上述“Baby Boy”讀來,我們應該也可了解須菩提何以會由佛前高弟變成全真道士口中的“祖師”。

至於“嬰兒”一詞,丹家初則取法老莊,形容青春永駐,如人之初始的嬰兒

① 吴光正,《苦行與試煉——全真七子的宗教修辭與文學創作》,《中國文哲研究通訊》第23卷第1期,中研院中國文哲研究所,2013年3月,39頁。

狀態[1]。《參同契》第八十章《法象成功》謂"升熬於甑山",必不使火妄動,必壓之而使如"嗷嗷聲悲"的"慕母"之"嬰兒"[2]。劉一明稱這裏的瘦辭所喻象火之"燥性消化",亦即修煉者已回復那出生之"專氣致柔"之態。"嬰兒"一詞,此後在丹道上因指氣功化境,相對於我上文曾經提及的"姹女",而後者自亦出乎《周易參同契》,是《二土全功》一章中所謂"老翁復丁壯,耆嫗成姹女"之所指(《注》,頁61)。在《西遊記》裏,"嬰兒姹女"這對名詞出現的次數甚多,多到由不得讀者不加注意。《周易參同契》借《易經》論命,"嬰兒姹女"遂變成全真道修命的關鍵字之一。"嬰兒"不用再贅;"姹女"者,"美女"也,余國藩從首版起就譯之爲"fair girl"(e.g., 1 *JW*, 4: 70 and 123; 2 *JW*, 4: 63 and 96),既合本意,也頗傳神,因爲《西遊記》中此詞多指漂亮的女妖,不但青春長駐,同時也指她們在修煉上所欲化至的理想境界。不過在丹道的技術上,"姹女"指"汞"。魏伯陽嘗謂"河上姹女,靈而最神"(《注》,頁69)。然而魏氏不作此說便罷,說了反而令後世論者衆說紛紜,隔代互駁:中野美代子認爲魏氏之説典出"黃河邊上"的仙人"河上公"的舊説[3]。清人劉一明乃全真教龍門派傳人,嘗融攝儒佛,另從《詩經》的《王風·大車》中那"淫奔之女"而詮解道:"姹女者,離中之陰,在人心爲人心之靈",但也因"靈"而演爲"識神"。其本色正如"水銀"而"流性不定",故而會"逐風揚波"而"喜動不喜静",只要"心火稍起,則飛揚騰空"[4]。方之中野的溯源,劉一明係就技術面而論"姹女",我以爲較合丹道之理。

"姹女"既然相對於"嬰兒",而"嬰兒"——即使不就比喻而就其實體言之——可想仍寓南宋道士蕭應叟(fl. 1226)所謂"陰剥陽純"的修練結果,受佛教戒欲觀影響大,而此即《西遊記》第八十三回"陽精月熾,真身化生"所成就的

---

① 李叔還,《道教大辭典》,杭州:浙江古籍出版社,1990,頁242"嬰兒"條。另參同書頁240"姹女"條。

② (東漢)魏伯陽著,無名氏注,《周易參同契注》,見張繼禹主編,《中華道藏》48册,北京:華夏出版社,2004,16: 73。《周易參同契注》以下簡稱《注》,引用頁碼隨文夾註。

③ 中野美代子,《西遊記の秘密》,東京:福武書店,1987,94頁。《西遊記の秘密》以下以作者姓簡稱"中野"。

④ 魏伯陽著,(清)劉一明解,《參同契直指》,台北:真善美出版社,1983,7及56—57頁。另見陳子展,《詩經直解》,台北:書林出版公司,1992,223—226頁;以及余國藩著,李奭學譯,《重讀石頭記:〈紅樓夢〉裏的情欲與虚構》,台北:麥田出版,2004,132—134頁。《參同契直指》以下簡稱《直指》。

"聖胎"更上層樓的表現[①]。"嬰兒"與"聖胎"這兩個名詞,余譯修訂版以爲"同"(none other than; 2 *JW*, 1: 86),理解和蕭應叟有異,不過相去並不遠,因爲後者長成後才會變成前者。所以修訂版《西遊記》的《導論》裏,余國藩並不以"嬰兒"今日通譯譯之,而是解爲"小娃兒"(baby boy)。類似之譯,書中例子還真不少(e.g., 2 *JW*, 1: 86-87 and 1: 115),比首版僅以"baby"說之强多了,可見三十年來余國藩對丹經的認識已臻化境,閱讀愈廣愈博,也愈精愈深。

有關"聖胎"一詞,第八十三回示範最佳。回中悟空化身爲桃子,計騙陷空山無底洞的地湧夫人吞下,從而在她腹内大肆淩虐了一翻。待他跳出口外,兩造當又是一陣廝殺。此時敍述者再賦一詩,其中有對句如下,吟唱目的分別爲鼠精及猴王修煉之所欲,而"聖胎"一詞出焉:

> 那個要取元陽成配偶,
> 這個要戰純陰結聖胎。(《西》,頁943)

地湧夫人是靈山鼠精化身的"姹女",在陷空山苦等唐僧三百年,因爲她本擬匹配這長保元陽的嬰兒。悟空代三藏出戰,但他可也不是個空殼的行者,當有意效法乃師,讓自己也變成是位剥陰存陽的嬰兒。上面這兩句詩,余譯本修訂版並未改動,連"聖胎"亦英譯爲"holy babe",把"胎"字給實體化了。修訂版注"聖胎"依舊,道是隱喻"内丹論述中修煉成仙之境"(the attainment of realized immortality in the discourse of internal alchemy; 2 *JW*, 4: 392)。不論譯文或注文,余國藩從之而作的翻譯詮釋,我以爲都無可挑剔,幾達爐火純青的化境。

話雖如此,譯文中有一處,我仍然覺得或可稍稍再議:

> That one desires primal *yang* to be her mate;
> This one fights pure *yin* to form the holy babe.
>
> (1 *JW*, 4: 123; 2 *JW*, 4: 113)

坦白說,這一聯譯得甚佳,不論平仄(metrics)或尾韻(rhyme),余國藩都押得

---

① 這裡有關"嬰兒"與"聖胎"之别,引自(宋)蕭應叟,《元始無量度人上品妙經内義》,見張繼禹主編,《中華道藏》,3: 552。

極其工整，漂亮之至，可比18世紀英國的新古典詩風(Neo-classicism)，尤其是蒲柏(Alexander Pope, 1688—1744)式的英雄雙行體(heroic couplet)。難怪修訂版一仍舊貫，未曾再修。兩句詩中相對的動詞一爲“取”，用“desire”而不借字典直譯之，高明得很。我吹毛求疵的是“戰”字的譯法。此字這裏未必全指“作戰”而言。此“戰”之目的乃在“結聖胎”，所以——除非我們以道教房中術看待這兩行詩——此“戰”反而有“力拒”(fight against)或王重陽所謂“戰退”狐魅之意。全真教係道教中的堅忍學派，色戒斷不容犯。詩中的“戰”字固有“性”的聯想，然而魏伯陽以來，丹詩中每每容許性意象，筆法一貫，也一致，而全真道用得更多，王重陽乾脆稱之爲“假名”[①]。此一“假名”，劉一明的説法是“托物寓言”[②]：全真道若不如此看待丹經中的性意象，則如此之“戰”怎又可能“結聖胎”或——延而申之——“結嬰兒”？全真典籍筆下的“嬰兒”乃“聖胎”化育而成，而清代全真道士閔一得(1758—1836)訂正的《尹真人東華正脈皇極闔辟證道仙經》對“嬰兒”的看法，則從《性命圭旨》而有另解(1: *JW*, 508n1)，謂之乃釋門的“法身”，猶言其性體清净無垢，一塵不染[③]。

《西遊記》中刻畫的“姹女”，陳士斌(fl. 1692)的《西遊真詮》踅回《易經》，稱之爲“坤象；坤得乾之中爻而成坎，乾易坤之中爻而成離”。職是之故，“姹女”才會變成丹術中“汞”的隱語。“文王後天八卦”形成於唐末，係僞託的《易》象之論。其中陰性化了“離卦”，陽性化了“坎卦”，因使“離女”對應於“坎男”，在四獸圖中且解之爲“青龍”，使之與“白虎”交媾而結成金丹(中野，頁79)。凡此種種，張伯端的《悟真篇》道得最精，而《西遊記》亦常引之。我們從上述名詞再看，張派全真當然大受《易經》的影響。其實也是：《西遊記》開書的宇宙觀便從《易經》而來(《西》，頁1—2)，走的是《悟真篇》一脈的思想。三藏十世修行，“元陽未泄”，而此亦《西遊記》因《周易参同契》而察悉的修命之道。

---

① 以上王重陽語，見(金)王喆，《重陽真人金關玉瑣訣》，載張繼禹主編，《中華道藏》，26: 395及397。

② 見(北宋)張伯端著，(清)知幾子(仇兆鰲)纂輯，《悟真篇集解》，載《秘書集成》第3册，北京：團結出版社，1994，27頁；另參2 *JW*, 1: 91—93。《悟真篇集解》以下簡稱《悟》。

③ (清)閔一得訂正，《尹真人東華正脈皇極闔辟證道仙經》，見高雅峰等整理、編校，《道藏男女性命雙修秘功》，瀋陽：遼寧古籍出版社，1994，125頁。這位“尹真人”，應指全真道第六代掌教尹志平(1169—1251)。我所用的現代版，係青羊宮傳抄本。《道藏男女性命雙修秘功》下文簡稱《秘功》。

余國藩的《導論》於此有深論(2 *JW*, 1: 82-95; 525n5 for chap 17; and 528n10—529n12),不過張伯端此處的指涉,反而源自佛門,並非《易經》。

所謂"修命",當然是要延年益壽,長生不死,《悟真篇》就此所論,多用隱語,此元人陳觀吾所以謂之"多假喻辭而不截然直指"(《悟》,頁20;另參見2 *JW*, 1: 91—93)。佛教以色爲戒,崇拜真陽,但在《悟真篇》中,張伯端處處反使之帶有性的暗示,"龍虎交媾"與"捉坎填離"等詞故而看似南宗偏支的房中術,從而又比爲"姹女嬰兒"與"金烏玉兔"的結合,也就是《西遊記》第二回繼之所稱的"龜蛇盤結"(《悟》,頁176)。張伯端還指出這些對比的喻詞非得"匹配成既濟"不可,蓋其"本是真陰陽",故而"夫妻同一義"(《悟》,頁117)。佛教和全真南宗的矛盾,《西遊記》借《悟真篇》如此調合,編次者顯然不含糊。儘管如此,這種調和卻生出一極端的弔詭,即西行道上多數的妖魔幾乎都不問原由,從第十三回三藏初離兩界山,進入西域那蠻陌之境開始(《西》,頁152),便設爲他們修煉上的終極目的,以爲吃了這十世修行的和尚的肉,便可正壽長生而金丹得度,化爲張伯端所説的紫微"太極之精"了[①]。在這種狀況下,内丹術在《西遊記》中卻諷刺地反轉成了葛洪(284—363)等外丹家提倡的服餌之術,亦即以玄奘爲還丹金液,而食其肉便可"得道飛升,畢天不朽"[②]。

我們認識《西遊記》,張伯端的《悟真篇》才是相對重要的理論鑰著,對小説成書影響尤大,而這相關種種,余譯本《導論》與各分册之章節附注,敘之甚詳,修訂本就不用贅述了。《西遊記》開書須菩提祖師所教的"逆修之法",係所謂"攢簇五行顛倒用"在小説中的實例解説(2 *JW*, 1: 509n10)。此等"逆修之法"無不攸關性命,正是《西遊記》令孫悟空自稱"齊天大聖"的諸多原因之一,蓋《悟真篇》繼之所述乃"温養十月,以成真人,與天齊壽"(《悟》,頁176)。悟空撐起"齊天大聖"這張大纛,並非太田辰夫就第四十一回故事表面悟空那"平天大聖"一干舊交六人的民間傳説所致[③],而是有其丹道意義上的成因。

---

① (宋)石泰著,閔一得授,閔陽林述,蔡陽倪訂,《還原篇闡微》,見《秘功》,299頁。另見吴光正,《神道設教:明清章回小説敘事的民族傳統》,武漢:武漢大學出版社,2012,81—86頁。

② 參見李著頁63—64。另見(東晉)葛洪著,李中華注譯,黄志民校閱,《抱朴子》,2册,台北:三民書局,1996,1: 83—126;以及柳存仁,《全真教和小説〈西遊記〉》,載於所著《和風堂文集》,3:1319—1391。

③ 太田辰夫,《西遊記の研究》,東京:研文出版,1984,80頁。另見《西》,471頁。

簡言之，這面大旗同樣涉及性命雙修。在“性”上，這旗幟乃因靈霄寶殿上玉帝輕賢，亦即悟空對“弼馬瘟”一職的極端反動煽起（《西》，頁41）；在“命”上，則是天上數日，下得凡界卻是年壽飛逝，而如此仙凡不一，悟空那裏心甘，是以務必討個公道所致。雙題迸發，猴王當然要來個“與天齊壽”，做個跳出輪迴，在天界爲人敬重的神仙了（《西》，頁40）。余譯本中，“齊天大聖”一詞均作“the Great Sage, Equal to Heaven”（e.g., 2 *JW*, 1: 151），確把上述寓意網羅殆盡，不失爲佳譯。

不過《悟真篇》還有令《西遊記》的作者更難釋懷的地方。用基督宗教的觀念講，“齊天大聖”就像興立一座巴別塔（The Tower of Babel），就像上帝座下反出天界的大天使露西弗爾（Lucifer），幾乎是“驕傲”（*hubris*）的代名詞，絕不容於天地間。《西遊記》中倒没有如此嚴重的罪宗首端，所以玉帝下令開打的那場“天上的戰爭”（heavenly war），就不比密爾頓（John Milton, 1608—1674）《失樂園》（*Paradise Lost*）裏的那一場[①]，而兵敗既成事實，“齊天大聖”一稱遂將就授與悟空了。通《悟真篇》全書，下面一詩或許才是“齊天大聖”一名靈感所自，至少有詮釋上的聯繫，英譯本《西遊記》兩版倒都忽略了：

了了心猿方寸機，
三千功行與天齊。
自然有鼎烹龍虎，
何必擔家戀子妻？

（《悟》，頁247—248）

我在他文中也曾提過這首詩，同樣認爲攸關“齊天大聖”這個名頭（李著，頁68）：“三千”之數，在此不過形容功行之多，而“心猿”一方面結合“方寸”，回應了《西遊記》首回的“靈臺方寸山”，另一方面又告訴我們修行得力，金丹練成，“心猿”在性命上都可以“與天齊”。是以上引《悟真篇》中的詩，不管從哪個角度看，在在教人都難以不想起孫大聖！全詩的後兩句，不論是丹家的喻詞或“行者”的自了漢性格，悟空又無一不能擔當。余譯本爲悟空做了許多源流考

① John Milton, *Paradise Lost* (Oxford and New York: Oxford University Press, 2008), VI.

證,“齊天大聖”的由來,上引詩或可供參考。

《西遊記》名爲“遊記”,實爲“心法”。《心經》在小説中的關鍵地位,因此而得。小説爲强調“心”的重要,除了文前提到的字謎外,不惜長篇抄録所謂《多心經》(即《心經》),而這點《悟真篇》早已有呼應,不但作《〈心經〉頌》以頌之,還吟出了《無心頌》以和之,而《即心是佛頌》在小説中雖經小幅改易(《悟》,頁302—306),幾乎句句也都抄爲第十四回的回前詩(《西》,頁153)。《悟真篇》在《西遊記》中之爲德也,由此又可窺一斑。凡此,余譯《西遊記》的修訂版當然不會放過(2 *JW*, 1: 13),而且闡述精甚,已非首版可比。

上述“太極之精”一語,我從石泰的《還原篇》引出。石泰嘗於驛中遇張伯端,知仙之可習,其書乃成(《秘功》,頁273)。張伯端走的既屬丹道的技術路線,《還原篇》當然不會有如全真七子而以道情爲内文主旨。石泰認爲修真者“只尋身内藥”便可,實則“不用檢丹書”(《秘功》,頁304),《還原篇》所傳者,因此可稱全真教内的禪宗。話説回來,《還原篇》中的口訣也最多,不檢丹書卻又以注解之,這弔詭要理解可不易。前文述及悟空變成桃子,計騙地湧夫人吞下,而他真身一現,在地湧夫人腹内左踢右踹,讓她疼得滿地打滚,最後只好告饒投降。《西遊記》第八十三回這段故事,看在全真道眼裏,當然也是寓言,陳士斌引《悟真篇》形容之曰:“果生枝上終期熟,子在胞中豈有殊?”這兩句詩,實爲《悟真篇》中張伯端另一首七律的下半部,其上另有兩句云:“虎躍龍騰風浪麤,中央正位産玄珠。”(《悟》,頁77—78)這兩行詩,我在他文同樣也談過(李著,頁67)。余譯修訂本的《導論》從李約瑟(Joseph Needham, 1900—1995)的《中國科技史》(*Science and Civilisation in China*)第五册取了一幅佛陀像,其中但見一嬰兒從其袈裟中探出頭來[1],而如此繪圖實乃以圖像喻丹道(2 *JW*, 1: 87),所謂“玄珠”,正是這裏的“嬰兒”或其前身的“聖胎”。質而再言,整首詩不啻前述“龍虎交媾”的隱喻,而《還原篇》此處説得更精,更近《西遊記》的丹道瘦辭:“姹女乘鉛虎,金翁跨汞龍,甲庚明正令,錬取一爐紅。”(《秘功》,頁281)修煉技術既如此,“姹女”和“汞”及“青龍”的關係在短短兩行

① Joseph Needham, *Science and Civilisation in China*, vol 5 (Cambridge: Cambridge University Press, 1954), pp. 81, 83-84, and 90.

的詩内便道盡，而“嬰兒”即“金翁”，正是丹術中的“鉛”或“白虎”。全真南宗的正統雖也不近女色，卻比北宗更好性意象，藉以隱喻修煉的技術。石泰這兩行詩講陰陽合和，看似合乎《易》道，卻不合丹道，比喻的仍爲《西遊記》開書須菩提祖師教給悟空的“逆修之法”，亦即所謂“攢簇五行顛倒用”在小説中的實例解説。《還原篇》是一套八十一章的五言詩，講“金液交結，聖胎圓成”（《秘功》，頁273），除可印證内丹修行的大、小周天之數外，也在回應唐僧在《西遊記》中的八十一難。若視爲《西遊記》的作者藉以形塑故事的丹術間架，應該也説得通。

《西遊記》中，三藏嘗借住烏雞國寶林寺。夜來他見銀輪當空，不禁賦詩懷歸。余譯本的《導論》對這段插曲特感興趣，從舊版起就頻見討論（e.g., 1 *JW*, 1: 49）。三藏才剛感歎何日可以取得真經返故園，悟空便近前爲他説明月家之意。悟空話中頂真所述，無非是晦朔間，月之上弦與下弦俱有其意，牽涉到所謂“先天采練”之術：“前弦之後後弦前，藥味平平氣象全”。這首詩變化自《悟真篇》所吟[①]，涉及養丹之道。悟空繼之所謂“采得歸來爐裏煉”，不過用了個“爐”字，指顧間就把外丹轉成了内丹。繼“煅成温養自烹煎”（《悟》，頁187）之後的“志心功果即西天”（《西》，頁419—420），有趣的卻又把佛、道綰爲一體（2 *JW*, 1: 67—68）。全真道雖稱“道”，生活中的實踐卻是三教一家，或將三教宗旨解爲自家大道。元初道士李道純（fl. 1306）著《中和集》，書題本身即出自《禮記》，亦因所習而以釋家之“圓覺”或“真如”、儒家之“太極”與道教的“金丹”視同一體，認爲唯“體同名異”罷了，不必强分[②]。《西遊記》中另可見諸上引詩後沙僧吟出的“三家同會無競争”，也可因悟空的議論而得窺其真義：“我等若能温養二八，九九成功，那時節，見佛容易，返故田亦易也。”（《西》，頁419—420）這些故事情節，余譯兩版都强調，可見重視，而就《西遊記》的詮釋來講，也是慧見。

---

① 在1 *JW*, 1: 49中，余國藩引的是南宋理宗淳祐己酉年（1249）王慶升的《爰清子至命篇》（《道藏》，84: 742），不過王氏卻道自己所從乃紫陽真人之作，亦即張伯端的《悟真篇》。

② （元）李道純，《中和集》，見《正統道藏》，4: 497。另參郭健，《取經之道與務本之道：〈西遊記〉内丹學發微》，成都：巴蜀書社，2008，84—86頁。《禮記·中庸》云：“喜怒哀樂，未發謂之中，發而皆中節謂之和。”見華義姜譯注，黃俊郎閲校，《新譯禮記讀本》，台北：三民書局，1997，737頁。

“温養二八”與“九九成功”俱屬煉丹術語，因旨隱意晦，歷來解法不一。中野美代子有驚人之論，她把《西遊記》中的丹詩視爲房中術的托喻，“二八”遂解爲年在“十六”的處女，是採陰補陽最佳的丹爐(中野，頁108)。由是中野又予引申，把“九九”當成至高聖數，隱喻“龍虎交媾”的性高潮。從好談房中術者的角度觀之，中野的説辭頗能服人，而《西遊記》不就變成一部房中術的寓言？唐三藏經歷的八十一難，經此觀照，跟着也變成了性交逐步走向高潮的長篇隱喻了(中野，頁118—119)。儘管如此，反對如此看待丹經者仍有其人，葛洪在《神仙傳》中態度清楚：如此閲讀《周易參同契》，他難以苟同，故謂“世之儒者不知神丹之事，多作陰陽注之”，正可謂“殊失其旨”[①]。後世《悟真篇》的注者，多數持論相同。閔一得(1758—1836)向以北宗龍門後人自居，所著《還原篇闡微》首開以《易》卦象數看待“二八”的風氣，是以若非視之爲日期，就是證爲鉛汞等的藥量。余國藩在舊譯中走的詮釋路線，原爲乾隆年間的全真道董德寧的《悟真篇正義》，把“二八”解爲月之前弦與後弦(1 *JW*, 2: 430n2)，不過董氏雖從時間解釋此一名詞，所解卻和接下來的詩重複而自相矛盾了。故而修訂版中，余國藩修訂自己的前見，同意以藥量解釋這裏的“二八”，符合動詞“温養”擬設的受詞。余國藩勇於挑戰、修正自己，確實令人佩服(2 *JW*, 2: 383n23—25)。至於上述的閔一得之説，當因《還原篇》有“但知行二八，便可煉金丹”之説，又有“姹女方二八，金翁正九三”使然(《秘功》，頁278及289)。我們只要想到第七十八回三藏在比丘國面聖之際，業已明示“採陰補陽，誠爲謬語”，連那“服餌長壽”，他也批爲“虛詞”(《西》，頁897)，則中野之説若然，豈非指取經五聖都得破那“色戒”不可，否則難以證成正果？如此論述，業已和全真丹術得保“元陽之身”方能“結聖胎”或“結嬰兒”的戒律抵牾，當然也跳不出佛門“一心不動，萬行俱全”的超越法門(《西》，頁897)，從而犯下今人所謂“詮釋過當”之病而致走火入魔。

我們倘無《還原篇》這類丹書的輔助，上面的數字就只能任人定義，甚至各取所需，而全真教也會變成孫思邈(541?—682)一脈的道教。中國宗教傳

① 葛洪，《神仙傳》，與(清)隱夫玉簡《疑仙傳》、(漢)劉向《列仙傳》合刊，台北：廣文書局，1988，11頁。

統中,“性命雙修”一向懸爲理想,但説其根本,則只有道教能夠成雙兼顧。就《西遊記》的文學一面而言,“修性”是刻畫人物的動力,而“修命”當然就是推動情節的力量。全真道士無不以“性命雙修”爲生命大任,所以特別講究祛色戒欲。既然如此,何以整部《西遊記》所借魏伯陽、張伯端、石泰,乃至其他丹書中的語句或語彙大多又充滿情色色彩,讓中野美代子一類的論者“想入非非”呢?

余譯本《西遊記》修訂版的《導論》,於上述問題新寫數頁,但求三致其意。煉丹名詞每常互稱“夫妻”,合爲“交媾”等等,上引劉一明的話把此一關係説得甚明:“托物寓言。”但上文我没再予徵引的是下一句劉文“旁引曲證”,亦即邏輯關係並非丹家書寫之所重。百回本《西遊紀》的序者乃陳元之(fl. 1592),他同樣認爲是書係以佛道設喻,評得甚至更絶:“此其書直寓言者哉!”陳氏這句話,余譯本新添之,繼而寫進了《西遊記》的《導論》中(2 *JW*, 1: 42-43),而且還以“宗教寓言”直呼《西遊記》的本質,似乎在向胡適等人開戰。所謂“寓言”(allegory/ extended metaphor),我們鬆散稱之爲“譬喻”(metaphor),亦無不可。修訂後的余譯本《導論》,最後另附“體象陰陽升降圖”一幀,其中即以“譬喻”代“寓言”(2 *JW*, 1: 92—93),而不論“譬喻”或“寓言”,這兩個名詞——容我再借王重陽的話——都是佛教所稱的“假名”。《西遊記》因此設爲托喻,假名成書,而凡此確需經驗老到的讀者或譯者一一勘破,方能中天見月明。

在道經丹書中,“體象陰陽升降圖”名氣大,所出乃前及蕭應叟著《元始無量度人上品妙經内義》[①]。蕭氏嘗試以圖代文,把内丹修煉示之以山,譬之以身,使人易知易懂。但也因此,《論語·雍也篇》所謂“能近取譬”的推己及人理想[②],全真道士才可謂念玆在玆。《西遊記》的作者當然也牢記在心,而余國藩更據之——此外,當然也涵容大量全真北宗如王重陽等人的道詩的研究——爲翻譯此一明代説部偉構的基礎,從而詳究之,細勘之,甚至視之爲戲語趣譚,以諧謔插曲而表現在小説中。故事因而益顯生動,活潑無比,活脱一幅巴

① 蕭應叟,《元始無量度人上品妙經内義》,見張繼禹主編,《中華道藏》,3: 551。
② (宋)朱熹,《四書集注》,台北:世界書局,1956,100頁。

克汀(Mikhail Bakhtin, 1895—1975)式的全真語言的嘉年華會。

當然,我們若退回《西遊記》的宗教傳統,則此書絶對是中國三教孕育而出的長篇説部,某個意義上亦有如一部宗教百科全書。三教中,道教的基礎無疑最稱雄厚,余國藩緣此而由丹經出發,予以重詮,下手傳譯,《西遊記》幾乎換上了新裝,重新西遊。余國藩眼光獨到,由此可窺一斑,而他筆下秀出班行,恐怕也是世無可疑。

美國漢學期刊《哈佛亞洲學報》(*HJAS*)論文目録(1936—2014)漢譯*

卞東波　何沁心　編譯

《哈佛亞洲學報》(*Harvard Journal of Asiatic Studies*,下簡稱*HJAS*)是美國著名的老牌漢學期刊,由哈佛燕京學社資助出版,從1936年出版第1卷,至2014年第74卷,至今已經有78年的辦刊史,歷史之悠久在美國漢學史上僅次於《美國東方學會學報》(*Journal of the American Oriental Society*)。除了二戰中的1943年未正常出版第8卷,1960—1961年、1962—1963年、1964—1965年合併出版第23、24、25卷之外,*HJAS*一直保持着連續出版,1936—1957年,基本上是每年一卷四期;1958—1979年基本上每年一卷一期;1980年至今則是每年一卷二期,分別在6月和12月發行。

根據*HJAS*官方網站上的説明,70多年來*HJAS*的辦刊宗旨是追求具有原創性的、優秀的有關亞洲的人文研究,刊載研究中國、日本、韓國各方面的論文,也發表關於中國蒙古、西藏歷史文化的論著,以及内陸亞洲(Inner Asia)阿爾泰語系民族的學術論文。*HJAS*的論文注重的是論文的學術性、論證的質量和研究方法的明晰度(strength of the scholarship, the quality of the argument, the clarity of the methodology)。即使關注的課題較爲狹窄,也要以某種方式

編譯者單位:南京大學文學院

* 本文爲江蘇省社科基金青年項目“北美中國古典文學研究名家研究”(12WWC014)及南京大學文科規劃項目“中國古典文學的新視鏡:新世紀海外中國文學研究之再研究與譯介”成果之一。毋澤暉、彭博、陳佳寧做了部分初譯工作,我們對譯稿進行了重譯和加工,對她們的幫助,特此致謝。

呈現給學術界，而不只是針對小圈子的少數專家。換句話說，發在*HJAS*上的論文不僅僅是"荒村野店，二三素心人的事"，不是追求所謂絶學、獨門之學，而是能讓整個學術共同體都能欣賞的論文。光有新的材料，新的觀點，而没有豐富的思想和充分的論述也達不到*HJAS*發表的要求。當然，*HJAS*實行匿名審稿制度，投稿者不能在論文及注釋中出現個人信息。如果通過評審，並同意發表後，投稿者要根據評審人的意見對文章進行修改，有時候這種修改幅度非常之大，以至於發表後的論文與原稿判若兩樣。

就漢學研究而言，*HJAS*發表的論文涉及中國的歷史、思想、宗教、文學、語言、藝術、政治、外交、建築等等各個方面，時間跨度從上古一直到近代，近年來也有關於中國現代文學的論文，但鮮有關於當代中國研究的文章，可能與*HJAS*定位以及漢學期刊的分工有關。*HJAS*的欄目比較固定，早期基本上由論文（article）、劄記（note）、書評（review）、書評論文（review article）四部分構成，現在基本上主要是論文和書評兩部分。每一期登載的論文很少，少則二三篇，最多五六篇。雖然*HJAS*要求每篇論文稿件不超過15000個英文單詞，而且歡迎短稿，但據筆者觀察，*HJAS*刊載的論文基本上都在20頁以上，接近或超過百頁的文章也不在少數，如1991年第51卷第2期發表了梅維恒、梅祖麟合著的《近體詩律的梵文來源》就幾乎長達百頁，而1994年第54卷第1期發表的蘇珊·徹尼亞克《宋代的書籍文化與文本傳播》長達120頁，早期還連載過黄仁宇的專著規模的論文《葉名琛總督與廣州事件（1856—1861）》（1941年第6卷第1期，37—127頁）。少而精，少而長，長而透，是*HJAS*發表論文的特色。另一個特色是注重書評，每一期都會發表十幾篇書評，數量上絶對超過論文。*HJAS*規定不接受書評投稿，書評一般由*HJAS*指定的學者撰寫。所評之書皆爲該領域新出的優秀之作，而撰寫書評者也是该領域的專家或權威人士。如果説，*HJAS*刊載的論文反映了美國漢學的高精尖，那麼書評則起到學術信息交流以及學術評論的作用。兩者相輔相成，反映了美國漢學期刊的基本特色。

*HJAS*設有一個主編（Editor），1936年以來，擔任過主編的有葉理綏（Serge Elisséeff, 1936—1957年在任）、John Bishop（1958—1974年在任）、Timothy Connor（1975年在任）、Donald Shively（1976—1983年在任）、艾朗諾

（Ronald Egan，1983—1987年在任）、Howard Hibbett（1988—2002年在任）、韓德琳（Joanna Handlin Smith，2003—2014年在任）。現任主編是哈佛大學東亞系研究日本歷史的教授David L. Howell，執行編輯（Managing Editor）是Melissa J. Brown。主編和編輯主要做的是事務性工作，學術上的工作主要依靠*HJAS*强大的編委會（Editorial Board），其成員基本上都是哈佛東亞系的教授，如宇文所安（Stephen Owen）、包弼德（Peter K. Bol）、歐立德（Mark C. Elliott）、普鳴（Michael J. Puett）、宋怡明（Michael Szonyi）、李惠儀（Wai-yee Li）、王德威（David Der-wei Wang）等人，是他們保證了*HJAS*論文的品質。

目前，*HJAS*是半年刊，一年出版兩本小16開紙質的期刊，其封面的顏色70多年如一日，一直是樸素的灰藍色。除了紙本之外，還可以在網上的"西文過刊系統"（JSTOR）看到*HJAS*距今5年前的所有內容；而在Project MUSE上則可以看到最近幾年內最新的內容。*HJAS*也有自己的專門網站（http://www.hjas.org），介紹了*HJAS*的基本情況，包括已經出版的部分年份期刊的目録、提要，編委會的成員，投稿須知等等。

鑒於*HJAS*在國際漢學史上的重要地位，以及其70多年來積累的論文的巨大學術價值，筆者一直發願將其揭載的論文目録譯爲中文，介紹給中國學界。筆者的目的在於：其一可以讓中國學者按圖索驥，根據論文題目找到原文加以參考，避免重復研究；其二也希望以此目録漢譯爲契機，比較系統地將美國漢學研究成果以目録的方式介紹給中國學者，加强中國學者與國際漢學界的對話。由於70多年的目録實在太豐厚，暫時無法全部翻譯，筆者先行將論文部分的題目譯爲中文，書評部分則留待他日。*HJAS*也登載了很多研究日本、韓國歷史文化的論文，由於術業有專攻，筆者主要關注中國研究，故日韓部分的論文題目也暫未譯出。

中國學者也早已注意到*HJAS*的巨大學術價值，不但著文加以介紹，如程章燦教授撰有《歲月匆匆六十年——由〈哈佛亞洲學報〉看美國漢學的成長》（上下篇，載《古典文學知識》1997年第1、2期），而且也積極將*HJAS*上重要的論文翻譯爲中文。爲了方便我國學者參考，筆者也儘量利用各種工具查找*HJAS*論文中譯的情況，備註其出處。本文的基本體例是，先列*HJAS*論文英文原題，下列漢譯，如果該論文有中譯本，則再於脚注中説明。

1936年第1卷第1期

Yuen Ren Chao, A Note on Lia, Sa, etc. (pp. 33-38)

趙元任:論漢語中的“倆”“仨”等[1]

Tschen Yinkoh and J. R. Ware, Han Yü and The T'ang Novel (pp. 39-43)

陳寅恪撰,魏楷譯:韓愈與唐代小説[2]

Unokichi Hattori and Serge Elisseeff, Confucius' Conviction of His Heavenly Mission (pp. 96-108)

服部宇之吉撰,葉理綏譯:孔子的天命觀

William Ernest Hocking, Chu Hsi's Theory of Knowledge (pp. 109-127)

霍金:朱熹的認識論

T'ang Yung-T'ung, The Editions of The *Ssŭ-shih-êrh-chang-ching* (pp. 147-155)

湯用彤撰,魏楷譯:《四十二章經》版本源流考

James R. Ware, Notes on *The Fan Wang Ching* (pp. 156-161)

魏楷:《梵網經》考

1936年第1卷第2期

Peter A. Boodberg, The Language of the T'o-Pa Wei (pp. 167-185)

卜弼德:拓跋魏的語言

Kaiming Chiu, The Introduction of Spectacles Into China (pp. 186-193)

裘開明:眼鏡傳入中國簡史

Wolfram Eberhard and Rolf Mueller, Contributions to The Astronomy of The Han Period III: Astronomy of The Later Han Period (pp. 194-241)

艾伯華、羅夫·穆勒:漢代天文學的貢獻(三):東漢的天文學

1936年第1卷第3–4期

H. Y. Fêng, The Origin of Yü Huang (pp. 242-250)

馮漢驥:“玉皇”起源考

---

① 參見趙元任《“倆”、“仨”、“四呃”、“八阿”》,載《東方雜誌》第24卷第12號,1927;又見趙元任著,吴宗濟、趙新那編《趙元任語言學論文集》,北京:商務印書館,2002。

② 參見程千帆先生的譯文,收入《陳寅恪集·講義及雜稿》,北京:生活·讀書·新知三聯書店,2001。

Peter A. Boodberg, Two Notes on The History of The Chinese Frontier (pp. 283-307)

卜弼德:中國邊疆史劄記二則

1937第2卷第1期

Rufus Suter, A Note About Ingenuousness in The Ethical Philosophy of Mencius (pp. 4-8)

蘇特:孟子倫理哲學中的"真"

1937第2卷第3–4期

Peter A. Boodberg, Some Proleptical Remarks on The Evolution of Archaic Chinese (pp. 329-372)

卜弼德:簡論古代漢語的演變

Hu Shih, A Criticism of Some Recent Methods Used in Dating *Lao Tzu* (pp. 373-397)

胡適:評論近人考據《老子》年代的方法[①]

1938第3卷第1期

Tschen Yinkoh, The *Shun-Tsung Shih-Lu* and *The Hsu Hsuan-Kuai Lu* (pp. 9-16)

陳寅恪:《順宗實録》與《續玄怪録》[②]

1938年第3卷第2期

Roswell S. Britton, A Horn Printing Block (pp. 99-102)

白瑞華:一片牛角雕版

Feng Han-Yi and J. K. Shryock, The Historical Origins of The Lolo (pp. 103-127)

馮漢驥、施瑞奥克:中國西南少數民族羅羅歷史淵源考

---

① 參見《胡適論學近著》第一集卷一中同名論文,濟南:山東人民出版社,1998 。

② 參見陳寅恪《金明館叢稿二編》中同名論文,北京:生活·讀書·新知三聯書店,2001。

Han Fei Tzu and W. K. Liao, Learned Celebrities: A Criticism of The Confucians and The Moists (pp. 161-171)

廖文奎:《韓非子·顯學篇》譯註

1938第3卷第3–4期

Ku Chieh-Kang and L. Carrington Goodrich, A Study of Literary Persecution During The Ming (pp. 254-311)

顧頡剛撰,傅路德譯:明代文字獄禍考略①

1939第4卷第1期

Cheng Te-k'un, The Excavation of T'ang Dynasty Tombs at Ch'uan-chou, Southern Fukien (pp. 1-11)

鄭德坤:福建泉州唐墓的發掘

J. K·Fairbank and S. Y. Teng, On The Transmission of Ch'ing Documents (pp. 12-46)

費正清、鄧嗣禹:清代公文的傳遞

Otto Maenchen-Helfen, The Ting-Ling (pp. 77-86)

奧托·曼森-黑爾芬:論丁零一族

1939第4卷第2期

Knight Biggerstaff, Some Notes on The *Tung-hua lu* and The *Shih-lu* (pp. 101-115)

畢乃德:《東華録》和《實録》叢劄

1939第4卷第3–4期

Peter A. Boodberg, Marginalia to The Histories of The Northern Dynasties (pp. 223-253)

卜弼德:北朝史瑣談

① 顧頡剛原文發表於《東方雜誌》第32卷第14期,1935年7月。

George A. Kennedy, Metrical "Irregularity" in The *Shih ching* (pp. 284-296)

金守拙:《詩經》中的不合律現象

1940年第5卷第1期

J. K. Fairbank and S. Y. Teng, On The Types and Uses of Ch'ing Documents (pp. 1-71)

費正清、鄧嗣禹:論清代檔案的種類與使用

Malcolm F. Farley, Some Mirrors of Supposed Pre-Han Date (pp. 72-94)

馬爾科姆·法利:疑似漢以前銅鏡年代考

1940年第5卷第2期

Paul K. Benedict, Studies in Indo-Chinese Phonology (pp. 101-127)

白保羅:印支語系研究

Peter A. Boodberg, Chinese Zoographic Names as Chronograms (pp. 128-136)

卜弼德:中國生肖紀年法

Edwin O. Reischauer, Notes on T'ang Dynasty Sea Routes (pp. 142-164)

賴肖爾:唐代海路考

Philip K. Reynolds, C. Y. Fang, The Banana in Chinese Literature (pp. 165-181)

菲力浦·K. 雷諾茲、房兆楹:中國文學中的"蕉"

James R. Ware, The So-called Final *Wei* (pp. 182-188)

魏楷:論句末助詞"爲"

1941年第5卷第3–4期

Yuen Ren Chao, Distinctions within Ancient Chinese (pp. 203-233)

趙元任:漢語上古音之別

Michael J. Hagerty, Comments on Writings Concerning Chinese Sorghums (pp. 234-260)

邁克爾·哈格蒂:漢語文獻中有關"黍"之記載

1941年第6卷第1期

Wilma Fairbank, The Offering Shrines of "Wu Liang Tz'ü" (pp. 1-36)

費慰梅:漢武梁祠建築原形考[①]

Huang Yen-yu, Viceroy Yeh Ming-ch'en and The Canton Episode (1856—1861): 1. Hsieh Fu-ch'eng (1838—1894) (pp. 37-45)

黃仁宇:葉名琛總督與廣州事件(1856—1861):1.薛福成(1838—1894)

Huang Yen-yu, Viceroy Yeh Ming-ch'en and The Canton Episode (1856—1861): 2. On Grand Secretary Yeh of Han-yang in Relation to The Episode of Canton (pp. 45-91)

黃仁宇:葉名琛總督與廣州事件(1856—1861):2.漢陽葉大學士及其與廣州事件之關係

Huang Yen-yu, Viceroy Yeh Ming-ch'en and The Canton Episode (1856—1861): 3. A Brief Biography of Yeh Ming-ch'en (1812—1859) (pp. 91-94)

黃仁宇:葉名琛總督與廣州事件(1856—1861):3.葉名琛傳略

Huang Yen-Yu, Viceroy Yeh Ming-Ch'en and The Canton Episode (1856—1861): 4. The Canton Episode (pp. 94-127)

黃仁宇:葉名琛總督與廣州事件(1856—1861):4.廣州事件

1941年第6卷第2期

Paul K. Benedict, A Cham Colony on The Island of Hainan (pp. 129-134)

白保羅:海南島上的占族僑民

J. K. Fairbank and S. Y. Têng, On The Ch'ing Tributary System (pp. 135-246)

費正清、鄧嗣禹:論清代的朝貢體制

1941年第6卷第3–4期

Yao Shan-Yu, The Chronological and Seasonal Distribution of Floods and Droughts in Chinese History, 206 B. C.-A. D. 1911 (pp. 273-312)

姚善友:中國歷史上周期性與季節性旱澇之分佈(公元前206—公元1911年)

① 參見王世襄中譯文,載《中國營造學社彙刊》第7卷第2期,1935。

Derk Bodde, Some Chinese Tales of The Supernatural: Kan Pao and His *Sou-shen chi* (pp. 338-357)

卜德:中國的志怪小説——干寶及其《搜神記》

1942年第7卷第1期

Fung Yu-Lan and Derk Bodde, The Philosophy of Chu Hsi (pp. 1-51)

馮友蘭撰,卜德譯注:朱熹的哲學①

Wilma Fairbank, A Structural Key to Han Mural Art (pp. 52-88)

費慰梅:漢代墓葬藝術的關鍵結構

1942年第7卷第2期

Fung Yu-Lan and Derk Bodde, The Rise of Neo-Confucianism and Its Borrowings From Buddhism and Taoism (pp. 89-125)

馮友蘭撰,卜德譯注:道學之初興及道學中二氏之成分②

Tenney L. Davis and Ch'En Kuo-Fu, Shang-Yang Tzu: Taoist Writer and Commentator on Alchemy (pp. 126-129)

戴維斯、陳國符:上陽子——道教作家與煉金術之研究者

L. Carrington Goodrich, The Revolving Book-Case in China (pp. 130-161)

傅路德:中國的輪藏

1943年第7卷第3期

Chu Shih-Chia, Tao-Kuang to President Tyler (pp. 169-173)

朱士嘉:道光皇帝致美國總統泰勒的信(即:耆英致顧盛函)

G. N. Kates, A New Date for The Origins of The Forbidden City (pp. 180-202)

G. N. 凱特:紫禁城溯源新考

K. T. Wu (Wu Kuang-Ch'ing), Ming Printing and Printers (pp. 203-260)

吴光清:明代的印刷業和刻工

① 參見馮友蘭《中國哲學史》第二篇《經學時代》第十三章《朱子》,北京:商務印書館,2011。

② 參見馮友蘭《中國哲學史》第二篇第十章。

1943年第7卷第4期

Ssu-yü Têng, Chinese Influence on The Western Examination System: I. Introduction (pp. 267-312)

鄧嗣禹:中國對西方考試制度的影響:I. 導言[①]

1944年第8期第1期

Sidney D. Gamble, Hsin Chuang: A Study of Chinese Village Finance (pp. 1-33)

西德尼·D. 甘博:辛莊:中國鄉村財政研究

Olov R. T. Janse, Notes on Chinese Influences in The Philippines in Pre-Spanish Times (pp. 34-62)

奧洛夫·詹斯:中國對西班牙殖民之前菲律賓的影響

1944年第8期第2期

Schuyler Cammann, The Development of the Mandarin Square (pp. 71-130)

斯凱勒·坎曼:官服補子之發展

Helen B. Chapin, Yünnanese Images of Avalokiteśvara (pp. 131-186)

海倫·查平:雲南人眼中的觀音菩薩

Chu Shih-chia, Chinese Local Histories at Columbia University (pp. 187-195)

朱士嘉:哥倫比亞大學所藏的中國方志

Yao Shan-Yu, Flood and Drought Data in The *T'u-shu chi-ch'êng* and the *Ch'ing shih kao* (pp. 214-226)

姚善友:《圖書集成》和《清史稿》中的洪澇旱災資料

1945年第8卷第3-4期

Cheng Te-k'un, The Royal Tomb of Wang Chien (pp. 235-240)

鄭德坤:王建皇陵考

---

① 參見王漢中譯文《中國考試制度西傳考》,載黄培、陶晉生主編《鄧嗣禹先生學術論文選集》,台北:食貨出版社,1980。又參見鄧嗣禹、彭靖《家國萬里:鄧嗣禹的學術與人生》中同名論文,上海:上海人民出版社,2014。

Chou Yi-liang, Tantrism in China (pp. 241-332)

周一良:中國的密教[1]

Li Fang-kuei, Some Old Chinese Loan Words in The Tai Languages (pp. 333-342)

李方桂:台語中若干古代漢語借詞[2]

Lo Ch'ang-p'ei, A Preliminary Study on The Trung Language of Kung Shan (pp. 343-348)

羅常培:貢山獨龍語初探

1945年第9卷第1期

Kenneth K. S. Ch'en, Buddhist-Taoist Mixtures in The *Pa-shih-i-hua T'u* (pp. 1-12)

陳觀勝:《八十一化圖》中的釋、道融合

Chou Yi-liang, Notes on Marvazi's Account on China

周一良:論馬爾瓦兹有關中國的記載

Chia Ssu-hsieh, Huang Tzu-ch'ing, Chao Yun-ts'ung and Tenney L. Davis, The Preparation of Ferments and Wines

賈思勰撰,黄子卿、趙雲從譯,騰尼·戴維斯導言:發酵與酒(由《齊民要術》選出)

Yang Chih-chiu and Ho Yung-chi, Marco Polo Quits China (p. 51)

楊志玖、何永佶:馬可波羅之離華

1946年第9卷第2期

Chêng Tê-k'un, The Slate Tomb Culture of Li-fan (pp. 63-80)

鄭德坤:理番版岩葬[3]

---

① 參見周一良著,錢文忠譯《唐代密宗》,上海:上海遠東出版社,2012。

② 參見張光宇、陳秀琪譯文,載《李方桂全集》第1册《漢藏語論文集》,北京:清華大學出版社,2012。

③ 參見《鄭德坤古史論集選》中同名論文,北京:商務印書館,2007。

J. Rahder, Portrait-Painting on a Chinese Porcelain Cup (pp. 101-106)

拉德麗:中國瓷器上的肖像畫

Lien-Sheng Yang, Notes on The Economic History of The Chin Dynasty (pp. 107-185)

楊聯陞:晉代經濟史釋論[1]

1947年第9卷第3–4期

Fung Yu-lan and Derk Bodde, A General Discussion of The Period of Classical Learning (pp. 195-201)

馮友蘭撰,卜德譯注:經學時代概論

Lien-sheng Yang, A Note on The So-called TLV Mirrors and The Game *Liu-po* (pp. 202-206)

楊聯陞:所謂TLV紋鏡與六博考

1947年第10卷第1期

Francis Woodman Cleaves, K'uei-k'uei or Nao-Nao? (pp. 1-12)

柯立夫:"夒"字讀音考

Karl A. Wittfogel, Public Office in The Liao Dynasty and The Chinese Examination System (pp. 13-40)

魏復古:遼代的官職與中國的科舉制度

Lien-Sheng Yang, A Theory about The Titles of The Twenty-Four Dynastic Histories (pp. 41-47)

楊聯陞:二十四史名稱試解[2]

L. Z. Ejdlin and Francis Woodman Cleaves, The Academician V. M. Alexeev as a Historian of Chinese Literature (pp. 48-59)

L. Z. 艾德林著,柯立夫譯:中國文學史研究者阿列克謝耶夫

① 參見楊聯陞《國史探微》中同名論文,沈陽:遼寧教育出版社,1998。

② 參見楊聯陞《國史探微》中同名論文。

1947年第10卷第2期

E. A. Kracke, Jr., Family Vs. Merit in Chinese Civil Service Examinations Under The Empire (pp. 103-123)

柯睿格:家庭與朝廷科舉中的功名

T'ang Yung-t'ung and Walter Liebenthal, Wang Pi's New Interpretation of The *I Ching* and *Lun-yü* (pp. 124-161)

湯用彤著,李華德譯:王弼之《周易》《論語》新義

Homer H. Dubs, A Canon of Lunar Eclipses for Anyang and China, -1400 to -1000 (pp. 162-178)

德效騫:公元前1400至公元前1000年安陽和中國的月食典

John De Francis, Biography of The Marquis of Huai-yin (pp. 179-215)

德範克:《史記·淮陰侯列傳》譯注

1947年第10卷第3–4期

John A. Pope, Sinology or Art History: Notes on Method in The Study of Chinese Art (pp. 388-417)

約翰·蒲柏:漢學抑或藝術史:論中國藝術研究的方法

Lien-Sheng Yang, A "Posthumous Letter" From The Chin Emperor to The Khitan Emperor in 942 (pp. 418-428)

楊聯陞:公元942年後晉皇帝致契丹皇帝的一封遺書

1948年第11卷第1–2期

Rhea C. Blue, The Argumentation of the *Shih-Huo Chih:* Chapters of the Han, Wei, and Sui Dynastic Histories (pp. 1-118)

布露:《漢書》《魏書》《隋書》之《食貨志》譯註

Tung Tso-pin and Lien-sheng Yang, Ten Examples of Early Tortoise-Shell Inscriptions (pp. 119-129)

董作賓著,楊聯陞譯注:武丁龜甲卜辭十例

Chung-han Wang, The Authorship of the *Yu-Hsien-K'u* (pp. 153-162)

王鍾翰:《游仙窟》著者考[1]

Alexander C. Soper, Some Technical Terms in the Early Literature of Chinese Painting (pp. 163-173)

梭柏:早期中國繪畫文獻中的術語

Richard C. Rudolph, Dynastic Booty: An Altered Chinese Bronze (pp. 174-180)

理查·魯道夫:國家的戰利品:一件重鐫銘文的青銅器

Benjamin Rowland, Jr., A Note on the Invention of the Buddha Image (pp. 181-186)

本傑明·羅蘭:佛教造像小考

Sidney M. Kaplan, Some Observations on Ch'i-Chia and Li-Fan Pottery(pp. 187-196)

西德尼·M. 卡普蘭:齊家、理番陶器叢論

Paul K. Benedict, Archaic Chinese g and d (pp. 197-206)

白保羅:論古漢語之g與d音

1948年第11卷第3–4期

James Robert Hightower, The *Han-shih wai-chuan* and the San chia shih (pp. 241-310)

海陶瑋:《韓詩外傳》與三家詩

Arthur Frederick Wright, Fo-t'u-têng: A Biography (pp. 321-371)

芮沃壽:佛圖澄傳

Michael J. Hagerty, Tai K'ai chih's *Chu-p'u* (pp. 372-440)

邁克爾·哈格蒂:戴凱之《竹譜》譯註

① 參見王鍾翰《清史補考》中同名論文,沈陽:遼寧大學出版社,2004。

1949年第12卷第1–2期

Francis Woodman Cleaves, The Sino-Mongolian Inscription of 1362 in Memory of Prince Hindu (pp. 1-133)

柯立夫：至元二十二年（1362）漢蒙合璧《忻都神道碑》考

Wang Yü-ch'üan, An Outline of The Central Government of The Former Han Dynasty (pp. 134-187)

王毓銓：西漢中央政府的組織

Achilles Fang, A Note on Huang T'ung and Tzŭ-wu (pp. 207-215)

方志彤："黄童"、"子烏"考釋

Lien-sheng Yang, Numbers and Units in Chinese Economic History (pp. 216-225)

楊聯陞：中國經濟史上的數目和單位[①]

L. Carrington Goodrich, Maternal Influence: A Note (pp. 226-230)

傅路德：論科舉制度中外家的影響

1949年第12卷第3–4期

John Hall, Notes on The Early Ch'ing Copper Trade With Japan (pp. 444-461)

約翰·霍爾：論清初中日間的銅貿易

Lien-sheng Yang , The Concept of "Free" and "Bound" in Spoken Chinese (pp. 462-469)

楊聯陞：漢語口語中"自由"與"限定"的觀念

1950年第13卷第1–2期

Francis Woodman Cleaves, The Sino-Mongolian Inscription of 1335 in Memory of Chang Ying-Jui (pp. 1-131)

柯立夫：至順三年（1335）漢蒙合璧《張應瑞神道碑》考

---

① 參見楊聯陞著，彭剛、程鋼譯《中國制度史研究》中同名譯文，南京：江蘇人民出版社，2007。亦見楊聯陞《國史探微》，題作《中國經濟史上的數詞與量詞》。

Achilles Fang, Bookman's Decalogue (pp. 132-173)

方志彤:葉德輝《藏書十約》譯註

Lien-sheng Yang, Buddhist Monasteries and Four Money-raising Institutions in Chinese History(pp. 174-191)

楊聯陞:佛教寺廟與中國歷史上的四種募錢制度[①]

Chaoying Fang, A Technique for Estimating The Numerical Strength of The Early Manchu Military Forces (pp. 192-215)

房兆楹:估算滿清早期軍力數量的方法

1950年第13卷第3–4期

Feng Han-Yi and J. K. Shryock, Marriage Customs in The Vicinity of I-Ch'ang (pp. 362-430)

馮漢驥、施瑞奥克:宜昌周邊地區的婚俗

Francis Woodman Cleaves, The Sino-Mongolian Edict of 1453 in The Topkapi Sarayi Müzesị (pp. 431-446)

柯立夫:土耳其伊斯坦布爾托普卡帕宮博物館所藏1453年漢蒙合璧詔書考

K. T. Wu, Chinese Printing under Four Alien Dynasties: (916-1368 A. D.) (pp. 447-523)

吴光清:中國遼夏金元時期(916—1368)的印刷業

Lien-sheng Yang, Notes on Dr. Swann's *Food and Money in Ancient China* (pp. 524-557)

楊聯陞:評孫念禮《漢書·食貨志》譯註[②]

① 參見楊聯陞著,彭剛、程鋼譯《中國制度史研究》中同名譯文。亦見楊聯陞《國史探微》,題作《佛教寺院與國史上四種籌措金錢的制度》。

② 參見楊聯陞著,彭剛、程鋼譯《中國制度史研究》,該文在此書中譯作《斯納博士〈中國古代的食物與貨幣〉注解》。

1951年第14卷第1–2期

Francis Woodman Cleaves, The Sino-Mongolian Inscription of 1338 in Memory of Jigüntei (pp. 1-104)

柯立夫：至元四年（1338）漢蒙合璧《温竹台神道碑》考

Edward H. Schafer, Ritual Exposure in Ancient China (pp. 130-184)

薛愛華：中國古代的儀式性的"露身"

Kenneth Ch'en, Notes on The Sung and Yüan Tripiṭaka (pp. 208-214)

陳觀勝：論宋元藏經

Sun Ts'ung-T'ien and Achilles Fang, Bookman's Manual (pp. 215-260)

方志彤：孫從添《藏書紀要》譯注

1951年第14卷第3–4期

William Hung, The Transmission of The Book Known as *The Secret History of The Mongols* (pp. 433-492)

洪業：《蒙古秘史》源流考[①]

Achilles Fang, Rhymeprose on Literature, The *Wên-Fu* of Lu Chi (A.D. 261-303) (pp. 527-566)

方志彤：陸機《文賦》譯注

1952年第15卷1–2期

Francis Woodman Cleaves, The Sino-Mongolian Inscription of 1346 (pp. 1-123)

柯立夫：至正六年（1346）漢蒙合璧碑銘考

Lien-sheng Yang, An Additional Note on The Ancient Game *Liu-po* (pp. 124-139)

楊聯陞：古代遊戲"六博"再考

Kenneth Ch'en, Anti-Buddhist Propaganda During The Nan-ch'ao (pp. 166-192)

陳觀勝：南朝的反佛宣傳

Arthur F. Wright, Biography of The Nun An-ling-shou (pp. 193-196)

芮沃壽：《安令首尼傳》譯注

---

① 參見黄時鑒譯文，載《中國元史研究通訊》，元史研究會編，1982年2期。

1952年第15卷3–4期

Lien-sheng Yang, Hostages in Chinese History (pp. 507-521)

楊聯陞:中國歷史上的人質[①]

1953年第16卷第1–2期

Glen William Baxter, Metrical Origins of the *Tz'u* (pp. 108-145)

白思達:詞律起源考

George A. Kennedy, Another Note on *Yen* (pp. 226-236)

金守拙:也論古漢語中的“焉”

1953年第16卷第3–4期

Wang Yi-t'ung, Slaves and Other Comparable Social Groups During The Northern Dynasties (386-618) (pp. 293-364)

王伊同:中國北朝(386-618)的奴隸與其他類似的社會群體

Lien-sheng Yang, The Form of The Paper Note Hui-tzu of The Southern Sung Dynasty (pp. 365-373)

楊聯陞:南宋朝紙幣會子的形式[②]

Y. P. Mei, The *Kung-sun Lung Tzu* With a Translation Into English (pp. 404-437)

梅貽寶:《公孫龍子》譯注

1954年第17卷第1–2期

Ping-ti Ho, The Salt Merchants of Yang-Chou: A Study of Commercial Capitalism in Eighteenth-Century China (pp. 130-168)

何炳棣:揚州鹽商:18世紀中國商業資本研究[③]

James Robert Hightower, The Fu of T'ao Ch'ien (pp. 169-230)

海陶瑋:論陶潛的賦

---

① 參見楊聯陞著,彭剛、程鋼譯《中國制度史研究》中同名譯文。亦見楊聯陞《國史探微》,題作《國史上的人質》。

② 參見楊聯陞著,彭剛、程鋼譯《中國制度史研究》中同名譯文。

③ 參見巫仁恕譯文,載《中國社會經濟史研究》1999年2期。

Immanuel C. Y. Hsü, The Secret Mission of The Lord Amherst on The China Coast, 1832 (pp. 231-252)

徐中約:阿美士德勳爵號1832年在中國沿海的秘密使命

Bernard S. Solomon, “One is No Number” in China and The West (pp. 253-260)

伯納德·S. 所羅門:中西方觀念中的“一非數”

Kenneth Ch’en, On Some Factors Responsible for The Antibuddhist Persecution Under The Pei-Ch’ao (pp. 261-273)

陳觀勝:論北朝滅佛運動的諸因素

1954年第17卷第3–4期

Lien-sheng Yang, Toward a Study of Dynastic Configurations in Chinese History (pp. 329-345)

楊聯陞:國史諸朝興衰芻論[①]

Francis A. Rouleau, The Yangchow Latin Tombstone as a Landmark of Medieval Christianity in China (pp. 346-365)

魯洛:揚州出土拉丁文墓碑:中國中古時期基督教的重要資料

Lao Kan, Six-Tusked Elephants on a Han Bas-Relief (pp. 366-369)

勞榦:漢浮雕上的六牙大象

Roy Andrew Miller, The Sino-Burmese Vocabulary of The *I-Shih chi-yu* (pp. 370-393)

羅伊·安德魯·米勒:《譯史紀餘》中的漢緬語詞彙

John L. Bishop, A Colloquial Short Story in The Novel *Chin p’ing mei* (pp. 394-402)

約翰·萊曼·畢曉普:《金瓶梅》中的白話短篇小説

1955年第18卷第1–2期

William Hung, Huang Tsun-Hsien’s Poem “The Closure of The Educational Mission in America” (pp. 50-73)

洪業:論黃遵憲之詩《罷美國留學生感賦》

---

① 參見楊聯陞《國史探微》中同名論文。

Thurston Griggs, The *Ch' ing shih kao*: A Bibliographical Summary (pp. 105-123)

瑟斯頓·格里格:《清史稿》文獻簡介

Yoshikawa Kōjirō and Glen W. Baxter, The *Shih-shuo hsin-yü* and Six Dynasties Prose Style (pp. 124-141)

吉川幸次郎著,白思達譯:《世説新語》與六朝散文風格

1955年第18卷第3–4期

Lien-sheng Yang, Schedules of Work and Rest in Imperial China (pp. 301-325)

楊聯陞:中華帝國的作息時間表[①]

Michael Sullivan, Notes on Early Chinese Landscape Painting (pp. 422-446)

蘇立文:論中國早期的山水畫

1956年第19卷第1–2期

William Hung, Three of Ch' ien Ta-hsin' s Poems on Yüan History (pp. 1-32)

洪業:論錢大昕的《元史雜詩》三首

Lien-sheng Yang, Marginalia to The *Yüan tien-chang* (pp. 42-51)

楊聯陞:《元典章》零拾

Kenneth Ch' en, The Economic Background of The Hui-ch' ang Suppression of Buddhism (pp. 67-105)

陳觀勝:會昌滅佛的經濟背景

1957年第20卷第1–2期

Lien-sheng Yang, Economic Justification for Spending: An Uncommon Idea in Traditional China (pp. 36-52)

楊聯陞:侈靡論——傳統中國一種不尋常的思想[②]

---

① 參見楊聯陞著,彭剛、程鋼譯《中國制度史研究》中同名譯文。亦見楊聯陞《國史探微》,題作《帝制中國的作息時間表》。

② 參見楊聯陞《國史探微》中同名論文。

William Hung, A Bibliographical Controversy at the T'ang Court A. D. 719 (pp. 74-134)

洪業:公元719年唐代朝廷的經學文獻真僞之争

Cheng Te-k'un, Yin-Yang Wu-Hsing and Han Art (pp. 162-186)

鄭德坤:陰陽五行與漢代藝術

H. F. Schurmann, On Social Themes in Sung Tales (pp. 239-261)

舒爾曼:宋代話本中的社會主題

So Kwan-wai, Eugene P. Boardman and Ch'iu P'ing, Hung Jen-Kan, Taiping Prime Minister, 1859—1864 (pp. 262-294)

蘇均煒、尤金·鮑德曼、邱炳:洪仁玕(1859—1864):太平天國的"總理"

Schuyler Cammann, Chinese Inside-Painted Snuff Bottles and Their Makers (pp. 295-326)

卡曼:内畫鼻煙壺及其製作者

1957年第20卷第3-4期

Francis Woodman Cleaves, The "Fifteen 'Palace Poems' " by K'o Chiu-ssu (pp. 391-479)

柯立夫:柯九思《宫詞十五首》譯注

J. K. Fairbank, Patterns Behind The Tientsin Massacre (pp. 480-511)

費正清:天津教案背後的模式[①]

James R. Hightower, The *Wen Hsüan* and Genre Theory (pp. 512-533)

海陶瑋:《文選》與文類理論[②]

Peter A. Boodberg, Philological Notes on Chapter One of The *Lao Tzu* (pp. 598-618)

卜弼德:《老子》第一章的文字考釋

---

① 參見杜繼東譯文,載《費正清集》,天津:天津人民出版社,1992。

② 參見史慕鳴譯文(周發祥校),載鄭州大學古籍所編《中外學者文選學論集》,北京:中華書局,2006。

Derk Bodde, Evidence for "Laws of Nature" in Chinese Thought (pp. 709-727)
卜德:中國思想中"自然法則"之證據

1958年第21卷

Charles O. Hucker, Governmental Organization of The Ming Dynasty (pp. 1-66)
賀凱:明代的政府組織

Kai-yü Hsü, The Life and Poetry of Wen I-to (pp. 134-179)
許芥昱:聞一多的生平與詩歌創作

1959年第22卷

Chang Kwang-Chih, Chinese Prehistory in Pacific Perspective: Some Hypotheses and Problems (pp. 100-149)
張光直:太平洋視界中的中國史前史:假説和問題

Henry Serruys, Mongols Ennobled During The Early Ming (pp. 209-260)
司律斯:明初受封的蒙古人

1960—1961年第23卷

W. A. C. H. Dobson, Towards a Historical Treatment of the Grammar of Archaic Chinese (pp.5-18)
杜百勝:古漢語語法的歷史考察

Lien-sheng Yang, Female Rulers in Imperial China (pp.47-61)
楊聯陞:國史上的女主[①]

Francis Woodman Cleaves, The Sino-Mongolian Inscription of 1240 (pp.62-75)
柯立夫:1240年的漢蒙合璧碑銘考

William Hung, The T'ang Bureau of Historiography before 708 (pp.93-107)
洪業:公元708年前的唐代史館

Fa-kao Chou, Certain Dates of the Shang Period (pp.108-113)
周法高:商代確年考

① 參見楊聯陞《國史探微》中同名論文。

Charles O. Hucker, An Index of Terms and Titles in "Governmental Organization of the Ming Dynasty" (pp.127-151)
賀凱:《明代的政府組織》中條目和標題索引

1962—1963年第24卷

Yu-kung Kao, A Study of the Fang La Rebellion (pp.17-63)
高友工:方臘起義之研究
E-tu Zen Sun, The Board of Revenue in Nineteenth-Century China (pp.175-228)
任以都:19世紀的中國户部
Chi Li, The Changing Concept of the Recluse in Chinese Literature (pp.234-247)
李祁:中國文學中隱逸觀念之變遷

1964—1965年第25卷

Tsuen-hsuin Tsien, First Chinese-American Exchange of Publications (pp.19-30)
錢存訓:中美出版品交換之肇始[①]
Ying-shih Yü, Life and Immortality in the Mind of Han China (pp.80-122)
余英時:漢代思想中的生與不朽[②]
Chauncey S. Goodrich, Two Chapters in the Life of an Empress of the Later Han (I) (pp.165-177)
顧傳習:東漢靈帝宋皇后生平二題(一)
Immanuel C. Y. Hsü, The Great Policy Debate in China, 1874:Maritime Defense Vs. Frontier Defense (pp.212-228)
徐中約:1874年中國重大政策辯論:海防還是塞防[③]

---

① 參見《錢存訓文集》中《中美書緣——紀念中美文化交換百周年》一文,北京:國家圖書館出版社,2012。

② 參見余英時著、侯旭東譯《東漢的生死觀》第一章,上海:上海古籍出版社,2005。

③ 參見徐中約著,計秋楓、朱慶葆譯《中國近代史》中《清代在新疆的統治與回民叛亂:海防與塞防之争》,北京:世界圖書出版公司,2008。

Fa-kao Chou, On the Dating of a Lunar Eclipse in the Shang Period (pp.243-247)
周法高:論商代月蝕的紀日法[1]

1966年第26卷

Ping-ti Ho, Lo-yang, A. D. 495-534: A Study of Physical and Socio-Economic Planning of a Metropolitan Area (pp. 52-101)
何炳棣:公元495—534年洛陽的城市與社會經濟規劃之研究
Chun-shu Chang, Military Aspects of Han Wu-ti's Northern and Northwestern Campaigns (pp.148-173)
張春樹:漢武帝時代塞北與西域戰爭中的軍事活動
Chauncey S. Goodrich, Two Chapters in the Life of an Empress of the Later Han (II) (pp.187-210)
顧傳習:東漢靈帝宋皇后生平二題(二)
Yu-kung Kao, Source Materials on the Fang La Rebellion (pp.211-240)
高友工:方臘起義的原始文獻

1967年第27卷

Silas Hsiu-liang Wu, The Memorial Systems of the Ch'ing Dynasty (pp.7-75)
吴秀良:清朝的奏摺體系
Francis Woodman Cleaves, The Sino-Mongolian Inscription of 1348 (pp.76-102)
柯立夫:至正八年(1348)的漢蒙合璧碑銘考
Patrick Hanan, The Early Chinese Short Story: A Critical Theory in Outline (pp.168-207)
韓南:中國早期的短篇小説:一個批評理論綱要

① 參見趙林譯文,載《大陸雜誌史學叢書·史學先秦史研究論集》第3輯第1册,台灣《大陸雜誌》編委會編,1970。

Philip Kuhn, The *T' uan-lien* Local Defense System at the Time of the Taiping Rebellion (pp.218-255)

孔飛力:太平天國時期地方"團練"的防禦體系①

1968年第28卷

Tsu-lin Mei and Yu-kung Kao, Tu Fu's "Autumn Meditations": An Exercise in Linguistic Criticism (pp.44-80)

梅祖麟、高友工:杜甫的《秋興》:語言學批評的嘗試②

Kenneth Ch'en, Filial Piety in Chinese Buddhism (pp.81-97)

陳觀勝:中國佛思想中的孝③

Willard J. Peterson, The Life of Ku Yen-wu (1613—1682) (I) (pp.114-156)

裴德生:顧炎武生平考(1613—1682)(一)

Laurence G. Thompson, The Junk Passage across the Taiwan Strait: Two Early Chinese Accounts (pp. 170-194)

勞倫斯·G. 湯普森:台灣海峽的舶船通道:兩份中國的早期史料

1969年第29卷

William Hung, A T'ang Historiographer's Letter of Resignation (pp.5-52)

洪業:《史通·忤時篇》譯注

Chia-ying Yeh Chao, Wu Wen-ying's *Tz'u* (pp.53-92)

葉嘉瑩:論吴文英的詞④

Yuen Ren Chao, Dimensions of Fidelity in Translation with Special Reference to Chinese (pp.109-130)

趙元任: 論翻譯中信達雅的信的幅度⑤

---

① 參見孔飛力著,謝亮生、楊品泉、謝思煒譯《中華帝國晚期的叛亂及其敵人:1796—1864年的軍事與社會結構》,北京:中國社會科學出版社,1990。

② 參見參見高友工、梅祖麟著,李世躍譯《唐詩三論》中同名譯文,北京:商務印書館,2013。

③ 參見許章真譯文,載《西域與佛教文史論集》,台北:學生書局,1989。

④ 參見葉嘉瑩《拆碎七寶樓臺——談夢窗詞之現代觀》,載《迦陵論詞叢稿》,石家莊:河北教育出版社,1997。

⑤ 參見羅新璋編《翻譯論集》中同名論文,北京:商務印書館,1984。

Průšek, Jaroslav, Lu Hsiün's "Huai Chiu" (pp. 169 -176)

普實克:論魯迅的《懷舊》[①]

Patrick Hanan, The Authorship of Some *Ku-chin hsiao-shuo* Storie (pp. 190-200)

韓南:《古今小說》中某些故事的作者問題[②]

Willard J. Peterson, The Life of Ku Yen-wu (1613—1682) (II) (pp. 201-247)

裴德生:顧炎武生平考(1613—1682)(二)

1970年第30卷

Kwang-ching Liu, The Confucian as Patriot and Pragmatist: Li Hung-chang's Formative Years, 1823-1866 (pp. 5-45)

劉廣京:愛國務實的儒家:李鴻章事業的形成階段(1823—1866)[③]

Mei Tsu-lin, Tones and Prosody in Middle Chinese and The Origin of The Rising Tone (pp. 86-110)

梅祖麟:中古漢語的聲調與上聲的起源

Patrick Hanan, Sung and Yüan Vernacular Fiction: A Critique of Modern Methods of Dating (pp. 159-184)

韓南:宋元白話小說:評近代繫年法[④]

Brian E. McKnight, Administrators of Hangchow Under The Northern Sung: A Case Study (pp. 185-211)

馬伯良:北宋杭州官員的個案研究

Fu-mei Chang Chen, On Analogy in Ch'ing Law (pp. 212-224)

陳張富美:清代法律中的類推[⑤]

---

① 參見普實克著,李歐梵編,郭建玲譯《抒情與史詩——中國現代文學論集》,上海:上海三聯書店,2010。

② 參見韓南著,王秋桂譯《韓南中國小說論集》中同名譯文,北京:北京大學出版社,2008。

③ 參見陳絳譯本,載劉廣京、朱昌峻編《李鴻章評傳》第二編"李鴻章的崛興",上海:上海古籍出版社,1995。

④ 參見韓南著,王秋桂譯《韓南中國小說論集》中同名譯文。

⑤ 參見陳新宇譯文,載《中西法律傳統》第5卷,北京:中國政法大學出版社,2006。

1971年第31卷

James R. Hightower, Allusion in The Poetry of T'ao Ch'ien (pp. 5-27)

海陶瑋:陶潛詩歌中的典故[①]

R. G. Wagner, The Original Structure of The Correspondence Between Shih Hui-Yüan and Kumārajīva (pp. 28-48)

瓦格納:釋慧遠與鳩摩羅什通信的原始構成

Kao Yu-kung and Mei Tsu-lin, Syntax, Diction, and Imagery in T'ang Poetry (pp. 49-136)

高友工、梅祖麟:唐詩的句法、用字與意象[②]

Lawrence D. Kessler, Chinese Scholars and The Early Manchu State (pp. 179-200)

凱思樂:漢族士大夫與早期滿清政權

Patrick Hanan, The Composition of The P'ing Yao Chuan (pp. 201-219)

韓南:《平妖傳》著作問題之研究[③]

1972年第32卷

Britten Dean, Sino-British Diplomacy in The 1860s: The Establishment of The British Concession at Hankow (pp. 71-96)

布里頓·丁: 19世紀60年代時的中英外交:漢口英租界的建立

Li-li Ch'en, Outer and Inner Forms of Chu-kung-tiao, with Reference to Pien-wen, Tz'u and Vernacular Fiction (pp. 124-149)

陳荔荔:諸宮調的内形與外形——兼及諸宮調與變文、詞及宋元白話小説的異同[④]

---

① 參見張宏生譯文,載莫礪鋒編《神女之探尋:英美學者論中國古典詩歌》,上海:上海古籍出版社,1994。

② 參見高友工、梅祖麟著,李世躍譯《唐詩三論》中同名譯文。

③ 參見韓南著,王秋桂譯《韓南中國小説論集》中同名譯文。

④ 參見陳淑英譯文,載《中外文學》第5卷第12期,1977年5月。

1973年第33卷

James R. Hightower, Yüan Chen and "The Story of Ying-Ying" (pp. 90-123)

海陶瑋:元稹和《鶯鶯傳》

Patrick Hanan, The Making of The Pearl-Sewn Shirt and The Courtesan's Jewel Box (pp. 124-153)

韓南:《蔣興哥重會珍珠衫》和《杜十娘怒沉百寶箱》撰述考[①]

Thomas L. Kennedy, Chang Chih-Tung and the Struggle for Strategic Industrialization: The Establishment of the Hanyang Arsenal, 1884—1895 (pp. 154-182)

康念德:張之洞及其工業戰略的探索:漢陽兵工廠之創建(1884—1895)

Ronald C. Miao, The "Ch'i ai shih" of the Late Han and Chin Periods (I) (pp. 183-223)

繆文傑:漢末及西晉時的《七哀詩》(一)

Li-li Ch'en, Some Background Information on The Development of Chu-kung-tiao (pp. 224-237)

陳荔荔:諸宮調形成的背景

1974年第34卷

John W. Dardess, The Cheng Communal Family: Social Organization and Neo-Confucianism in Yüan and Early Ming China (pp. 7-52)

達第斯:義門鄭氏:元代明初的社會組織與理學

Patrick Hanan, The Technique of Lu Hsün's Fiction (pp. 53-96)

韓南:魯迅小説的技巧[②]

Ho Peng Yoke, Goh Thean Chye and David Parker, Po Chü-i's Poems on Immortality (pp. 163-186)

何丙郁等:白居易的神仙詩

① 參見韓南著,王秋桂譯《韓南中國小説論集》中同名譯文。

② 參見韓南著,王秋桂譯《韓南中國小説論集》中同名譯文。

George A. Hayden, The Courtroom Plays of The Yüan and Early Ming Periods (pp. 192-220)

喬治·海頓:元及明初的公堂劇[①]

Albert E. Dien, The Use of The *yeh-hou chia-chuan* as a Historical Source (pp. 221-247)

丁愛博:歷史文獻《鄴侯家傳》之利用

Hans H. Frankel, The Chinese Ballad "Southeast Fly The Peacocks"(pp. 248-271)

傅漢思:中國民謡《孔雀東南飛》

1975年第35卷

Chia-ying Yeh Chao, The Ch'ang-chou School of Tz'u Criticism (pp. 101-132)

葉嘉瑩:常州詞派的詞學批評[②]

Y. W. Ma, The Textual Tradition of Ming Kung-an Fiction: A Study of The *Lung-t'u Kung-an* (pp. 190-220)

馬幼垣:明代公案小説的版本傳統——《龍圖公案》考[③]

1976年第36卷

William T. Graham, Jr., Yü Hsin and "The Lament for the South" (pp. 82-113)

葛藍:庾信與《哀江南賦》

Ronald Egan, On The Origin of The Yu hsien k'u Commentary (pp. 135-146)

艾朗諾:《遊仙窟》注者考源[④]

Francis Woodman Cleaves, A Chinese Source Bearing on Marco Polo's Departure From China and a Persian Source on His Arrival in Persia (pp. 181-203)

柯立夫:關於馬可波羅離華的漢文資料和抵達波斯的波斯文資料

---

① 參見梁曉鶯譯文,載王秋桂編《中國文學論著譯叢》,台北:學生書局,1985。

② 參見葉嘉瑩《常州詞派比興寄託之説的新檢討》,載《清詞論叢》,石家莊:河北教育出版社,2002。

③ 參見静宜文理學院中國古典小説研究中心編《中國古典小説研究專集》(二),台北:聯經出版事業有限公司,1980。

④ 參見卞東波譯文,載《域外漢籍研究集刊》第9輯,北京:中華書局,2013。

1977年第37卷

David Johnson, The Last Years of a Great Clan: The Li Family of Chao chün in Late T'ang and Early Sung (pp. 5-102)

姜士斌：世家大族的没落——唐末宋初的趙郡李氏[①]

James R. Hightower, The Songs of Chou Pang-yen (pp. 233-272)

海陶瑋：論周邦彥的詞

Ronald C. Egan, Narratives in *Tso Chuan* (pp. 323-352)

艾朗諾：《左傳》的記敘文論[②]

1978年第38卷

David M. Farquhar, Emperor as Bodhisattva in The Governance of The Ch'ing Empire (pp. 5-34)

法誇爾：清朝統治時期皇帝的菩薩扮相

Victor H. Mair, Scroll Presentation in The T'ang Dynasty (pp. 35-60)

梅維恒：唐代投卷考

Yu-Kung Kao and Tsu-Lin Mei, Meaning, Metaphor, and Allusion in T'ang Poetry (pp. 281-356)

高友工、梅祖麟：唐詩的語義、隱喻和典故[③]

David N. Keightley, The Bamboo Annals and Shang-Chou Chronology (pp. 423-438)

吉德煒：《竹書紀年》與商周繫年

1979年第39卷

Pei-Yi Wu, Self-Examination and Confession of Sins in Traditional China (pp. 5-38)

吴百益：傳統中國的自省與悔罪

---

① 參見耿立群譯文，載《唐史論文選集》，台北：幼師文化事業公司，1990。

② 參見許章真譯文，載《中國古典小説研究專集》，台北：聯經出版事業有限公司，1983。

③ 參見高友工、梅祖麟著，李世躍譯《唐詩三論》中同名譯文。

William T. Graham, Jr., Mi Heng's "Rhapsody on a Parrot" (pp. 39-54)

葛藍:論禰衡的《鸚鵡賦》

Christopher C. Rand, Li Ch'üan and Chinese Military Thought (pp. 107-137)

蘭德:李筌與中國軍事思想

Derk Bodde, Chinese "Laws of Nature": A Reconsideration (pp. 139-155)

卜德:再論中國的"自然法則"

Stephen Owen, Transparencies: Reading The T'ang Lyric (pp. 231-251)

宇文所安:透明度:解讀唐代的抒情詩[①]

Ronald C. Egan, The Prose Style of Fan Yeh (pp. 339-401)

艾朗諾:范曄的散文風格

Hoyt Cleveland Tillman, Proto-Nationalism in Twelfth-Century China? The Case of Ch'en Liang (pp. 403-428)

田浩:中國12世紀的"原型民族主義"? ——以陳亮爲例

1980年第40卷第1期

Chia-Ying Yeh Chao, On Wang I-sun and His Yung-wu Tz'u (pp. 55-91)

葉嘉瑩:王沂孫及其詠物詞[②]

David Johnson, The Wu Tzu-hsü Pien-wen and Its Sources: Part I (pp. 93-156)

姜士斌:伍子胥變文及其來源(一)[③]

1980年第40卷第2期

Patrick Hanan, Judge Bao's Hundred Cases Reconstructed (pp. 301-323)

韓南:《百家公案》考[④]

---

① 參見陳引馳譯文(黃寶華校),載倪豪士編選《美國學者論唐代文學》,上海:上海古籍出版社,1994。

② 參見作者同名論文,載《文學遺産》1987年6期。

③ 參見蔡振念譯文,全文載《中華文化復興月刊》第16卷第7—9期,1983年7—9月。

④ 參見韓南著,王秋桂譯《韓南中國小說論集》中同名譯文。

Patricia Ebrey, Later Han Stone Inscriptions (pp. 325-353)
伊佩霞:東漢的石刻
John D. Langlois, Jr., Chinese Culturalism and The Yüan Analogy: Seventeenth-Century Perspectives (pp. 355-398)
蘭德璋:中華文化主義以及與元朝的類比:17世紀的視角
David Johnson, The Wu Tzu-hsü Pien-wen and its Sources: Part II (pp. 465-505)
姜士斌:伍子胥變文及其來源(二)

1981年第41卷第1期

David R. Knechtges, Ssu-ma Hsiang-ju' s "Tall Gate Palace Rhapsody" (pp. 47-64)
康達維:司馬相如的《長門賦》①
Dennis Grafflin, The Great Family in Medieval South China (pp. 65-74)
格拉夫林:中古中國南方的世家大族
Katrina C. D. McLeod and Robin D. S. Yates, Forms of Ch' in Law: An Annotated Translation of The *Feng-chen shih* (pp. 111-163)
卡崔娜·麥克里爾德、葉山:秦律的形式:《封診式》譯注
John D. Langlois, Jr., "Living Law" in Sung and Yüan Jurisprudence (pp. 165-217)
蘭德璋:宋元法學中的"活法"②

1981年第41卷第2期

James R. Hightower, The Songwriter Liu Yung: Part I (pp. 323-376)
海陶瑋:詞家柳永(一)
Edward H. Schafer, Wu Yün' s "Cantos on Pacing The Void" (pp. 377-415)
薛愛華:吴筠的《步虚詞》

① 參見康達維著,蘇瑞隆譯《漢代宫廷文學與文化之探微——康達維自選集》中同名譯文,上海:上海譯文出版社,2013。

② 參見李明德、李涵譯文,載高道蘊、高鴻鈞、賀衛方主編《美國學者論中國法律傳統》,北京:中國政法大學出版社,1994。

Benjamin E. Wallacker, The Poet as Jurist: Po Chü-i and a Case of Conjugal Homicide (pp. 507-526)

班傑明·華立克:作爲法理學家的詩人:白居易和一樁婚姻悲劇

K. C. Chang, The Animal in Shang and Chou Bronze Art (pp. 527-554)

張光直:商周青銅器上的動物紋様[1]

1982年第42卷第1期

James R. Hightower, The Songwriter Liu Yung: Part II (pp. 5-66)

海陶瑋:詞家柳永(二)

Willard J. Peterson, Making Connections: "Commentary on The Attached Verbalizations" of The *Book of Change* (pp. 67-116)

裴德生:創建聯繫:《易經·繫辭傳》

Shan Chou, Beginning with Images in The Nature Poetry of Wang Wei (pp. 117-137)

周姍:從王維山水詩中的意象談起

Lynn A. Struve, The Hsü Brothers and Semiofficial Patronage of Scholars in The K'ang-hsi Period (pp. 231-266)

司徒琳:徐乾學三兄弟與康熙朝對學者的半官方扶持

1982年第42卷第2期

Robert M. Hartwell, Demographic, Political, and Social Transformations of China, 750-1550 (pp. 365-442)

郝若貝:750—1550年中國的人口、政治與社會的轉型[2]

Jonathan W. Best, Diplomatic and Cultural Contacts Between Paekche and China (pp. 443-501)

喬納森·百思德:百濟與中國的外交和文化交流

---

① 參見《中國青銅時代》中同名論文,北京:生活·讀書·新知三聯書店,2013。

② 參見林岩譯文,載《新宋學》第三輯,上海:復旦大學出版社,2014。

Roger V. Des Forges, The Story of Li Yen: Its Growth and Function From The Early Ch'ing to The Present (pp. 535-587)
戴福士:李岩故事的產生及其從清初到當下的功用
Glen Dudbridge, Miao-shan on Stone: Two Early Inscriptions (pp. 589-614)
杜德橋:兩方早期石刻上的妙善故事
Daniel L. Overmyer, The White Cloud Sect in Sung and Yüan China (pp. 615-642)
歐大年:宋元時期的白雲教

1983年第43卷第1期

William T. Graham, Jr. and James R. Hightower, Yü Hsin's "Songs of Sorrow" (pp. 5-55)
葛藍、海陶瑋:論庾信的《哀江南賦》
John D. Langlois, Jr. and Sun K'o-K'uan, Three Teachings Syncretism and The Thought of Ming T'ai-tsu (pp. 97-139)
蘭德璋、孫克寬:三教合一與明太祖的思想
Susan W. Chen, The Personal Element in Mao Tun's Early Fiction (pp. 187-213)
陳蘇珊:茅盾早期小說中的個人色彩

1983年第43卷第2期

Pauline R. Yu, Allegory, Allegoresis, and The *Classic of Poetry* (pp. 377-412)
余寶琳:諷寓、諷寓闡釋與《詩經》①
Ronald C. Egan, Poems on Paintings: Su Shih and Huang T'ing-chien(pp. 413-451)
艾朗諾:蘇軾和黃庭堅的題畫詩②
David S. Nivison, The Dates of Western Chou (pp. 481-580)
倪德衛:西周年曆

① 參見曹虹譯文《諷喻與〈詩經〉》,載莫礪鋒編《神女之探尋——英美學者論中國古典詩歌》。
② 參見藍玉、周裕鍇譯文,載莫礪鋒編《神女之探尋——英美學者論中國古典詩歌》。

Arthur N. Waldron, The Problem of The Great Wall of China (pp. 643-663)

林霨:中國的長城問題

1984年第44卷第1期

James R. Hightower, Han Yü as Humorist (pp. 5-27)

海陶瑋:韓愈的幽默

Daniel K. Gardner, Principle and Pedagogy: Chu Hsi and The *Four Books* (pp. 57-81)

賈德納:原則和教條:朱熹與《四書》

Tom Fisher, Loyalist Alternatives in the Early Ch'ing (pp. 83-122)

湯姆·費舍:清初遺民的選擇

Victor H. Mair, Li Po's Letters in Pursuit of Political Patronage (pp. 123-153)

梅維恒:李白的干謁書信

1984年第44卷第2期

Chauncey S. Goodrich, Riding Astride and The Saddle in Ancient China (pp. 279-306)

顧傳習:古代中國的"騎"和馬鞍

Daniel L. Overmyer, Attitudes Toward The Ruler and State in Chinese Popular Religious Literature: Sixteenth and Seventeenth Century Pao-chüan (pp. 347-379)

歐大年:中國民間宗教文學對君主和國家的態度:以十六七世紀的寶卷爲例

Pai Hua-Wen and Victor H. Mair, What is Pien-wen? (pp. 493-514)

白化文著,梅維恒譯:什麽是變文[1]

Allan Barr, The Textual Transmission of *Liaozhai zhiyi* (pp. 515-562)

白亞仁:《聊齋志異》文本源流考

---

① 參見周紹良、白化文編《敦煌變文論文録》,上海:上海古籍出版社,1982。

1985年第45卷第1期

Robert L. Thorp, The Growth of Early Shang Civilization: New Data From Ritual Vessels (pp. 5-75)

杜樸:從新見禮器資料看早商文明的興起

Shan Chou, Allusion and Periphrasis as Modes of Poetry in Tu Fu's "Eight Laments" (pp. 77-128)

周姍:杜甫《八哀詩》中作爲詩歌程式的典故和曲喻

Allan Barr, A Comparative Study of Early and Late Tales in *Liaozhai zhiyi* (pp. 157-202)

白亞仁:《聊齋志異》中早期與晚期故事比較研究

James T. C. Liu, Polo and Cultural Change: From T'ang to Sung China (pp. 203-224)

劉子健:唐宋間的馬球和文化的變遷[①]

Francis Woodman Cleaves, The Eighteenth Chapter of an Early Mongolian Version of the *Hsiao Ching* (pp. 225-254)

柯立夫:早期蒙文譯本《孝經》第十八章

Grace S. Fong, Wu Wenying's Yongwu Ci: Poem as Artifice and Poem as Metaphor (pp. 323-347)

方秀潔:吴文英的詠物詞:作爲隱喻與技巧的詩歌

1985年第45卷第2期

David Johnson, The City-God Cults of T'ang and Sung China (pp. 363-457)

姜士斌:唐宋時期的城隍崇拜

Donald Harper, A Chinese Demonography of the Third Century B. C. (pp. 459-498)

夏德安:公元前3世紀的睡虎地日書研究

① 參見劉子健《南宋中葉馬球衰落和文化的變遷》,載《歷史研究》1980年2期。

Andrew H. Plaks, After the Fall: *Hsing-shih yin-yüan chuan* and the Seventeenth-Century Chinese Novel (pp. 543-580)

浦安迪:墜落之後:《醒世姻緣傳》與17世紀中國小説[1]

Patricia Ebrey, T'ang Guides to Verbal Etiquette (pp. 581-613)

伊佩霞:唐代的書儀

Don J. Wyatt, Chu Hsi's Critique of Shao Yung: One Instance of the Stand Against Fatalism (pp. 649-666)

韋棟:朱熹對邵雍的批評:對宿命論批判之一例

1986年第46卷第1期

Donald Holzman, The Cold Food Festival in Early Medieval China (pp. 51-79)

侯思孟:中國早期中古時代的寒食節[2]

John B. Henderson, Ch'ing Scholars' Views of Western Astronomy (pp. 121-148)

韓德森:清代學者對西方天文學的看法

Edward L. Shaughnessy, On The Authenticity of the Bamboo Annals (pp. 149-180)

夏含夷:《竹書紀年》的真僞[3]

1986年第46卷第2期

Tao-Chung Yao, Ch'iu Ch'u-chi and Chinggis Khan (pp. 201-219)

姚道中:丘處機與成吉思汗

Kang-i Sun Chang, Symbolic and Allegorical Meanings in the Yüeh-fu pu-t'i Poem Series (pp. 353-385)

孫康宜:《樂府補題》中的象徵與托喻[4]

---

① 參見劉倩等譯《浦安迪自選集》中同名譯文,北京:生活·讀書·新知三聯書店,2010。

② 參見楊玉君譯文《中世紀早期的寒食節》,載台灣《民俗曲藝》100期,1996。

③ 參見夏含夷《也談武王的卒年——兼論〈今本竹書紀年〉的真僞》,載《古史異觀》,上海:上海古籍出版社,2005。

④ 參見孫康宜《文學經典的挑戰》中同名論文,南昌:百花洲文藝出版社,2002。

Jonathan Chaves, Moral Action in the Poetry of Wu Chia-chi (1618-84) (pp. 387-469)

齊皎瀚:吴嘉紀(1618—1684)詩中的忠與孝

Dore J. Levy, Constructing Sequences: Another Look at the Principle of *Fu* "Enumeration" (pp. 471-493)

李德瑞:詩歌次序之構建:六義中"賦"之新論[①]

Elizabeth Endicott-West, Imperial Governance in Yüan Times (pp. 523-549)

韋絲特:元帝國的統治

1987年第47卷第1期

Eric Henry, The Motif of Recognition in Early China (pp. 5-30)

艾瑞克·亨利:上古中國"知"的母題

Hoyt Cleveland Tillman, Consciousness of T'ien in Chu Hsi's Thought (pp. 31-50)

田浩:朱熹思想中關於"天"的意識[②]

Theodore Huters, From Writing to Literature: The Development of Late Qing Theories of Prose (pp. 51-96)

胡志德:從文到文學:晚清散文理論的發展

Cynthia Brokaw, Yüan Huang (1533—1606) and The Ledgers of Merit and Demerit (pp. 137-195)

包筠雅:袁黄(1533—1606)與功過格[③]

Robin D. S. Yates, Social Status in The Ch'in: Evidence From The Yün-meng Legal Documents. Part One: Commoners (pp. 197-237)

葉山:雲夢法律文書中所反映的秦代社會狀況(一):普通民衆

① 參見吴伏生譯文,載《國際漢學》第4輯,鄭州:大象出版社,1999。

② 參見田浩《論朱熹和天——跟隨史華兹老師研究宋代思想史》,載《華東師範大學學報》(哲學社會科學版),2008年1期。

③ 參見包筠雅著,杜正貞、張林譯《功過格——明清社會的道德秩序》,杭州:浙江人民出版社,1999。

Donald Harper, Wang Yen-shou's "Nightmare Poem" (pp. 239-283)

夏德安:王延壽的《夢賦》

1987年第47卷第2期

Ying-Shih Yü, "O Soul, Come Back!" A Study in The Changing Conceptions of The Soul and Afterlife in Pre-Buddhist China (pp. 363-395)

余英時:"魂兮歸來!"——論佛教傳入中國以前靈魂與來世觀念的轉變[①]

Anthony C. Yu, "Rest, Rest, Perturbed Spirit!": Ghosts in Traditional Chinese Prose Fiction (pp. 397-434)

余國藩:安魂:傳統中國筆記小説中的鬼魂

Peter K. Bol, Seeking Common Ground: Han Literati Under Jurchen Rule (pp. 461-538)

包弼德:求同:女真統治下的漢族文人

Donald Harper, The Sexual Arts of Ancient China as Described in a Manuscript of The Second Century B. C. (pp. 539-593)

夏德安:2世紀馬王堆帛書中所見古代中國的房中術

Stephen H. West, Cilia, Scale and Bristle: The Consumption of Fish and Shellfish in The Eastern Capital of The Northern Song (pp. 595-634)

奚如谷:纖毛、魚鱗和豬鬃:北宋東京的魚類和貝類消費

1988年第48卷第1期

Hung-Lam Chu, The Debate Over Recognition of Wang Yang-ming (pp. 47-70)

朱鴻林:王陽明從祀孔廟的争議[②]

Susan Wilf Chen, Mao Tun The Translator (pp. 71-94)

陳蘇珊:翻譯家茅盾

---

① 參見余英時著,侯旭東等譯《東漢生死觀》中同名譯文,上海:上海古籍出版社,2005。

② 參見朱鴻林《〈王文成公全書〉刊行與王陽明從祀争議的意義》,載楊聯陞、全漢升、劉廣京主編《國史釋論——陶希聖先生九秩榮慶祝壽論文集》(下),台北:食貨出版社,1988。

Thomas Shiyu Li and Susan Naquin, The Baoming Temple: Religion and The Throne in Ming and Qing China (pp. 131-188)

李世瑜、韓書瑞:保明寺:明清時期的宗教與皇權①

Edward L. Shaughnessy, Historical Perspectives on The Introduction of The Chariot Into China (pp. 189-237)

夏含夷:中國馬車的起源及其歷史意義②

1988年第48卷第2期

Kang-I Sun Chang, The Idea of The Mask in Wu Wei-yeh (1609—1671)

孫康宜:隱情與"面具"——吴梅村詩試説③

Joseph Roe Allen III, From Saint to Singing Girl: The Rewriting of The Lo-fu Narrative in Chinese Literati Poetry (pp. 321-361)

周文龍:從貞女到歌伎:中國文人詩歌中對羅敷故事的改寫

Yenna Wu, The Inversion of Marital Hierarchy: Shrewish Wives and Henpecked Husbands in Seventeenth-Century Chinese Literature (pp. 363-382)

吴燕娜:婚姻中的乾坤倒置——17世紀中國文學中的潑悍之婦與懼内之夫的形象家庭與社會④

Hoyt Cleveland Tillman, Ch'en Liang on Statecraft: Reflections From Examination Essays Preserved in a Sung Rare Book (pp. 403-431)

田浩:《圈點龍川水心二先生文粹》中所見陳亮的治國方略

Stephen F. Teiser, "Having Once Died and Returned to Life": Representations of Hell in Medieval China (pp. 433-464)

太史文:起死回生:中國中古時期對地獄的表現

Jonathan Chaves, The Yellow Mountain Poems of Ch'ien Ch'ien-i (1582—1664): Poetry as Yu-chi (pp. 465-492)

齊皎瀚:作爲遊記的詩歌:錢謙益(1582—1664)的黄山詩

---

① 參見李世瑜本文的初稿《順天保明寺考》,載《北京史苑》1985年3期。

② 參見夏含夷《古史異觀》中同名論文。

③ 參見孫康宜《文學經典的挑戰》中同名論文。

④ 參見張國剛編《家庭與社會》中同名論文,北京:清華大學出版社,2009。

1989年第49卷第1期

Zbigniew Slupski, Three Levels of Composition of the *Rulin Waishi* (pp. 5-53)

史羅甫:《儒林外史》創作的三層結構

Anthony C. Yu, The Quest of Brother Amor: Buddhist Intimations in The Story of The Stone (pp. 55-92)

余國藩:情僧的索問:《紅樓夢》的佛教隱意[①]

Charles Holcombe, The Exemplar State: Ideology, Self-Cultivation, and Power in Fourth-Century China (pp. 93-139)

何肯:塑造範型:4世紀中國的意識形態、修身與權力

Daniel K. Gardner, Transmitting The Way: Chu Hsi and His Program of Learning (pp. 141-172)

賈德納:傳道:朱熹和他的致知論

1989年第49卷第2期

John W. Dardess, A Ming Landscape: Settlement, Land Use, Labor, and Estheticism in T'ai-ho County, Kiangsi (pp. 295-364)

約翰·W. 達第斯:一幅明朝景觀:從文學看江西泰和縣的居民點、土地使用和勞動力[②]

Ronald C. Egan, Ou-yang Hsiu and Su Shih on Calligraphy (pp. 365-419)

艾朗諾:歐陽修與蘇軾論書法

Kidder Smith, Jr., *Zhouyi* Interpretation From Accounts in The *Zuozhuan* (pp. 421-463)

蘇德愷:《左傳》中所見的《周易》闡釋

Timothy Brook, Funerary Ritual and The Building of Lineages in Late Imperial China (pp. 465-499)

卜正民:中國明清時期的喪禮與世系之建構

---

① 參見林國華譯文,載樂黛雲、陳珏編《北美中國古典文學研究名家十年文選》,南京:江蘇人民出版社,1996。

② 參見王波譯文,載《農業考古》1995年1期、3期。

Allan Barr, Disarming Intruders: Alien Women in *Liaozhai zhiyi* (pp. 501-517)

白亞仁:去妖入凡:《聊齋志異》中的奇女子

1990年第50卷第1期

Christoph Harbsmeier, Confucius Ridens: Humor in The *Analect*s (pp. 131-161)

何莫邪:孔子的樂:《論語》中的幽默

Pauline Yu, Poems in Their Place: Collections and Canons in Early Chinese Literature (pp. 163-196)

余寶琳:詩歌的定位:早期中國文學的選集與經典[①]

Daniel L. Overmyer, Buddhism in The Trenches: Attitudes Toward Popular Religion in Chinese Scriptures Found at Tun-Huang (pp. 197-222)

歐大年:與世隔絶的佛教:敦煌所見漢文抄本中的民間宗教

Joe Eng, Laughter in a Dismal Setting: Humorous Anecdotes in The *Kuei-ch' ien chih* (pp. 223-238)

喬·英格:苦中作樂:《歸潛志》中的奇聞趣事

Pei Huang, New Light on The Origins of The Manchus (pp. 239-282)

黄培:滿族起源新探

1990年第50卷第2期

Stephen Owen, Place: Meditation on the Past at Chin-ling (pp. 417-457)

宇文所安:地:金陵懷古[②]

Grace S. Fong, Persona and Mask in the Song Lyric (*Ci*) (pp. 459-484)

方秀潔:詞中的代言與面具

Ronald C. Egan, Su Shih's "Notes" as a Historical and Literary Source (pp. 561-588)

艾朗諾:作爲歷史和文學研究資料的蘇軾書簡

---

① 參見何鯉譯文,載樂黛雲、陳珏編《北美中國古典文學研究名家十年文選》。

② 參見陳躍紅、王軍譯文,載樂黛雲、陳珏編《北美中國古典文學研究名家十年文選》。

Jennifer W. Jay, Memoirs and Official Accounts: The Historiography of the Song Loyalists (pp. 589-612)

謝慧賢:回憶録與正史:南宋遺民的編史工作

Angelina Yee, Counterpoise in *Honglou meng* (pp. 613-650)

余珍珠:《紅樓夢》中的復筆與互現

1991年第51卷第1期

Shan Chou, Tu Fu's Social Conscience: Compassion and Topicality in His Poetry (pp. 5-53)

周姍:杜甫的社會責任感:杜詩中的惻隱之心與時事

Yenna Wu, Repetition in *Xingshi yinyuan zhuan* (pp. 55-87)

吴燕娜:《醒世姻緣傳》中的復疊錯綜

Scott Pearce, Status, Labor, and Law: Special Service Households under The Northern Dynasties (pp. 89-138)

裴士凱:地位、工作與刑律:北朝特殊的役户

1991年第51卷第2期

Victor H. Mair and Tsu-Lin Mei, The Sanskrit Origins of Recent Style Prosody (pp. 375-470)

梅維恒、梅祖麟:近體詩律的梵文來源[①]

Harold D. Roth, Psychology and Self-Cultivation in Early Taoistic Thought (pp. 599-650)

羅浩:早期道家思想中的心理學和自我修煉

Richard Von Glahn, The Enchantment of Wealth: The God Wutong in the Social History of Jiangnan (pp. 651-714)

萬志英:財富的法術:江南社會史中的"五通神"[②]

---

① 參見王繼紅譯文,載張西平主編《國際漢學》第16輯,鄭州:大象出版社,2007。

② 中譯本可參劉永華編《中國社會文化史讀本》,北京:北京大學出版社,2011;陳仲丹譯《中國大衆宗教》,南京:江蘇人民出版社,2006。

1992年第52卷第1期

Wu Hung, What is Bianxiang? On The Relationship Between Dunhuang Art and Dunhuang Literature (pp. 111-192)

巫鴻:何爲變相?——兼論敦煌藝術與敦煌文學的關係[①]

Meir Shahar, The Lingyin Si Monkey Disciples and The Origins of Sun Wukong (pp. 193-224)

夏維明:靈隱寺猴徒與孫悟空的原型

Paula M. Varsano, Immediacy and Allusion in The Poetry of Li Bo (pp. 225-261)

方葆珍:李白詩歌中的直言和典故

Stuart H. Sargent, Colophons in Countermotion: Poems by Su Shih and Huang T'ing-chien on Paintings (pp. 263-302)

薩進德:另類的題跋:蘇軾和黄庭堅的題畫詩

1992年第52卷第2期

Eric Henry, Chu-ko Liang in The Eyes of His Contemporaries (pp. 589-612)

艾瑞克·亨利:同時代人物眼中的諸葛亮

Hok-Lam Chan, The Organization and Utilization of Labor Service under The Jurchen Chin Dynasty (pp. 613-664)

陳學霖:金朝勞動力的組織和利用

1993年第53卷第1期

Michael A. Fuller, Pursuing The Complete Bamboo in The Breast: Reflections on a Classical Chinese Image for Immediacy (pp. 5-23)

傅君勱:追求胸中之成竹:對中國古典意象直接性的反思

Leo Tak-Hung Chan, Narrative as Argument: The *Yuewei caotang biji* and The Late Eighteenth-Century Elite Discouse on The Supernatural(pp. 25-62)

陳德鴻:志怪爲證:《閱微草堂筆記》與18世紀後期士大夫的鬼神論述

---

① 參見巫鴻著,鄭岩等譯《禮儀中的美術》(下),北京:生活·讀書·新知三聯書店,2005。

Pamela Kyle Crossley and Evelyn S. Rawski, A Profile of The Manchu Language in Ch'ing History (pp. 63-102)

柯嬌燕、羅友枝:清朝滿語概觀

1993年第53卷第2期

Karen Turner, War, Punishment, and The Law of Nature in Early Chinese Concepts of The State (pp. 285-324)

高道蘊:早期中國國家觀念中的戰爭、懲罰與自然法

Alice W. Cheang, Poetry, Politics, Philosophy: Su Shih as The Man of The Eastern Slope (pp. 325-387)

鄭文君:詩歌,政治,哲理:作爲東坡居士的蘇軾[①]

1994年第54卷第1期

Susan Cherniack, Book Culture and Textual Transmission in Sung China (pp. 5-125)

蘇珊·徹尼亞克:宋代的書籍文化與文本傳播

Judith T. Zeitlin, Shared Dreams: The Story of The Three Wives' Commentary on *The Peony Pavilion* (pp. 127-179)

蔡九迪:一人同夢:三婦合評牡丹亭[②]

1994年第54卷第2期

Wai-Yee Li, The Idea of Authority in the *Shih chi (Records of the Historian)* (pp. 345-405)

李惠儀:《史記》中的權威觀

Paula M. Varsano, The Invisible Landscape of Wei Yingwu (pp. 407-435)

方葆珍:韋應物詩中的無形山水

① 參見卞東波、鄭瀟瀟、劉傑譯文,載《中國蘇軾研究》第5輯,北京:學苑出版社,2015。

② 參見解芳譯文,載徐永明、陳靝沅主編《英語世界的湯顯祖論文集》,杭州:浙江大學出版社,2013。

Grace S. Fong, Inscribing Desire: Zhu Yizun's Love Lyrics in *Jingzhiju qingqu* (pp. 437-460)

方秀潔:刻寫欲望:《静志居情趣》中朱彝尊的豔詞

1995年第55卷第1期

Nathan Sivin, State, Cosmos, and Body in The Last Three Centuries B. C. (pp. 5-37)

席文:公元前三個世紀觀念中的國家、宇宙和身體

Stephen Owen, The Formation of The Tang Estate Poem (pp. 39-59)

宇文所安:唐代别業詩的形成[①]

1995年第55卷第2期

Laura Hua Wu, From *Xiaoshuo* to Fiction: Hu Yinglin's Genre Study of *Xiaoshuo* (pp. 339-371)

吴華:從"小説"到虛構作品:胡應麟關於"小説"文類之研究

Angelina C. Yee , Self, Sexuality, and Writing in *Honglou meng* (pp. 373-407)

余珍珠:《紅樓夢》中的自我、性事與書寫

Philip A. Kuhn, Ideas Behind China's Modern State (pp. 295-337)

孔飛力:中國現代國家的觀念

1996年第56卷第1期

Ellen Widmer, The Huanduzhai of Hangzhou and Suzhou: A Study in Seventeenth-Century Publishing (pp. 77-122)

魏愛蓮:蘇杭還讀齋:17世紀出版業之個案研究

Xiaoshan Yang, Having It Both Ways: Manors and Manners in Bai Juyi's Poetry (pp. 123-149)

楊曉山:其道兩全:白居易詩歌中的園林與生活方式[②]

---

① 參見陳磊譯文,載《古典文學知識》1997年6期、1998年1期。

② 參見楊曉山著,文韜譯《私人領域的變形》,南京:江蘇人民出版社,2009。

1996年第56卷第2期

Paula M. Varsano, Getting There From Here: Locating the Subject in Early Chinese Poetics (pp. 375-403)

方葆珍:由此及彼:尋找早期中國詩學中的主體[①]

James M. Hargett, Song Dynasty Local Gazetteers and Their Place in The History of *Difangzhi* Writing (pp. 405-442)

何瞻:宋代方志及其在中國地方志史上的地位

1997年第57卷第1期

Ronald Egan, The Controversy Over Music and "Sadness" and Changing Conceptions of The *Qin* in Middle Period China (pp. 5-66)

艾朗諾:中國中古時期關於音樂與"悲"的争論以及琴學觀念之變遷

Stephen H. West, Playing With Food: Performance, Food, and The Aesthetics of Artificiality in The Sung and Yuan (pp. 67-106)

奚如谷:食之戲:宋元演劇與食物的"假扮"美學[②]

Allan H. Barr, The Wanli Context of The "Courtesan's Jewel Box" Story (pp. 107-141)

白亞仁:萬曆文化背景下的《杜十娘怒沉百寶箱》

Jeffrey Riegel, Eros, Introversion, and The Beginnings of *Shijing* Commentary (pp. 143-177)

王安國:情欲、内省與《詩經》的早期注釋

1997年第57卷第2期

Catherine Vance Yeh, The Life-Style of Four *Wenren* in Late Qing Shanghai (pp. 419-470)

葉凱蒂:四個晚清上海文人的生活方式

① 參見張萬民、張楣楣譯文,載《古代文學理論研究》第35輯,上海:華東師範大學出版社,2013。

② 參見王敦、張舒然譯文,載《文化遺産》2013年1期。

Michael Puett, Nature and Artifice: Debates in Late Warring States China Concerning The Creation of Culture (pp. 471-518)

普鳴：自然與機巧：中國戰國後期有關文化創生的争論

Haun Saussy, Repetition, Rhyme, and Exchange in *The Book of Odes* (pp. 519-542)

蘇源熙：《詩經》中的復沓、韻律和互換[①]

1998年第58卷第1期

Charles Hartman, The Making of a Villain: Ch'in Kuei and Tao-hsüeh (pp. 59-146)

蔡涵墨：一個邪惡形象的塑造：秦檜與道學[②]

Alice W. Cheang, Poetry and Transformation: Su Shih's Mirage (pp. 147-182)

鄭文君：詩與變：蘇軾的幻想

1998年第58卷第2期

Patrick Hanan, *Fengyue Meng* and the Courtesan Novel (pp. 345-372)

韓南：《風月夢》與青樓小説[③]

Wei Shang, Ritual, Ritual Manuals, and the Crisis of the Confucian World: An Interpretation of *Rulin waishi* (pp. 373-424)

商偉：儒家世界的禮儀、禮書和危機：解讀《儒林外史》[④]

Michael Puett, Sages, Ministers, and Rebels: Narratives from Early China Concerning the Initial Creation of the State (pp. 425-479)

普鳴：聖人、衆臣和逆賊：上古中國關於國家初建的敘事

---

① 參見卞東波、許曉穎譯文，載蘇源熙著《中國美學問題》附録，南京：江蘇人民出版社，2009。

② 參見楊立華譯文，載田浩編，楊立華、吴艷紅等譯《宋代思想史論》，北京：社會科學文獻出版社，2003。

③ 參見宋莉華譯文，載《上海師範大學學報》(哲學社會科學版)2004年1期。

④ 參見商偉《禮與十八世紀的文化轉折：〈儒林外史〉研究》，北京：生活·讀書·新知三聯書店，2012。

1999年第59卷第1期

Sally K. Church, Beyond the Words: Jin Shengtan's Perception of Hidden Meanings in *Xixiang ji* (pp. 5-77)

程思麗：弦外之音：金聖歎對《西廂記》言外之意的解讀

David M. Robinson, Politics, Force and Ethnicity in Ming China: Mongols and the Abortive Coup of 1461 (pp. 79-123)

魯大维：明代中國的政治、武力和民族：蒙古和天順五年的未遂政變

Eric Henry, "Junzi Yue" versus "Zhongni Yue" in *Zuozhuan* (pp. 125-161)

艾瑞克·亨利：《左傳》中"君子曰"和"仲尼曰"之比較

1999年第59卷第2期

David Schaberg, Song and the Historical Imagination in Early China (pp. 305-361)

史嘉柏：歌與上古中國的歷史想象

Wai-Yee Li, Heroic Transformations: Women and National Trauma in Early Qing Literature (pp. 363-443)

李惠儀：英雄的變形：清初文學中的女性與國族創傷

Emma Teng, Taiwan as a Living Museum: Tropes of Anachronism in Late-Imperial Chinese Travel Writing (pp. 445-484)

鄧津華：作爲活的博物館的台灣：明清時期遊記中"時光倒錯"之隱喻

2000年第60卷第1期

Bryna Goodman, Being Public: The Politics of Representation in 1918 Shanghai (pp. 45-88)

顧德曼：走向共和：1918年上海的代議政治

Xiangyun Wang, The Qing Court's Tibet Connection: Lcang skya Rol pa'i rdo rje and the Qianlong Emperor (pp. 125-163)

王祥雲：清廷與西藏的紐帶：章嘉·若必多吉與乾隆皇帝

Eva Shan Chou, Tu Fu's "General Ho" Poems: Social Obligations and Poetic Response (pp. 165-204)

周姗:杜甫的《陪鄭廣文遊何將軍山林十首》:社會責任與詩意的回應

2000年第60卷第2期

Colin Hawes, Meaning beyond Words: Games and Poems in the Northern Song (pp. 355-383)

柯霖:言外之意:北宋的遊戲與詩歌①

Allan H. Bar, The Early Qing Mystery of the Governor's Stolen Silver (pp. 385-412)

白亞仁:清初官銀被盜之謎

Patrick Hanan, The Missionary Novels of Nineteenth-Century China (pp. 413-443)

韓南:19世紀中國的傳教士小説②

2001年第61卷第1期

Anthony DeBlasi, Striving for Completeness: Quan Deyu and the Evolution of the Tang Intellectual Mainstream (pp. 5-36)

鄧百安:"全"之追求:權德輿和唐代思想主潮的演進

Peter K. Bol, The Rise of Local History: History, Geography, and Culture in Southern Song and Yuan Wuzhou (pp. 37-76)

包弼德:地方史的興起:南宋至元代婺州的歷史、地理與文化③

Sophie Volpp, Classifying Lust: The Seventeenth-Century Vogue for Male Love (pp. 77-117)

袁書菲:规範欲望:17世紀的男色風尚

① 參見張麗華譯文,載《西南交通大學學報》(社會科學版)2004年5期。

② 參見韓南著,徐俠譯《中國近代小説的興起》,上海:上海教育出版社,2004。

③ 參見吴松弟譯文,載《歷史地理》第21輯,2006。

2001年第61卷第2期

Stephen Owen, Reproduction in the *Shijing (Classic of Poetry)* (pp. 287-315)

宇文所安:《詩經》中的繁殖與再生[①]

Charles Hartman, Li Hsin-ch'uan and the Historical Image of Late Sung Tao-hsüeh (pp. 317-358)

蔡涵墨:李心傳與晚宋道學的歷史形象

Meir Shahar, Ming-Period Evidence of Shaolin Martial Practice (pp. 359-413)

夏維明:明代少林武事考

2002年第62卷第1期

Beverly Bossler, Shifting Identities: Courtesans and Literati in Song China (pp. 5-37)

柏文莉:身份變化:中國宋朝藝妓與士人[②]

Steven B. Miles, Rewriting the Southern Han (917-971): The Production of Local Culture in Nineteenth-Century Guangzhou (pp. 39-75)

麦维哲:重寫南漢(917—971):19世紀廣州的本土文化之産生

Vincent Goossaert, Starved of Resources: Clerical Hunger and Enclosures in Nineteenth-Century China (pp. 77-133)

高萬桑:匱於食用:中國19世紀僧道的忍饑與坐關

2002年第62卷第2期

Scott Cook, The *Lüshi chunqiu* and the Resolution of Philosophical Dissonance (pp. 307-345)

顧史考:《吕氏春秋》與哲學分歧的調和

---

① 參見宇文所安著,田曉菲譯《他山的石頭記》中同名譯文,南京:江蘇人民出版社,2003。

② 參見伊沛霞、姚平主編《當代西方漢學研究集萃·婦女史卷》中同名譯文,上海:上海古籍出版社,2012。

Xiaofei Tian, A Preliminary Comparison of the Two Recensions of *Jinpingmei* (pp. 307-345)

田曉菲:《金瓶梅》詞話本與繡像本的初步比較

2003年第63卷第1期

Li Feng, "Feudalism" and Western Zhou China: A Criticism (pp. 115-144)

李峰:"Feudalism"和西周時期的中國:一種批判

François Louis, The Genesis of an Icon: The *Taiji* Diagram's Early History (pp. 145-196)

弗朗索瓦·路易:聖圖的起源:太極圖的早期史

Patrick Hanan, The Bible as Chinese Literature: Medhurst, Wang Tao, and the Delegates' Version (pp. 197-239)

韓南:作爲中國文學的《聖經》:麥都思、王韜與《聖經》委辦本①

2003年第63卷第2期

Liu Xiaogan, From Bamboo Slips to Received Versions: Common Features in the Transformation of the *Laozi* (pp. 337-382)

劉笑敢:《老子》演變中的趨同現象——從簡帛本到通行本②

Martin Kern, Western Han Aesthetics and the Genesis of the *Fu* (pp. 383-437)

柯馬丁:西漢美學與賦體的起源③

2004年第64卷第1期

Tobie Meyer-Fong, Packaging the Men of Our Times: Literary Anthologies, Friendship Networks, and Political Accommodation in the Early Qing (pp. 5-56)

梅爾清:天下英才,盡入吾彀:清初的文學總集、交遊網絡與政治妥協

---

① 參見段懷清譯文,載《浙江大學學報》(人文社會科學版)2010年2期。

② 參見陳静譯文,載《文史》2004年1輯。

③ 參見復旦大學文史研究院編《着壁成繪》(復旦文史講堂之二)中同名譯文,北京:中華書局,2009。

Xun Liu, Visualizing Perfection: Daoist Paintings of Our Lady, Court Patronage, and Elite Female Piety in the Late Qing (pp. 57-115)
劉訊:圖繪完美:晚清道教的聖母畫、朝廷扶持與上層女性的虔信
Man-Houng Lin, Late Qing Perceptions of Native Opium (pp. 117-144)
林滿紅:晚清土產鴉片觀念[①]

2004年第64卷第2期

Margaret Baptist Wan, The *Chantefable* and the Novel: The Cases of *Lü mudan* and *Tianbao tu* (pp. 367-397)
包美歌:唱本與小説:以《緑牡丹》和《天豹圖》爲例[②]

2005年第65卷第1期

Xiaofei Tian, Illusion and Illumination: A New Poetics of Seeing in Liang Dynasty Court Literature (pp. 7-56)
田曉菲:幻與照:梁代宫廷文學中新的觀看詩學[③]
Jack W. Chen, The Writing of Imperial Poetry in Medieval China (pp. 57-98)
陳威:中國中古時代的帝王詩歌創作
Sophie Volpp, The Gift of a Python Robe: The Circulation of Objects in *Jin Ping Mei* (pp. 133-158)
袁書菲:蟒袍之禮:《金瓶梅》中物的流轉

2005年第65卷第2期

K. E. Brashier, Symbolic Discourse in Eastern Han Memorial Art: The Case of the Birchleaf Pear (pp. 281-310)
白瑞旭:東漢墓葬畫像石中的象徵性話語:以甘棠爲例

---

① 參見唐博譯文(董建中、林滿紅校),載《清史譯業》第8輯,北京:中國人民大學出版社,2010。
② 參見《國際漢學》第22輯中譯文,題作《唱本對小説的影響》,鄭州:大象出版社,2012。
③ 參見田曉菲《烽火與流星:蕭梁王朝的文學與文化》,北京:中華書局,2010。

Michael A. Fuller, Aesthetics and Meaning in Experience: A Theoretical Perspective on Zhu Xi's Revision of Song Dynasty Views of Poetry (pp. 311-355)

傅君勱:經驗中的美學與意義:朱熹對宋代詩歌觀念改造的理論考察

Ling Hon Lam, The Matriarch's Private Ear: Performance, Reading, Censorship, and the Fabrication of Interiority in *The Story of the Stone* (pp. 357-415)

林凌瀚:賈母的耳朵:《紅樓夢》中的演劇、讀戲本、审查與情之滋生

Edward L. Shaughnessy, The Guodian Manuscripts and Their Place in Twentieth-Century Historiography on the *Laozi* (pp. 417-457)

夏含夷:郭店楚簡及其在20世紀《老子》學史中的地位

2006年第66卷第1期

Sarah M. Allen, Tales Retold: Narrative Variation in a Tang Story (pp. 105-143)

艾文嵐:重講故事:一篇唐傳奇的不同敘述

Hilde De Weerdt, Byways in the Imperial Chinese Information Order: The Dissemination and Commercial Publication of State Documents (pp. 145-188)

魏希德:中國帝制時期情報秩序中未開拓的一面:政府文書的傳播與商業出版

Bruce Rusk, Not Written in Stone: Ming Readers of the *Great Learning* and the Impact of Forgery (pp. 189-231)

阮思德:未刻於石:明代《大學》的讀者與僞書《石經大學》的衝擊

2006年第66卷第2期

Anna M. Shields, Remembering When: The Uses of Nostalgia in the Poetry of Bai Juyi and Yuan Zhen (pp. 321-361)

田安:記住彼時:懷舊在白居易和元積詩歌中的表現

Paul Jakov Smith, *Shuihu zhuan* and the Military Subculture of the Northern Song, 960-1127 (pp. 363-422)

史樂民:《水滸傳》和北宋(960—1127)的軍事亞文化

Charlotte Furth, The Physician as Philosopher of the Way: Zhu Zhenheng (1282-1358) (pp. 423-459)

費俠莉:作爲理學家的醫者:朱震亨(1282—1358)

Chang Woei Ong, The Principles Are Many: Wang Tingxiang and Intellectual Transition in Mid-Ming China (pp. 461-493)

王昌偉:"理"者有萬:王廷相與明清時期的思想轉型

2007年第67卷第1期

Paize Keulemans, Listening to the Printed Martial Arts Scene: Onomatopoeia and the Qing Dynasty Storyteller's Voice (pp. 51-87)

古柏:聆聽刻本中武打場面:擬聲法與清代説書人的聲音

Reiko Shinno, Medical Schools and the Temples of the Three Progenitors in Yuan China: A Case of Cross-Cultural Interactions (pp. 89-133)

秦玲子:中國元代時的醫學校與三皇廟:跨文化互動的個案

Paul R. Goldin, Xunzi and Early Han Philosophy (pp. 135-166)

金鵬程:荀子與漢初哲學

2007年第67卷第2期

Rania Huntington, Memory, Mourning, and Genre in the Works of Yu Yue (pp. 253-293)

韓瑞亞:俞樾著作中的記憶、悼亡與文體

Stephen Owen, The Manuscript Legacy of the Tang: The Case of Literature (pp. 295-326)

宇文所安:唐代的手抄本遺産:以文學爲例[1]

Khee Heong Koh, Enshrining the First Ming Confucian (pp. 327-374)

許齊雄:第一位從祀孔廟的明代儒學家

[1] 參見卞東波、許曉穎譯文,載《古典文獻研究》第15輯,南京:鳳凰出版社,2012。

2008年第68卷第1期

Andrea S. Goldman, Actors and Aficionados in Qing Dynasty Texts of Theatrical Connoisseurship (pp. 1-56)

郭安瑞:清代"花譜"中的演員與戲迷

James A. Benn, Another Look at the Pseudo-Śūraṃgama sūtra (pp. 57-89)

貝劍銘:從另一個角度看僞本《楞嚴經》

Hok-Lam Chan, The "Song" Dynasty Legacy: Symbolism and Legitimation from Han Liner to Zhu Yuanzhang of the Ming Dynasty (pp. 91-133)

陳學霖:"宋"朝的遺産:從韓林兒到明朱元璋的"國號"與政權合法性

2008年第68卷第2期

Yuming He, Difficulties of Performance: The Musical Career of Xu Wei's: *The Mad Drummer* (pp. 77-114)

何予明:表演的困境:徐渭《狂鼓史》的音樂特質

Seunghyun Han, Bandit or Hero? Memories of Zhang Shicheng in Late Imperial and Republican Suzhou (pp. 115-162)

韓承賢:强盜還是英雄?——明清及民國時期蘇州人記憶中的張士誠

2009年第69卷第1期

Pauline Lin, Rediscovering Ying Qu and His Poetic Relationship to Tao Qian (pp. 37-74)

林葆玲:重審應璩與陶潛之間的詩學聯繫[1]

Xiaorong Li, Gender and Textual Politics during the Qing Dynasty: The Case of the *Zhengshi ji* (pp. 75-107)

李小榮:清代的性别與文本政治:以《正始集》爲例

① 參見卞東波譯文,載《古典文獻研究》第16輯,南京:鳳凰出版社,2013。

2009年第69卷第2期

Natasha Heller, The Chan Master as Illusionist: Zhongfeng Mingben's *Huanzhu Jiaxun* (pp. 271-308)

賀納嫻:作爲幻術師的禪師:中峰明本的《幻住家訓》

Carrie Reed, Parallel Worlds, Stretched Time, and Illusory Reality: The Tang Tale *Du Zichun* (pp. 309-342)

卡麗·里德:平行的世界、延展的時間和虚幻的現實:論唐傳奇《杜子春》

Lynn A. Struve, Self-Struggles of a Martyr: Memories, Dreams, and Obsessions in the Extant Diary of Huang Chunyao (pp. 343-394)

司徒琳:殉道者的自我掙扎:黄淳耀現存日記中的記憶、夢境與癡迷

Judith T. Zeitlin, The Cultural Biography of a Musical Instrument: Little Hulei as Sounding Object, Antique, Prop, and Relic (pp. 395-441)

蔡九迪:樂器、古董、道具、遺物——小忽雷文化傳記[①]

2010年第70卷第1期

Xiaoshan Yang, Tradition and Individuality in Wang Anshi's *Tang bai jia shixuan* (pp. 105-145)

楊曉山:王安石《唐百家詩選》中的傳統與個性

Matthew W. Mosca, Empire and the Circulation of Frontier Intelligence: Qing Conceptions of the Ottomans (pp. 147-207)

馬世嘉:帝國和邊疆情報的流通:清人觀念中的奥斯曼帝國

2010年第70卷第2期

Tamara T. Chin, Defamiliarizing the Foreigner: Sima Qian's Ethnography and Han-Xiongnu Marriage Diplomacy (pp. 311-354)

秦大倫:異族人的陌生化:司馬遷的《匈奴列傳》與漢匈和親

① 參見宋巧燕譯文,載《戲曲研究》第84、85輯,2012。

Wendy Swartz, Naturalness in Xie Lingyun's Poetic Works (pp. 355-386)

田菱:謝靈運詩歌中的自然

Cho-ying Li,Charles Hartman, A Newly Discovered Inscription by Qin Gui: Its Implications for the History of Song *Daoxue* (pp. 387-448)

李卓穎、蔡涵墨:新近面世之秦檜碑記及其在宋代道學史中的意義[1]

Norman A. Kutcher, Unspoken Collusions: The Empowerment of Yuanming yuan Eunuchs in the Qianlong Period (pp. 449-495)

柯啓玄:私下的勾結:乾隆時期圓明園宦官的掌權

2011年第71卷第1期

Ao Wang, Poetry Matters: Interpretative Community, *pailü*, and "Yingying zhuan" (pp. 1-34)

王敖:詩之重要性:解釋團體、排律與《鶯鶯傳》

Linda Rui Feng, Chang'an and Narratives of Experience in Tang Tales(pp. 35-68)

馮令晏:長安及其在唐人小説敘事中的呈現

Elisabeth Kaske, Fund-Raising Wars: Office Selling and Interprovincial Finance in Nineteenth-Century China (pp. 69-141)

白莎:融資戰爭:捐納制度與19世紀中國的省際財政

2011年第71卷第2期

Yiqun Zhou, Temples and Clerics in *Honglou meng* (pp. 263-309)

周軼群:《紅樓夢》中的寺觀與僧侶道士

2012年第72卷第1期

Beverly Bossler, Vocabularies of Pleasure: Categorizing Female Entertainers in the Late Tang Dynasty (pp. 71-99)

柏文莉:娛悦的字眼:晚唐歌妓的分類

① 參見姜錫東編《宋史研究論叢》(第12輯)中同名論文,保定:河北大學出版社,2011。

2012年第72卷第2期

Jie Shi, "My Tomb Will Be Opened in Eight Hundred Years": A New Way of Seeing the Afterlife in Six Dynasties China (pp. 217-257)

石介:"天度八百而後開吾墓":中國六朝"來世"觀念新探

Jeffrey Moser, The Ethics of Immutable Things: Interpreting Lü Dalin's *Illustrated Investigations of Antiquity* (pp. 259-293)

孟絜予:永恒之物的倫理:吕大臨《考古圖》考

Wai-Yee Li, Gardens and Illusions from Late Ming to Early Qing (pp. 295-336)

李惠儀:從明末到清初的園林及幻像

2013年第73卷第1期

Sukhee Lee, Cooperation and Tension: Revisiting Local Activism in the Southern Song Dynasty (pp. 43-82)

李蘇姬:合作與緊張:重探南宋的地方能動性

2013年第73卷第2期

David Brophy, The Junghar Mongol Legacy and the Language of Loyalty in Qing Xinjiang (pp. 231-258)

大衛·布羅菲:清朝新疆地區準格爾蒙古人的遺産與語言選擇

Lawrence Zhang, Legacy of Success: Office Purchase and State-Elite Relations in Qing China (pp. 259-297)

張樂翔:成功的遺産:捐納與清代的"國家—精英"關係

2014年第74卷第1期

Xiaoqiao Ling, Law, Deities, and Beyond: From the *Sanyan* Stories to *Xingshi yinyuan zhuan* (pp. 1-42)

凌筱嶠:律條、神明與超越:從"三言"到《醒世姻緣傳》

2014年第74卷第2期

Sookja Cho, Within and Between Cultures: The Liang-Zhu Narrative in Local Korean Cultures (pp. 207-248)

曹淑子:文化之中與文化之間:韓國本土文化中的梁祝敘事

Richard von Glahn, The Ningbo-Hakata Merchant Network and the Reorientation of East Asian Maritime Trade, 1150-1350 (pp. 249-279)

萬志英:寧波-博多的商人網絡與東亞海上貿易的重組(1150—1350)

基地紀事

國際漢學系列講座紀要(2015.01—2015.06)

國際漢學系列講座·第七十三講

題　目: 西方儀式研究(Ritual Studies)與中國禮學異同辨

主講人: 英國牛津大學聖安學院　羅伯特·恰德(Robert Chard)　教授

主持人: 北京大學中國古文獻研究中心　劉玉才　教授

時　間: 2015年1月5日(星期一)下午

地　點: 北京大學國際漢學家研修基地學術報告廳

羅伯特·恰德(Robert Chard)教授爲牛津大學聖安學院副院長、日本東京大學的客座教授和《國際亞洲研究學刊》的執行主編,長期致力於中國文化的研究,尤其關注傳統禮學。

中國禮學和西方Ritual Studies之間的異同問題,直接關涉到中西方學者之前的對話交流。目前在很多國際會議上,"禮學"被徑直譯爲Ritual Studies,羅伯特先生認爲這種作法是不合適的。西方的Ritual Studies理論常被用來分析中國的社會、政治或者宗教,對西方漢學有一定的影響。羅伯特先生認爲我們首先應該弄明白Ritual Studies是一門什麽樣的學問,其學科自身又有着怎樣的歷史。

西方的Ritual Studies是一門跨學科的學問,它並不是如歷史、地理一類有着清晰界定的學科。其研究對象涉及全人類的衆多儀式禮儀,研究者主要用人類學和史學的方法,兼及心理學等學科。這門學科的主要特點是非常的理論化,很抽象,倚重方法。羅伯特教授認爲Ritual Studies在西方學術界是一個非常"時髦"的學問,頗多不需要提及的地方,也會以之來分析問題。近幾十年來,這門學問廣受關注,出現了專門期刊,如*Journal of Ritual Studies*等。實際上,Ritual Studies研究從一開始就存在這樣的問題,那就是Ritual Studies研究源自西方,天然帶着西方學術自身的偏見。

關於Ritual Studies的歷史,治歐洲上古禮儀之學的Philippe Buc教授,認爲西方的Ritual的觀念並不强,像古希臘羅馬本没有關於抽象化的Ritual概念。直到16世紀,耶穌會從天主教中分了出來,批評天主教"没有意義的儀式太多,脱離了上帝,没有神聖的内容",所以天主教徒出於護教的考慮,開始研究古書,包括《聖經》和其他古書中關於儀式的記載,其目的是爲了證明禮儀是非常神聖且有價值的。自此,基督教和天主教對抽象概念上的Ritual開始有了比較清楚的認識。

19世紀下半葉,Ritual Studies的研究開始從基督教中區分出來,轉换了其原本從基督教神學角度來看定義真理的標準。學者們開始將關注點放在其他宗教上,也不再依基督教的標準去作惟一的價值判斷。理雅各(James Legge,1815—1897)是一個轉折點,他雖然是一個基督徒,但不同于其他傳教士,他認爲有必要研究中國語言和中國文化。雖然他仍然認同基督教的優越性。但他認爲中國的文化亦有值得欣賞的地方,所以他下定了决心,將所有的"五經"翻譯成了英文。虽然他翻譯的《禮記》知道和使用的人不多,但从西方汉学的角度来看,他翻译的《礼记》是最好的。Max Müller是德國的世界宗教學學者,和James Legge一樣受聘于牛津大學,他想將世界各個宗教的經典翻譯成英文。這樣,在分析Ritual的時候,學者們逐漸擺脱了基督神學的束縛,試圖找出宗教和Ritual的一些普遍性的特征。他們研究的對象廣涉早期文明中的儀式和神話,認爲整個部族的信仰是藴含在其神話當中,而神話又是通過儀式表現出來的。真正現代意義上的Ritual研究,當肇端於此。

關於Ritual的特點,主要有兩派意見:一派認爲Ritual具有社會功能和效果,它能夠强化群體内部的自我認同。Ritual具有象征意義,這種象征意義非常抽象。另外一派認爲Ritual只是一種行動,關涉到整個社會的權利層次及其之間的鬥争,而不認爲其中的象征意義很重要。

Ritual Studies和中國的禮學雖然有相同的地方,但必須區分出來。因爲西方學者重視的只是Ritual的層面,而忽略其中的倫理性和政府結構等問題。禮學範圍比較廣，用Ritual Studies來提出和研究中國禮學就縮小了其範圍。

羅伯特教授主張，禮學的概念最好是不譯，尤其是在專業的學術論著中不用翻譯。但西方漢學家有一個很重要的責任，他們必須把中國的文明和文化解釋給西方人，這又需要有一個翻譯。就這種情況而言，羅伯特教授認爲禮學可以翻譯成Chinese Ritual Studies或Traditional Chinese Ritual Studies.

講座由北京大學古文獻研究中心劉玉才教授主持，中國社會科學院吴麗娱老師及北京大學文史哲部分師生參加了本次報告會。

（趙培　執筆）

國際漢學系列講座·第七十四講

題　目： 氣候與内亞史研究

主講人： 普林斯頓高等研究院　狄宇宙（Nicola Di Cosmo）　教授

主持人： 北京大學歷史學系暨中國古代史研究中心　榮新江　教授

評議人： 北京大學歷史學系暨中國古代史研究中心　陸揚　教授

時　間： 2015年4月20日（星期一）上午

地　點： 北京大學中國古代史研究中心報告廳

狄宇宙教授是西方研究東亞古代史的著名學者，在内亞史、中國史以及古代内亞政權與中原王朝關係史等領域取得了豐碩的研究成果。本次講演，他著眼於氣候、環境與人類社會三者之間的相互作用，力圖揭示遊牧社會内部的動力及其與古代中原地區的關係。蒙古高原有兩種常見的氣候災害：乾旱與嚴寒，嚴寒比較迅猛，乾旱比較緩慢而持久，它們都可能對脆弱的遊牧社會造成沉重打擊。今天，牧民可以在政府的幫助下搬遷，或者得到國際援助，失去畜群的牧民甚至可以搬進城市，但在歷史上，遇到這類災難會發生什麼呢？

狄宇宙首先闡述了氣候與遊牧民族遷移之關係。歷史學的研究，特别是19世紀末至20世紀初的研究，往往會把氣候與遊牧民族的遷移聯繫到一起，而遊牧民族的遷移又跟語言、技術的傳播以及遊牧政權的形成密切相關。目前,對遊牧民族史的進一步深入研究需要借重新的技術手段,新技術對這類研究有重要影響,會帶來新的課題,比如,基因研究對遊牧遷移的研究就非常有幫助。遊牧民族的遷移主要有兩種類型。一種是推進式遷移,即有一種力量將遊牧民族從某個區域往外驅離,從氣候的角度來看,這種方式主要是突然降温的災害造成的。突然降温使得土地資源喪失,牲畜缺乏草料,從而引起一連串的反應,造成部族内部的衝突和不穩定。另一種是拉進式遷移,即外部力量誘導遊牧民族進入到某個新的地區。造成這種遷移有多種原因,包括掠奪、商業貿易、農産品需求,以及完全隨意的遷移。

然後,狄宇宙介紹了内亞邊界與氣候關係之主要理論學説。早期的著名學者湯因比(A. J. Toynbee)把遊牧民族的入侵和征服理解爲一種自然現象,他認爲遊牧民族跟外界的關係大致有兩種:或者是農業地區的人群進入到遊牧地區,促使他們作出反應,或者是遊牧部族自己遷移到别的地方。古氣候學的先驅之一亨廷頓(E. Huntington)是一個氣候決定論者,他認爲草原地區的氣候災害,嚴寒與乾旱,促使遊牧民族週期性遷移。另一位早期研究古代内亞史與氣候的重要學者拉鐵摩爾(O. Lattimore),反對氣候決定論,而提出了氣候依賴理論。他認爲環境當然對遊牧民族有影響,但是隨着遊牧社會的發展,遊牧民族的自發選擇越來越强,而不是消極被動地接受環境的影響。有關内亞遊牧民族與農耕民族關係有兩個重要理論學派:一是功能學派,認爲遊牧社會仰賴農耕地區的産品,因此他們需要通過征服等手段來獲取這些産品,而在征服過程中必然産生更高級的政治形態;另一個是進化學派,認爲遊牧民族和農耕社會平行發展,由於中原帝國的形式變得越來越複雜,那麽遊牧地區也需要複雜的政權來適應這種變化。狄宇宙反對所有這些理論,强調遊牧民族内部非常多樣化,不能簡單地用一兩種理論來概括。他希望用其他辦法來重新審視遊牧社會的興衰變化,因此他注意到了古氣候學,不斷關注這個領域的方法和成果。

作爲講演的主要部分,狄宇宙教授介紹了他近期的一個重要研究課題,即氣候與蒙古帝國崛起之關係。1974年,詹金斯(G. Jenkins)就曾研究過氣候對成吉思汗崛起的影響,他認爲在成吉思汗時期出現過非常迅疾的降温,這減緩了蒙古部族内部的争鬥,從而促使成吉思汗權威的形成。狄宇宙教授三年前也開始關注這個課題,與三位科學家申請了美國的國家科學基金,重新用氣候學的方法來探索蒙古部族在鄂爾渾河流域的歷史。他們的研究跟過去的有所不同——他們不僅關注災害性的氣候,也關注適宜的氣候對蒙古帝國的影響,更强調好的氣候對遊牧社會内部産生的作用,探究這種作用是否對遊牧政權産生過積極的影響。初步的研究成果表明,在1180—1205年間,這一地區出現了持續很長時間的嚴重乾旱;1211—1225年,這裏的氣候又變得比較温暖,降雨量比較充沛。正是在1180—1205年這個時期,蒙古各個部落的内部争鬥非常激烈,部落軍事化現象非常突出;而1206年是鐵木真被推爲大汗的時間;成吉思汗的第一次西征則發生在1211—1225年這段氣候非常適宜的時期。

最後,狄宇宙提出了草原高生産力假説,主要包括以下幾點:第一,好的氣候有利於草原經濟迅速復原,這一點對遊牧地區的影響非常顯著;第二,好的氣候造成牲畜數量,特别是馬匹的增加,這一點很重要,因爲大規模的軍事征服没有充足的馬匹是不可能成功的;第三,古代草原地區的農業生産不可忽視,這種生産力會促進中央集權形態的出現;第四,草原地區的城市化對於遊牧政權的穩定也有很重要的作用。此外,除了研究蒙古帝國的中心區域,狄宇宙還呼籲對金帳汗國等周邊政權的關注。有學者指出,金帳汗國之所以選擇伏爾加河流域作爲政權的核心區域,就是因爲當地的氣候比較宜人。關於氣候與遊牧社會關係的研究,現在已經受到了較多的關注,這確實是一個有價值的途徑,值得進一步關注和深入。

（羅帥　執筆）

國際漢學系列講座·第七十五講

題　目:《源氏物語》的"古注釋書"與中國古文獻

主講人: 日本早稻田大學文學學術院　河野貴美子　教授

主持人: 北京大學中國古文獻研究中心　劉玉才　教授

時　間: 2015年5月6日(星期三)上午

地　點: 北京大學人文學苑6號樓B124室

《源氏物語》是古代日本最負盛名的文學名著，成書於11世紀初。作者紫式部，爲當時日本宫廷内服侍天皇、皇后的女官。整部小説共五十四卷，圍繞着主人公光源氏的私人生活展開，描寫了宫廷内的各種場景。這部用假名寫成的作品問世不久，就陸續有引用中國古文獻爲之作注釋的“注釋書”出現，主要包括藤原伊行《源氏釋》（？—1175?）、藤原定家《奥入》（1233年前後）、四辻善成《河海抄》（1362—1368前後）、一條兼良《花鳥余情》（1472）等，今日則形成《源氏物語古注集成》（25册）和《源氏物語古注釋叢刊》（10册）這樣兩種大型叢書。在河野貴美子教授看來，《源氏物語》的注釋者引用中國古文獻的記載來注釋這部和文著作的故事内容，本身就是和漢文學史上一件值得關注並應引起思考的事情；而探討《源氏物語》的“古注釋書”與中國古文獻的關係問題，也有助於推進古代的和漢語言研究與文學研究。

在介紹了《源氏物語》及其“古注釋書”的基本情況之後，河野教授主要從三個方面考察了它們和中國古文獻之間的關係：一是《源氏物語》和白居易詩；二是《河海抄》所引的中國古文獻；三是《花鳥余情》所引的中國古文獻。

衆所周知，838年，白居易詩就已東傳日本，随後受到了普遍熱烈的歡迎。而《源氏物語》的“須磨”卷記述光源氏由於政治鬥争的失敗左遷須磨之後心情鬱鬱寡歡，面對一輪圓月，憶及京都宫廷的美好時光，他禁不住地吟誦起“二千里外故人心”的詩句，這正是白居易《八月十五日夜，禁中獨直，對月憶元九》詩中的一句。同樣地，“桐壺”卷記載桐壺天皇因思念自己故去的愛人而吟誦無法入眠的和歌，《源氏物語》的注釋者在此就以《長恨歌》中的“夕殿螢飛思悄然，秋燈挑盡未能眠”兩句詩來加以解説。還值得注意的是，現在通行的白詩文本均作“孤燈挑盡未成眠”，但金澤文庫所藏古寫本《白氏文集》卷十二的文字和古注釋書所引一致，這種異文資料自然非常珍貴。

到了日本中世，最有成就的《源氏物語》古注釋書，當推四辻善成的《河海抄》和一條兼良的《花鳥余情》，二者旁徵博引，内容豐富。

首先，對於《源氏物語》中的らいし和わらび，《河海抄》的解釋都引用了《毛詩音義》。這就説明，直至公元14世紀，日本學者學習《毛詩》，仍以《經典釋文》爲範本，並且《河海抄》的部分文字與宋刻宋元遞修本《經典釋文》的對應之處也相同。其次，唐宋文獻方面，《河海抄》的徵引對象既有《金谷園記》

這樣的佚書佚文,也有《增注唐賢絶句三體詩法》這種在日本歷史上流傳廣泛的文學選集及相關抄物資料。再次,如果從作品内容和成書時代上加以考量,《源氏物語》中並没有杜甫詩和蘇軾詩的影響痕跡,但《河海抄》卻在注釋時引用了南宋的杜詩注本(《集千家注分類杜工部詩》)和蘇詩注本(《王狀元集注分類東坡先生詩》),這一方面説明了杜甫、蘇軾這些名家的别集在日本的早期流傳主要憑藉宋人注本的形式,另一方面也反映了它們當時的主要作用在於爲作詩(漢詩、和歌、和漢聯句等)提供方法幫助。凡此種種,都是河野教授通過《河海抄》與中國古文獻的仔細比對而加以揭示出來的。

較之《河海抄》,《花鳥餘情》體現出的則是15世紀學者一條兼良在注釋《源氏物語》時對中國古文獻的採擇情況。這主要包括:第一,引用元代類書《新編排韻增廣事類氏族大全》;第二,引用杜甫詩句及南宋阮閱《詩話總龜》中的故事;第三,引用蘇軾詩詞、詩注及《詩話總龜》等,特别是像《四河入海》這樣的蘇詩集注性質的抄物資料,時代更早的《河海抄》則尚未涉及。

講座最後,河野貴美子教授再次由《源氏物語》的"古注釋書"與中國古文獻的關係出發,表達了對於在漢字文化圈内從事文學、文化、文獻研究的個人理解與思考,即單純的中國文獻、日本文獻、韓國文獻研究之外,更需要在開闊的東亞視野下進行整體性的觀照。

北京大學中文系杜曉勤教授、顧永新教授,歷史學系井上亘教授以及北京大學的部分研究生參加了此次講座。

(趙昱　執筆)

國際漢學系列講座·第七十六講

題　目: 關公、媽祖在越南的流傳與演變

主講人: 台灣成功大學中文系、人文社會科學中心　陳益源　教授

主持人: 北京大學中國語言文學系　潘建國　教授

時　間: 2015年6月4日(星期四)下午

地　點: 北大人文苑6號樓中文系B-124會議室

陳益源教授在中國古典小説、民俗學、民間文學、越南漢文學、東南亞閩南文化等領域均有卓越的研究成果。此次講座,陳益源教授以關公、媽祖爲題,關注源自於中國的人物與信仰,如何隨着移民而盛行海外,並從文學作品、文獻材料、民俗活動、現存廟宇建築等各方面深入探討,分别講述兩者在越南的流傳與演變。

首先,陳益源教授從關公談起,《三國志演義》對關公形象塑造相當重要,透過文學藝術的潛移默化,關公及其忠義精神不僅影響了中國本地民衆,亦受到越南人民的認同,加上越南各地關帝廟的建立,以及關帝信仰的蓬勃發展,更進一步讓關公深入人心。開始談關公在越南的流傳與演變之前,陳教授特别强調進行田野調查之前需充分把握已見之文獻,帶着文獻跑田野調查,方能事半功倍。目前越南現存的文獻裏,與關帝信仰有關者有三,一是方志中載及關帝廟者,二是現存之關帝廟碑銘拓片,三是與關帝信仰有關的漢喃書籍,從這些方志、碑銘拓片、書籍等文獻中的大量文字記載,可見當時關帝信仰之興盛蓬勃。陳教授以會安及順化地區的關帝信仰爲例,説明目前越南關帝信仰概況。早期越南關帝信仰只在當地的華人間流通,後來逐漸成爲在地信仰,並廣設廟宇祭祀關帝,然而目前廟宇祭祀現況(例如關帝由主殿移往後殿或偏殿)、建築用途(廟宇挪用爲小學)、居民信仰重心轉移(例如祭祀其他神祇)等皆與昔時不同。從這些現象來看,關公信仰已逐漸式微,並被其他神祇所取代。最後,陳教授認爲,關公雖作爲一個小説人物,有其鮮明獨特的人物個性,但是研究關公,不可只局限在小説文本,更需重視民間信仰的部分,方能理解關公信仰何以深入民心,並能在海外流播甚廣。

其次,媽祖信仰隨着中國移民向外遷徙而傳向海外,例如台灣、港澳地區和越南、馬來西亞、新加坡、印度尼西亞、菲律賓等東南亞國家,幾乎都建有"媽祖廟"或"天后宫",這些廟宇往往成爲當地信仰的中心。其中,影響深遠,但却鮮爲人知者,應屬越南。越南民間稱媽祖廟爲"婆寺",一般廟名仍以"天后宫"或"天妃廟"居多。早期在越南北部、中部、南部,海邊或江畔,都有華人(尤其是福建人)建立的天后宫,爲數衆多;如今隨着華人在越南的去留與移動,目前的天后宫主要集中在越南南部湄公河流域,天后信徒也由早期的華人,擴及到越南的一般百姓(越族)。陳教授指出,初期越南天后宫的興建與廣東、福建等地居民移居在此密切相關，他們多將此廟宇與會館結合，並以此爲中心建設學校、醫院、義地等。根據現今留存天后宫的重修碑文拓片或捐贈芳名録，可以發現當中有若干閩商並未回到故鄉，而是留在當地参與地方事務及興建廟宇，這些閩商對於推廣天后信仰起到一定程度的影響。此後天后信仰逐漸跨越地域、族群的界綫，成爲越南普遍的信仰，以越南南部湄公河流域爲例，現存天后宫數量超過六十座以上。雖然部分地區天后信仰已不復當時盛況，但是天后信仰在越南仍具影響力，甚至成爲市政發展的考慮因素之一。

陳益源教授强調小説文本中的關帝與媽祖雖具影響力，但深刻地影響普羅大衆的部分却體現在民間信仰之中，因此不可忽視信仰的凝聚力與影響力。在此次講座中，陳益源教授將越南的關帝信仰與媽祖信仰發展興衰做了仔細梳理，展示了中國信仰在異國他鄉如何被接受，並與當地信仰相互影響、融合的過程。

（蔡芷瑜　執筆）

徵稿啓事

一、《國際漢學研究通訊》是北京大學國際漢學家研修基地主辦的綜合學術刊物，辦刊宗旨爲報導國際漢學界在中國傳統人文學科領域的研究動態，搭建中外學者溝通交流的學術平臺。本刊分設漢學論壇、文獻天地、漢學人物、論著評介、研究綜覽、基地紀事等欄目，歡迎海内外學人賜稿或提供信息。

二、本刊暫定爲半年刊，分别在三月、九月底截稿。

三、本刊以中、英文爲主。來稿篇幅以中文一萬五千字内爲宜，特約稿件不在此限。除經本刊同意，不接受已刊發稿件。論著評介欄目原則上不接受外稿，但可以推薦。

四、來稿請提供Word文檔和PDF文檔，同時寄送打印紙本。中文稿件請提供繁体字文本。如附有插圖，請提供原圖圖片格式（JPG之類）的電子文件。具体撰稿格式請參照文稿技術規範。因編輯人員有限，恕不退稿，請自留底稿。咨詢稿件處理事宜，請儘量通過電子郵件。

五、來稿如涉及著作權、出版權方面事宜，請事先徵得原作者或出版者之書面同意，本刊不負相關責任。本刊有權對來稿進行删改加工，如不願删改，請事先注明。

六、來稿刊出之後，即致贈稿酬、樣刊。本刊享有已刊文稿的著作財産權和數據加工、網絡傳播權，如僅同意以紙本形式發表，請在來稿中特別注明。

七、來稿請注明中英文姓名、工作單位、職稱，並附通信地址、郵政編碼、電話傳真、電子郵件等項聯絡信息。

八、來稿請寄：

北京市海淀區頤和園路5號　100871

北京大學國際漢學家研修基地
《國際漢學研究通訊》編輯委員會
E-mail:sinology@pku.edu.cn

附:

文稿技術規範

一、來稿請以Word文檔(正文五號字,1.5倍行距)打印紙本,同時提供電子文檔。

二、來稿正文請按“一、(一)、1.、(1)”的序號設置層次,其中“1.”以下的章節段落的標題不單獨占一行;文稿層次較少時可略去“(一)”這一層次;段内分項的可用①②③等表示。

如:一、XXXX

(一) XXXX

1. XXXX

(1) XXXX。① XXX;② XXX;③ XXX。

三、來稿中的中文譯名,除衆所熟知的外國人名(如馬克思、愛因斯坦)、地名(如巴黎、紐約)、論著名(如《聖經》、《資本論》)按照通用譯名外,其他人名、地名、論著名在文中首次出現時,請括注外文原名,如沃爾特·福克斯(Walter Fuchs),地名、論著名照此處理。

四、來稿中的注釋,請採用頁下注、每頁各自編號,注號置於句末的標點符號之前,如孔子已有“六藝”之説①,“……將邊界查明來奏”②。但引文前有冒號者,句號在引號内,則注號置於引號之外,如《釋名》云:“經者,徑也,常典也。”③

五、頁下注釋文字的具體格式如下:

1. 著作類:著作者名,《書名》,出版地:出版者,出版年(不加“年”字),X—X頁。又:著作者名,《書名》卷X,X年X本。

2. 雜誌類:著作者名,《論文名》,《期刊名》X年X期,X-X頁。又:著作者名,《論文名》,《期刊名》X卷X號,X-X頁。

3. 西文書名與雜誌名均用斜體，文章名加引號。日文、韓文參考中文樣式。

4. 重複出現的注釋不用"同上"簡略，但標注文獻出處只列著作、論文名和頁碼即可。

例：① 郭紹虞，《宋詩話考》，北京：中華書局，1979，75頁。

② 張裕釗，《濂亭文集》卷四，清光緒八年查氏木漸齋刊本。

③ 袁行霈，《〈新編新注十三經〉芻議》，《北京大學學報》2009年2期，7頁。

④ 池田秀三著，金培懿譯，《韋昭之經學——尤以禮爲中心》，《中國文哲研究通訊》第15卷3期，141-155頁。

⑤ Ad Dudink, "The Chinese Christian Books of the Former Beitang Library", *Sino-Western Cultural Relations Journal* XXVI (2004), pp. 46—59.

六、圖表按先後順序編號，在文中應有相應文字説明，如見圖X，見表X。

七、數字用法：

1. 公曆世紀、年代、年、月、日用阿拉伯數字，如18世紀50年代。

2. 中國清代和清代以前的歷史紀年、其他國家民族的非公曆紀年，用中文數字表示，且正文首次出現時需用阿拉伯數字括注公曆。如秦文公四十四年(公元前722)，清咸豐十年(1860)，日本慶應三年(1867)。

3. 中文古籍卷數均用中文數字表示，如作卷三四一，不作三百四十一。